ENVIRONMENTAL ANAEROBIC TECHNOLOGY

Applications and New Developments

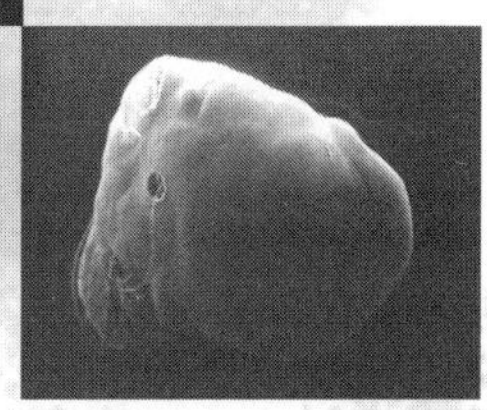

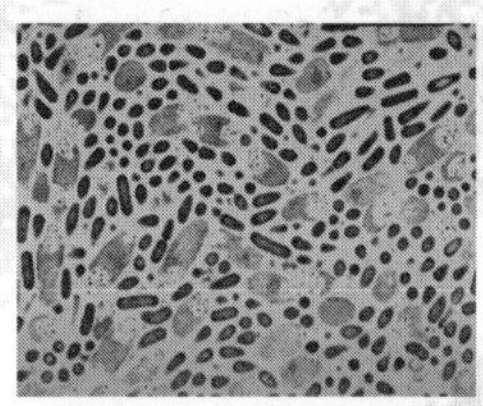

ENVIRONMENTAL ANAEROBIC TECHNOLOGY

Applications and New Developments

Editor

Herbert H P Fang

The University of Hong Kong, Hong Kong

ICP

Imperial College Press

Published by

Imperial College Press
57 Shelton Street
Covent Garden
London WC2H 9HE

Distributed by

World Scientific Publishing Co. Pte. Ltd.
5 Toh Tuck Link, Singapore 596224
USA office: 27 Warren Street, Suite 401-402, Hackensack, NJ 07601
UK office: 57 Shelton Street, Covent Garden, London WC2H 9HE

British Library Cataloguing-in-Publication Data
A catalogue record for this book is available from the British Library.

ENVIRONMENTAL ANAEROBIC TECHNOLOGY
Applications and New Developments

ISBN-13 978-1-84816-542-7
ISBN-10 1-84816-542-0

Typeset by Stallion Press
Email: enquiries@stallionpress.com

Printed in Singapore by World Scientific Printed.

Contents

Preface

Compared to the more conventional aerobic process, anaerobic process for waste and wastewater treatment has three intrinsic advantages, i.e., energy saving, reduced sludge yield, and production of biofuel. Because of these advantages, in addition to the technical development in recent years, anaerobic treatment technology has become a cost-effective means for pollution control, resulting in the installations of thousands of full-scale treatment facilities worldwide in the past two decades. These advantages also make anaerobic process a sustainable environmental technology of enormous potential for the years to come.

Full-scale application of anaerobic process was first conducted in about 1860 for the treatment of domestic wastewater by Louis M. Mouras and Abbé Moigno in France, more than 50 years before the invention of aerobic activated sludge process. The first biological treatment of wastewater process was patented by W. D. Scott-Moncrieff in 1891 in Great Britain, in which organic matters were first degraded anaerobically and the products were then further degraded and mineralized aerobically, producing a clear effluent and inoffensive gases. Thereafter, anaerobic process had been continually improved and applied in many cities in Europe and the United States; the value of the generated methane was generally recognized for heating and power generation. Unfortunately, anaerobic process was treated at the time as a black box, with little understanding of its microbiology and biochemistry, and with little improvement in process technology. It was gradually replaced in the 1930s by the aerobic treatment process, which was perceived by many as better understood and more effective in operation.

Anaerobic technology was re-vitalized in the 1960s in the United States by Perry McCarty's pioneering research on the anaerobic filter. However, the first successful full-scale application of anaerobic technology for high-strength industrial wastewater treatment was not demonstrated until the late 1970s by Gatze Lettinga in the Netherlands, using a new reactor design known as upflow anaerobic sludge blanket (UASB). The success of UASB relies on the effective gas–liquid–solid separation by a built-in device at the top of the reactor, resulting in better retention of biomass in the reactor and the formation of granular sludge with superb settleability and bioactivity. Since then, UASB and its modifications, including expanded granular sludge bed (EGSB) and internal circulating (IC) reactors, have been successfully applied worldwide for the treatment of high-strength industrial wastewaters containing readily biodegradable pollutants, such as effluents from food, beverage, brewery, and distillery industries; information related to these reactors and their applications is abundant in literature.

Today, anaerobic technology continues to flourish in many aspects, such as development of new types of reactors, uses of molecular techniques for microbial studies and kinetic modeling, production of bio-hydrogen by fermentation and microbial electrolysis cell, and broadening applications to municipal wastewater, effluents from chemical industry and agricultural wastes with high lignocellulose content. This book compiles state-of-the-art information related to these new developments in anaerobic technology. It consists of 18 chapters contributed by 40 renowned experts from 13 countries/regions. The first chapter is authored by Gatze Lettinga, whose contributions to this technology have won him the Tyler Prize (2007) and the Lee Kuan Yew Water Prize (2009). In this chapter, Lettinga shares with the readers the historic development of anaerobic technology and his concept of de-centralized municipal wastewater treatment in which anaerobic process plays the primary role.

The remaining chapters are divided into three categories: full-scale applications, emerging technologies, and new developments. Discussions of full-scale applications include those in France (Chapter 2) and Japan (Chapter 3), and the treatment of municipal wastewaters in India, Colombia, and Brazil (Chapters 4 and 5). Several new designs of anaerobic reactor (Chapters 2 and 3) and experiences in the design and operations of municipal wastewater treatment (Chapters 3 and 4) are also reviewed. Discussions of emerging technologies include granulation mechanism and reactor systems (Chapter 6), anaerobic membrane bioreactor (AnMBR) (Chapter 7), and anaerobic baffled reactor (ABR) (Chapter 8), along with the applications of anaerobic process for chemical effluents (Chapter 9), the uses of molecular techniques for the identification of crucial microbes and their roles in the treatment process (Chapter 10), and mathematical models for process control (Chapter 11). Discussions of new developments include the pretreatment processes which are required for breaking down the recalcitrant lignocellulosic wastes prior to anaerobic degradation, either chemically (Chapter 12) or enzymatically (Chapter 13). Also discussed are studies of fermentative production of bio-hydrogen in Taiwan (Chapter 14), China (Chapter 15), Korea (Chapter 16), and Thailand (Chapter 17), as well as production of hydrogen by microbial electrolysis cells (Chapter 15), bio-ethanol by fermentation (Chapter 17), and chemicals from synthetic gas fermentation (Chapter 18).

I would like to thank all the authors for their efforts and prompt submissions, as well as The University of Hong Kong's Strategic Research Theme for its support.

It is my hope that this book will not only be useful for those engaged in research, development, design, and operation of anaerobic technology, but that it will also arouse interests of postgraduate and undergraduate students to dedicate themselves to this environmental technology of enormous potential for the future.

April, 2010

Herbert H. P. Fang
Chair Professor of Environmental Engineering
The University of Hong Kong

Chapter 1

The Route of Anaerobic Waste (Water) Treatment toward Global Acceptance

Gatze Lettinga
Emeritus Professor, LeAF-Foundation,
Sub-Department of Environmental Technology
Wageningen University, 6700 EV Wageningen, Netherlands
Gatze.Lettinga@wur.nl

Although the feasibility of anaerobic wastewater treatment (AnWT) has been successfully demonstrated over 100 years ago, its development and implementation had been hampered for decades until the re-introduction of the anaerobic filter process in the 1960s and the technological breakthrough in the 1970s of the upflow anaerobic sludge blanket (UASB) process. This has led to the further developments of a number of modern high-rate AnWT systems, such as expanded granular sludge bed (EGSB) reactor, and anaerobic baffled reactor (ABR), for numerous types of industrial wastewaters. Furthermore, UASB reactor systems have been successfully applied for the treatment of raw domestic sewage, followed by an aerobic polishing post-treatment, if needed; the enormous potentials of the innovative micro-aerobic post-treatment will soon be demonstrated at full scale. Moreover, based on improved understanding of these processes, e.g., in the microbiology, biochemistry, the immobilization of required organisms and consortia, and reactor and process technology, and in control and steering, undoubtedly substantial further progress in the applicability of these anaerobic systems will be achieved in near future. These treatment systems, greatly based on the Natural Biological Mineralization route (NBM), are going to act as a crowbar to force Environmental Protection toward a significantly more sustainable tackle. They enable closing of water and substance loops, viz. valorization of pollutants, reuse of treated water, and will stimulate minimization of wasting of clean water in wastewater collection and transport. Along with the implementation of these systems a major step will be made toward more sustainability in society.

1. Introduction

Consciousness of the humanity, including that of environmental protection (EP), is growing in all domains of society. The question is, however, whether or not it grows fast enough, regarding the exponential developments in science and technology. Is mankind becoming sufficiently capable to employ its enormous scientific, technological, and cultural achievements optimally for the well-being of future citizens, all of them, and wherever living? In essence, and in line with the ideas of the Brundtland Commission (1987), it is in that way how the notion sustainability should be interpreted with three main pillars, namely:

— socialization of human society, i.e., maximization of social security for all world citizens,
— resource conservation, both fossil and living, and
— optimizing development of humanity, with emphasis on education and talent development.

It is the moral task for individual humans to contribute to these issues. The question is whether they are capable, becoming sufficiently aware in time about the "whys and hows". During their short lifetime, consciousness of individual humans develops somewhat similarly to mankind over millions of years, urging them to contribute. Are we controlling or just following the process of social consciousness? Regarding the exponential developments in science and technology, but at the same time comparably slow progress made in public awareness of all the three sustainability pillars, we face risks for dangerous unbalances in society. Science and technology should always remain well rooted in human society, i.e., not become self-controlling, leading to a disastrous robotization.

Back-casting how things in various sectors in society developed, for instance, in the domain of public sanitation (PuSan) and environmental protection (EP), it is clear that many decisions made in the past were far from optimal, neither for society as a whole nor for separate communities. Illustrative is how things evolved with respect to the development and implementation of anaerobic digestion (AnDi) and anaerobic wastewater treatment (AnWT) systems. Why is that AnDi and AnWT didn't get the core-position in wastewater, why is that so much emphasis in the thirties has been put on the development and implementation of the aerobic wastewater treatment (AeWT) concept, why is that the implementation of modern AnWT proceeded so much smoother in the industrial sector than in the PuSan sector, why is that so much emphasis has been put on extremely centralized off-site sanitation concepts and so tiny on optimally de-centralized on-site concepts? It looks that in the public sector decision makers tend to select costly conventional, proven concepts. Citizens are still insufficiently aware of the "why and how" of developments in many public domains even though they have to pay the bill; but with respect to making the decisions they generally have to rely on the insights and interests of the "specialists" and (local) politicians. Citizens are so far from sufficiently informed and consequently involved in decision-making, and are generally unable to oppose, particularly when not well organized. Many disastrous decisions in the past can be attributed to that.

The further going back into our past, the more difficult, if not horrifying, were conditions of life for majority of humans; most of the man-made disastrous events can be due to an almost complete lack on scientific insight, catastrophic governance and consequently disastrous inequity. However, mankind becomes slowly aware of the origins of the evil, gradually understands that most of the man-made tragedies can be due to a serious inherent lack of social security for citizens, entrepreneurs, companies, etc., because it leads to self-interested decision-making, to misleading of citizens, to prevention of their involvement in decision-making.

About one and half a century ago, citizens lived in great ignorance with respect to many important aspects for satisfactory conditions of life; science was still in its very initial stages of development, the knowledge in the field of microbiology, biochemistry, chemistry, physics, etc., was negligible. Citizens did not have any idea of the public health risks of wastes, wastewaters and polluted surface water. Residents of Dutch cities at that time disposed excreta in the canals, which at the same time frequently were used as source of potable water. As biological treatment methods for wastewaters and wastes did not exist, hygienic conditions were dramatic. Only in the second part of nineteenth century, things started to improve as a result of implementation of practice directed to collecting human excreta in residences and conveying them out from the neighborhood. Although these activities initially were primarily conducted for use as soil-conditioner and fertilizer to farmers, instead of improving public health conditions, the net result was, however, positive. Only in the second half of nineteenth century the first biological, viz.

anaerobic treatment, systems were developed and implemented. It was the commencement of PuSan and EP; both nowadays are major issues. The consciousness about the crucial importance of a clean environment, the need to attain sustainability in society grew considerably in the last half century, particularly since the publication of the report *Our Common Future* of the Brundtland Commission (1987), with respect to society in general and to the various specific domains, such as PuSan and EP. Gradually the opinion seems to evolve that in these domains sustainability is within reach. A growing group of "specialists" in the field of waste and wastewater treatment is convinced that AeWT as core wastewater treatment method needs to be substituted by AnDi and AnWT treatments, supplemented with the adequate complementary methods. However, developing sustainable EP concepts/systems is just the first step; the second comprises their proper and timely implementation, and this is at least as difficult as it implies a drastic change. From practice it is learnt that the substitution of well-established concepts/technologies, if they are very expensive, complex and in essence non-sustainable, for innovative more sustainable alternatives is extraordinarily difficult. The reason for that generally lays not in any deficits of the innovative technology, but is sociological in nature. It can be attributed to the fact that established institutions, companies, and groups commercializing the traditional established systems generally do not want to give up their privileged position. A major bottleneck on the route to sustainability is that private interests generally prevail over those of the society as a whole. On the other hand, as will be elucidated below, sometimes the private (industrial) sector takes the lead in the implementation of innovative sustainable systems/concepts, like AnWT systems as primary wastewater treatment; although the public sanitation sector is far behind, it irrevocably needs to follow.

2. Roots of Modern High Rate AnWT

2.1. *Historical aspects of PuSan with AnDi system (e.g., McCarty, 2001)*

The first recognition that anaerobic biological processes could represent an effective method of converting waste organic matter into methane can be attributed to Volta (Barker 1956). He showed in 1776 the formation of "combustible air" from sediments in lakes, ponds, and streams. Eighty years later Reiset observed methane formation in decomposing manure piles and he proposed to study this process in order to understand the decomposition of organic matter (Buswell and Hatfield 1938). The first full-scale application of the AnDi process, i.e., the Mouras' Automatic Scavenger, was used for treating domestic wastewater in about 1860 (Moigno 1881, 1882). In 1890–1891 W. D. Scott Moncrieff constructed a tank with an empty space below and a bed of stones above, consequently a hybrid of a tank digester and an anaerobic filter (AF). The wastewater of 10 people entered the tank and then passed up through the AF system, resulting in a great reduction in sludge volume. Unfortunately, at that time it was almost completely a black-box system, although the researchers observed the development of a thin bacterial film covering the stone; they indicated, and likely the first to do so, the advantages of retaining wastewater solids for a period of time in order to achieve hydrolytic or bacteriolytic action on waste solids for the conversion of a portion of the organic matter into inoffensive gases or soluble compounds that pass out with the wastewater. Already in 1893!

The first septic tank (patented) was installed in 1895 by Donald Cameron in Exeter, England, treating successfully about 230 $m^3 \cdot d^{-1}$ of wastewater initially and the entire city's wastewater two years later. Moreover the gas was collected and used for heating and lighting. Similar systems were designed in 1894 by A. N. Talbot for Urbana and Champaign, Illinois in the USA in 1897 (Buswell and Hatfield 1938).

In 1904, a new two-stage anaerobic process was put into operation at Emscher, Germany, in which the suspended solids settled into a separate digestion compartment, the so-called Travis tank (Metcalf and Eddy 1915). In 1905, Karl Imhoff modified the system in order to prevent wastewater flowing through the hydrolyzing chamber, where the sludge was allowed to stay for a period, from a few weeks to several months, in order to make it inoffensive upon withdrawal. The Imhoff tank greatly reduced the cost of sludge disposal and became quite popular (Metcalf and Eddy 1915). However, since the system appeared to be not the complete solution to wastewater treatment, efforts began with separate digestion of sludge (Imhoff 1916); but the system became practically successful only around 1927 with a (first) sludge-heating apparatus in a separate digestion tank at Ruhrverband installed at Essen-Rellinghausen (Imhoff 1938). From then onwards the application of separate digestion grew rapidly in popularity, particularly in larger cities.

The value of generated methane was generally recognized. In addition to its use for heating digesters, applications were found for other purposes as well. In 1923, methane was collected on large scale by the Emschergenossenschaft and delivered to the municipal gas system at the Essen-Rellinghausen plant (Imhoff 1938). In 1927, the Ruhrverband utilized the sludge gas in Iserlohn and then in Essen-Rellinghausen in order to generate power for a biological treatment plant, and the cooling water from the motors was used for heating the digestion tanks. Such use of digester gas is now common practice at wastewater treatment plants throughout the world. By the 1930s, many cities in Germany added compressors in order to store the gas in steel cylinders for use as a motor fuel (Imhoff 1938). This practice also has been used on and off in modern times. Along with these applications, there were several studies then dealing with the full-scale application of separate anaerobic sludge treatment process which resulted in sufficient understanding to accept its wide scale application by the end of the 1930s. However, the basic understanding of the anaerobic digestion process, including the microbiology, biochemistry, and consequently the process technology, was still so poor that the application of anaerobic wastewater treatment (AnWT) could not take off in these years. Interest on AnWT, despite the initial satisfactory results of the AF systems, was almost completely lost in the époque around 1930 in favor of the first-generation aerobic wastewater treatment (AeWT) systems, which were better understood and therefore considered as much more feasible, effective, and robust in operation.

2.2. *Historical aspects of high-rate AnWT*

The development of modern high-rate AnWT systems for treatment of low and medium strength wastewaters began in the sixties with the anaerobic filter (AF) system at the Stanford University by McCarty and co-workers (McCarty 1964; Young and McCarty 1969). Famous is his statement: *Anaerobic wastewater treatment didn't enjoy the popularity it truly deserves* (McCarty 1964). Besides the re-introduction of the AF system, McCarty made great contributions to the improved understanding and applicability of AnWT, and to the recognition of the crucial importance of bacterial immobilization, i.e., their ability to aggregate in flocs or biofilms. Although McCarty's contributions were crucial for the development and implementation of AnWT, the USA somehow has not found real acceptance of modern AnWT technology, and its funding of research in this

Fig. 1.1. The reversed flow Dorr-Oliver Clarigester in Stellenbos treating wine distillery wastewater, which was full with granular sludge (1980).

field has remained marginal. The main reason for that lies in the extremely reluctant attitude of the established water pollution control sector; any alternative of the conventional AeWT tackle is rejected.

In South Africa, AnWT had already had some "credit" in the fifties as a result of the rather successful application of the so-called "reversed flow Dorr-Oliver Clarigester" system (Fig. 1.1) for the treatment of high-strength industrial wastewaters (Stander 1950, 1966). However, the process was still very poorly understood in all aspects, including microbiology, biochemistry, and particularly reactor operation. In the late sixties and early seventies attempts were made to improve the situation. Fundamental research of microbiology and biochemistry was initiated (e.g., Tourien *et al.* 1967; Thiel 1969; Kotzé *et al.* 1969), along with applied technological research (Simpson 1971; Pretorius 1971). However, like in the USA, the AnWT technology did not take off; in fact, almost all the emphasis in South Africa was soon put on the development of modern AeWT systems, such as the successful biological phosphate removal process.

In Europe, contrary to the USA and South Africa, the interest in AnWT remained marginal until the early seventies. In April 1970, the great breakthrough came resulting from the comprehensive investigations at the Agricultural University in Wageningen and the development of the upflow anaerobic sludge bed (UASB) reactor. Since then, the status of AnWT changed dramatically (Lettinga *et al.* 1980).

3. Development of Anaerobic Upflow Sludge Bed Systems

3.1. *Conventional UASB (Lettinga 1995)*

Inspired by McCarty's work but completely ignorant of the "ins and outs" of biological wastewater treatment methods, I moved into the field of AnWT when I started my career at the Wageningen University. With a chemical engineering background and specialization in physicochemical treatment of radioactive wastewaters, unprejudiced with respect to any biological treatment system, I started investigating the feasibility of the AF system, copying from McCarty's design, for the treatment of wastewater of the potato starch industry, which had since many decades seriously polluted surface waters in the North-East of the Netherlands. The results obtained in the laboratory with diluted freshly prepared potato juice solutions were very promising and challenging. First of all, high treatment efficiency was achieved at volumetric loads exceeding 10 kg-COD·m^{-3}·day^{-1}, and the startup proceeded smoothly. Furthermore, quite importantly, like Young and McCarty (1969) we observed the formation of a highly active and extremely well-settling sludge. We hoped these findings would convince the potato starch company to move into further research directed toward the implementation of AnWT at their factories, more particularly to the even more promising UASB reactor developed in our laboratory (Lettinga *et al.* 1980). Like the AF system, the UASB reactor has a simple design, but it looked applicable for a much wider range of wastewater (including some containing partially insoluble particulates) with substantially higher loading potentials and is likely cheaper in investment and operation. Although at first sight the UASB reactor resembles the South African "reversed flow Clarigester" system, it in essence deviates considerably in design and in mode of operation. Following the laboratory and bench-scale investigations in 1971–1972, the feasibility of the UASB reactor concept was successively demonstrated at 6 and 30 m^3 pilot plants for (although not the potato starch wastewater as we hoped) sugar beet wastewater, which was substantially less complex than the potato starch wastewater. This work was made in close co-operation with a Dutch sugar beet company (CSM) with involvement of researchers of Wageningen Agricultural University, Amsterdam University and Delft Technical University, and with substantial financial support from the Dutch government. The first 200 m^3 demonstration reactor was put in operation in 1976 at a sugar beet factory near Amsterdam, and at the same location one year later the first 1000 m^3 full-scale UASB reactor and the following year the second at the sugar beet factory in Breda (Fig. 1.2). Scaling up of the UASB system to 1000 m^3 was achieved within only 5–6 years. Starting from the mid-seventies, the feasibility of the system was assessed for numerous types of effluents of agro-industries, including the complex potato-starch wastewater during 1979–1981. Based on the successful demonstration and implementation of the system for a number of agro-industrial wastewaters, the Dutch government forced the board of potato Starch Company to try out the system (the alternative treatments tried out by the industry before then appeared to be far too complex and expensive). However, contrary to our results of laboratory experiments conducted with diluted freshly prepared potato juice solutions, the treatment of real wastewater appeared quite problematic. It was in fact barely feasible only after the continuous supply of a trace element cocktail to the raw wastewater; we discovered the significant effect of trace elements in preliminary experiments in preceding years and in comprehensive continued still on-going investigations (Zandvoort *et al.* 2005).

The sudden development of this niche for AnWT in Europe can be attributed to a coincidence of reasons: (i) the implementation of strict legislation of effluent discharge in 1970 in the Netherlands, which led to a strong increase in the interest from the polluting industries in low-cost treatment systems, (ii) the established water pollution control industry was then completely directed to AeWT and quite overloaded, and was also ignorant, and in fact prejudiced regarding the potentials of AnWT, (iii) the very smooth and successful demonstration of the UASB system, (iv) the considerable financial support from the Dutch government to improve the insight in various aspects of AnWT, i.e., the microbiology, biochemistry, and reactor and

Fig. 1.2. Full-scale UASB treatment of effluent of a CSM (company) sugar beet factory in the Netherlands (1976).

process technology, and (v) the establishment of new contractors moving into the field of modern AnWT. The interest in application of high-rate AnWT systems for the treatment of industrial wastewaters was not just restricted to the traditional mesophilic conditions but also developed for thermophilic (van Lier *et al.* 1992, 2001) and psychrophilic conditions (Rebac *et al.* 1998). Moreover, along with the steeply increasing interest in application of the UASB concept for the traditional anaerobic treatment, other applications of the UASB concept came in picture, such as biological sulfate reduction, de-nitrification, and for innovative micro-aerobic (post-)treatment systems, and more recently even for aerobic treatment systems, for final polishing of the effluent and/or nitrification.

Following the success in the industrial sector, the next big challenge for us was the issue to assess the feasibility of the UASB system for raw domestic sewage pre-treatment. Based on preliminary research conducted in sixties and early seventies in the USA and South Africa, we started in 1976 comprehensive laboratory, bench and (0.03–6 m^3) pilot plant investigations for sewage first at temperatures exceeding 20 °C (tropical conditions), and later since the late seventies for temperatures in the range of 10–20 °C. Since the investigations directed to applications under tropical conditions were very promising, the Dutch government provided financial support to continue the comprehensive feasibility studies at a 64-m^3 demonstration plant in Cali, Colombia during 1981–1986. The results clearly demonstrated the big potentials of the UASB system for that purpose, and also for application in decentralized settings. These positive findings (Louwe Kooijmans *et al.* 1985) led to the construction and startup of a number of big full-scale UASB installations in Colombia (Schellinkhout and Collazos 1992) and India (Wiegant 2001), once again by the financial support of the Dutch government.

The efforts to demonstrate the feasibility of the one-step UASB system for raw sewage pre-treatment under the moderate Dutch climatic conditions were less successful, i.e., not sufficiently attractive to promote its full-scale implementation, but on the other hand, for us, not too "poor" to terminate these investigations. It ultimately led to the development of an integrated UASB-AnDi system, in which the complementary AnDi step serves for: (i) the mineralization of the poorly stabilized sludge accumulating in the UASB reactor; and (ii) maintaining a sufficiently high methanogenic activity in the UASB reactor by returning part of the digested sludge into it (Mahmoud *et al.* 2004a, b). This integrated UASB-AnDi system, though slightly more complex in operation and design, offers attractive potentials for applications in moderate climates. It may be particularly attractive as a pre-treatment for toilet and kitchen wastewaters collected with vacuum systems in so-called decentralized sanitation and reuse (DESAR) settings, i.e., the optimally sized on-site settings, such as separate communities, hospitals, hotels, schools, apartment buildings, etc. This DESAR concept is directed not just to protect the living environment from pollution, but also to particularly recover useful byproducts (biogas, fertilizers, soil conditioners, etc.), and to prevent the installation of expensive transport systems (Kujawa-Roeleveld and Zeeman 2006; Zeeman *et al.* 2008). The interest in this innovative concept has gradually been growing in Europe. In China substantial progress has been made in the last decades with the implementation of community on-site AnWT systems in the public sector, especially in rural areas (Wang 2004).

Undoubtedly one of the very challenging features of the UASB system comprises its ability to generate granular anaerobic sludge, i.e., a sludge consisting of mechanical robust, very well settleable and biologically active bacterial aggregates up to about 3 mm in size. This peculiar phenomenon, which had in fact been noticed by Young and McCarty (1969) in their AF systems, became since mid-seventies a major joint research topic at our university between the Departments of Environmental Technology, Microbiology, and Physical-Colloid Chemistry (e.g., Hulshoff Pol *et al.* 1981, 1983; Grotenhuis *et al.* 1991; Visser *et al.* 1993; Alphenaar *et al.* 1993). Its immense potential impact for practice and scientific challenges was soon recognized in a wide circle, and became since the eighties a major issue of research at many other universities worldwide. It has led to significantly improved insight in the "whys and hows", although not yet to a complete general consensus (McHugh *et al.* 2003; Hulshoff Pol *et al.* 2004). Still numerous challenging questions remain to be elucidated (Lettinga and Hulshoff Pol 2002).

3.2. *The EGSB system*

Soon after the "discovery" of the renowned features of granular sludge, comprehensive investigations were directed to the development of the EGSB system. The EGSB reactor relies on: (i) the exclusive use of granular sludge, (ii) operation under conditions of a slight bed expansion, and, particularly, (iii) the expectation that granules remain stable and will smoothly augment. (Lettinga *et al.* 1980; Lettinga 2010; Kato *et al.* 1999; Rebac *et al.* 1998). The EGSB system offers great potentials: it can (i) accommodate very high organic and hydraulic loadings, (ii) treat very low-strength ($COD < 150\,mg{\cdot}l^{-1}$) wastewater, (iii) be feasible for low-strength acidified wastewaters under psychrophilic conditions (even at 4–10 °C), and (iv) be capable of treating wastewaters containing lipids and toxic/inhibitory compounds like formaldehyde, lauric acid, etc.

The EGSB reactor concept became popular in the last decades for treatment of a variety of industrial effluents. However, despite the many satisfactory full-scale experiences, still substantial improvements can be realized in order to improve its stability and to extend its applicability. Particularly further research needs to be done to improve the insight into the mechanisms underlying specific aspects of the "granular sludge phenomenon", viz., maintaining a sufficiently high stability, the smooth augmentation, and the possibilities to manipulate the bacterial composition, i.e., how to incorporate new organisms (Lettinga 2010).

3.3. *Other high rate reactor designs*

Two systems comparably very similar to EGSB were developed in the eighties: the anaerobic fixed-film bed system developed in the USA (Jewell *et al.* 1981; Switzenbaum and Jewell 1980) and the expanded-bed version of the anaerobic fluidized bed developed in Belgium (Binot *et al.* 1983) and in Spain (Iza *et al.* 1987). Both these systems rely: (i) on bacterial film attachment to small carrier materials supplied to the reactor, (ii) on the agglomeration of these attached biofilm grains, and (iii), like the EGSB system, on their augmentation.

Attempts (e.g., Heijnen 1988) to develop a completely fluidized anaerobic bed system failed; as a matter of fact, this system inherently cannot be successful, because maintaining a completely fluidized bed requires uniform (in shape, size, and density) bio-film particles, or agglomerates. That is impossible; irrevocably sludge bed segregation will occur.

Other innovative developments on anaerobic sludge bed systems in the eighties and nineties comprised the anaerobic sequencing batch system developed by Dague (Zhang *et al.* 1997), the anaerobic baffled reactor system which originally was developed in the USA (Barber and Stuckey 1999), and various hybrid reactor systems.

4. Place of AnWT in EP

Since the mid-seventies, anaerobic treatment has received significantly increased acceptance worldwide, although countries like the USA and South Africa have so far remained reluctant. The idea gradually finds acceptance that we urgently need a much more sustainable tackle in PuSan and EP, meaning that we need much more economical systems and concepts which enable a maximum extent of waste valorization, and which in due time can lead to all kinds of modern urban agricultural practices. The insight is growing that AnWT and AnDi processes indeed should be the first, and in essence the major, treatment step in what we consider as a really sustainable EP-tackle (EP_{sus}); they are the core step in the mineralization route of organic matter in nature. This *Natural Biological Mineralization* (NBM) route is exactly the tackle to be followed in waste and wastewater treatment; it virtually leads to the approach needed for EP_{sus}. After the AnDi/AnWT step, the subsequent treatment steps in the NBM tackle are successively the micro-aerobic and aerobic polishing steps, which may include, if necessary, the full oxidation of reduced S- and N-compounds. The innovative micro-aerobic steps, which rely on unique bio-conversions occurring under conditions of very low oxygen supply or concentrations, are effective in the conversion of highly reduced compounds, like H_2S and volatile organic S-compounds (Buisman *et al.* 1990; Janssen *et al.* 1997; van Leerdam 2007; Sipma *et al.* 2003) in the raw wastewater or generated in the preceding AnWT step (Fig. 1.3).

The NBM tackle offers the following advantages over the conventional AeWT-based approach:

— Energy generating, instead of energy demanding;
— Resource recovery, i.e., waste valorization and urban agricultural practices;

Fig. 1.3. Full-scale UASB plants (rectangular and circular installations in the center part of the photo) treating the combined reuse paper wastewater of three factories, and conventional aerobic post-treatment (right upper part of photo) and micro-aerobic biogas-treatment facilities (in a small separate building positioned between the AnWT-installations) in the city Eerbeek, Netherlands.

— Very low volumes of excess sludge which are concentrated and well stabilized;
— High organic and hydraulic volumetric loads, resulting in very low space requirements;
— Simple in operation and maintenance, and consequently hardly depending on specialists;
— Low in capital investment, as well as operational and maintenance costs;
— Use of conventional, plain equipment which can be generally manufactured locally;
— Long life-time of installations and auxiliary equipment;
— Anaerobic sludge can be stored unfed for long periods of time without losing its activity;
— Applicable at almost any scale, and at almost any location;
— Very robust, independent of any complex infrastructures, e.g., for power supply;
— Almost absence of any malodorous problems;
— Avoid the formation of recalcitrant compounds (e.g., humic acids); and
— Efficient in degrading various recalcitrant compounds, such as azo dyes, polyaromatic compounds, and nitro-aromatics.

These advantages enable industries to accomplish an almost complete closing of water and matter loops, especially when the NMB methods are implemented in settings which focus on clean water saving practices and minimization of transport, consequently on optimal de-centralized settings. But this is true not merely for the industrial domain, but also for the agricultural and PuSan sanitation domain. And this fortunately is gradually happening. Following the success in the industrial sector, the successful demonstration of AnWT for sewage pre-treatment in tropical regions, and the proven potentials of innovative NBM-based post-treatment meth-

Fig. 1.4. Full-scale UASB-Plant at Belo Horizonte, Brazil, consisting of 12 recator modules, (pre)treating raw municipal wastewater for 1 M inhibitants at 1.8 $m^3 \cdot s^{-1}$. (photo obtained from C. Chernicharo).

ods, this approach irrevocably will find worldwide acceptance in the PuSan sector. At present Brazil is in the frontline of such approach (Chernicharo 2006; Foresti *et al.* 2006) (Fig. 1.4). The potentials of domestic sewage pre-treatment by modern AnWT systems indeed are enormous, and certainly this is not merely true for tropical regions; the NBM tackle also can profitably be applied in subtropical regions (Mahmoud *et al.* 2004b), especially during summer season in touristic regions where the population during the season can reach tenfold higher than other seasons. Implementation of the NBM concept, along with adequate wastewater collection practices and transport minimization can and will lead to very substantial savings, both in investment and in operational costs.

5. What Could the Future Bring?

With the NBM-based treatment concept, most of the tools for attaining a sustainable tackle in PuSan and EP lay ready on the shelf. Combined with adequate measures and technologies directed to prevention of pollution, e.g., application of clean water saving collection practices for liquid and solid residues as well as minimization of transport, the optimal conditions for a really sustainable environmental protection and public health care in communities can be realized in near future. By putting the focus on a step-wise implementation of optimal DESAR concepts, which are directed to closing loops of water and other matters in residential communities, at the industrial sites, in separate buildings or clusters of buildings, or at farm-scale, a maximum of valorization of residues can be achieved at a minimum of input of resources and it can promote urban agricultural practices. A very important step made in that respect comprises the raising interest of hospitals, at least in the Netherlands, to become completely self-sufficient in treating their hazardous waste and wastewaters. Very likely, hotels, restaurants, schools, etc. will soon follow, if it can be proven that these systems are economically profitable and may lead to a high extent of self-sufficiency, as is the case for industries.

Moreover, these NBM methods, especially when supplemented with adequate physicochemical methods, are also ideally suitable for treatment of polluted water, solid residues, and gases, as well as for the decontamination of polluted soils. Contemporary and future generations can make a major contribution toward a much higher sustainability in society in various domains.

A question coming up is: Can anything prevent us from advancing in that direction of EP_{sus}? Judging from the significant progress made in the industrial sector in the last decades, the answer is simply "No". Politicians, decision makers, consultants, and engineers have gradually become much more confident that the issue of EP_{sus} is no longer a dream, but a well achievable option. The bottlenecks certainly lay not in technological deficits or lack of appropriate EP_{sus}-concepts, but predominantly can be found in all kinds of sociological factors, i.e., too much tunneling in the approach, monopolization of decision-making in the sector by the civil engineering discipline, too dominant interests of established private business in existing methods and concepts. But certainly there exists also a general lack on long-term vision and/or courage from the side of public policy makers. And last but not least, the involvement of citizens in decision-making is still far too little, and the need to be adequately informed about what is going on in the development of anaerobic technology. Then, the extreme reluctance of the established pollution control would ultimately vanish, and these concepts irrevocably will find their way into the public domains of sanitation.

In various countries the first steps have been made in the PuSan domain, particularly in Brazil where the federal government stimulates the implementation via the so-called PROSAB program. This to some extent also happens in India, Mexico and slowly even in European countries like Germany and the Netherlands (Kujawa-Roeleveld and Zeeman 2006). However, despite that, it will be a long and difficult way to go. The expulsion of all kinds of evil heritages needs prolonged periods of time and tremendous effort. But, it will happen; the benefits of EP_{sus} concepts simply are too pregnant and too challenging. This issue is not only just a clean living environment, but also a realization of innovative forms of urban agricultural practices, more integration between urban and rural regions, and achievement of a significant higher extent of self-sufficiency in many other domains (Lettinga 2004). Consequently, there is hope that gradually the general insight will establish that Brundtland's envisaged drastic social changes for achieving "real" sustainability in society at global scale, merely can be attained when our enormous scientific and technological achievements are optimally employed for the well-being of all citizens.

References

Alphenaar, P. A., Visser, A., and Lettinga, G. (1993) The effect of liquid upward velocity and hydraulic retention time on granulation in UASB reactors treating waste water with a high sulphate content. *Biores. Technol.* **43**(1), 249–258.

Barker, H. A. (1956) Biological formation of methane. In *Bacterial Fermentations*, John Wiley and Sons, Inc., New York, pp. 1.

Barber, W. P. and Stuckey, D. C. (1999) The use of the anaerobic baffled reactor (ABR) for wastewater treatment: A review. *Wat. Res.* **33**(7), 1559–1578.

Binot, B. A., Bol, T., Naveau, H. T., and Nyns, E. J. (1983) Biomethanation by immobilized fluidized cells. Preprints of IAWPR, Specialized seminar anaerobic treatment of wastewaters in fixed film reactors, 16–18 June, Copenhagen, 211–223.

Brundtland Commision (1987) *Our Common Future*. The World Commission on Environment and Development (WCED), Oxford University Press.

Buisman, C. J., Yspeert, P., Janssen, A., and Lettinga, G. (1990) Kinetics of chemical and biological sulphide oxidation in aqueous solutions. *Wat. Res.* **24**(5), 667–671.

Buswell, A. M. and Hatfield, W. D. (1938) *Anaerobic Fermentions*. Bulletin No. 32. State Water Survey.

Chernicharo, C. A. L. (2006) Post-treatment options for the anaerobic treatment of domestic wastewater. *Rev. Environ. Sci. Biotechnol.* **5**(1), 73–92.

Foresti, E., Zait, M., and Vallero, M. (2006) Anaerobic processes as the core technology for sustainable domestic wastewater treatment: Consolidated applications, new trends, perspectives and challenges. *Rev. Environ. Sci. Biotechnol.* **5**, 3–19.

Grotenhuis, J. T. C., van Lammeren, A. A. M., Plugge, C. M., Smit, M., Stams, A. J. M., Xu, Y. S., and Zehnder, A. J. B. (1991) Bacteriological compounds and structure of granular sludge adapted to different substrates. *Appl. Environ. Microbiol.* **57**, 1942–1949.

Heijnen, J. J. (1988) Large scale anaerobic aerobic treatment of complex industrial wastewater using immobilized biomass in fluid bed and air-lift suspension reactors. *Verfahrenstechnik Abwasserreinigung*, GVC, Baden-Baden, Band 2, 203–218, 17–19 October.

Hulshoff Pol, L. W., de Zeeuw, W. J., Velzeboer, C. T. M., and Lettinga, G. (1981) Granulation in UASB-reactors. *Wat. Sci. Technol.* **15**, 291–304.

Hulshoff Pol, L. W., de Zeeuw, W. J., Velzeboer, C. T. M., Wiegant, W., and Lettinga, G. (1983) Granulation in UASB reactors. *Wat. Sci. Technol.* **15**, 291–304.

Hulshoff Pol, L. W., Castro Lopes, S. I. de, Lettinga, G., and Lens, P. N. L. (2004) Anaerobic sludge granulation. *Wat. Res.* **38**(6), 1376–1389.

Imhoff, K. (1916) Separate sludge digestion improves Imhoff tank operation by keeping sewage fresh. *Eng. Record* **74**, 101–102.

Imhoff, K. (1938) *Sedimentation and digestion in Germany*. In: L. Pearse ed., *Modern Sewage Disposal*, Lancaster Press, Lancaster, Pennsylvania, pp. 47.

Iza, J., Garcia, P. A., Sanz, I., and Fdz-Polanco, F. (1987) Granulation results in anaerobic fluidized bed reactors, in *Granular Anaerobic Sludge: Microbiology and Technology*, In: *Proceedings of the GASMAT Workshop*, 25–27, October, 1987, Lunteren, the Netherlands.

Janssen, A. J. H., Ma, S. C., Lens, P., and Lettinga, G. (1997) Performance of a sulphide-oxidizing expanded-bed reactor supplied with dissolved oxygen. *Biotechnol. Bioeng.* **53**, 32–40.

Jewell, W. J., Switzenbaum, M. S., and Morris, J. W. (1981) Municipal wastewater treatment with the anaerobic attached microbial film expanded bed process. *J. Wat. Pollut. Control Fed.* **53**(4), 482–490.

Kato, M. T., Rebac, S., and Lettinga, G. (1999) The anaerobic treatment of low-strength brewery wastewater in expanded granular sludge bed. *Appl. Biochem. Biotechnol.* **76**, 15–32.

Kotzé, J. P., Hattingh, W. H. J., Siebert, M. L., Thiel, P. G., and Toerien, D. F. (1969) A biological and chemical study of several anaerobic digester. *Wat. Res.* **2**, 195–213.

Kujawa-Roeleveld, K. and Zeeman, G. (2006) Anaerobic treatment in decentralised and source-separation-based sanitation concepts. *Rev. Environ. Sci. Biotechnol.* **5**, 115–139.

Lettinga, G., van Velsen, A. F. M., Hobma, S. W., de Zeeuw, W. J., and Klapwijk, A. (1980) Use of the upflow sludge blanket (USB) reactor. *Biotechnol. Bioeng.* **22**, 699–734.

Lettinga, G. (1995). Anaerobic digestion and wastewater treatment systems. *Anthony van Leeuwenhoek* **67**, 328.

Lettinga, G. and Hulshoff Pol, L. W. (2002) The sludge granulation phenomenon: Its augmentation, manipulation and challenges (part 2). In: *Granulation and Auto-Immobilisation Processes in Wastewater Treatment*, pp. 97–109, Wageningen University, The Netherlands.

Lettinga, G. (2004) A good life environment for all, *Environ. Sci. Eng.* **1**, 77–82.

Lettinga, G. (2010) Challenges of a feasible route towards sustainability in environmental protection, *Frontiers of Environmental Science & Engineering in China, no. 4* (accepted for publication). Springer.

Louwe Kooijmans, J., Lettinga, G., and Rodrquez-Parra, G. (1985) The UASB-process for domestic waste water treatment in developing countries. *J. Ind. Wat. Eng. Sci.* **39**(5), 437–451.

Mahmoud, N., Zeeman, G., Gijzen, H., and Lettinga, G. (2004a) Anaerobic sewage treatment in a one-stage UASB reactor and a combined UASB-Digester system. *Wat. Res.* **38**(9), 2348–2358.

Mahmoud, N., Zeeman, G., Lettinga, G., and Gijzen, H. (2004b) Perspectives for integrated sewage management in Palestine and the Middle East. *Water* **21**, 24–29.
McCarty, P. L. (1964) *Anaerobic Waste Treatment Fundamentals.* Public Works, September, 107–112, October, 123–126, November 91–94, December, 95–99.
McCarty, P. L. (2001) The development of anaerobic treatment and its future. *Wat. Sci. Technol.* **44**(8), 157–176.
McHugh, S., O'Reilly, C., Mahony, T., Colleran, E., and O'Flaverty, V. (2003) Anaerobic granular sludge bioreactor technology. *Rev. Env. Sci. Biotechn.* **2**, 225–245.
Metcalf, L. and Eddy, P. (1915) *American Sewerage Practice, III. Disposal of Sewage.* McGraw-Hill, New York.
Moigno, A. F. (1881) Mouras' Automatic Scavenger. *Cosmos*, p. 622.
Moigno, A. F. (1882) Mouras' Automatic Scavenger. *Cosmos*, p. 97. Monroy, O., Famá, G., Meraz, M., Montoya, L., Marcarie, H. (2000) Anaerobic digestion for wastewater treatment in Mexico: A state of the technology. *Wat. Res.* **34**(6), 1803–1816.
Pretorius, W. A. (1971) Kinetics of anaerobic fermentation. *Wat. Res.* **5**, 681.
Rebac, S., van Lier, J. B., Lens, P., van Capellen, J., Vermeulen, M., Stams, A. J. M., Dekkers, F., Swinkels, K. Th. M., and Lettinga, G. (1998) Psychrophilic (6–15°C) high-rate anaerobic treatment of malting wastewater in two-module expanded granular sludge bed. *Biotechnol. Prog.* **14**, 856–864.
Schellinkhout, A. and Collazos, C. J. (1992) Full scale application of the UASB technology for sewage treatment. *Wat. Sci. Technol.* **25**(7), 159–166.
Simpson, D. E. (1971) Investigations on a pilot plant contact digester for the treatment of a dilute urban waste. *Wat. Res.* **5**, 523.
Sipma, J., Janssen, A. J. H., Hulshoff Pol, L. W., and Lettinga, G. (2003) Development of a novel process for the biological conversion of H_2S and methanethiol to elemental sulfur. *Biotechnol. Bioeng.* **82**(1), 1–11.
Stander, G. J. (1950) Effluents from fermentation industries: Part IV. A new method for increasing and maintaining efficiency in the anaerobic digestion of fermentation effluents. *J. Inst. Sew. Purif.* **4**, 438.
Stander, G. J. (1966) Water pollution research — A key to wastewater management. *J. Wat. Pollut. Control Fed.* **38**, 774.
Switzenbaum, M. S. and Jewell, W. J. (1980) Anaerobic-attached film expanded-bed reactor treatment. *J. Wat. Pollut. Control Fed.* **52**, 1953.
Thiel, P. G. (1969) The effect of methane analogues on methanogenesis in anaerobic digestion. *Wat. Res.* **3**, 215.
Tourien, D. F., Siebert, M. L., and Hattingh, W. H. J. (1967) The bacterial nature of the acid-forming phase of anaerobic digestion. *Wat. Res.* **1**, 497.
van Leerdam, R. C. (2007) *Anaerobic Degradation of Methanethiol in a Process for Liquefied Petroleum Gas (LPG) Biodesulfurization*, PhD thesis, Wageningen University.
van Lier, J. B., van Grolle, K. C. F., Stams, A. J. M., de Macario, E. C., and Lettinga, G. (1992) Start-up of a thermophilic upflow anaerobic sludge bed (UASB) reactor with mesophilic granular sludge. *Appl. Microbiol. Biotechnol.* **37**, 130–135.
van Lier, J. B., van der Zee, F. P., Tan, N. C. G., Rebac, S., and Klerebezedm, R. (2001) Advances in high rate anaerobic treatment: staging of reactor systems. *Wat. Sci. Technol.* **44**(8), 15–25.
Visser, A., Alphenaar, P. A., Gao, Y., and Lettinga, G. (1993) Granulation and immobilization of methanogenic and sulfate-reducing bacteria in high rate anaerobic reactors. *Appl. Microbiol. Biotechnol.* **40**(4), 575–581.
Wang, K. (2004) The development and application of anarobic biotechnology in the industrial and agricultural sector in China. In: *Proceedings of 10th Anaerobic Digestion Conference*, Montreal, 962–969.
Wiegant, W. M. (2001) Experiences and potential of anaerobic wastewater treatment in tropical regions, *Wat. Sci. Technol.* **44**(8), 107–113.
Young, J. C. and McCarty, P. L. (1969) The anaerobic filter for wastewater treatment. *J. Wat. Pollut. Control Fed.* **41**(5), R160–173.

Zandvoort, M. H., Hullebusch, E. D., van, Peerbolte, A., Golubnic, S., Lettinga, G., and Lens, P. N. L. (2005) Influence of pH shocks on trace metal dynamics and performance of methanol fed granular sludge bioreactors. *Biodegradation* **16**(6), 549–567.

Zeeman, G., Kujawa, K., de Mes, T., Hernandez, L., de Graaff, M., Abu-Ghunmi, L., Mels, A., Meulman, B., Temmink, H., Buisman, C., van Lier, J. B., and Lettinga, G. (2008) Anaerobic treatment as a core technology for energy, nutrients and water recovery from source-separated domestic waste(water). *Wat. Sci. Technol.* **57**(8), 1207–1212.

Zhang, R. H., Yin, Y., Sung, S., and Dague, R. R. (1997) Anaerobic treatment of swine waste by the anaerobic sequencing batch reactor, *Trans. ASAE* **40**(3), 761–767.

Chapter 2

Developments of New Anaerobic Treatment Technology in France

Rene Moletta

INRA/Savoie university,
Parc Technolac 73376 Le Bourget du lac, France
rene.moletta@yahoo.fr

In France, treatment of industrial effluents by anaerobic treatment has been well established especially in the agricultural and food industries. With the increasing cost of green electricity, we see an increase of anaerobic reactors constructions for farm and municipal solid wastes. Here we report five developments currently underway. The first is a modified moving bed process, called "anaerobic moving bed biofilm reactor" (AMBBR), in which a carrier is moving inside the reactor. It is easy to manage, and we can reach an organic load of 24 kg-COD·m^{-3}·d^{-1}. The second technology, called "PROVEO", is for wastewater treatment. The process uses support media for trapping microbial flocs and for biofilm growth. The third technology is combining an anaerobic filter with ultrafiltration (UF) membranes that have been applied to the treatment of dairy wastewater. The treated wastewater is free of suspended solids and the organic load reaches 10 kg-COD·m^{-3}·d^{-1}. The fourth technology, named ERGENIUM, is about the treatment of solid wastes in a two-stage process, i.e., acidogenesis followed by methanogenesis. The leachate from the first reactor is separated from the solid fraction and further treated in an anaerobic filter. The fifth technology development is on the pH control of the anaerobic reactor so that the biogas may have a steady methane concentration.

1. Introduction

France has around 200 anaerobic reactors treating industrial wastewater and one-third of sludge produced by municipal wastewater treatment plants. Now we implement sludge digester to 15,000–20,000 IE capacity plants. Anaerobic treatment of wastewater can produce up to 30% of the total energy needed for agricultural and food industries, and reduce two-thirds of the sludge produced by municipal wastewater treatment plants. But we have installed since 2006 only two digesters for municipal solid wastes and four digesters for the farm. For comparison, Germany has several thousand digesters installed on farm at the same period. This is due to the very low price of green energy in our country! As a result, little active research was conducted in France until recently, except the National Institute for Agricultural Research (INRA) laboratory in Narbonne.

With the new price of green energy in France, we have seen since 2006 a great deal of development on implementation of anaerobic treatment for solids wastes and wastewater. I will present in this chapter some new technologies in development in France. Some have already been in full-scale operation; others are in pilot plant tests. In addition, I will also describe strategies developed to improve the performance of anaerobic treatment.

2. New Anaerobic Treatment Technologies

2.1. *Anaerobic moving bed biofilm reactor (AMBBR)*

Anaerobic moving bed biofilm reactor (AMBBR) has been proved to be very reliable because of its high volumetric loading rates and low sludge accumulation in the reactor (Jahren *et al.* 2002). Based on our previous researches (Garcia-Calderon *et al.* 1998; Steyer *et al.* 1999; Arnaiz *et al.* 2003), the idea of developing anaerobic treatment of single-phase AMBBR originated from the view that it is cost-effective and might be feasible to apply with cautions in treating undiluted high-strength dairy wastewater. For balanced digestion, supplement of alkalinity is required to prevent accumulation of volatile fatty acids (VFAs) in the reactors (Rajeshwari *et al.* 2000; Jantsch and Mattiasson 2004). However, although various anaerobic reactor configurations have been investigated in treating dairy wastewaters (Hwang and Hansen 1992; Imbeault *et al.* 1998), none has used the AMBBR design. The purpose of this research was to evaluate the single-phase AMBBR process for its stability and efficiency on the anaerobic treatment of milk permeate. The aim was to develop this technology for small reactors with a daily loading capacity of 50–500 kg-COD.

2.1.1. *Reactor configuration*

The design of AMBBR is illustrated in Fig. 2.1. The bench-scale AMBBR was 32 l in volume (diameter 25.0 cm; height 90.0 cm; effluent outlet height 86.5 cm). It was filled with 10 l of seeding sludge, 18 l of biofilm carriers and tap water to make up its working volume of 30 l. The reactor was operated continuously by intermittent feeding controlled by programmed timers.

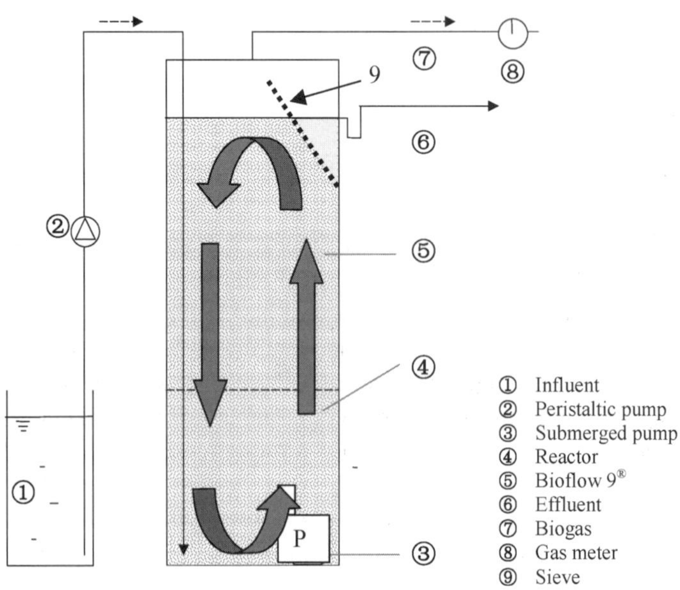

Fig. 2.1. Schematic diagram of bench-scale anaerobic moving bed biofilm reactor (AMBBR).

The mixing is generated by a submerged pump situated at the bottom of the reactor. The reactor was placed in a plastic chamber equipped with air heaters so as to operate at stable mesophilic condition of 35 °C.

Mineral Vithane® solution provided by Biothane Company was added to the influent at 1.0 $ml{\cdot}l^{-1}$. The influent pH was adjusted at around pH 7.0 with 2.5 $mol{\cdot}l^{-1}$ NaOH solution. The organic load (express as kg-$COD{\cdot}m^{-3}{\cdot}d^{-1}$) of the reactor was gradually increased until the soluble COD removal efficiency dropped below 80%.

2.1.2. *Wastewater characteristics*

The wastewater was from a cheese factory which used ultrafiltration (UF) membrane to recover proteins and other materials from milk for cheese making. The wastewater was a mixture of milk permeate, which is poor in protein and nitrogen, and COD-concentrated washing water (10% of the wastewater total volume). The washing water was from cleaning of transport lines and equipments. The COD came mainly from lactose in the permeate. The wastewater was stored at -20 °C after collection and defrosted at 4 °C for about 24 h before feeding.

2.1.3. *Biofilm carriers*

Low-density polyethylene Bioflow 9® (diameter 10 mm; height 8 mm; density 0.94; specific surface 530 $m^2{\cdot}m^{-3}$) was used as biofilm carriers. Each such carrier consists of small cylindrical elements with longitudinal fins that protrude on the outside surface and an internal cross member that divides each element into three circular sectors. The objective was that the density of carrier with biofilm may reach about 1.0 in normal operation.

2.1.4. *Influence of organic loading rate on the COD removal efficiency*

Figure 2.2 illustrates the total chemical oxygen demand (TCOD) and soluble chemical oxygen demands (SCOD) at the inlet and outlet of the reactor, respectively. Organic loading rate (OLR; expressed in kg-$COD{\cdot}m^{-3}{\cdot}d^{-1}$) plus TCOD and SCOD removal efficiencies are plotted in Fig. 2.3. The treated wastewater contained more suspended solids (SS) than the raw wastewater did. No severe VFA accumulation was observed during the whole period even when the TCOD removal efficiency varied significantly. This means that, for this kind of wastewater and process, the SCOD removal efficiency is a better criterion in deciding the stability of the system.

Although the TCOD removal efficiency fluctuated substantially during the whole period, the SCOD removal efficiency decreased only mildly with the increase of OLR, from near 95% to 80–90% at OLR of 28.0 kg-$COD{\cdot}m^{-3}{\cdot}d^{-1}$. The maximum OLR reached in this study was 28.0 kg-$COD{\cdot}m^{-3}{\cdot}d^{-1}$.

2.2. *Floating anaerobic filter reactor PROVEO*

An anaerobic fixed bed reactor consists of a vertical bed containing inert media as supports for microbial attachment. This allows the reactor to maintain a much higher quantity of biomass than those in suspended-growth processes, resulting in higher volumetric loading rates, shorter hydraulic retention times (HRTs) and better performance stability. One of the problems associated

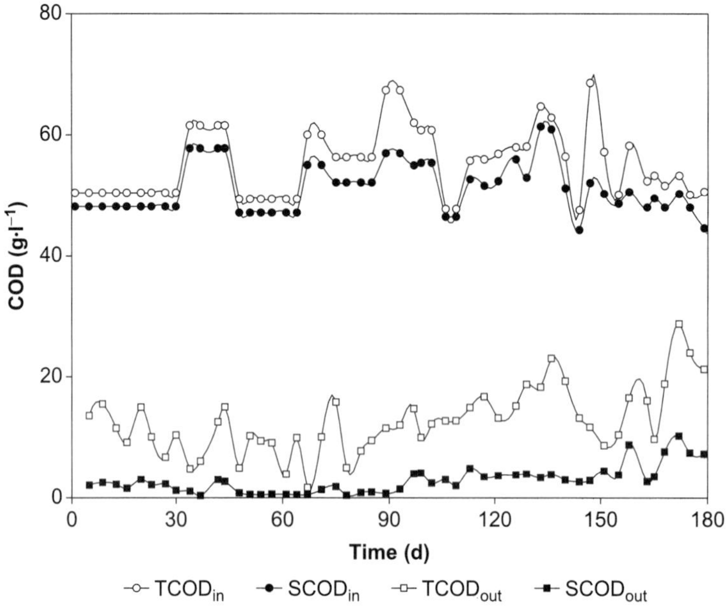

Fig. 2.2. Total and soluble chemical oxygen demands (TCOD and SCOD) measured at inlet and outlet of the reactor.

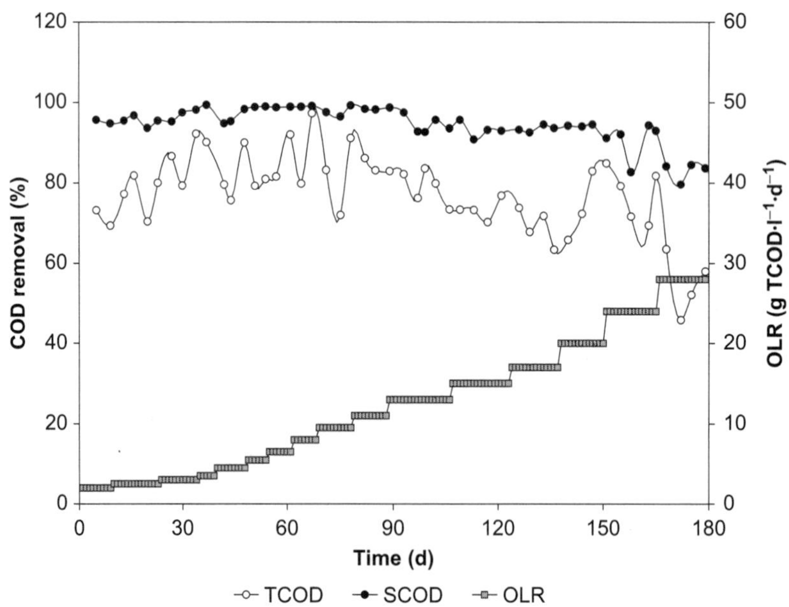

Fig. 2.3. Total and soluble COD removal efficiencies at various OLRs.

with anaerobic fixed beds is the risk of clogging and channeling due to the dense accumulation of biomass in the media. This needs sometimes a strong agitation in order to remove clogging and channeling. This problem can be acute and severe when the reactors are operated at high loading rates or with an effluent containing high levels of solids.

2.2.1. *Description of the developed process*

In the new process developed, called PROVEO, the problem of clogging has been resolved by the use of small loose floating supports. Unclogging of the reactor could be easily done by fluidization of the supports using liquid- or gas recirculation. The supports used are made of polyethylene with a diameter and a height both of 3 cm and a density of 0.93.

In normal operation, the supports will remain floating thus creating a fixed bed in the upper part of the reactor (see Fig. 2.4). When solids have accumulated in the reactor, the floating media can be fluidized by an up-flow current of liquid or by recirculation of biogas to fluidize the bed and to allow the removal of the excess solids accumulated in the media. Fluidization is done only when it is necessary to unclog the bed, that is to say generally once every 2–4 months. This process has been patented by INRA and is currently commercialized by the French company Naskeo.

2.2.2. *Assessment of process performances*

The performances of the PROVEO process have been investigated in 10 l bench-scale reactors for low-strength wastewater (1.9 g-COD·l^{-1}) and effluents from various agricultural and food industries, including fruit canning (10 g-COD·l^{-1}), winery (20 g-COD·l^{-1}) and cheese dairy (30 g-COD·l^{-1}). The reactors were operated for about 6 months. At the beginning of the experiments, the reactors were fed at a low OLR of 0.4 kg-COD·m^{-3}·d^{-1} and long HRT. The OLR was then increased stepwise, by increasing the feeding rate, by about 20% per week until the COD removal efficiency decreased to 80%, which was considered as the threshold level for a safe and economical operation of the reactors. At the end of the experiments, all the reactors were dismantled and all the supports were removed from the reactors in order to quantify the biomass inside the reactors.

An example of the PROVEO reactor performance in treating winery effluent is illustrated in Fig. 2.5. During the 14-day start-up period, the COD in the effluent remained quite high even at a low OLR of 0.4 kg-COD·m^{-3}·d^{-1}. The effluent COD values were then decreased gradually from 2.7 to 0.8 g-TCOD·l^{-1} and 2.1 to 0.4 g-SCOD·l^{-1}. At the maximum OLR of 35 kg-COD·m^{-3}·d^{-1},

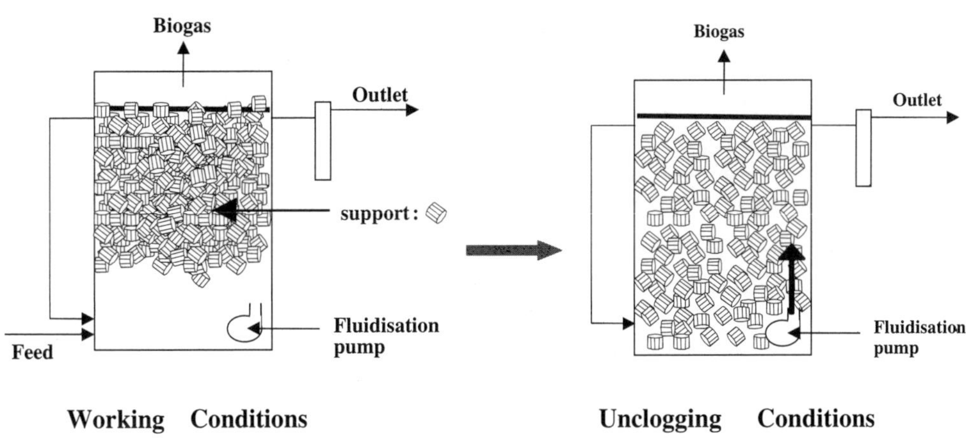

Fig. 2.4. PROVEO reactor in working conditions and during fluidization.

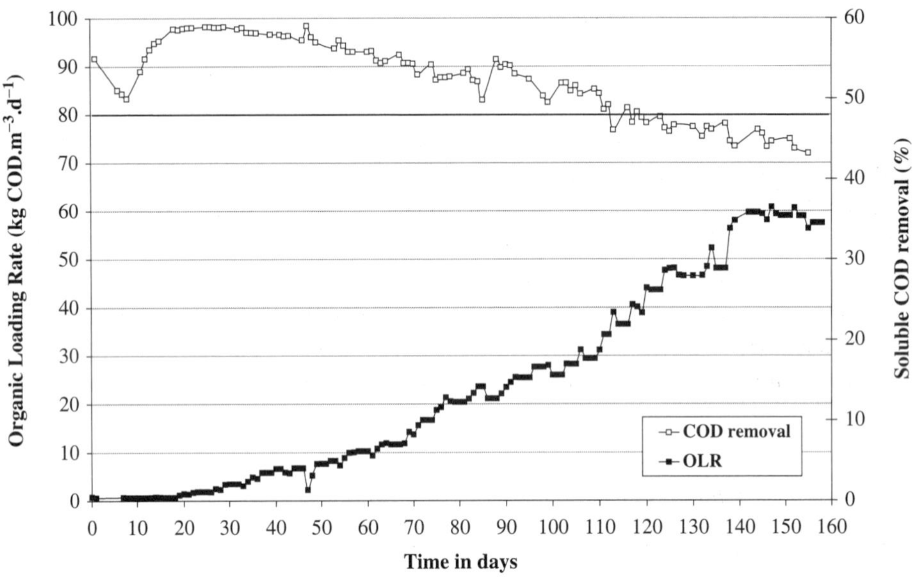

Fig. 2.5. Changes of OLR and soluble COD removal efficiency in the PROVEO reactor.

the process removed 70% of TCOD and 74% of SCOD (Fig. 2.5). As the SCOD removal efficiency of 80% was considered as the threshold level, results showed that the reactor should not be operated at OLR higher than 27 kg-COD·m^{-3}·d^{-1} and HRT lower than 19 h. The SS concentration at outlet was always low, averaging 0.86 ± 0.03 g·l^{-1} for the entire study period. Similar trends were observed for other reactors but with different threshold limits.

2.2.3. *Design parameters*

For low-strength wastewater with COD below 2 g·l^{-1}, results showed that HRT is the limiting factor for the PROVEO reactor with a minimum requirement of 4 h. For effluents with higher COD, the OLR becomes the controlling parameter, and is used in the design of the reactor volume. Results of treating three wastewaters at respective COD of 10, 20 and 30 g·l^{-1}, showed that the PROVEO reactor performed effectively at OLR ranging 17–27 kg-COD·m^{-3}·d^{-1}. However, there was no logical relationship found between the wastewater COD level and the maximum OLR obtained; instead, the nature of the organic matters in the wastewater seemed more important than the overall COD level. At the end of the experiment, all the supports were removed from the reactors to measure the overall quantity of biomass inside the reactors. It was observed that the supports were full of biomass, mostly entrapped by the support media instead of attached to the support surface. Each support was able to retain 0.7–1.6 g of TS, showing an excellent retention of biomass. Furthermore, the fraction of biomass entrapped inside the supports represented 60–70% of total biomass inside the reactor, indicating that the packing support had a dual role in the retention of the biomass, namely entrapping biomass within the support and filtering the biomass in suspension.

2.2.4. *Examples of industrials applications*

Two examples of PROVEO reactors are as follows. One is using a 30-m^3 reactor to treat the effluents of a plant which produces pharmaceutical coatings (Fig. 2.6). The wastewater had a daily volume of 3-4 m^3 and a COD of 60 $g{\cdot}l^{-1}$. The other is to treat a mixture of whey and cleaning wastewater from a cheese dairy plant (Fig. 2.7). The COD of the mixture is 20 $g{\cdot}l^{-1}$.

2.3. *Anaerobic membrane bioreactor*

Most of the dairy plants are often far away from sewage collection systems. To comply with environmental regulations, each production unit must be associated with an autonomous and efficient system for the removal of its pollutants. The combination of anaerobic treatment and ultrafiltration is a technique that has attracted a global interest. The anaerobic process allows

Fig. 2.6. PROVEO treatment plant for the effluent from a pharmaceutical industry.

Fig. 2.7. PROVEO treatment plant for the effluent from a cheese dairy.

the treatment of organic wastes with a low impact on the environment with both low energy consumption and low sludge production. The membrane treatment increases the soluble COD removal efficiency and delivers an effluent free of SS, bacteria and pathogens. The effluent can then be either recycled by the dairy plant, or used for irrigation or safely discharged into the environment. An industrial pilot plant has been in operation for three years to validate this treatment technology previously developed at laboratory scale.

2.3.1. *Process overview*

The anaerobic membrane bioreactor principle is schemed in Fig. 2.8. This design focuses on a specific coupling approach that makes possible to dissociate the dynamics of each process, thus offering a better flexibility to the whole system. The industrial pilot plant was installed on an industrial site and fed with real wastewaters. Two wastewaters with different loads were treated. The mean compositions of these wastewaters are summarized on Table 2.1.

The pilot plant consists of an anaerobic fixed bed reactor ($9\,m^3$) coupled with an external cross-flow ultrafiltration unit equipped with inorganic membranes. The reactor is filled with $5\,m^3$ of PVC media (Flocor). The unit of ultrafiltration is a module X UNIT 14 provided by Pall-Exekia. The membranes used have the average pore sizes of 20 and 100 nm. The membrane area installed is $0.72\,m^2$. The pressure is 5 bar and the maximum flow of circulation is $5\,m{\cdot}s^{-1}$.

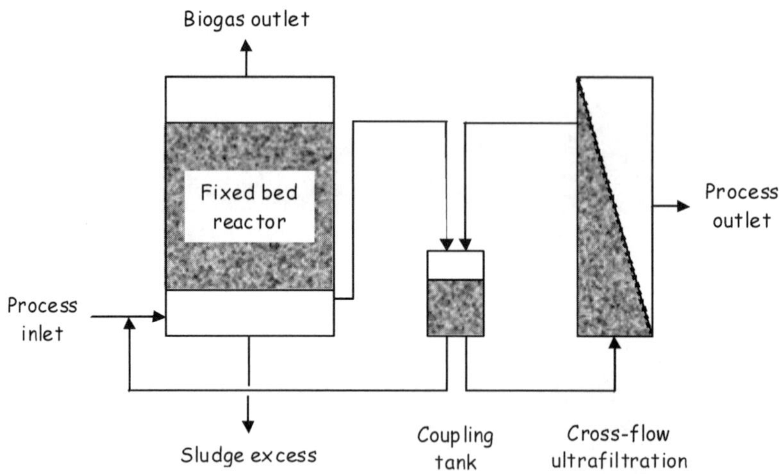

Fig. 2.8. The anaerobic fixed bed membrane bioreactor principle.

Table 2.1. Characteristics of dairy wastewaters.

	Wastewater 5–10 g-COD·l^{-1}	Wastewater 25–30 g-COD·l^{-1}
COD (mg·l^{-1})	5935 ± 2362	26784 ± 9821
TS (mg·l^{-1})	4088 ± 1988	11952 ± 4064
SS (mg·l^{-1})	1472 ± 1271	5360 ± 2417
Fat (mg·l^{-1})	1400	6300
pH	4.97 ± 1.07	3.71 ± 0.32

A 1-m^3 high-density polyethylene (HDPE) holding tank allows the hydraulic dissociation of reactor and the ultrafiltration unit.

Agitation in the anaerobic reactor is ensured by an external liquid recirculation loop. The heating of the reactor (37 °C) is carried out via a plate heat exchanger placed on the reactor recirculation loop. The following operational parameters are recorded electronically: temperatures (circulation loop and reactor); flow rates (feed and recirculation); biogas composition (CH_4 and CO_2); and recirculation loop pH. The ultrafiltration unit is also equipped with its own data recording system.

2.3.2. *Performance*

Figures 2.9 and 2.10 illustrate the COD removal efficiencies by the combined process (anaerobic treatment plus UF), the anaerobic treatment alone and the increased efficiencies due to the UF treatment for two wastewaters: one with 5–10 g-COD·l^{-1} and the other 25–30 g-COD·l^{-1}. Both final effluents of after the UF treatment were free of SS. Using a 20-nm membrane treating wastewater of 5–10 g-COD·l^{-1} at HRT ranging 15–30 h, we could reach an OLR of 10 kg-COD·m^{-3}·d^{-1} with 95–98% COD removal efficiency and a membrane permeate of 270 l·m^{-2}·h^{-1}. On the other hand, treating a wastewater of 25–30 g-COD·l^{-1} at HRT ranging 84–134 h, we could reach an OLR of 6.5 kg-COD·m^{-3}·d^{-1} with 96–99% COD removal efficiency and a membrane permeate of 110 l·m^{-2}·h^{-1}.

2.4. *Organic solid waste digestion: the ERGENIUM^{TM} technology*

Industrial process ERGENIUM^{TM} is developed from laboratory studies. This two-step process is of particular interest for the treatment of organic solid wastes. The hydrolysis rate is enhanced by the recirculation of treated effluent. The recirculation lowers the VFA concentrations and thus stabilizes the pH, which may otherwise inhibit the degradation.

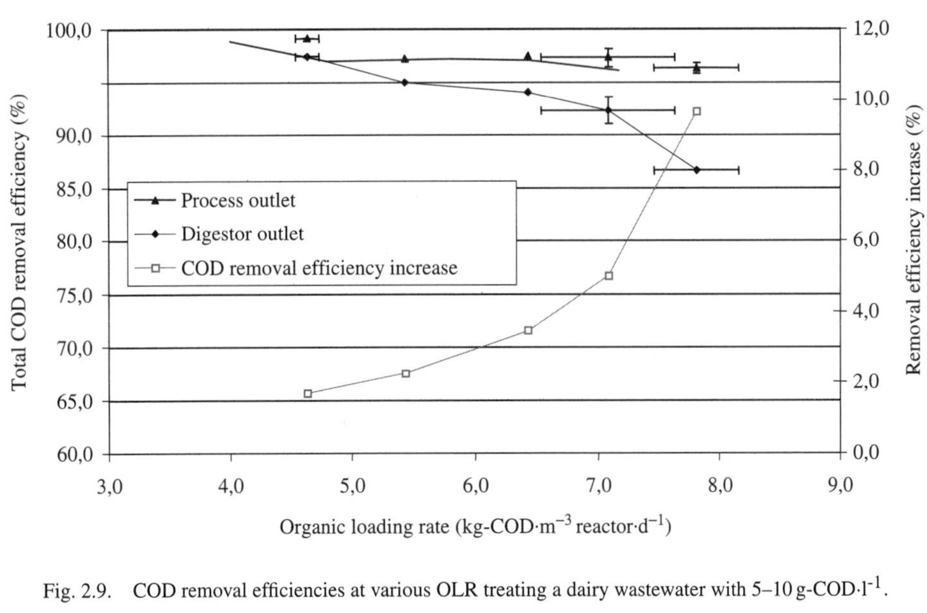

Fig. 2.9. COD removal efficiencies at various OLR treating a dairy wastewater with 5–10 g-COD·l^{-1}.

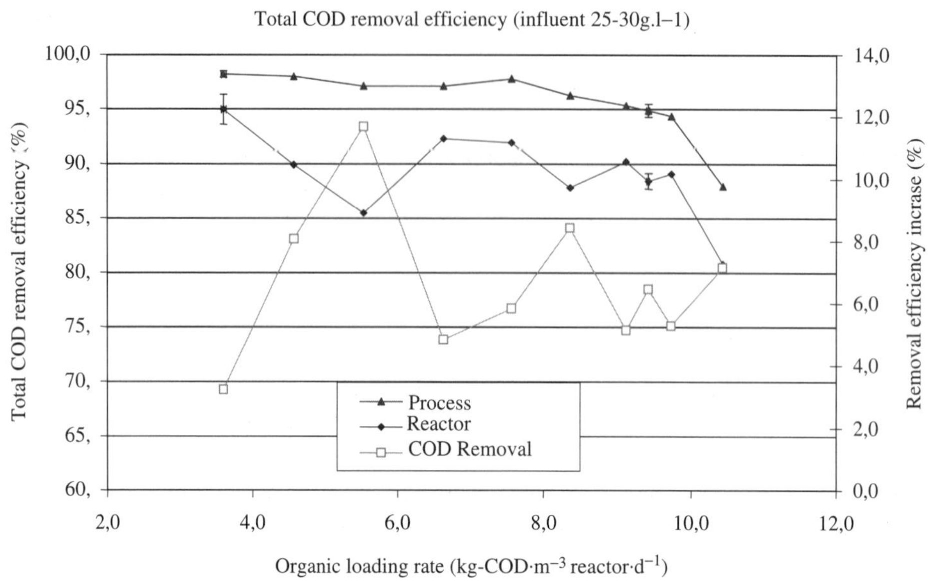

Fig. 2.10. COD removal efficiencies at various OLR treating a dairy wastewater with 25–30 g-COD·l^{-1}.

2.4.1. *Description of the process*

The ERGENIUM^{TM} process (Fig. 2.8) is based on the principle of decoupling the residence times of solid and liquid organic matters. In this system, organic solid wastes are hydrolyzed and acidified in a reactor. Part of the effluent of this reactor is recycled, and the rest is treated in a separate methanogenic reactor. Therefore this process is a two-stage system in which leachate recycling ensures the transfer of dissolved organic matter from the hydrolysis to the methanogenic step. This flushing effect is already described by Banks and Wang (1999). In addition, the solid residence time is also adjusted through solid recycling after liquid–solid separation.

Both steps are performed in separate optimal conditions (physical and chemical), which also justifies a separation of the two operations. Separation of soluble matters from solids has two advantages: (a) dedicated treatment of VFAs from the hydrolysis and acid-forming steps; and (b) improved organic matter degradation, especially hydrolysis, avoiding the accumulation of potentially inhibitory compounds and the reduction of pH.

The ERGENIUM^{TM} process has been patented in France (French deposed number N° FR0757492). The solid–liquid separation makes possible to control the residence time of the solids in the process. This is interesting and innovative for the following reasons:

- adjustment of the solid content in the reactor;
- adjustment of the content of active biomass to reach optimal conditions for hydrolysis; and
- management of the flow of organic matters (and particularly avoiding the accumulation of inert organic matters in the hydrolysis reactor).

2.4.2. *Model and simulation*

A model has been developed for the design of the ERGENIUM^{TM} process. This model accounts for the two stages, each with a phase separation, and a double recirculation (R_1 and R_2 in

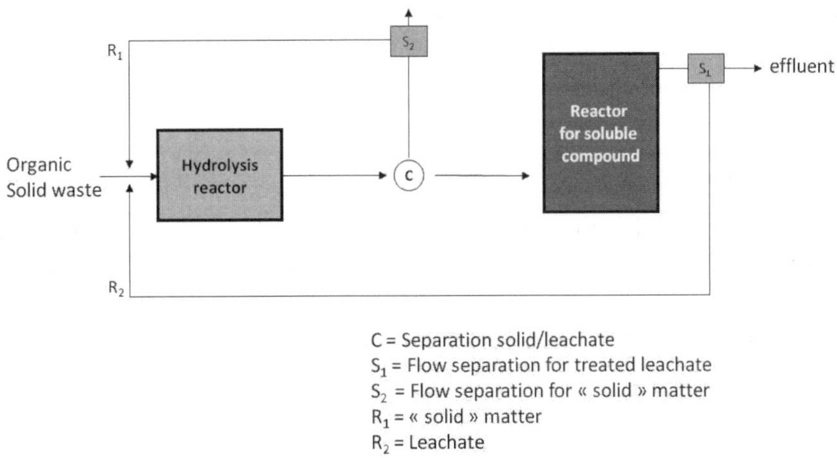

Fig. 2.11. Flowsheet of the ERGENIUM^{TM} process.

Fig. 2.11). In this model, the organic matter is fractionated into four parts as shown in Fig. 2.12. This description is done according to the analytical methodology described by Buffière and Frédéric (2008).

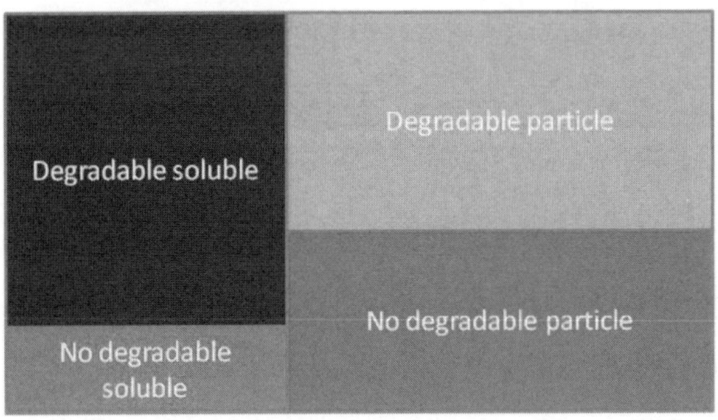

Fig. 2.12. Composition of organic matter.

The design of ERGENIUM^{TM} is thus performed according to the desired parameters, namely:

- The rate of degradability of the organic substrate;
- The VFA concentration in the hydrolysis reactor;
- The solid content in the hydrolysis reactor; and
- The concentration of inerts (soluble and particulate).

The model is based on the mass balance of soluble and solid matters written in stationary state in order to assess the flows of solids, volatile solids and methane under different operating conditions. Three design parameters are tested:

- The organic loading rate;
- The ratio of flow separation for "treated leachate" S1 (see Fig. 2.8); and
- The ratio of flow separation for "solid matter" S2 (see Fig. 2.8).

The influence of those parameters on the efficiency of the system can thus be assessed according to several criteria, such as particulate solid removal efficiency, soluble compound concentration, soluble inert content in the hydrolysis reactor.

2.4.3. *Implementation of ERGENIUM^{TM} with a total recycle of C, N and P*

To prove the concept of this technology, a prototype of this technology will be implemented in the suburb of Paris. The main features of this project, called ErgeO (Fig. 2.13), are the following:

- The system consists of a two-stage anaerobic treatment according to the ERGENIUM^{TM}.

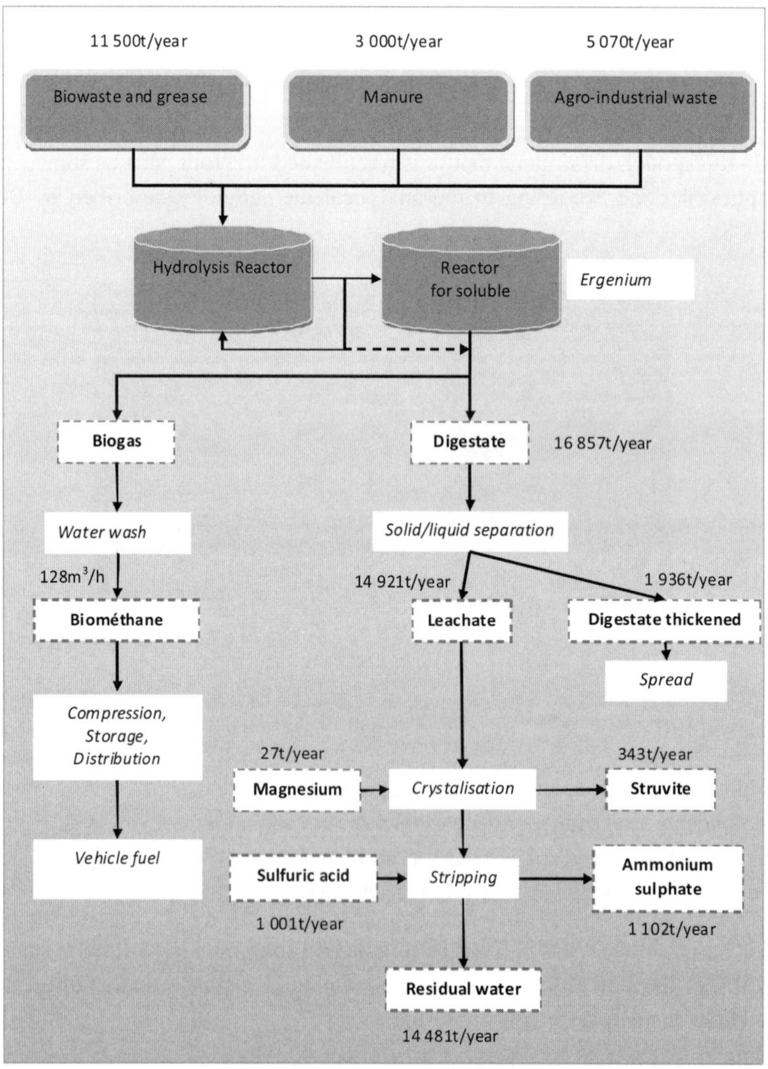

Fig. 2.13. Flow of different reagents and products for the project ErgeO.

- Leachate obtained after phase separation undergoes a crystallization and a stripping step to recover nitrogen and phosphorus. The effluent containing only traces of minerals and organic matters, is sent to either sewage treatment or irrigation.

Finally, the recovered biogas will be purified and compressed to power a fleet of waste-collecting vehicles.

2.5. *Control of biogas quality*

Presently the world derives most of its energy from fossil fuels that are expected to be depleted in the near future. It becomes thus crucial to find alternative sources of renewable energy. In addition to the conventional sources of renewable energy, such as solar, wind and hydroelectric energy, biomass offers great prospects. Among the biofuels, biogas produced from anaerobic treatment is very promising, as the product methane can be readily used for electricity and/or heat production. However, the cost reduction that could be generated from the energy recovery might be counterbalanced by the fluctuating and inconsistent biogas quality. The energy value is indeed directly linked to the CH_4 content in the biogas. It is thus crucial for the plant manager to operate the process at maximal biogas production rate and with a constant composition.

Yet, the biogas quality is hard to predict and even harder to control. In a previous work (Hess and Bernard 2006), a mechanistic model was developed for the gas–liquid transfer coefficient $K_L a$ in order to accurately determine the dissolved CO_2 in the liquid phase. This model took into account the pressure distribution along the height of the reactor and included bubbles nucleation and rise in the water column to assess more precisely the surface of exchange. A positive relationship between the average $K_L a$ and the biogas flow rate at the top of the column was pointed out, which corroborated the relationship mentioned by Kalyuzhnyi *et al.* (2001).

Here we introduce a simplified model for the liquid–gas exchange surface. This leads to a new model of gas transfer which will be used within a control strategy developed for the optimization of the biogas quality. The controller is tested and validated on a 0.5-m^3 pilot plant. Details are described in the following sections.

2.5.1. *Description of the developed process*

The influent is a wine distillery wastewater obtained from a local winery in Narbonne, France. The process is a 528 l up-flow anaerobic fixed bed reactor (Fig. 2.14). The following parameters are measured on-line: input and recirculation liquid flow rates, pH in the reactor and the influent, temperature, biogas production rate, contents of CO_2, CH_4 and H_2 in the biogas, VFA and bicarbonate concentrations, plus total alkalinity (TA) and partial alkalinity (PA) in the liquid phase. More details on the process can be found elsewhere (Steyer *et al.* 2002).

2.5.2. *Simplified model of* $K_L a$

The mechanistic model presented in Hess and Bernard (2006) pointed out a linear correlation between the biogas flow rate and the average gas–liquid transfer coefficient $K_L a$ without a major significance of the pressure distribution, given the small height of our pilot plant. In order to assess more easily the total surface of exchange, average properties will be here considered for all the bubbles inside the water column: a mean size, a mean composition (*i.e.*, no exchange

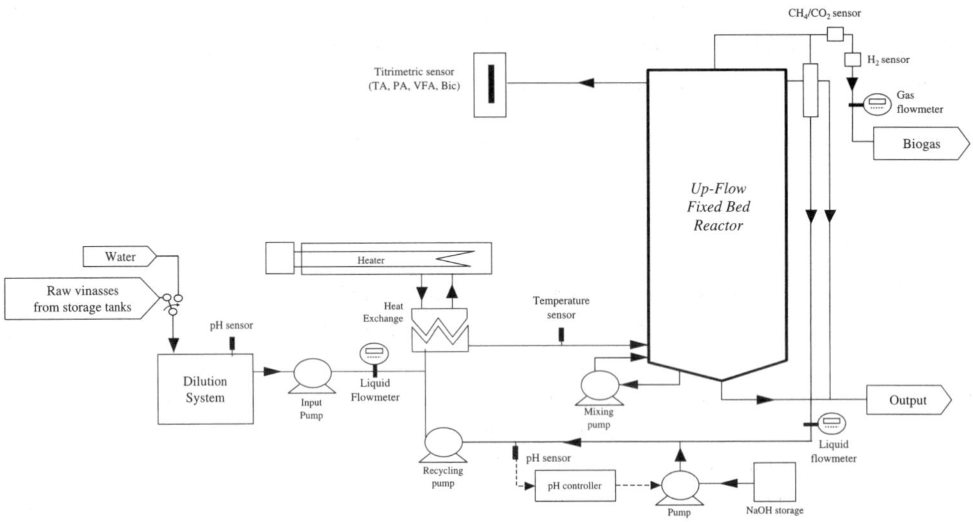

Fig. 2.14. Schematic layout of the up-flow anaerobic fixed bed treatment plant.

during the bubbles elevation) and a mean rise velocity depending on the bubble radius. A linear trend between $K_{\mathrm{L}}\mathrm{a}$ and the gas flow rate q_{R} can be expressed as:

$$K_{\mathrm{L}}\mathrm{a} = \gamma \cdot q_{\mathrm{R}} \tag{2.1}$$

where γ is a proportional constant. Moreover, the average molar CO_2 transfer (from gas to liquid) rate can be expressed as $q_{\mathrm{C}} = K_{\mathrm{L}}\mathrm{a}(C_{\mathrm{CO_2}}\text{-}K_{\mathrm{H}} \cdot P_{\mathrm{C}})$ where $C_{\mathrm{CO_2}}$ is the dissolved concentration of CO_2, K_{H} the Henry's constant for the CO_2 and P_{C}, the partial pressure of CO_2. If P_{C} is expressed using CH_4 and CO_2 flow rates and by replacing $K_{\mathrm{L}}\mathrm{a}$ with Eq. (2.1), we finally get the following relation between the biogas quality and the dissolved CO_2:

$$\%C_{\mathrm{CO_2}} = \frac{RT\gamma}{P_{\mathrm{T}} + RT\gamma K_{\mathrm{H}} P_{\mathrm{T}}} C_{\mathrm{CO_2}} \tag{2.2}$$

where P_{T} is the headspace pressure. It thus appears that controlling the biogas composition can be performed through the control of the dissolved CO_2 in the liquid phase.

2.5.3. *Control algorithm*

As previously highlighted, the biochemical equilibrium between the dissolved and the gaseous CO_2 in the anaerobic reactor is driven by the pH; for a fixed concentration of total inorganic carbon, the higher the pH, the higher the fraction of dissolved CO_2. Addition of caustic soda increases the reactor alkalinity and along with it the pH. This could be a way to regulate the dissolved CO_2 and, as a consequence, the biogas quality. Simple control systems (*i.e.*, a proportional (P) and proportional-integral (PI)) and regulation scheme based on the measurement of the CO_2 content in the biogas and alkali addition directly inside the reactor can thus be implemented.

2.5.4. *Assessment of process performances*

Soda was added into the recycling flow (from the top to the bottom of the reactor) just in front of the pH meter (see Fig. 2.14). The pH measurements are thus highly affected by the controller,

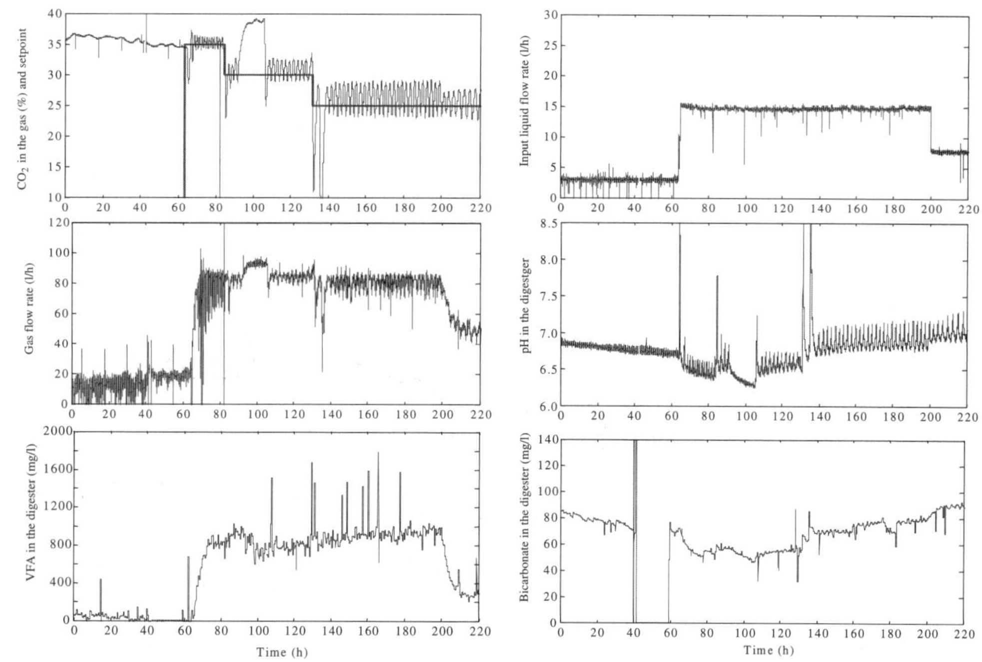

Fig. 2.15. Experimental validation of the control of CO_2 content in biogas from base addition in the liquid phase — controller is a variable gain P-controller.

and fluctuations would have been less pronounced if the pH was measured inside the reactor. However, from a practical point of view, this was unfortunately impossible. The pH shown in Figs. 2.3 and 2.4 were thus those in the recycling loop, the pH in the reactor being the base line.

The first implemented controller is a basic controller (*i.e.*, a P-controller with variable gain) that serves as a reference to evaluate the benefits of implementing more advanced controllers. The controller was switched on at $t = 62$ h with a set-point equal to 35% (see Fig. 2.15). At $t = 83$ h and $t = 131$ h, the set-point was changed to 30% and 25%, respectively. Feed flow was changed from $3\,l\cdot h^{-1}$ to $15\,l\cdot h^{-1}$ when the controller was switched on to evaluate its ability to maintain constant biogas composition. It was later (at $t = 200$ h) reduced to $8\,l\cdot h^{-1}$ to induce an additional disturbance. Two main technical problems occurred during this experiment. Firstly, from $t = 90$ to 105 h, the pump in charge of the base addition was clogged and no control could be performed during this period, leading the CO_2 to deviate from the set-point. Then, between $t = 131$ and 136 h, a similar operational problem was noticed twice, inducing high fluctuations of the CO_2 for few hours.

Small oscillations can be seen on the C_{CO_2}, mainly due to the process configuration that induces a delay between base addition and its effect in the gas phase. Nevertheless, the controller behaved very well, rejecting the main imposed disturbances, and no effect of the changes of the input flow rate (at $t = 63$ and 200 h) could be noticed on the CO_2 content. Moreover, the controller was able to track the different set-points imposed. From a more general perspective, the increase of liquid feed flow leading to an increase of biogas produced together with a VFA accumulation up to about $1\,g\cdot l^{-1}$ which was not detrimental to the process operation.

A second controller was implemented few days later to include a predictive term in order to minimize the oscillations of the CO_2 content in biogas. This controller is a "mixed" controller: the

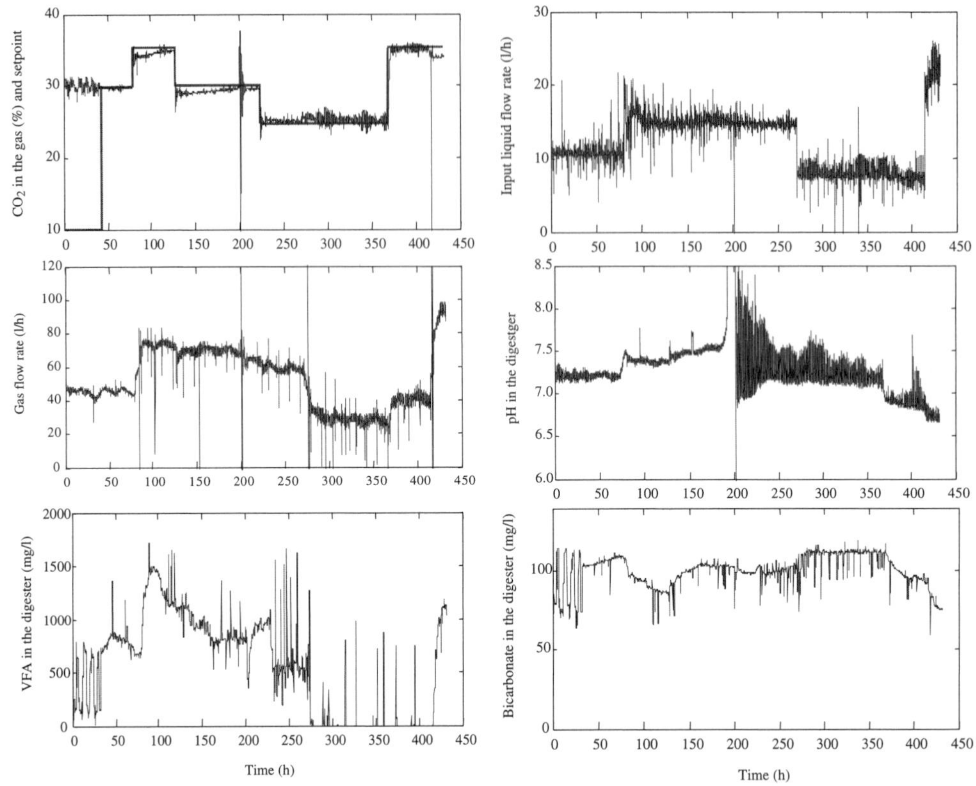

Fig. 2.16. Experimental validation of the control of CO_2 content in biogas from base addition in the liquid phase — controller is a mixed PI + linearizing controller.

regulation of the alkalinity is performed through a linearizing control and a PI-controller based on the measurement of the partial pressure of CO_2 to account for potential modeling errors. The controller was switched on at $t = 42$ h with a set-point equal to 30% (see Fig. 2.16). At $t = 80$, 128, 223 and 369 h, the set-point was, respectively, changed to 35%, 30%, 25% and again 35%.

Besides small technical problems (*e.g.*, faulty pH meter at $t = 200$ h for few hours), experiments demonstrated that the biogas quality could be very efficiently regulated around different desired set-points, despite changes of the organic loading rate. No oscillations are noticed on the CO_2 content in biogas, illustrating the benefits of the mixed controlled over the simple P-controller previously applied. The performance is even improved after $t = 200$ h when tuning parameters were modified (*i.e.*, increase of the integral term leading to a more active control — see pH evolution in Fig. 2.16) and the set-points are then perfectly followed without any offset. Moreover, the control did not induce any detrimental effects on the overall functioning of the plant (see additional variables in Fig. 2.16).

As a conclusion we can say that the control of the biogas quality is an important issue for a better recovery of bioenergy and thus for the better prospect of anaerobic technology. This objective was addressed in the present study through soda addition in the liquid phase and these results demonstrated the feasibility of the approach. Moreover, the amount of soda used was similar to the quantity commonly used to regulate the inlet pH and no additional soda consumption was noticed during the experiments. Although the proposed control could not overcome usual biogas purification techniques to achieve a very high biogas quality, it can help improve things through a smarter use of the caustic soda.

References

Arnaiz, C., Buffière, P., Elmaleh, S., Lebrato, J., and Moletta, R. (2003) Anaerobic digestion of dairy wastewater by inverse fluidization: The inverse fluidized bed and the inverse turbulent bed reactors. *Environ. Technol.* **24**, 1431–1443.

Banks, C. J. and Wang, Z. (1999) Development of a two phase anaerobic digester for the treatment of mixed abattoir wastes. *Wat. Sci. Technol.* **40**(1), 69–76.

Buffière, P. and Frédéric, S. (2008) A comprehensive method for organic matter characterization in solid wastes in view of assessing their anaerobic biodegradability. *Wat. Sci. Technol.* **58**(9), 1783–1788.

Garcia-Calderon, D., Buffière, P., Moletta, R., and Elmaleh, S. (1998) Anaerobic digestion of wine distillery waste-water in downflow fluidized bed. *Wat. Res.* **32**(12), 3593–3600.

Hess, J. and Bernard, O. (2006) Mechanistic K_La modeling in an up-flow anaerobic digester. In: *Proceedings of MATHMOD'06 Conference.* Vienna, Austria, 8 pages on CDROM.

Hwang, S. H. and Hansen, C. L. (1992) Performance of upflow anaerobic sludge blanket (UASB) reactor treating whey permeate. *Trans. ASAE* **35**, 1665–1671.

Imbeault, N., Paquet, M., and Cote, R. (1998) Volatile fatty acids production by anaerobic whey permeate biodegradation in continuous bioreactor. *Wat. Qual. Res. J. Can.* **33**, 551–563.

Jahren, S. J., Rintala, J. A., and Odegaard, H. (2002) Aerobic moving bed biofilm reactor treating thermo-mechanical pulping whitewater under thermophilic conditions. *Wat. Res.* **36**(4), 1067–1075.

Jantsch, T. G. and Mattiasson, B. (2004) An automated spectrophotometric system for monitoring buffer capacity in anaerobic digestion processes. *Wat. Res.* **38**(17), 3645–3650.

Kalyuzhnyi, S., Fedorovich, V., and Lens, P. (2001) Novel dispersed plug flow model for UASB reactors focusing on sludge dynamics. In: *Proceedings of the 9th International IWA Conference on Anaerobic Digestion*, Antwerpen, Belgium.

Rajeshwari, K. V., Balakrishnan, M., Kansal, A., Lata, K., and Kishore, V. V. N. (2000) State-of-the-art of anaerobic digestion technology for industrial wastewater treatment. *Renew. Sustain. Energy Rev.* **4**(2), 135–156.

Steyer, J. P., Buffière, P., Rolland, D., and Moletta, R. (1999) Advanced control of anaerobic digestion process through disturbances monitor. *Wat. Res.* **33**(9), 2059–2068.

Steyer, J. P., Bouvier, J. C., Conte, T., Gras, P., and Sousbie, P. (2002) Evaluation of a four year experience with a fully instrumented anaerobic digestion process. *Wat. Sci. Technol.* **45**(4–5), 495–502.

Chapter 3

Applications and New Developments of Biogas Technology in Japan

Yu-You Li[1,2,*] and Takuro Kobayashi[2]

[1]*Department of Environmental Science, Graduate School of Environmental Studies, Tohoku university, 6-6-06 Aza-Aoba, Aramaki, Aoba-ku, Sendai 9808579, Japan*

[2]*Department of Civil and Environmental Engineering, Graduate School of Engineering, Tohoku University, 6-6-06 Aza-Aoba, Aramaki, Aoba-ku, Sendai 9808579, Japan*

**yyli@epl1.civil.tohoku.ac.jp*

This chapter is to summarize the history and applications of anaerobic technology in Japan for decades to industrial wastewater treatment, sewage sludge digestion and methane fermentation of food/agricultural wastes, based on literature available only in Japanese. Over 300 upflow anaerobic sludge bed (UASB) and expanded granular sludge bed (EGSB) full-scale plants are now in operation for the treatment of beer, soft drink, distillery, food and chemical wastewaters. Anaerobic sludge digestion began in 1932, and is now used in over 300 sewage treatment plants with a total digester volume of 2.1×10^6 m^3, producing enough methane to heat the digester or to generate biofuel or electricity. Since the enactment of the Basic Act for the Promotion of the Recycle-Oriented Society, a national program called Biomass Nippon Strategy started in 2002 has attracted much public attention to the concept of converting the annual 89×10^6 tons of livestock wastes to resources. The estimated potentials of methane recovery are 1.6×10^9 $m^3 \cdot y^{-1}$ from the livestock wastes and 1.7×10^9 $m^3 \cdot y^{-1}$ from the municipal solid wastes (MSW). Since 2006, over seventy methane fermentation plants have been in operation treating livestock wastes, and another fifty plants for the treatment of food wastes and the organic fraction of MSW. Recent developments of biogas technology include the application of anaerobic membrane bioreactor and bio-desulfurization of biogas.

1. Introduction

Japan has for decades applied anaerobic technology for the treatment of industrial wastewater, night soil, sewage sludge and food/agricultural wastes. In recent years, anaerobic technology has aroused much public interest because of its potential to help offset global warming, establish a recycling-oriented society, and effectively utilize biomass. The anaerobic treatment of organic solid wastes and industrial wastewater is becoming increasingly common in Japan. Over 300 industrial wastewater plants are using UASB and EGSB treatment processes. In addition, fermentative methane production from organic solid wastes is considered to be one of the dominant technologies for recovering biofuel from the biomass stock in Japan. Based on the stoichiometry shown in Table 3.1 for methanogenic fermentation, the estimated potential of methane production from various potential feedstocks in Japan is shown in Table 3.2. It is noteworthy that the total amount of annual potential methane production corresponds to about 3,950 Ml of crude oil, comparable to the 2010 national goal of overall biomass utilization (4,000 $Ml \cdot y^{-1}$). It is thus

Table 3.1. Stoichiometry of methane fermentation (Li and Nishimura 2007).

Feedstock	Stoichiometry of methane fermentation	Biogas yield ($m^3 \cdot kg^{-1}$-VS)	Methane (%)
Carbohydrate	$(C_6H_{10}O_5)_n + nH_2O \rightarrow 3nCH_4 + 3nCO_2$	0.830	50.0
Protein	$C_{16}H_{24}O_5N_4 + 14.5H_2O \rightarrow 8.25CH_4 + 3.75CO_2 + 4NH_4^+ + 4HCO_3^-$	0.764	68.8
Lipid	$C_{50}H_{90}O_6 + 24.5H_2O \rightarrow 34.75CH_4 + 15.25CO_2$	1.425	69.5
Cattle manure	$C_{22}H_{31}O_{11}N + 10.5H_2O \rightarrow 11.75CH_4 + 9.25CO_2 + NH_4^+ + HCO_3^-$	0.970	56.0
Municipal solid waste	$C_{46}H_{73}O_{31}N + 14H_2O \rightarrow 24CH_4 + 21CO_2 + NH_4^+ + HCO_3^-$	0.888	53.3
Sewage sludge	$C_{10}H_{19}O_3N + 5.5H_2O \rightarrow 6.25CH_4 + 2.75CO_2 + NH_4^+ + HCO_3^-$	1.003	69.4

Table 3.2. Potential of biogas production in Japan.

Feedstock	Amount ($Mt \cdot y^{-1}$)	Biogas yield ($m^3 \cdot t^{-1}$)	Methane (%)	Methane production ($Gm^3 \cdot y^{-1}$)	Crude oil equivalent ($Ml \cdot y^{-1}$)
Municipal solid wastes	22	130	60	1.72	1,720
Livestock wastes	89	30	60	1.60	1,600
Sewage sludge	75	14	60	0.63	630
Total	186	174	60	3.95	3,950

fair to state that the effort made over the last few decades to develop new biogas technologies and to promote their applications stands to bear considerable fruit. In this chapter, the new developments and recent applications of biogas technology in Japan are outlined, with a focus on industrial wastewater treatment, digestion of sewage sludge, methanogenic fermentation of food/agricultural waste, and new developments in biogas plant technology. This overview, based on mostly literature available in Japanese, shall enable readers to understand the trends and the issues affecting the biogas technology in Japan.

2. Applications of Biogas Technology in Japan

2.1. *Sewage sludge*

2.1.1. *History*

The history of anaerobic digestion of sewage sludge in Japan dates back to 1932 when the nation's first anaerobic digester treating sewage sludge from wastewater treatment plants (WWTPs) was built in Nagoya. In the subsequent years, digesters were built in big cities such as Tokyo, Osaka, and Kyoto. During the World War II (WWII), there was a considerable interest in using recovered biogas as fuel for automobiles. During the 1960s and 1970s, the establishment of sewage lines throughout Japan accelerated the pace of construction of anaerobic digesters in WWTPs; over 180 plants were installed with anaerobic digesters. In more recent years, the high level of interest in biomass energy resulted in a fresh look into the anaerobic digestion of sewage sludge by the government and researcher institutes alike. This attention spurned an interest in high-rate methane-producing processes for treating sewage sludge, with research emphasis on physicochemical pretreatment, thermophilic digestion, and temperature-phased digestion.

2.1.2. *Applications*

In 2001, there were 305 anaerobic digesters in WWTPs throughout Japan, treating 34.5% of the sewage sludge. There are two typical process configurations of sludge treatment: both start with thickener followed by anaerobic digestion and dewatering, with one process finishing up with composting and the other with incineration. The average conditions of all the 305 methanogenic plants operated in Japan, as shown in Fig. 3.1, are summarized as follows (Japan Sewerage Association 1998): (1) digester volume, 7,600 m^3; (2) hydraulic retention time (HRT), 34 d; and (3) temperature, 34 °C. Because of the low total solids (TS) concentration of influent, the biogas production is only averaging 11 $m^3 \cdot m^{-3}$. The average TS and volatile solids (VS) reductions are 51% and 57%, respectively. The biogas is typically composed of 60% of CH_4, 36% of CO_2, and 1,164 ppm of H_2S.

2.1.3. *Example 1: Sludge treatment centers in Yokohama city (Li and Nishimura 2007)*

In Yokohama, there are two sludge treatment centers (Hokubu and Nanbu) collecting the sewage sludge produced from eleven WWTPs using pipelines. The process flow of the Hokubu sludge treatment center is shown in Fig. 3.2. In this plant, the sludge collected from five WWTPs is treated in twelve egg-shaped mesophilic anaerobic digesters ($12 \times 6, 800\, m^3$) with 30 days of HRT after concentrating by centrifugation to 5% TS. The recovered biogas is mainly converted into electricity and heat energy being used a gas generator and fuel cells, as well as being used as fuel for the sludge incinerator. Before anaerobic digestion, the TS concentration of the influent sludge can be controlled at 4.5–5.0% by centrifugation process. The average pH and alkalinity in the digesters are 7.25 and 4,800 $mg \cdot l^{-1}$, respectively, and both remain stable throughout the year. Reductions of TS, VS and COD_{Cr} are 44.2%, 54.2% and 60.0%, respectively. Biogas production

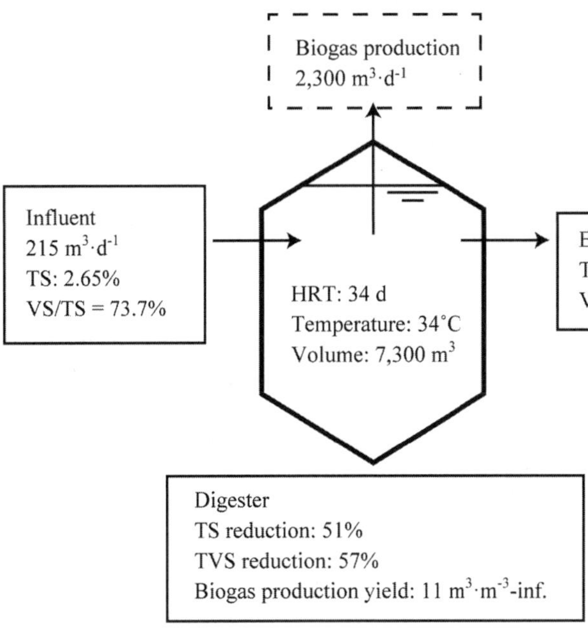

Fig. 3.1. Average performance of the 305 methanogenic plants in Japan in 2001.

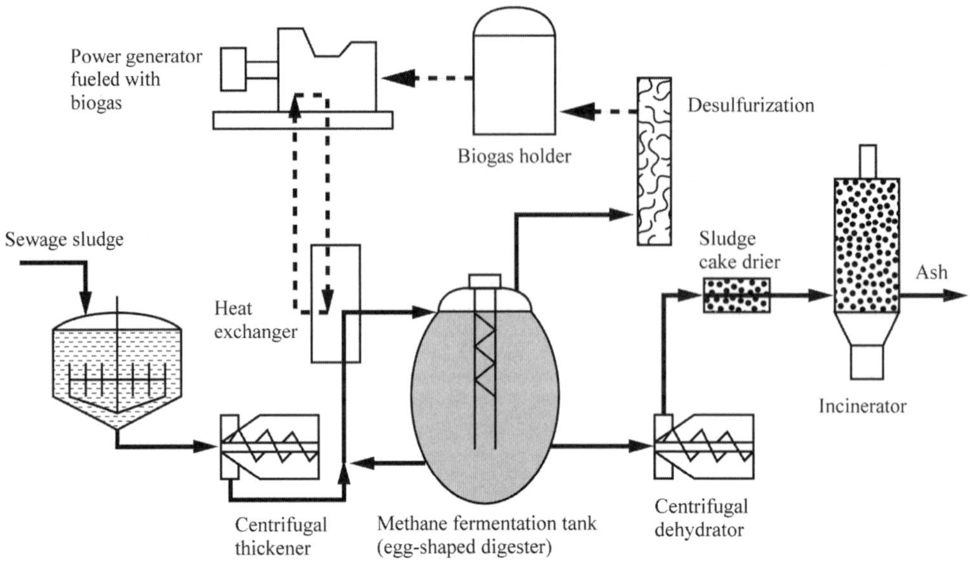

Fig. 3.2. Flow chart of the Hokubu sludge treatment center in Yokohama.

magnifications range from 20 to 30 $m^3 \cdot m^{-3}$, which are about twice the average of other WWTPs in Japan.

2.1.4. *Example 2: Process configuration of the WWTP in Yamagata (Noike 2008)*

The WWTP in Yamagata, with a treating capacity of 40,000 $m^3 \cdot d^{-1}$ of wastewater, successfully treats sewage sludge (TS 4.69%) of 227 $m^3 \cdot d^{-1}$ using four mesophilic anaerobic digesters. Table 3.3 shows the energy balance of this plant. The digesters produce 4,082 $m^3 \cdot d^{-1}$ of biogas containing 58% methane, and the digested sludge (TS 2.21%) is dewatered and then composted. The biogas generates 7,226 $kWh \cdot d^{-1}$ of electricity, corresponding to 48.7% of the total energy consumption of this plant. This is much higher than what was expected before the plant went into operation.

2.2. *Night soil (human waste)*

2.2.1. *History*

Since ancient times, night soil was generally stored as fertilizer for agricultural use in Japan. However, the use of night soil has been gradually replaced by chemical fertilizer after the WWII.

Table 3.3. Energy balance in the wastewater treatment plant of Yamagata.

Sewage sludge ($m^3 \cdot d^{-1}$)	227
Biogas production ($m^3 \cdot d^{-1}$)	4,082
CH_4 (%)	58
CO_2 (%)	40
Energy consumption ($kWh \cdot d^{-1}$)	14,823
Electric generation from biogas ($kWh \cdot d^{-1}$)	7,226
Supply of plant electricity consumption (%)	48.7

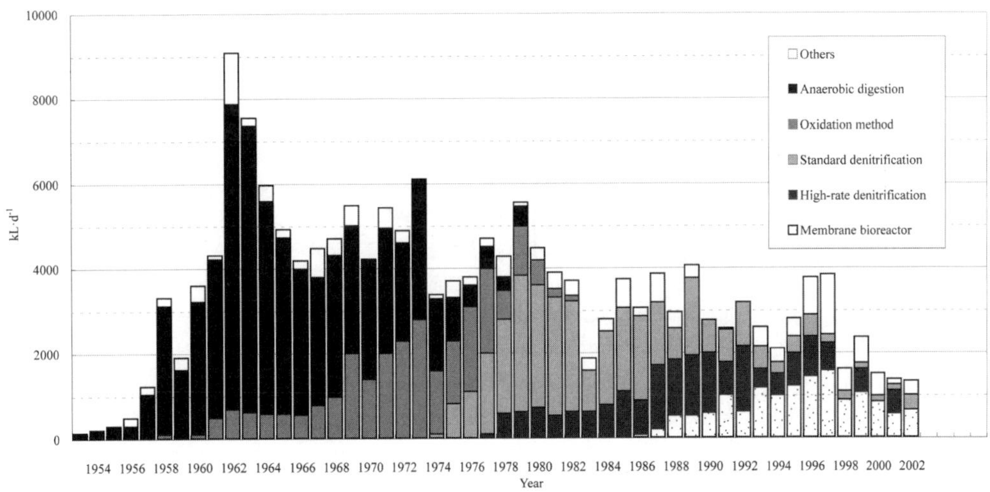

Fig. 3.3. The history of night soil treatment methods in Japan.

Advances in community hygiene in Japan came with awareness that the storage of night soil was responsible for a number of significant health problems, such as parasite and bacterial infections. The treatment of night soil in Japan started in the 1950s (Fig. 3.3). In 1950, the Economic Stabilization Agency, supported by the General Headquarters of the allied forces recommended the treatment of stored night soil using anaerobic digestion. Years later, in the 1960s, fundamental studies into the anaerobic digestion of night soil were carried out (Matsumoto 1962; Endo 1965; Matsumoto and Onuma 1965), and technical guidelines were proposed. During the period of 1953–1975, anaerobic digestion was one of the major methods for treating night soil in Japan, but the tightening of water quality standards required the BOD/N ratio of the effluent be kept below 3. As a result, most plants began to employ biological nitrogen removal processes for the treatment of night soil, instead of anaerobic digestion. In recent years, new technologies including high-rate denitrification and membrane bioreactor have been developed and applied in many plants. Now, the prevalence of the sewer system has resulted in a decrease in the treatment of stored night soil, as shown in Fig. 3.3.

2.2.2. *Applications*

The typical characteristics of night soil produced in Japan are summarized as follows: pH 7.6, BOD 7,800 mg·l^{-1}, COD_{Mn} 4,700 mg·l^{-1}, SS 8,300 mg·l^{-1}, T-N 2,700 mg·l^{-1}, T-P 350 mg·l^{-1}, and coliform 10^6–10^7 CFU·ml^{-1}. Only 6% of the total amount of night soil produced (5,856 m^3·d^{-1}) was treated by anaerobic digestion in 2006 (Morikawa 2009). Figure 3.4 illustrates a typical night soil treatment process using anaerobic digestion. The process consists of a two-stage anaerobic digester, a standard activated sludge tank and a settling tank with a volume ratio of 30:6.25:2.5. During 20–30 days of anaerobic digestion, about 50% of influent VS is converted to biogas (300–400 m^3·t-VS^{-1}) containing 45–65% methane. The digester effluent is diluted before being treated with an activated sludge process with a BOD loading rate of 0.4 kg·m^{-3}·d^{-1} or less. The final treated effluent contains 13 mg·l^{-1} of SS, 3.8 mg·l^{-1} of BOD, 120 mg·l^{-1} of T-N and 4.5 mg·l^{-1} of T-P.

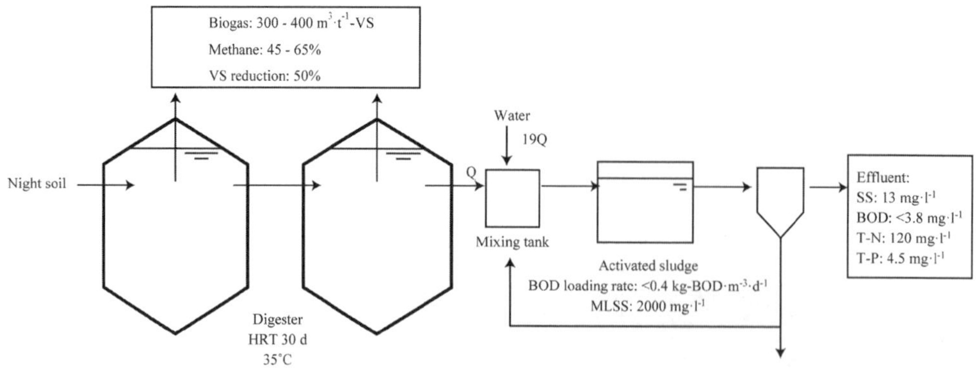

Fig. 3.4. Typical flow and mass balance in the anaerobic digestion of night soil (Morikawa 2009).

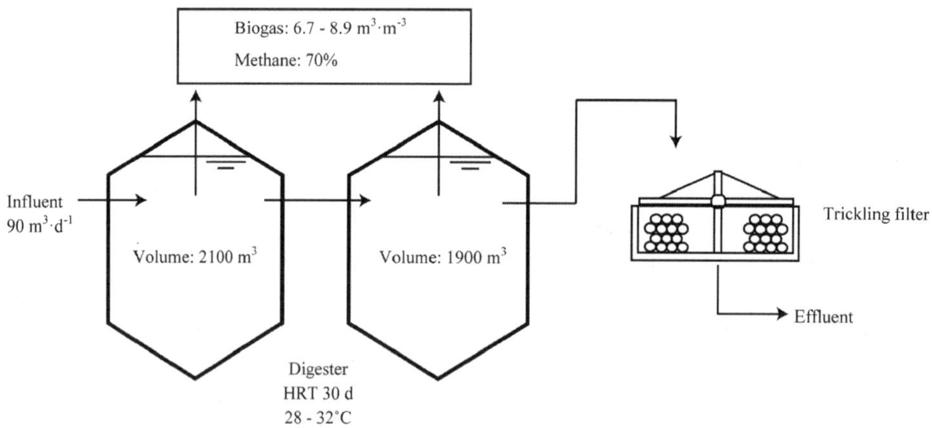

Fig. 3.5. Flow diagram of the Hokko plant.

2.2.3. *Example 1: Hokko plant (Inoue 2009)*

The Hokko plant in Hokkaido began operating in 1956 to treat the night soil (90 m^3·d^{-1}) produced from Sapporo. A flow diagram of this plant is shown in Fig. 3.5. The process consists of a two-stage mesophilic anaerobic digester and a trickling filter. Thirty days of anaerobic digestion produces 6.7–8.9 m^3·m^{-3} of biogas containing 70% of methane. The supernatant of the digested sludge is treated by a trickling filter packed with 6–10 cm pebbles.

2.3. *Livestock waste*

2.3.1. *History*

Over the years, there have been three movements in the anaerobic digestion of livestock waste. The first was in the 1950s and 1960s when farmers thought it was a smart new technology for energy production. During this time, many technological developments were being heavily promoted, but it failed to take off due to both the low levels of gas produced during the winter and the prevalence of liquefied petroleum gas. The energy crises in 1973 and 1980, however, led to renewed attention to anaerobic digestion. By the early 1980s, anaerobic digestion had been the focus of a great number of studies globally, and, by 1981, there were thirty-four digesters demonstrably in operation in Japan. In the period that followed, however, oil prices went down and stayed low for quite a while, effectively making those digesters redundant. The third wave of interest in anaerobic digestion has come more recently with the effort to create a recycling-oriented society. In Japan, the Basic Act for the Promotion of the Recycling-Oriented Society was enacted in 1999, and renewable energy sources, waste recycling and sustainable agriculture became issues that society as a whole began to take a wide interest in. The Biomass Nippon Strategy, which was first implemented in 2002, brought livestock wastes as a biomass resource to the public's attention. In Japan, the yearly production of 89 million tons of livestock waste makes it the largest biomass resource in the organic wastes category. The estimated potential of annual methane recovery from livestock waste (1.6 billion m^3) is the second largest following that from municipal solid wastes (MSW, 1.7 billion m^3) in Japan. A number of biogas plants treating livestock waste have been established in cities across Japan since 2000. In these plants, the biogas recovered from digesters has been converted into heat or electric energy and successfully utilized. The digested sludge tends to have been used as liquid fertilizer, or otherwise dewatered being composted. Developing better treatment methods for digested sludge is one of the most important challenges for the better management of biogas plants in Japan. There are two reasons for this. One is the lack of agricultural landscape available to farmers to spread digested sludge because farms in Japan are considerably smaller than those in many other countries. The estimated amount of nitrogen from livestock waste is around 710 thousand ton/y, which is approximately 1.5 times larger than the nitrogen requirement of agriculture in Japan. Another reason is the poor framework for utilizing digested sludge for agriculture in Japan. This means that the cost of treating redundant sludge weighs on the biogas plant operator. To summarize, the biogas plants treating livestock waste are focused on two issues: the promotion of the use of fertilizer and cost reduction of wastewater treatment.

2.3.2. *Applications*

In 2005, there were seventy biogas plants treating livestock waste throughout Japan. As shown in Fig. 3.6, 66% of these plants are in Hokkaido. This can be explained by the larger area of farmland in Hokkaido which can make use of digested sludge than those in other areas of Japan. Table 3.4 illustrates the process mode of biogas plants treating livestock waste in Japan. Half of the seventy plants treat only livestock waste, with cattle manure being the most utilized followed by swine manure. Nine plants treat mixture of livestock waste and MSW, and another four plants treat a mixture of livestock waste, MSW and sludge. Plants are divided into three scales in terms of their treatment capacity: twenty-seven small-scale plants ($<10\,t{\cdot}d^{-1}$), nineteen middle-scale plants ($10–50\,t{\cdot}d^{-1}$) and twelve large-scale plants ($>50\,t{\cdot}d^{-1}$). Most livestock waste (over 90%) is treated in intensive middle- or large-scale plants. Being focused on the effective utilization of biogas energy, thirty-two plants use generators to convert biogas into electricity to supply the plants and adjacent farms, and utilize the heat waste warm the digester. Five plants sell their redundant electricity. Twenty-eight plants produce heat energy using boilers, and ten plants utilize only electric energy. Thirty-five of the plants directly distribute effluent from the digesters to agricultural fields, and in five plants the effluent is dewatered followed by composting and wastewater treatment. Table 3.5 summarizes the operating data of the biogas plants, which treat

Fig. 3.6. Distribution of biogas plants treating livestock waste in Japan (New Energy and Industrial Technology Development Organization 2005).

Table 3.4. Processing mode of biogas plants in Japan.

		Number of plants
Feedstock	Cattle manure	28
	Swine manure	9
	Mixed manure	6
	Manure + MSW	4
	Manure + MSW + sludge	9
	Unidentified	15
	Total	70
Capacity	$<10\,t{\cdot}d^{-1}$	27
	$10–50\,t{\cdot}d^{-1}$	19
	$>50\,t{\cdot}d^{-1}$	12
	Unidentified	12
	Total	70
Biogas utilization	Electric	10
	Heat	28
	Electric + heat	32
	Total	70
Effluent	Liquid fertilizer	35
	Wastewater treatment	5
	Unidentified	30
	Total	70

Table 3.5. Operating data of biogas plants treating livestock waste in Japan (Japan Association of Rural Resource Recycling Solutions 2009).

Process	Feedstock	Capacity ($t{\cdot}d^{-1}$)	Temperature (see note)	Biogas production ($Nm^3{\cdot}t^{-1}$)	VS reduction (%)	Year
CSTR	Cattle manure	32.8	T	19.8		2004
CSTR	Cattle manure	18.0	M	21.1		2002
CSTR	Cattle manure	15.0	M	33.3		2000
CSTR (Multi-stage)	Cattle manure	94.8	M	41.1	35–40	2007
CSTR (Multi-stage)	Cattle manure + Agricultural waste	3.2	M	22.2		2004
CSTR	Mixed manure + Agriculture waste	85.3	M	38.1		2006
CSTR	Cattle manure + MSW	13.3	M	13.2		2003
CSTR	Cattle manure	13.5	M	35.6	30	2005
CSTR	Cattle manure + Agricultural waste	5.0		20		2005
CSTR	Cattle manure + MSW	4.9	M	61.2		2005
CSTR	Mixed manure	4.7	T	38.9		2002
CSTR	Mixed manure + MSW	55.9	M	37.4	52 (COD_{Cr})	1998
BIMA	Mixed manure + MSW + sludge	67.8	M	17.4		2005
CSTR	Swine manure + MSW + sludge	58.6	M	53.9		2006
CSTR (Fixed bed)	Cattle manure	5.0	M	37.8		2004
CSTR	Mixed manure	10.0	M	25.2		2006

Note: T represents thermophilic digestion; M represents mesophilic digestion.

livestock waste in Japan. Most of these plants were built after 2000, and use the completely stirred tank reactor (CSTR) process operated under mesophilic conditions. To summarize, biogas plants have largely been put to practical use in agriculture in Japan. Biogas is mainly used to supply the plants with their electricity needs and their heat energy. Once the issue of how to best treat digested sludge is resolved, biogas plants are expected to appear all over Japan.

2.3.3. *Treatment of digested sludge*

The effluent of anaerobic digestion is either distributed directly to farm area or clean up after dewatering. Generally speaking, in Japan, only Hokkaido has sufficient farmland to handle the effluent produced by anaerobic digestion. A typical flow diagram of effluent treatment in biogas plants is shown in Fig. 3.7. First, digested sludge is dewatered using a mechanical dehydrator with the addition of inorganic (such as ferric and aluminum salts) or polymeric flocculants. The addition of flocculants can significantly increase in the running cost, presenting a serious problem for many plants. Table 3.6 shows examples of the characteristics of dewatered supernatant, suggesting a small amount of biodegradable organic content and high concentration of nitrogen.

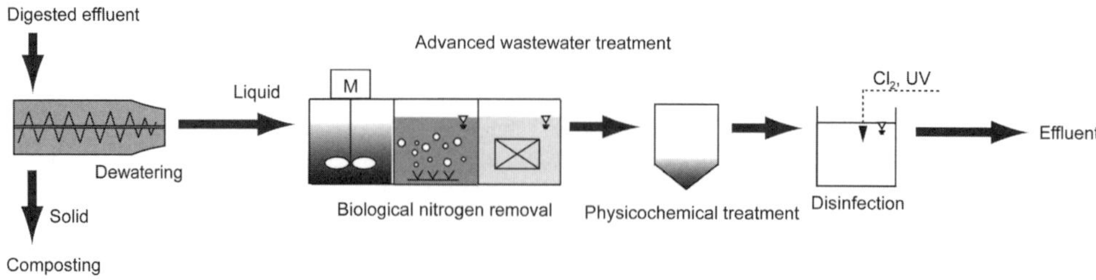

Fig. 3.7. Basic flow diagram on the treatment of digested effluent (Li 2007).

Table 3.6. Examples of characteristics of dewatered supernatant (Li 2007).

Item		Cattle manure	Swine manure
COD_{Cr}	($mg \cdot l^{-1}$)	62,000	—
BOD	($mg \cdot l^{-1}$)	4,900	1,000–8,000
T-N	($mg \cdot l^{-1}$)	4,900	2,000–5,000
T-P	($mg \cdot l^{-1}$)	—	1,000–2,000

The effluent is typically treated biologically, followed by solid–liquid separation, advanced treatment and disinfection.

Whether advanced treatment is required depends on the standard of the effluent. Before being released into the public water environment, nitrogen and phosphorus in the effluent need to be removed by biological and chemical means. Biological treatment uses the following three processes: (1) typical activated sludge process, (2) activated sludge process with membrane separation (membrane bioreactor), and (3) oxidation ditch.

In Japan, the running cost of biogas plants is predominantly the cost of effluent treatment. Utilizing the sludge as fertilizer is the best way economically. However, because this is not possible in many areas, the effluent treatment process must be made more efficient by improving efficiency for chemical addition and nitrogen removal. Our research group has been investigating an improved anaerobic digestion system, which consists of a mesophilic anaerobic digester, a dewatering system without chemical addition, and an effluent treatment process using oxidation ditch (Fig. 3.8), to decrease both the cost and the space required for a farm-scale biogas

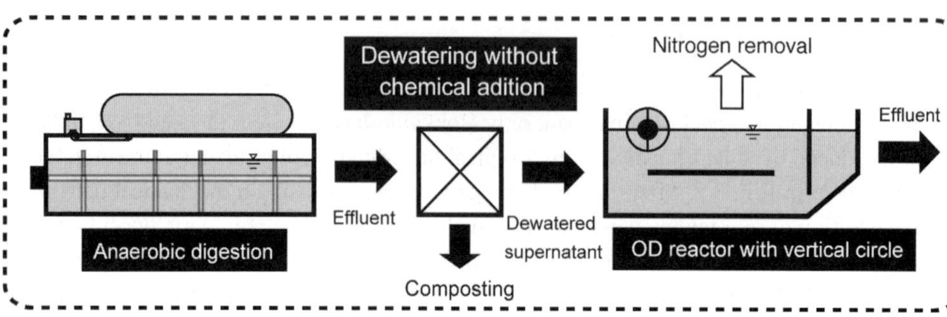

Fig. 3.8. Biomethanation system with purification units treating anaerobic digester effluent (Usami *et al.* 2009).

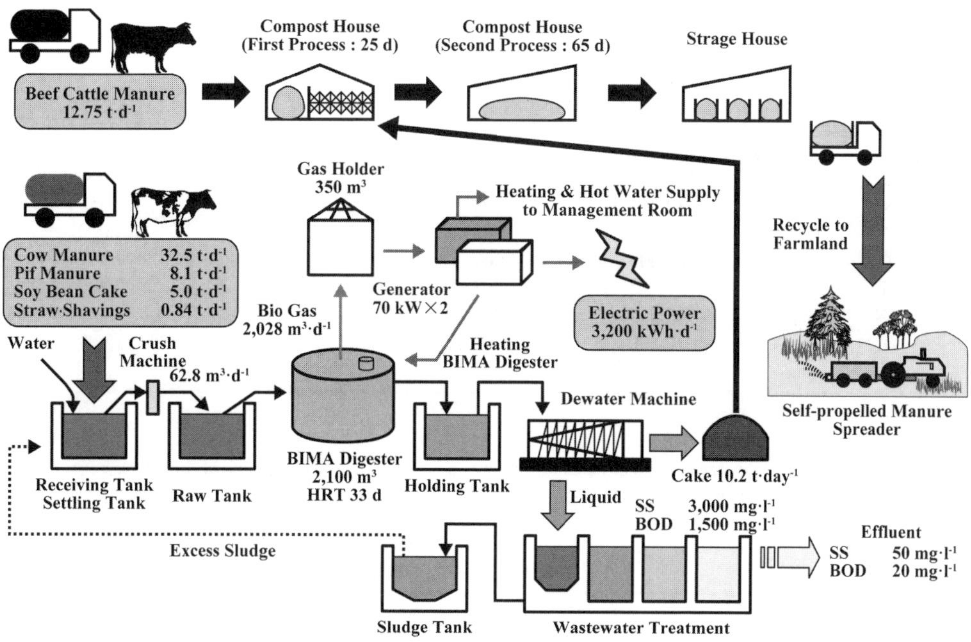

Fig. 3.9. Flow diagram of YBEC.

plant. A pilot plant has been operating successfully with the effluent water treatment requiring considerably less water for dilution and no chemicals.

2.3.4. *Example: Yagi Bio-ecology center (Ogawa et al. 2005)*

The Yagi Bio-ecology center (YBEC) is a typical biogas plant in successful operation. A flow diagram of the YBEC is shown in Fig. 3.9. This plant treats animal and food wastes collected from farms and food factories in Yagi, a small town with a population of 9,000, producing 42,000 tons of biomass resource yearly. This biomass resource was composed of 25,171 tons of livestock waste and 4,827 tons of food waste. Sixty-five percent of the biomass produced is treated by the anaerobic digesters in YBEC. As shown in Fig. 3.9, the animal and food wastes are mixed in a storage tank, and then fed either to a 2,100 m^3 mesophilic (37 °C) digester or a 600 m^3 thermophilic (55 °C) digester. The HRT and treatment capacity are, respectively, 33 d and 62.8 $t{\cdot}d^{-1}$ for the mesophilic digester, and 26 d and 23.2 $t{\cdot}d^{-1}$ for the thermophilic digester. The characteristics of this biomass resource are summarized in Table 3.7, and those of the influents and effluents of the digesters are in Table 3.8. The TS and COD_{Cr} in the influents are 6.6% and 72.5 $g{\cdot}l^{-1}$, and both digesters removed 51–53% of COD_{Cr}. The volumetric biogas production rate of the thermophilic digester averaged 1.00 $Nm^3{\cdot}d^{-1}{\cdot}m^{-3}$, approximately 1.1 times higher than that of the mesophilic digester. The average methane concentration in the biogas from both digesters was 55.4%. The thermophilic digester reduced substantially more *Escherichia.coli* and *Fical coliform bacteria* than the mesophilic digester. Biogas is converted into electricity, and waste heat is recovered as hot water. In YBEC, the electric power produced from the biogas is about enough to supply all that is required by the plant. Any excess electricity produced is sold

Table 3.7. Characteristics of biomass resource.

Item		Manure from 650 dairy farms: Raw manure	Manure from 650 dairy farms: Dewatered liquid	Manure from 1,500 pigs	Tofu reuse
TS	($mg \cdot l^{-1}$)	136,700	47,500	65,700	233,000
VTS	($mg \cdot l^{-1}$)	115,000	37,500	55,000	220,600
SS	($mg \cdot l^{-1}$)	89,000	34,500	44,500	24,850
BOD	($mg \cdot l^{-1}$)	24,000	6,500	25,500	840,000
COD_{Cr}	($mg \cdot l^{-1}$)	130,000	64,000	82,000	7
pH		7	7	6	7,290
T-N	($mg \cdot l^{-1}$)	5,530	1,950	4,370	2,400
NH_4^+-N	($mg \cdot l^{-1}$)	2,500	1,200	4,900	1,698
T-P	($mg \cdot l^{-1}$)	1,290	395	1,470	

Table 3.8. Average characteristics of influent and effluent and performance of digesters.

Characteristics		Influent	Effluent from mesophilic digester	Effluent from thermophilic digester
pH		6.5	7.9	8.1
TS	($mg \cdot l^{-1}$)	66,250	44,250	46,880
VTS	($mg \cdot l^{-1}$)	54,880	32,880	34,250
COD_{Cr}	($mg \cdot l^{-1}$)	72,500	34,250	35,250
VFA	($mg \cdot l^{-1}$)	8,520	50	353
T-N	($mg \cdot l^{-1}$)	3,050	3,040	3,160
NH_4^+-N	($mg \cdot l^{-1}$)	1,330	1,840	2,090
Alkalinity	($mg \cdot l^{-1}$)	6,280	12,120	12,630
T-P	($mg \cdot l^{-1}$)	515	490	510
Coliform bacteria	($CFU \cdot ml^{-1}$)	722,500	5,300	248
Fecal coliform	($CFU \cdot ml^{-1}$)	439,200	1,332	28
Performance			Mesophilic digester	Thermopilic digester
COD_{Cr} reduction	(%)		52,8	51.4
Biogas production	($Nm^3 \cdot d^{-1} \cdot m^{-3}$)		0.9	1.0
Methane content	(%)		Average 55.4	

to the WWTP. After the digested sludge is dewatered, the wastewater is treated and the solid is composted.

2.4. *Municipal solid wastes*

2.4.1. *History*

In Japan, the anaerobic digestion of municipal solid waste (MSW) attracted social attention in the latter half of the 1970s. A national research project entitled "Stardast'80" was initiated to promote experimental pilot studies of anaerobic digestion of MSW. However, the project failed due to the low energy recovery efficiency and the high cost of wastewater treatment, making it economically uncompetitive. Moreover, the poor removal of inappropriate matters from the influent became a barrier for operation. The anaerobic digestion of MSW was therefore not put into practical use in the 1980s.

The fully fledged application of the anaerobic digestion of MSW in Japan started from the "resource recycling centers of excess sludge" project in 1997, which promoted the anaerobic digestion of mixtures of food waste and excess sludge of night soil treatment. Thirteen subsidized biogas plants were built as resource recycling centers of excess sludge by 2005, ten of which employed a wet thermophilic digestion process (Moebius system), and the other three employed a wet mesophilic digestion process (REM system). Additionally, the Biomass Nippon Strategy, which started in 2002, positioned MSW as a principal biomass resource. It was estimated that 2,200 $t{\cdot}y^{-1}$ of food waste generated in Japan could produce by anaerobic digestion 1.72 billion $m^3{\cdot}y^{-1}$ of methane, more than that produced from livestock waste or sewage sludge. In recent years, not only are there more biogas plants treating only MSW but also there is an increase in the number of farmer's biogas plants treating mixtures of livestock waste and MSW. To enhance economic efficiency, a dry thermophilic process (Compogas System) and a waterless high-solids wet two-phase process (WTM system) have been developed. Hydrogen-producing processes are also being investigated. The many technical developments in anaerobic digestion of MSW briefly introduced here give rise to expectations that it could become widespread throughout Japan in the near future.

2.4.2. *Applications*

Figure 3.10 shows the distribution of biogas plants, which mainly treat MSW and food-processing wastes in Japan. In 2005, there were 47 plants in operation in Japan, the processing modes of which are summarized in Table 3.9. There are 16 plants treating only MSW and 17 treating only food-processing wastes. The others treat mixtures of MSW, night soil and sludge. Similar

Fig. 3.10. The distribution of biogas plants which mainly treat MSW in Japan (New Energy and Industrial Technology Development Organization 2005).

Table 3.9. The mode of processing used in biogas plants in Japan.

		Number of plants
Feedstock	MSW	16
	MSW + sludge	7
	MSW + sludge + human waste	4
	Food-processing waste	17
	Food-processing waste + sludge	1
	Unidentified	3
	Total	47
Capacity	$<10\,t{\cdot}d^{-1}$	24
	$10–50\,t{\cdot}d^{-1}$	12
	$>50\,t{\cdot}d^{-1}$	5
	Unidentified	6
	Total	47
Biogas utilization	Electric	10
	Heat	14
	Electric + heat	16
	Unidentified	7
	Total	47
Effluent	Liquid fertilizer	1
	Wastewater treatment	13
	Unidentified	33
	Total	47

to plants treating livestock waste, the treatment capacity can be divided into three categories: 24 small plants ($<10\,t{\cdot}d^{-1}$), 12 middle-sized plants ($10–50\,t{\cdot}d^{-1}$) and five large plants ($>50\,t{\cdot}d^{-1}$). Most small plants and half of the middle-sized ones treat only MSW or food-processing wastes. The other half of the middle-sized plants treat mixtures of MSW and sludge. The large plants treat only food-processing wastes or mixtures of MSW, night soils and sludge. When focusing on the utilization of biogas energy, 17 plants utilize both electric and heat energy, 14 plants utilize only heat energy, and ten plants use only electric energy. Most of the recovered energy is consumed inside the plants themselves. A few plants use digested sludge as fertilizer, but the majority of the plants are equipped with dewatering and wastewater treatment facilities. Dewatered solids are either incinerated or composted. The operating data of some of the biogas plants, which treat organic fraction of MSW, are summarized in Table 3.10. Unlike those run by farmers, thermophilic digestion processes are often employed.

Various types of reactors, such as the CSTR, the dry plug-flow reactor, the fixed-bed reactor and the anaerobic membrane reactor are used, as shown in Table 3.10.

2.4.3. *Example 1: Campo Recycle Plaza (Kawamura et al. 2005)*

The Campo Recycle Plaza is a typical example of the biogas plant treating MSW using a Compogas system, the flow diagram of which is shown in Fig. 3.11. The collected MSW is directly crushed to the size of 100 mm, and then further fragmented to about 10 mm with all inappropriate matter removed by a secondary sorting machine. The prepared influent mainly contains food and paper wastes, accounting for about 80% of the total weight (paper: 46% and MSW: 33%). The TS concentrations ranged from 15 to 25% (averaging 20%), and the T-N concentrations

Table 3.10. Operating data of biogas plants treating livestock waste in Japan (Japan Association of Rural Resource Recycling Solutions 2009).

Process	Feedstock	Capacity ($t{\cdot}d^{-1}$)	Temperature (see note)	Biogas production ($Nm^3{\cdot}t^{-1}$)	VS reduction (%)	Year
METAKLES	MSW + food processing waste	24.0	T	125	78	2003
Compogas	MSW	6.0	T	335	73–85	2004
CSTR	MSW	4.8	M	190		2003
CSTR	MSW + food-processing waste + sludge	3.9	T	318		2003
BIMA	MSW	22.0	M	123		2003
Fixed-bed	MSW	11.6	T	129		2003
Sirius	MSW	3.0	M	160		2003
Membrane	MSW	10.5	T	137	75–78 (COD_{Cr})	2003

Note: T represents thermophilic digestion; M represents mesophilic digestion.

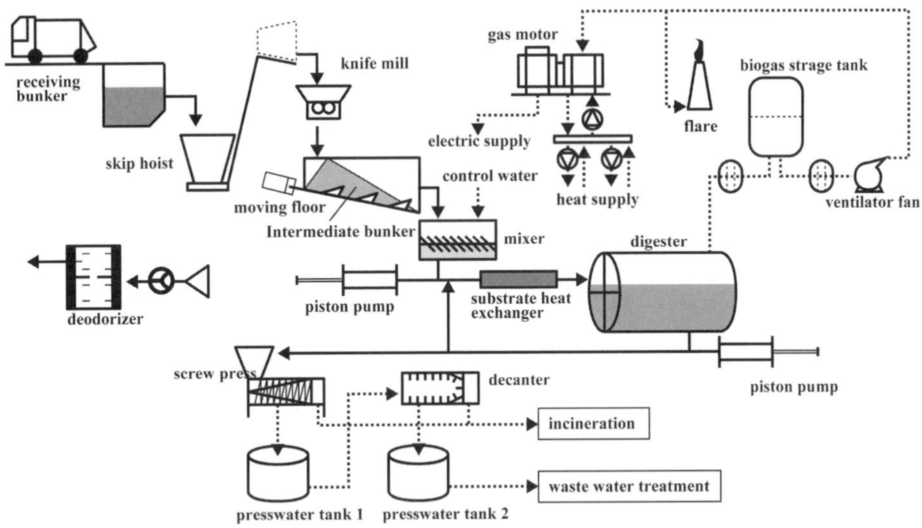

Fig. 3.11. Flow diagram of the Compogas system used in Campo Recycle Plaza.

ranged from 3,000 to 8,000 $mg{\cdot}kg^{-1}$ (averaging 4,000 $mg{\cdot}kg^{-1}$). The prepared influent is added into the dry thermophilic digester using the horizontal plug-flow system shown in Fig. 3.11, and treated for 20–25 days. In the Campo Recycle Plaza, two 1,150 m^3 digesters treat 50 ton/d of MSW daily. On average, the digester reduces 76.8% of VS, produces 205.8 m^3/t of biogas containing 57.8% methane. The recovered biogas is partly converted into electricity, heat, and utilized as fuel for natural gas automobiles after purification. Electric generation from the biogas is 14,880 kWh/d, 40% of which is consumed by the plant. After dewatering of the digested sludge, the wastewater is treated and the solid is composted.

2.4.4. *Example 2: Operating biogas plants in Hokkaido (Tanikawa et al. 2008)*

An evaluation of the cost and energy balance for the operation of three biogas plants with wet mesophilic digesters treating MSW in Hokkaido is discussed in the section. Flow diagram of

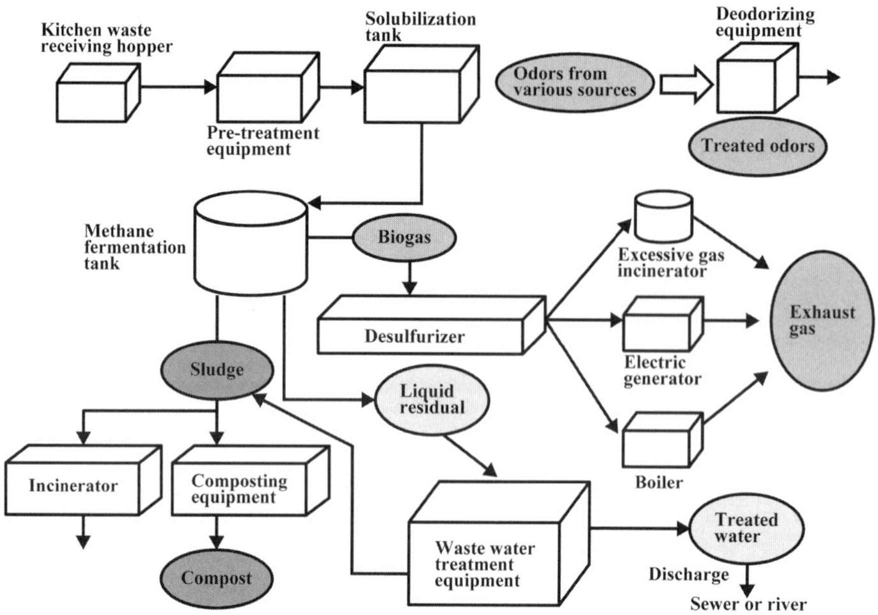

Fig. 3.12. Flow diagram of typical process of three biogas plants.

Table 3.11. The process modes of the three biogas plants.

Item		Plant A	Plant B	Plant C
Treatment capacity	(t-wet MSW·d^{-1})	16	55	22
Amount of MSW	(t-wet MSW·d^{-1})	9.7	25	10.1
Temperature	(°C)	55	35	55
Feedstock		MSW	MSW	MSW
Biogas production	(Nm^3·t^{-1})	111	113	146
Methane content	(%)	63.4	58.5	63.1

Table 3.12. Energy utilization in three biogas plants.

Item			Plant A	Plant B	Plant C
Amount kitchen waste treated		t·d^{-1}	9.0	22.9	10.0
Electricity	Amount used	kWh·t^{-1}	263	266	247
	Amount generated	kWh·t^{-1}	147	194	207
	Amount of used fuel	l·t^{-1}	0	8	0
Amount of used water		m^3·t^{-1}	2.1	1.3	1.6

typical process of three biogas plants is illustrated in Fig. 3.12. The process modes of three biogas plants are summarized in Table 3.11. In these plants, MSW is fed to digesters with a HRT of 10–15 days after removing inappropriate matters by sorting machines. The recovered biogas is converted into electricity and heat, supplying 56%, 73% and 84% of the plant consumptions. Table 3.12 shows the amount of energy utilization in each plant. The administrative and maintenance expenses of each plant are shown in Table 3.13. The cost of treating 1 ton

Table 3.13. Administrative and maintenance expenses (yen·$(t\text{-}MSW)^{-1}$).

Item	Plant A	Plant B	Plant C
Labor cost (by contract)	8,260	5,700	3,180
Electricity (buying cost)	1,860	1,370	720
(sale gain)	0	-30	0
Water rate	350	0	0
Sewage rate	260	0	460
Fuel cost	60	1,140	0
Chemicals and supplies expenses	1,880	5,600	930
Residue disposal cost	3,210	2,200	1,800
Inspection, repair, and subcontract cost	1,930	4,190	3,580
Total	17,840	20,170	10,670
Total except labor cost	9,580	14,470	7490

of MSW varies from 10,700 to 20,200 yen. Administrative and maintenance expenses account 30–46% of the cost. The cost difference among the three plants is due to their differences in the management control systems and the type of wastewater treatment process employed. The administrative and maintenance expenses excluding the labor cost are 9,600 yen·t^{-1}, 14,470 yen·t^{-1}, and 7,500 yen·t^{-1}, respectively. Of the three plants, the one with the largest capacity, plant B, is most expensive to operate, because it uses an advanced wastewater treatment process, while plants A and C discharge wastewater into a sewer system.

2.5. *Wastewater treatment*

2.5.1. *History*

The history of anaerobic industrial wastewater treatment in Japan dates back to the 1950s. Ono was the first to research the anaerobic treatment of alcohol on butanol distillery wastewater (Ono 1962a, b). He reported that a combination of methane fermentation and an activated sludge process for alcohol distillery wastewater was used at seven facilities (thermophilic fermentation and activated sludge: 6, mesophilic fermentation and activated sludge: 1) in Japan before 1962. Ono demonstrated that a loading rate (OLR) of 2.5–13.2 g-VS·l^{-1}·d^{-1}resulted in 80% BOD removal using thermophilic anaerobic process to treat various alcohol distillery wastes, and 98% BOD removal after following the activated sludge process (Ono 1962b). In the 1970s and 1980s, the waste and wastewater treatment of pulp and paper factories by methane fermentation was actively studied (Ono 1978; Fujimura *et al.* 1980). Ono (1978) successfully used an anaerobic contact process to treat pulp mill wastewater with an OLR of 4.2–4.3 g-VS/L/d in a laboratory experiment and with a COD_{Cr} loading rate of 5.0–5.1 g·l^{-1}·d^{-1} (HRT 2–5 days) in a pilot-scale experiment (Fujimura *et al.* 1980). After 1980, the UASB and EGSB have become the dominant processes in the treatment of industrial wastewater. Sumitomo Heavy Industries Ltd. (Chiku 2008) alone has installed over 70 UASB and EGSB plants for chemical and paper industries. In total, there are over 300 of UASB and EGSB full-scale plants now in operation.

2.5.2. *Applications*

The UASB and EGSB processes are applied for the treatment of brewery, soft drink, food and chemical wastewaters. Since the 1980s, Kurita Water Industries Ltd., Fujikasui Engineering

Table 3.14. Example of performances of UASB and EGSB treatment.

	UASB (10–15 kg-COD$\cdot$m$^{-3}\cdot$d^{-1})		EGSB (20–30 kg-COD$\cdot$m$^{-3}\cdot$d^{-1})
	Vegetable processing wastewater	Confectionery wastewater	Brewery wastewater
Inf. COD_{Cr} (mg$\cdot$l^{-1})	700–1700	5,400	2,000–3,000
Eff. COD_{Cr} (mg$\cdot$l^{-1})	120–170	440	800
Inf. BOD (mg$\cdot$l^{-1})	320–700	3,060	n.a.
Eff. BOD (mg$\cdot$l^{-1})	45–90	155	n.a.
Inf. SS (mg$\cdot$l^{-1})	150–650	330	200–300
Eff. SS (mg$\cdot$l^{-1})	30–60	95	100–500

Co. Ltd., and Sumitomo Heavy Industries Ltd. have made large improvements in these systems and commercialized each of their own technologies in Japan. Table 3.14 summarizes the performances of these UASB and EGSB treatment systems (Kurusu and Nakajima 1998). In 1988, Asahi Breweries was the first to use UASB treatment, followed by Kirin Breweries, Sapporo Breweries, and Suntory. At present, all of the total 33 brewery factories in Japan use either UASB or EGSB, plus an aerobic post-treatment process, to treat their wastewaters (Yoda 2004). Typical characteristics of brewery wastewater are as follows: 1,500–5,000 mg$\cdot$l^{-1} of COD_{Cr}, 1,000–3,000 mg$\cdot$l^{-1} of BOD_5 and 500–2,000 mg$\cdot$l^{-1} of SS. In these factories, the UASB and EGSB processes typically achieved 10–20 kg$\cdot$m$^3\cdot$d^{-1} of COD_{Cr} loading rate and over 90% COD_{Cr} removal efficiency.

The anaerobic treatment of shochu distillery waste water is also widely applied in Japan. Figure 3.13 illustrates a typical flow diagram of an anaerobic treatment process of shochu distillery wastewater (Tanpo 1997). The UASB process is also popular in the treatment of food-processing wastewater. Fujikasui Industries has applied UASB to treatment of rice processing wastewater (COD_{Cr}: 4,500 mg$\cdot$l^{-1}) and miso (bean paste) processing wastewater (Shiraishi 2004), in which over 80% and 98% of COD_{Cr} are successfully removed, respectively. For the treatment of chemical industrial wastewater, it is reported that three UASB and seven UAF processes have been in operation (Tagawa 2004). Figure 3.14 shows an example of such a facility.

3. New Developments in Biogas Technology in Japan

3.1. *Anaerobic membrane reactor*

A recent advance in membrane technology has made it possible to use membrane bioreactors for high-rate methane fermentation. In Japan, progressive approaches to developing a methane fermentation system using an anaerobic membrane reactor have been developed in the past decade. Figure 3.15 shows an example of the system being successfully used in many biogas plants, consisting of a methane fermentation reactor and a membrane separation unit. Digested sludge is recycled between the reactor and the membrane separation unit. This system has been successfully operated at an organic load of 18.7 kg-$COD_{Cr}\cdot$m$^3\cdot$d^{-1} with a cross membrane flux of 0.1–0.15 m$^3\cdot$m$^{-2}\cdot$d^{-1}. At present, there are 16 biogas plants using anaerobic membrane reactors in Japan, including many treating shochu distillery wastewater (Table 3.15).

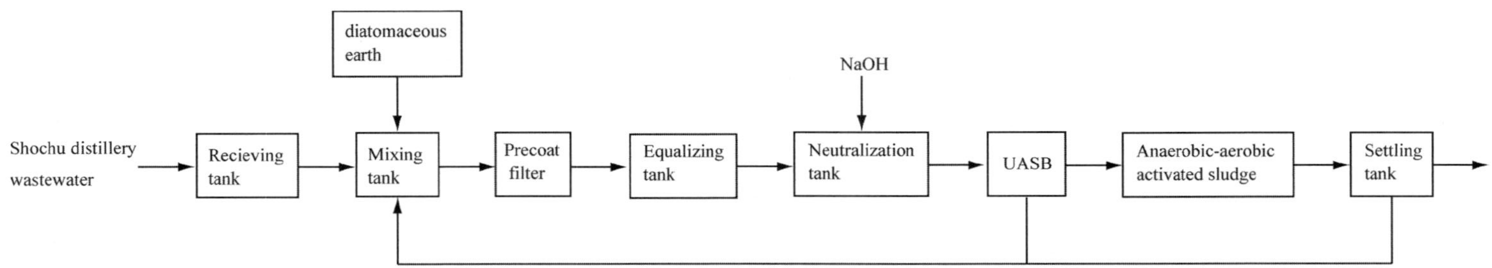

Fig. 3.13. Flow diagram of anaerobic treatment process of shochu distillery waste water.

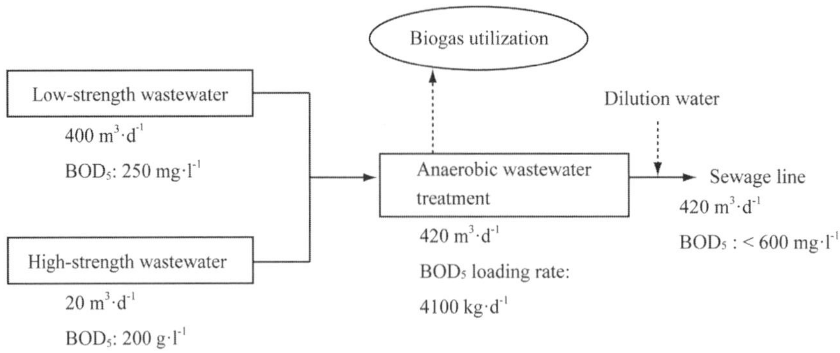

Fig. 3.14. A facility planned for treating chemical industrial wastewater.

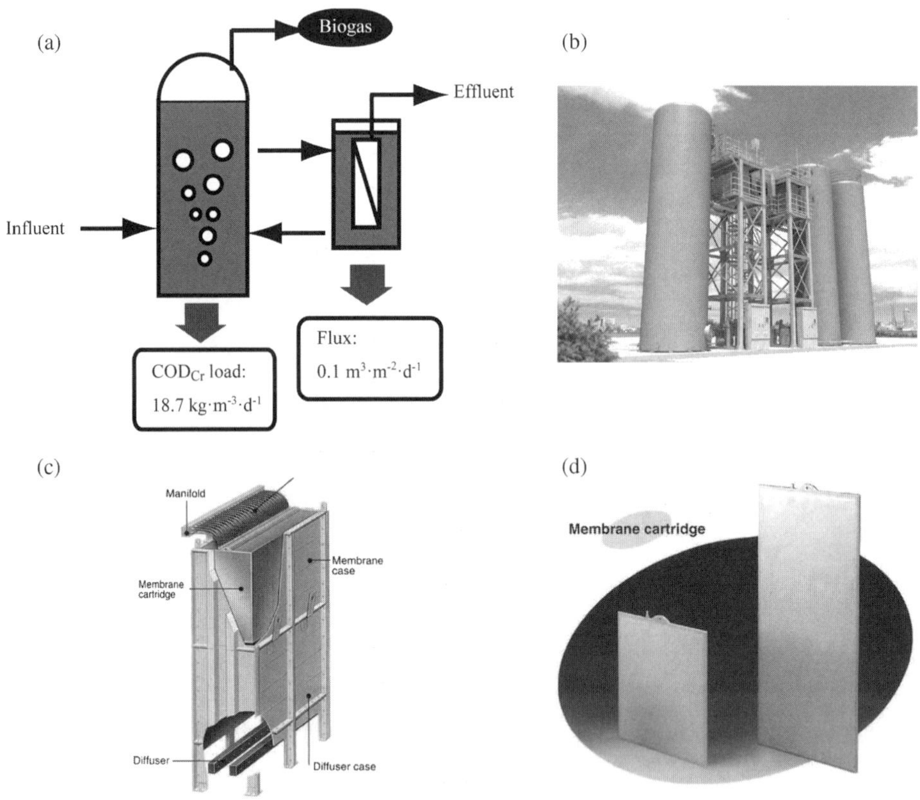

Fig. 3.15. Kubota anaerobic membrane system: (a) schematic diagram, (b) photo of a full-scale plant, (c) a membrane separation unit, and (d) membrane cartridge.

3.2. *Two-phase fermentation for hydrogen and methane production*

An effective two-phase anaerobic process producing hydrogen and methane has been developed at the Tohoku University (Ohba *et al.* 2005). As shown in Fig. 3.16, the process consists of a mixing tank, a thermophilic acidogenic hydrogen-producing reactor (55 °C) and a mesophilic methanogenic reactor (36 °C). Influent to the reactors is a 1:1 mixture of potato-processing

Table 3.15. Biogas plants employing anaerobic membrane reactor in Japan.

Number	Month/year	Waste/capacity
1	04/2000	MSW 8 $t{\cdot}d^{-1}$
2	04/2002	Food-processing waste 2 $t{\cdot}d^{-1}$
3	03/2003	MSW 16 $t{\cdot}d^{-1}$
4	03/2003	MSW 5 t/d and sludge 20 $kL{\cdot}d^{-1}$
5	09/2003	Food-processing waste 5.2 $t{\cdot}d^{-1}$
6	03/2004	Food-processing waste 6.5 $t{\cdot}d^{-1}$
7	01/2006	Awamori distillery waste 15 $t{\cdot}d^{-1}$
8	09/2006	Shochu distillery waste 60 $t{\cdot}d^{-1}$
9	03/2007	Shochu distillery waste 20 $t{\cdot}d^{-1}$
10	03/2007	Shochu distillery waste 20 $t{\cdot}d^{-1}$
11	03/2008	Shochu distillery waste 60 $t{\cdot}d^{-1}$
12	09/2008	Shochu distillery waste 12 $t{\cdot}d^{-1}$
13	11/2008	Food wastes 65 $t{\cdot}d^{-1}$
14	03/2009	Shochu distillery waste 65 $t{\cdot}d^{-1}$
15	07/2010	Awamori distillery waste 6 $t{\cdot}d^{-1}$
16	07/2010	Shochu distillery waste 35 $t{\cdot}d^{-1}$

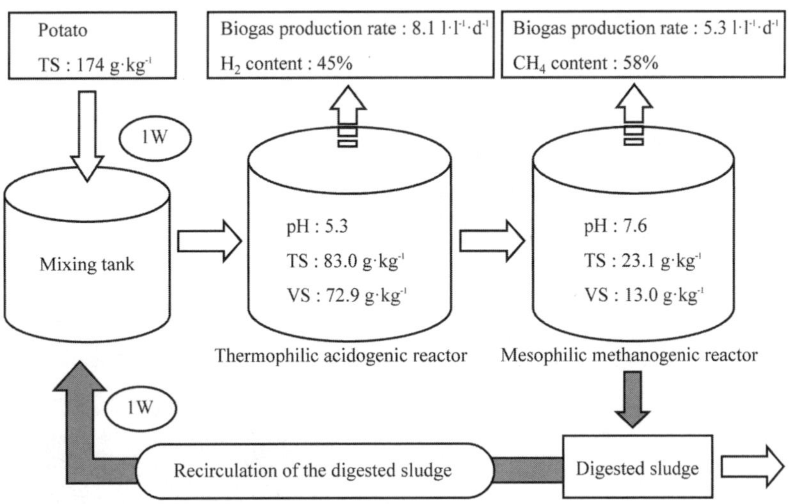

Fig. 3.16. Two-phase circulating high-concentration hydrogen–methane fermentation process (Ohba *et al.* 2005).

waste (TS 17.4%) and recycled digested sludge from a second mesophilic reactor (TS 2.3%). Successful continuous operation was achieved at an HRT of 20 d and a COD_{Cr} loading rate of 10.9 $kg{\cdot}m^{-3}{\cdot}d^{-1}$. In this operation, TS, VS and COD_{Cr} reductions are 89%, 93% and 87%, respectively. The first acidogenic reactor produced 8.1 $l{\cdot}l^{-1}{\cdot}d^{-1}$ of biogas containing 45% H_2, and the second methanogenic reactor produced 5.3 $l{\cdot}l^{-1}{\cdot}d^{-1}$ biogas containing 58% CH_4.

3.3. *Biological desulfurization of biogas*

In recent years, the biological desulfurization of biogas is widely used in many biogas plants because it is more cost-effective than the common chemical method using Fe_2O_3. The biological

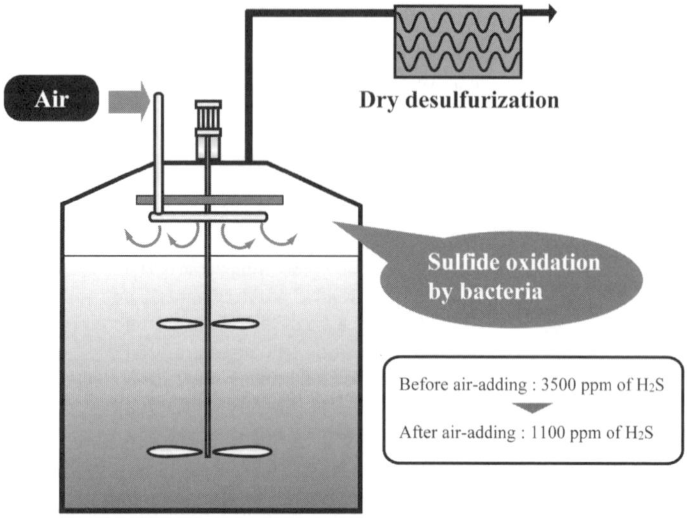

Fig. 3.17. Biological desulfurization system inside an anaerobic digester (Kobayashi *et al.* 2009a).

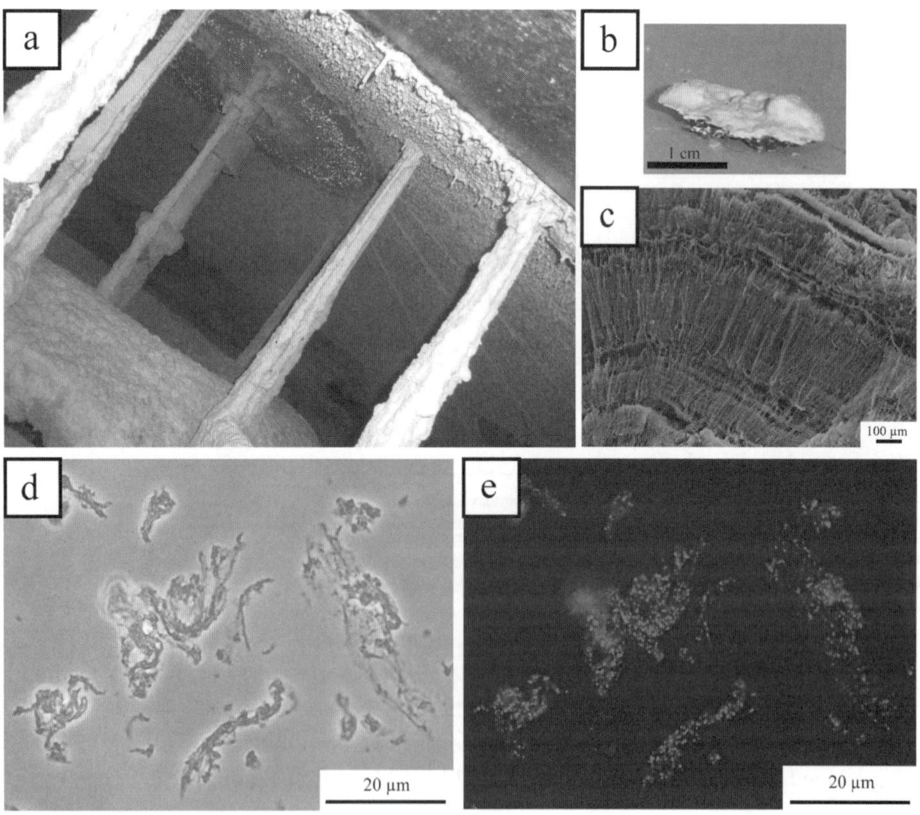

Fig. 3.18. Photographs of the headspace of the digester (a) and a fresh sulfur mat (b). SEM images of the vertical cross-section of sulfur mat composed of filamentous sulfur bundles (c). Phase contrast (d) and fluorescence images (e) using specific CARD-FISH probe targeting at sulfide-oxidizing bacteria of an identical microscopic field including some sulfur filaments. (Kobayashi *et al.* 2009b).

desulfurization processes in use can be classified into two patterns. One is the micro aeration in the headspace of a digester, and the other is using an external reactor to cultivate sulfur-oxidizing bacteria (SOB). Figure 3.17 shows an example of our pilot biological desulfurization system (Kobayashi *et al.* 2009a). The 338 m^3 digester treats 13.5 $t \cdot d^{-1}$ of daily cattle manure, and produces 481 $m^3 \cdot d^{-1}$ of biogas containing 3,500 ppm of H_2S. Biological desulfurization was carried out by micro-aeration with a flow rate equal to 5% of the biogas production rate. After aeration, the H_2S concentration in the biogas was reduced to 0–1,900 ppm; the remaining H_2S in the biogas must be further removed using a dry desulfurization technique. About half of the biogas plants using biological desulfurization use a chemical process for post-treatment.

To understand the biological reaction and the conditions inside a digester, we tried to identify the microorganisms which were responsible for desulfurization. We found microbial mats with sulfur deposits inside the digester (Figs. 3.18a and 3.18b). The mats contained high concentrations of S^o, accounting for 90% of the TS. This indicates that H_2S was converted into S^o inside the digester. Figure 3.18c illustrates the structure of a mat with bundles of filamentous sulfur. A catalyzed reporter deposition — fluorescence in situ hybridization (CARD-FISH) experiment demonstrated that sulfide-oxidizer could be found in aggregated cells around the sulfur filaments (Figs. 3.18d and 3.18e). The results of microbial analysis suggested that *Halothiobacillus neapolitanus* and *Sulfurimonas* sp. related bacteria are responsible for the production of sulfur from H_2S in the digester (Kobayashi *et al.* 2009b).

References

Chiku, H. (2008) UASB processing wastewater treatment systems. *Sumitomo Heavy Industries Technical Report* 167, 13–16.

Endo, I. (1965) Fundamental studies on the anaerobic digestion of night soil. *J. Japan Soc. Civil Eng.* **118**, 17–47.

Fujimura, E., Takeshita, N., and Mimoto, N. (1980) Energy recovery by methane fermentation of pulp mill waste water and sludge. *J. Japan Technical Assoc. Pulp Paper Ind.* **34**(10), 650–660.

Inoue, Y. (2009) Anaerobic digestion supported public health in Japan after WWII (2). *JOHKASOU* **395**, 26–32.

Japan Association of Rural Resource Recycling Solutions (2009) Reference system for information of biomass technology. http://www2.jarus.or.jp/biomassdb/.

Japan Sewerage Association (1998) *Sewerage Statistics*. Japan Sewerage Association, Tokyo, Japan.

Kawamura, K., Nakanishi, F., and Irie, N. (2005) A process report on bio-recycle facility of Kampo recycle plaza. *Takuma Technical Reports* **13**(1), 31–37.

Kobayashi, T., Li, Y.Y., Kubota, K., Harada, H., Maeda, T., Kawai, K., and Yoshida, H. (2009a) Hydrogen sulfide removal from biogas under microaerophilic environment in the anaerobic digester. *J. Japan Soc. Civil Eng. (G)* **65**(2), 104–113.

Kobayashi, T., Li, Y.Y., Kubota, K., and Harada, H. (2009b) Environmental and microbial ecology in sulfur mats responsible for the biological removal of hydrogen sulfide from biogas. *J. Environ. Conserv. Eng.* **38**(9), 42–51.

Kurusu, H. and Nakajima, A. (1998) Industrial wastewater treatment technology — biological methods and following advanced treatment. *Ind. Environ.* **27**(9), 80–85.

Li, Y.Y. (2007) Utilization and purification of digested sludge. In: *Production and Energy Utilization of Gaseous Fuel from Biomass*. NTS, Tokyo, Japan, pp. 188–203.

Li, Y.Y. and Nishimura, O. (2007) Recycle of waste biomass by methane fermentation technology. *Japanese J. Multiphase Flow*, **21**(1), 29–38.

Matsumoto, J. (1962) Anaerobic digestion treatment of night soil. *J. Japan Waste Manage. Assoc.* **15**(4), 2–11.

Matsumoto, J. and Onuma, M. (1965) Experimental studies on the treatment of night soil digestion tank supernatant by activated sludge process. *J. Japan Soc. Civil Eng.* **116**, 8–15.

Morikawa, N. (2009) Trend of the faculties for treatment of night soil. *JOHKASOU* **398**, 10–14.

New Energy and Industrial Technology Development Organization (2005) *Handbook for Introduction of Biomass Energy*, 2nd ed. New Energy and Industrial Technology Development Organization, Tokyo, Japan.

Noike, T. (2008) The importance of methane fermentation for the prevention of global warming. In: *Proceedings of the Seminar on Effective Utilization of Sewage Sludge*, 91–110.

Ogawa, Y., Fujita, M., and Nakagawa, Y. (2005) Comparison between mesophilic and thermophilic methane fermentation treatments of manure and food waste. *J. Japan Soc. Waste Manage. Experts* **16**(1), 44–54.

Ohba, M., Li, Y.Y., and Noike, T. (2005) Analysis of microbial community in methane fermentation by a two-phase circulating high-concentration process. *J. Japan Soc. Wat. Environ.* **29**(7), 399–406.

Ono, H. (1962a) Industrial wastewater treatment by methane fermentation I. *J. Wat. Waste* **4**(3), 51–58.

Ono, H. (1962b) Industrial wastewater treatment by methane fermentation II. *J. Wat. Waste* **4**(4), 23–31.

Ono, H. (1978) On the waste and waste-water treatment of pulp and paper factories by methane fermentation. *J. Japan Technical Assoc. Pulp Paper Ind.* **32**(3), 71–78.

Shiraishi, K. (2004) Anaerobic wastewater treatment in food industry. *J. Environ. Conserv. Eng.* **33**(6), 427–431.

Tagawa, T. (2004) Application of anaerobic treatment to chemical wastewater. *J. Environ. Conserv. Eng.* **33**(6), 432–436.

Tanikawa, N., Furuichi, T., Ishii, K., and Nishigami, K. (2008) Study on methane generation, environmental burdens and cost effectiveness of kitchen-waste biogasification facilities. *J. Japan Soc. Waste Manage. Experts* **19**(3), 182–190.

Tanpo, M. (1997) New systems of the treatment of shochu distillery waste. *Industo* **12**(12), 27–30.

Usami, S., Akutsu, Y., Li, Y.Y., Harada, H., Kanno, S., and Moriyama, H. (2009) A pilot study on cattle manure treatment system combining methane fermentation, dewatering without chemicals and nitrogen removal. *Environ. Eng. Res.* **46**, 673–680.

Yoda, M. (2004) Anaerobic treatment of brewery wastewater by the UASB process. *J. Environ. Conserv. Eng.* **33**(6), 417–421.

Chapter 4

Anaerobic Sewage Treatment using UASB Reactors: Engineering and Operational Aspects

Jules B. Van Lier[1], Anand Vashi[2], Jeroen Van Der Lubbe[3] and Barry Heffernan[3]

[1]*Faculty of Civil Engineering and Geosciences, Department of Water Management, Delft University of Technology, The Netherlands*
j.b.vanlier@tudelft.nl

[2]*Enviro Control Associates (I) PVT.LTD. Opp. Bank of Maharashtra, Ghod Dod Road, Surat-395 001, Gujarat, India*

[3]*Biothane Systems International, Tanthofdreef 21, 2600 GB Delft, The Netherlands*

Anaerobic pre-treatment of domestic sewage using UASB reactor systems offers a number of advantages, e.g., system compactness, negligible or no energy consumption, stabilised excess sludge production, potential for energy recovery, low-cost accessibility of sewage for agricultural reuse purposes, etc. Research on high-rate anaerobic sewage treatment started in the early 1980s with the 64 m^3 UASB pilot plant in Cali, Colombia and various other initiatives in Brazil. Hereafter, the technology was adapted to full-scale conditions and slowly introduced in the market. This chapter describes the anaerobic treatment process for domestic sewage and evaluates the performance of current large-scale reactor systems located in (sub-)tropical areas. Although the perspectives from the early 1980s were confirmed in bench-scale studies, a considerable number of the recently constructed treatment plants deliver disappointing performance results. In many cases, inadequate designs, improper reactor operation and insufficient control are responsible for these poor performances. A brief survey is made on the preconditions that must be met prior to starting and operating sewage treatment plants based on the UASB concept. Adequate maintenance and basic system knowledge at the level of plant manager seem to be indispensable for full-scale success.

1. Introduction

Since the mid-1990s, approximately 10 years after the installation of the first pilot-scale upflow anaerobic sludge bed (UASB) reactors for the treatment of municipal sewage in Cali Colombia (Lettinga *et al.* 1993; Schellinkhout *et al.* 1985) and Brazil (Vieira and Souza 1986), there has been a rapid increase in the number of large-scale municipal UASB sewage treatment plants (STPs) installed in sub-tropical countries. In particular, Brazil and India have extensively adopted the UASB technology. For example, the authors were able to identify over 45 municipal UASB STPs in India with an average daily flow rate of 10,000 m^3 or more (the largest STP was designed for a flow rate of 338,000 $m^3 \cdot d^{-1}$) while 15 such STPs were identified in Brazil. The main benefits associated with UASB-based STPs compared to conventional aerobic treatment are a significant reduction in energy consumption and the potential for energy self-sufficiency. Constraints include

Table 4.1. Main advantages and constraints of high-rate anaerobic sewage treatment systems.

Advantages (compared to aerobic processes)

- Substantial (reaching 90%) savings in operational costs as no energy is required for aeration.
- Significant reductions in investment costs as less treatment units are required.
- The produced methane (CH_4) is of interest for energy recovery or electricity production.
- The technologies do not make use of high-tech equipment, except for main headwork pumps and fine screens. The treatment system is less dependent on imported technologies.
- The process is robust and can handle periodic high hydraulic and organic loading rates.
- The system is compact with hydraulic retention times (HRTs) of 6–9 h, and is, therefore, suitable for application in urban areas, minimising conveyance costs.
- Small-scale applications allow decentralised treatment, making sewage treatment less dependent on the extent of sewage networks.
- The sludge production is low, well stabilised and easily dewatered; consequently, it does not require extensive post-treatment.
- The valuable nutrients (N and P) are conserved which give the treated wastewater a high potential for crop ferti-irrigation.
- A well-designed UASB filters effectively Helminth's eggs from the influent, allowing the treated effluent for agricultural reuse.

Constraints

- Anaerobic treatment, although converting most of influent COD into CH_4, is a partial treatment, requiring adequate post-treatment to meet the discharge or reuse criteria.
- The produced CH_4 is partially dissolved in the effluent (depending on the influent COD concentration). So far no measures are applied to prevent CH_4 escaping to the atmosphere.
- The collected CH_4 is often not utilised for energy generation and in some cases not even flared.
- There is little experience with full-scale application at moderate to low temperatures.
- Reduced gases, like H_2S, that are dissolved in the effluent may escape causing odour problems.
- High influent sulfate concentrations may limit the applicability of anaerobic sewage treatment as it results in the conversion of organic BOD/COD to inorganic BOD/COD.

potential odour problems and difficulties associated with, including nutrient removal in the treatment scheme. Table 4.1 lists the most important benefits and constraints associated with the implementation of anaerobic sewage treatment.

The rapid uptake of the municipal UASB technology was stimulated by the excellent performance of the Cali pilot plant and some other early municipal UASB plants such as Pedregal, Bucaramanga and Kanpur (van Haandel and Lettinga 1994; Schellinkhout and Collazos 1992; Schellinkhout and Osorio 1994; Draaijer *et al.* 1992 and 1994). The obtained maximum removal efficiencies for chemical oxygen demand (COD), biological oxygen demand (BOD) and total suspended solids (TSS) reached 70–80%, 75–85% and 70–80%, respectively. Table 4.2 lists average values from the early plants obtained in Colombia, Brazil and India.

The above mentioned reduction in investment costs (Table 4.1) is attributed to the omission of electro-mechanical units, such as aerators, as well as the reduced need for volumetric unit capacity, owing to the so-called 'shared functionality' of the anaerobic treatment tank. In fact,

Table 4.2. Treatment performance of the first full-scale UASB plants treating municipal sewage. COD refers to total COD of the raw wastewater (van Haandel and Lettinga 1994).

Country	Volume (m^3)	Temperature (°C)	HRT (h)	COD_{Inf} ($mg \cdot l^{-1}$)	$COD^{\$}_{eff}$ ($mg \cdot l^{-1}$)	COD_{rem} (%)
Colombia	64	24–26	4–6	267	110	65
Colombia	6,600	25	5.2	380	150	60–80
Brazil	120	23	4.7–9	315–265	145	50–70
Brazil	67.5	23	7	402	130	74
Brazil	810	30	9.7	563	185	67
India	1,200	20–30	6	563	146	74

$Calculated from the influent COD and removal efficiency.

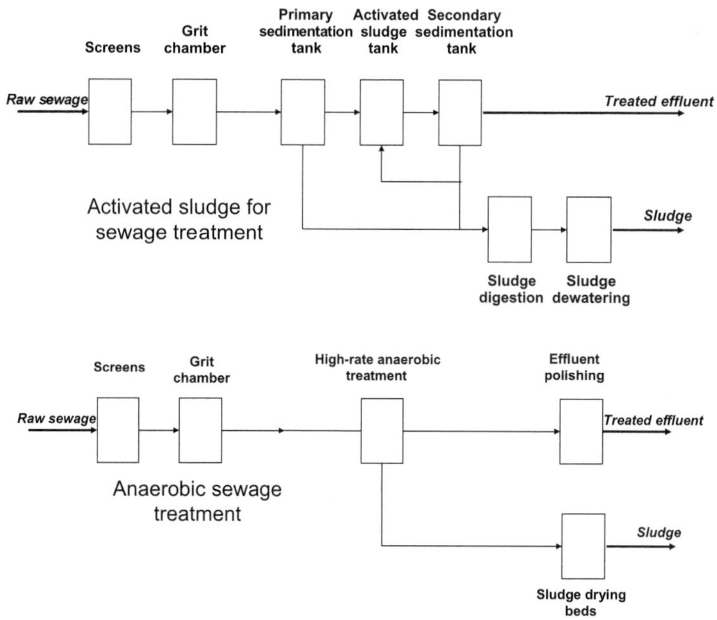

Fig. 4.1. Functional units of an STP, comparing activated sludge (up) and UASB technology (down).

a UASB reactor comprises four functional units (Fig. 4.1):

(1) *Primary clarifier*: removal/entrapment of (non)biodegradable suspended solids from the influent.
(2) *Biological reactors (secondary treatment)*: Removal of biodegradable organic compounds by converting them into methane.
(3) *Secondary clarifier*: clarifying the treated effluent in the settler zone at the top part of the UASB reactor.
(4) *Sludge digester*: stabilisation (digestion) and improving the dewatering characteristics of the retained influent primary sludge.

Obviously, the headworks, i.e., influent pumps if gravity cannot be used, screens and sand/grit removal chambers are needed for any treatment system. For anaerobic sewage treatment, though,

Table 4.3. Performance of the more recently installed full-scale UASB plants treating municipal sewage in different parts of the world. COD refers to total COD of the raw wastewater.

	Effluent content			Removal efficiency		
STP	COD ($mg{\cdot}l^{-1}$)	BOD ($mg{\cdot}l^{-1}$)	TSS ($mg{\cdot}l^{-1}$)	COD (%)	BOD (%)	TSS (%)
India[1]	364	137	357	43	55	18
India (survey)	285	121	107	46	41	49
Brazil[2]	251	98	85	65	74	71
Brazil (survey)	247	97	112	62	67	54
Middle East[3]	221	83	63	71	70	85

[1]Sato *et al.* (2006); [2]Oliveira and von Sperling (2009); and [3]Nada *et al.* (2006).

special care must be taken with the pre-treatment units as they determine reactor stability and long-term performance successes. For instance, clogging of influent pipes can be minimised by applying step screens. Several contractors advise fine (mesh or bar) widths not more than 6 mm clear distance. Such step screens should immediately follow from the coarse bar rack screens which generally have openings of 3–8 cm. With a width of 6 mm, screen automation is indispensable. Therefore, the step screen is often the most expensive part of the treatment system.

Notwithstanding the aforementioned excellent treatment results, recent literature reports and our own surveys suggest that the performance of the more recently constructed large-scale municipal UASB STPs is inferior to the results reported for the early pilot and full-scale plants (Table 4.3). For the sake of convenience, Table 4.3 lists averaged values only. Standard deviations, and particularly those referring to the data from India, have a very wide range, even yielding negative removal efficiencies. Very low average removal efficiencies are also obtained with a recently commissioned UASB reactor treating SO_4^{2-} (sulfate) rich sewage in the Gulf Area. Obviously, in the latter case the organic BOD is converted into inorganic BOD (H_2S/HS^-) leading to extremely disappointing results. Compared to the overall results of the initial installations, the average treatment efficiencies of the more novel systems dropped from 67% to 57% for COD removal, from 78% to 60% for BOD removal and from 63% to 60% for TSS removal.

Flaws in hardware engineering, combined with often underestimated shortcomings in operational and managerial aspects, contribute to the sometimes rather disappointing treatment results, which are particularly observed in India. For instance, very strikingly, half of the monitored UASBs by Sato *et al.* (2006) had more suspended solids in the effluent of the UASBs compared to the influent values. Since in India, discharge of untreated sewage is the single most important cause for pollution of surface and ground water (CPCB 2007), major attention has been put on the implementation of sewage treatment plants in the past decades. The programme of pollution abatement of rivers was started by the Ministry of Environment and Forests (MoEF), Government of India, with the launching of the Ganga Action Plan (GAP) Phase-I in June 1985. The UASB technology for sewage treatment was introduced in India under GAP-I in late 1980s. A pilot plant of 5,000 $m^3{\cdot}d^{-1}$ capacity, using UASB technology, followed by a one-day hydraulic retention time (HRT) polishing pond was constructed for the treatment of combined wastewater, i.e., domestic and tannery waste (Wiegant *et al.* 1999) and later exclusively for domestic sewage only. The results of these initial reactor systems were in agreement or even better than the expected performance data, although start-up procedures took longer than had been predicted

(Draaijer *et al.* 1994). Later in April 1993, Yamuna and Gomti Action Plans were approved under a new scheme of GAP Phase-II. This was followed by approval of pollution abatement programmes of other polluted rivers in the country in July 1995 under the National River Conservation Plan (NRCP). The National River Conservation Directorate (NRCD) under MoEF identified a good number of towns and cities within the states of Uttar Pradesh, Haryana and Delhi where sewage treatment facilities were urgently needed. The aforementioned states form the major part of the Yamuna River catchment, which is the largest tributary of the Ganges River and therefore the target area of GAP. Notwithstanding the limited experience obtained with the 5,000 $m^3 \cdot d^{-1}$ capacity pilot project at Kanpur city, the UASB technology with almost the same design was selected for about 16 STPs, with a combined capacity of almost 600,000 $m^3 \cdot d^{-1}$, under the Yamuna Action Plan (YAP) (CPCB 2003). Subsequent to this up-scaling and wide replication, a large number of STPs based on UASB technology have been constructed in the country based on low capital cost, low energy consumption and low cost of operation and maintenance. A large number of the installed plants do not perform as expected and site visits elucidated both engineering flaws and shortcoming in operational matters.

Disappointing treatment results (Sato *et al.* 2006) together with unrealistic expectations about biogas production and treatment capacity give rise to criticism and open opposition to UASB technology (Sharma 2002; Tare and Nema 2006). Apparently in India, the understanding of anaerobic pre-treatment is only limitedly diffused and hardly backed by academic institutions at all, unlike in countries like Brazil which have invested in fundamental and applied research and enhanced the cooperation between academics and consultants. This resulted in a large variety of STPs in which anaerobic technology is combined with adequate post-treatment systems offering a cost-effective solution that is complying with the national standards. Nonetheless, there are also a number of treatment plants in Brazil that are performing below expectations. Below various engineering and operational factors which impact the functionality and treatment performance of UASB-based plants are discussed. If these engineering and operational factors are not identified and remedied, the further application and development of UASB technology may be hindered. A review of existing literature is complemented with observations made during on-site visits to 12 large-scale municipal UASB STPs located in Brazil, India and the Middle East. The design capacities and parameters of these STPs vary largely, having flow rates ranging 30,000–164,000 m^3d^{-1} and reactor volumes ranging 8,000–55,000 m^3. Full-scale evaluations based on the authors observation are supported by questionnaire evaluations addressing operation managers, operators and operational records.

The more or less standardised concept of a UASB STP is depicted in Fig. 4.2. Exact configurations and dimensions of the functional units are site specific, depending on amongst others' local experience, contractor's experience, available funds, effluent requirements, etc.

In the past three decades the UASB reactor design has been more or less 'standardised', and is based on the original 64 m^3 pilot plant design of Cali, Colombia as described in the book of van Haandel and Lettinga (1994). The site visits of full-scale STPs revealed that there were a number of minor variations in plant design among the various reactors systems. It is not yet clear to what extent these differences in design determine the reactor's process efficiency; however, it is likely that a multiple combination of factors determines the final performance, thereby masking the impact of an isolated factor. When designing an UASB system a number of parameters need to be evaluated. In addition to the UASB tank itself, pre-treatment functional units, such as grit removal systems and screens, as well as post-treatment units are of crucial importance for the

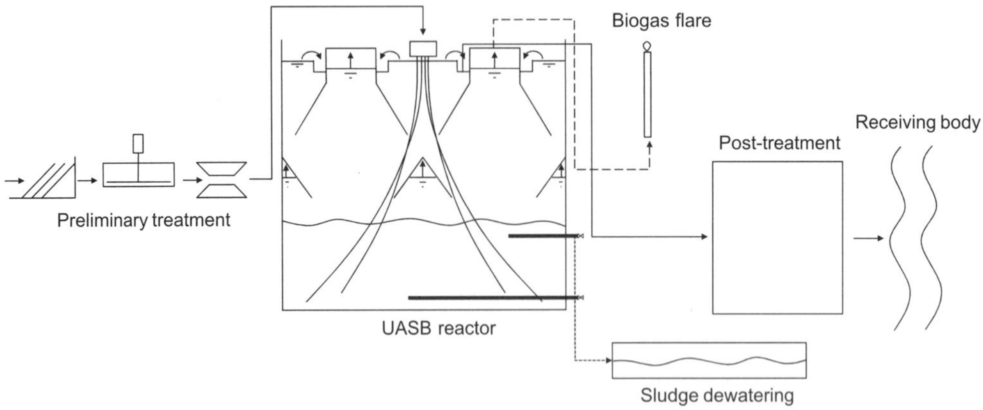

Fig. 4.2. General process configuration of a UASB based STP.

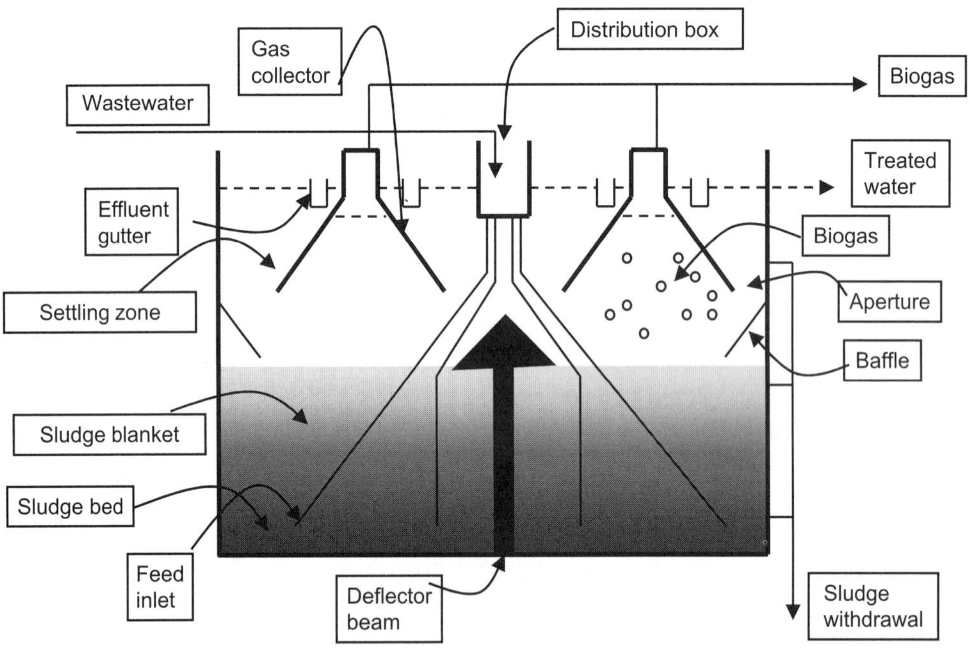

Fig. 4.3. Most important UASB design aspects, schematically indicated.

overall performance of the plant. The overall hydraulic profile requires special attention, as this impacts flow distribution to each corner of the reactor. Figure 4.3 indicates the most important design aspects to be considered for a UASB unit, and Table 4.4 lists the design criteria and basic assumptions used in the Indian UASB reactor designs.

It is interesting to note that, out of twelve full-scale installations visited, only two delivered satisfactory treatment results, and seven were classified as poor. For the remaining systems, data were missing and operational control was hardly available. To what extent that process and hardware design aspects are responsible for the poor performance results is not fully understood.

Table 4.4. Design criteria adopted in most of the Indian UASB reactors.

Parameter	Unit	Value
Temperature of sewage (minimum)	°C	20
HRT at average flow	hours	8–12
HRT at peak flow (minimum)	hours	4
Maximum aperture velocity	$m \cdot h^{-1}$	5
Feed inlet density (maximum)	m^2 per feed point	4
Angle of gas collector	degrees	50
Centre-to-centre distance between gas domes	m	4.0
Clear distance between gas domes	m	3.0
Gas hood width	m	0.44
Overlap of gas collector over deflector beam	m	0.15
Angle of deflector	degrees	45
Settling zone surface	% of total surface	75
Upflow velocity at average flow	$m \cdot h^{-1}$	0.5–0.6
Biogas loading (maximum)	$m^3 \cdot m^{-2} \cdot h^{-1}$	1.0
Biogas production	$m^3 \cdot kg\text{-}COD_{rem}^{-1} \cdot d^{-1}$	0.08–0.12
Sludge bed concentration	$kg\text{-}TSS \cdot m^{-3}$	65–70
Sludge bed height	% of height to gas collector	80–90
Sludge retention time (SRT)	days	32–45
Degradation non-sludge VSS	%	50
Bacterial yield coefficient (total)	$kg\text{-}VSS \cdot kg\text{-}BOCD^{-1}$	0.08
Sludge ash content	%	55–57
Volumetric loading rate	$kg\text{-}COD \cdot m^{-3} \cdot d^{-1}$	1.15–1.45
COD removal efficiency	%	65–70
BOD removal efficiency	%	75–80
TSS removal efficiency	%	75–80

However, from these site visits it became clear that the problems were not only due to inadequacy of treatment capacity alone but also due to operation and maintenance of the sewage treatment plants.

The following sections address several aspects which were found to have an impact on reactor performance or were found to be striking in a particular design. Optimising these design aspects will certainly result in improved reactor performance but will not guarantee good treatment results as this, obviously, is also dependent on managerial and operational attendance. Not needless to stress is the fact that a UASB reactor unit is a cost-effective pre-treatment step to remove the bulk of organic pollutants. Complying with effluent standards can only be achieved in conjunction with an appropriate post-treatment step that takes the functionality of the UASB reactor into account. In this respect, the choice for a one-day HRT polishing pond to upgrade the UASB effluent quality to discharge standards can be regarded as a highly questionable compromise, which should be discouraged under the Indian field conditions. In fact, after some years of operation, none of the inspected ponds reached their design treatment capacity, owing to the accumulation of suspended solids, contained in the effluent of the UASB units, turning the entire pond into an anaerobic, odorous, solids accumulation basin. Considering the limited attention given to the UASB reactors, e.g., the lack of sludge withdrawal, the subsequent post-treatment unit is likely to receive a much higher load than anticipated in the original design. In addition, the absolutely necessary major cleaning exercises such as the removal of excess solids from these one-day HRT polishing ponds are simply not taking place. As a result, none of these UASB plus polishing pond combinations reach the effluent standards that are set by the Indian regulations, thereby giving the UASB technology a doubtful and even negative impression. A thorough research programme from pilot to full-scale, with the objective of identifying the most appropriate combinations of UASB reactor units and post-treatment steps, was performed in Brazil, largely supported by the Brazilian Government under the name Prosab, documented by Chernicharo (2006). Lessons learnt include the need for increased operators input with the higher degree of compactness of the chosen combination.

As mentioned above, thorough screening and extended grit removal are indispensable for the long-term operational success of any compact wastewater treatment system. For an UASB reactor upfront pre-treatment is in fact crucial. Application of step screens complementing coarse screens is already indicated. The importance of removing sand and grit from the sewage before it enters the UASB reactor cannot be overemphasised. Given the high hydraulic loading rate typically applied to a UASB reactor, the rate of grit accumulation in the UASB tank can be significant. Grit accumulation reduces the available digestion volume of the UASB reactor, and it can potentially result in blockages of the inlet pipes. Grit chambers, also referred to as sand traps, are generally installed in order to reduce the accumulation of sand and grit inside the UASB reactors. Full-scale UASBs are generally preceded by square or rectangular horizontal flow grit chambers, with a few making use of an aerated grit chamber. The grit quantity in the sewage is location specific and amounts about 75 $cm^3 \cdot m^{-3}$ for typical Brazilian sewage (van Haandel and Lettinga 1994). Using this value and the obtained grit removal efficiency, it is possible to estimate the rate at which grit accumulates in the reactor. For some full-scale reactors the filling rate has been estimated to reach about 5% of the total reactor volume per year. Without efficient pre-treatment, the design, operation and maintenance of UASB reactors become an impossible job.

2. Reactor Size and Shape

The design of a UASB reactor treating a low-strength wastewater such as municipal sewage is governed by the hydraulic rather than the organic load (van Haandel and Lettinga 1994; von Sperling and Chernicharo 2005; van Lier *et al.* 2008). In fact, the required reactor volume (V, in m^3) is linearly related to the flow (Q, in $m^3 \cdot h^{-1}$) and the applied HRT (θ, in h):

$$V_{reactor} = \theta \cdot Q \tag{4.1}$$

Volumetric sizing of a UASB reactor fed with low-strength sewage of 500 g-COD·m^{-3} can be estimated using Eq. (4.1), applying an HRT of, say, 8 h. By taking a height of 5 m the required area can be roughly estimated. When the hydraulic loading is limiting, the applied up-flow velocity in the digestion and sedimentation compartments, as well as in the apertures between the gas hoods, is arguably the most critical design criteria: excessive velocities can result in biomass washout and thereby destabilise the biological treatment process.

UASB reactors can be constructed in two basic geometric shapes: circular or rectangular. A circular shape offers a number of advantages including increased structural stability and a smaller perimeter than a rectangle of the same surface area. Hence less building materials (concrete, steel reinforcements) will be required. However, when treating a large flow, which is often the case for UASB STPs, it is better to use a modular design. In this case a rectangular or square shape may be more advantageous as the UASB reactors can share walls. However, one should bear in mind that there is a maximum to the number of units that can be joined together as expansion from temperature variations may result in cracks in the civil structure. In general, the size of a single civil structure should be limited to a maximum of approximately 50 × 50 m. On the other hand, the construction of a circular gas–liquid–solid (GLS) separator is technically more difficult than that of a rectangular or square unit. When mounting a square or rectangular GLS separator in a circular reactor, a significant part of the reactor surface area will be lost for sludge settling. In addition, full surface coverage by the GLS separator near the reactor wall will be very difficult. For these reasons circular UASB reactors are generally not applied for UASB STPs. The majority of reactors and/or reactor modules are square or very close to square. A square reactor generally has lower civil construction costs than a rectangular one, because the perimeter of a square is smaller than a rectangle of the same area.

2.1. *Size*

Most UASB STPs are built using the modular design. The volume of individual UASB modules generally varies between 1,500 and 3,000 m^3, even though a UASB module of nearly 8,000 m^3 has been constructed in India (Sato *et al.* 2006). The choice of more reactors versus larger volume reactors is partly an economic one. Increasing the volume of individual reactors decreases the total perimeter which may decrease costs. For instance, when a 40 × 40 m reactor replaces four 20 × 20 reactors, almost 50% of the wall perimeter (i.e., the 'shared' walls) can be deleted. However, from an operational perspective it is advantageous to have more but smaller reactors over less but larger reactors, as this allows a greater deal of flexibility in operation. For example, it will be possible to take a single reactor out of operation for maintenance or repair, without significantly reducing overall treatment capacity.

2.2. *Basic shape*

In general, UASB reactors have a square or rectangular design, both in the cross-sectional area and the superficial area. Nonetheless, earlier Brazilian designs had a larger surface area of the settler than that of the digestion section (Fig. 4.4b). This might improve sludge retention compared to the conventional reactor due to the reduction of the upflow velocity and surface velocity in the settling compartment, which is important for dilute wastewaters (van Haandel and Lettinga 1994). However, the construction of vertical walls is far simpler and cheaper than that of inclined walls. In order to support the inclined walls, UASB reactors are often constructed completely subterranean. This, however, increases costs for earth excavation and removal. Also, a full subterranean construction is vulnerable to buoying forces when empty, potentially brought about by high ground water levels. In addition, access to sludge withdrawal valves and drain

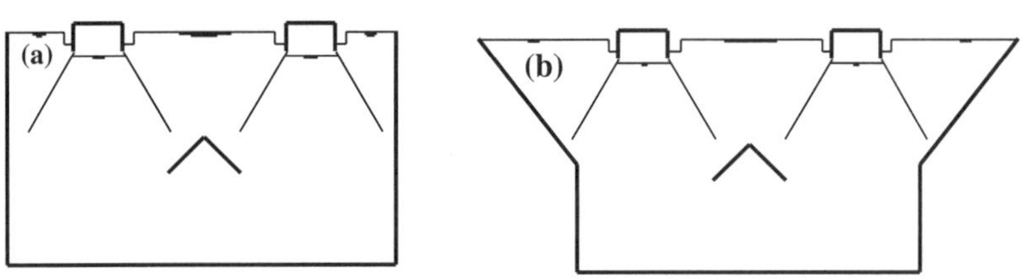

Fig. 4.4. Cross-sections of the two basic configurations available for constructing UASB reactors.

valves for the complete removal of sludge from the reactor for maintenance is more difficult with subterranean reactors. Clever mounting of the GLS separator and reduced upflow velocities prevent the need for enlarged surfaces at the top.

3. Influent Distribution System

A balanced distribution of influent over the bottom surface of a municipal UASB reactor is essential for optimum performance of the unit. A properly designed and constructed influent distribution system helps to guarantee maximum contact between the influent organic material and the sludge mass contained in the reactor. Furthermore, short circuiting between the influent distribution and effluent collection system is minimised. In contrast to industrial UASB reactors, where the influent is pumped through a bottom influent device, in a municipal UASB the influent is typically gravity fed from the top of the reactor through pipes that enter the municipal UASB reactor from above. It was demonstrated that an area coverage of 2–4 m^2 per inlet point was sufficient to obtain satisfactory treatment efficiencies (Royal Haskoning Consulting Engineers 1989). In general, most of the reactors took this criterion into practice with very few exceptions.

The selected density of inlet pipes and the total surface area of the STP determine the total number of flow splits required. For instance, a full-scale STP with a total surface area of 11,796 m^2, i.e., 38.4 m (l) $\times$ 12.8 m (w) $\times$ 24 (modules), and an inlet pipe density of one pipe per 2.5 m^2 requires an even flow division over 4,608 (i.e., 11796/2.5) equal parts. Reducing the density to one per 4 m^2 reduces the number of influent pipes drastically. It is noteworthy that no correlation could be made between the feed inlet density and the overall treatment performance of the inspected UASB systems. Therefore, it seems likely that overall performance is affected by multiple parameters, which mask the contribution of an individual parameter.

To facilitate proper influent distribution, distribution boxes are used to achieve the necessary division of flow. At this point, a distinction will be introduced between the final distribution box (FDB) located on top of the UASB from where the inlet pipes will enter into the reactor and all other distribution boxes located upstream of the UASB reactor. Distribution of the FDBs on top of a typical Indian-designed UASB reactor is depicted in Fig. 4.5.

3.1. *Final distribution box*

For optimal performance of a municipal UASB it is essential that the FDB is correctly designed and installed. Although the design of the FDB varies greatly in full-scale systems, ideally, the FDB should be symmetrical in shape to ensure equal flow to all inlet pipes. An FDB serving a large number of inlet pipes is easier to construct in a circular mode, which is frequently observed in full-scale plants in Brazil. The largest number of inlet pipes per FDB was 96, whereas rectangular FDBs are designed serving up to 24 inlet pipes.

The decision whether to select a limited number of FDBs with more inlet pipes or more FDBs with fewer inlet pipes has interesting repercussions. A single FDB with a high number of inlet pipes per reactor greatly simplifies the upfront influent distribution system, as the number of splits in the flow upstream of the FDB is greatly reduced. However, there are also some disadvantages associated with this configuration. Primarily, it results in a very large FDB; in the case of the 96 inlet pipes the diameter was 2.5 m. This resulted in a conflict between the FDB

Fig. 4.5. Location of final distribution boxes on top of a UASB reactor.

and the gas hoods as the FDB was too large to fit in between two gas hoods. The alternative of installing more FDBs with fewer inlet pipes ensures that there will be no conflict with the gas hood, but on the other hand it will increase construction costs and require more FDBs feed line connections through the reactor wall. A low number of inlet pipes per FDB also requires more hydraulic engineering skills to equally split the incoming sewage flows upfront over the various distribution boxes. Small errors upfront can have huge impacts down stream in the individual reactor modules. Upfront division using open distribution channels without V-notched overflow weirs should be prevented as equal flow splitting cannot be guaranteed in this way.

Van Haandel and Lettinga (1994) have suggested a number of important design principles that should be adhered to when designing an FDB, including the ability to:

- verify the flow rate at each split;
- adjust the flow rate at each split;
- easily detect if an inlet pipe has become obstructed; and
- easily eliminate any obstructions that do occur

To verify the uniform division of flow to all inlet pipes in an FDB, it is necessary that each inlet pipe connects to a single outlet only. However, in a large number of the STPs inspected, the inlet pipes have multiple outlets (Fig. 4.6), in an effort to save on construction costs. However, over time one or more of these outlets eventually become blocked, causing the influent to be distributed over the remaining outlets. The blocking of these outlets is unavoidable. If there is

(a) (b)

Fig. 4.6. Recommended inlet distribution systems: (a) each inlet pipe has a single outlet point, and (b) each inlet pipe has several outlet points.

only one outlet per pipe the blockage can immediately be identified and remedied. With multiple outlets blockages cannot be identified and unblocking of the outlet will be very difficult.

Adjustable triangular freefall overflow weirs (V-notches) are advised to ensure the even distribution of flow to each individual inlet pipe. During commissioning the weir level can be adjusted to ensure that each pipe receives an equal part of the flow. In the majority of inspected STPs, the V-notch weirs were indeed installed in the FDB but a significant number were not adjustable. Obviously, if the FDB is installed properly, i.e., levelled, there is no need for these weirs to be adjustable, but this depends on the skill of the civil contractor. In several full-scale reactors, the FDB consisted of a concrete rectangular box with a number of inlet pipes, where each pipe was connected to two outlets (Fig. 4.7). Influent enters from the centre of the FDB and is subsequently distributed over the surface of the FDB, because of this arrangement the UASB inlet pipes nearest to the FDB outlet likely receive more flow than those UASB inlet pipes farthest from the FDB outlet. Covering these boxes had unexpected consequences in one of the full-scale plants as a layer of scum developed in the FDB (Fig. 4.7b). Presently it is not clear whether the scum build-up resulted from the cover or from negligence of the operator to open the covered box. Two alternative improved FDB designs are shown in Figs. 4.7c and 4.7d.

3.2. *Blockages to inlet pipes*

Inlet pipe blockages can occur as a result of either a failure to remove debris, such as plastic, upstream of the UASB, or due to sludge and/or grit clogging the outlet point of the inlet pipes. In most full-scale STPs the problem of debris clogging the influent distribution system is prevented by installing step screens with a bar distance of 6 mm upstream of the UASB reactor. The clogging rate of the inlet pipes due to grit accumulation can be greatly reduced by the installation of a proper grit removal system. However, even with a properly functioning fine screen and grit removal system, clogging of inlet pipes will still occur occasionally. Thus an effective method to identify and remove blockages is always required. The identification of inlet pipes that have become blocked is obvious, once a freefall overflow weir is installed, as the liquid level after the weir will increase as shown in Fig. 4.8a.

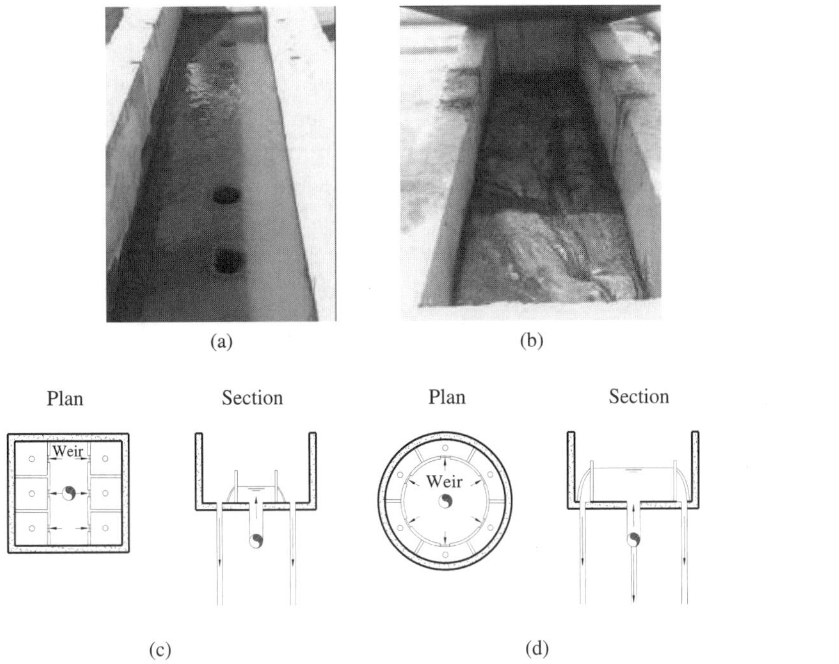

Fig. 4.7. (a) Images of a poorly designed final distribution boxes and (b) the scum formation resulting from the uneven flow distribution; and proper schematic designs of distribution boxes of: (c) rectangular (square) shape and (d) circular shape.

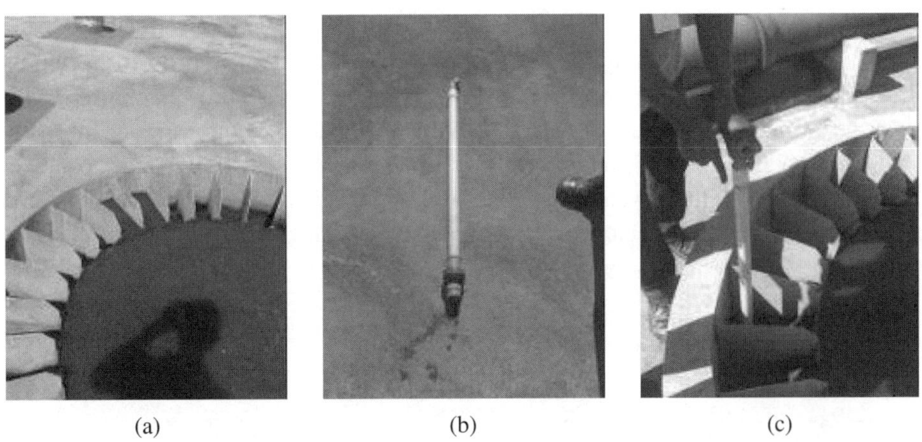

Fig. 4.8. Pressurised water procedure for unblocking inlet pipes.

Inlet pipe clogging can be further minimised by creating sufficient static head between the inlet of the individual influent distribution pipe, i.e., bottom surface of the FDB, and the reactor liquid surface. A difference of 50 cm water column is regarded as a minimum for regular automatic de-clogging of the tube. Obviously, increase in static head will result in more efficient self de-clogging. Nonetheless, from time to time, persistent inlet clogging will manifest. Several systems were employed to unclog these blockages, including flexible piping, inflexible piping, wire rods, flushing with pressurised water and the application of simple plungers that are used to de-clog water closet systems. Of these systems, the most effective results have been obtained using pressurised water (Chernicharo *et al.* 2009). Briefly the method involves the following steps: identifying the blocked inlet point (Fig. 4.8a), insert a PVC tube with a rubber seal attached on one end into the blocked inlet pipe (Fig. 4.8b), the rubber seal allows the application of pressurised water (5 bar gauge) which removes any blockages (Fig. 4.8c).

Insufficient static head and/or too narrow influent pipes may even result in complete flooding of the FDB, particularly during peak flows. Under such conditions, the shortest inlet pipes will receive more flow than longer pipes as the pressure drop in these pipes will be lower.

3.3. *Inlet pipes diameter*

Two conflicting design requirements must be reconciled when the diameter of the inlet pipes is selected: (1) the velocity at the top of the pipe should be low enough to allow any entrained air to escape back up the pipe; and (2) the velocity should be high enough at the outlet of the pipe to prevent heavy inert suspended solids from settling in or around the outlet. As to the first design requirement, entrainment of air bubbles with the influent is unavoidable because of the turbulent free fall that occurs when the sewage passes over the influent weir in the FDB and can have a number of negative effects. In general, the impact on the biological conversion process will be minimal as the entrapped O_2 will be immediately used by the facultative aerobic bacteria present in the sludge bed. An important detrimental effect is that the small air bubbles that are entrained with the sewage have a tendency to accumulate into large air pockets in the inlet pipes, especially at bends. These pockets can lead to temporary blockages of the inlet pipes. The air pockets eventually escape back up the inlet pipe but until, and as they do so, they prevent influent from entering into the blocked inlet pipe. If the FDB is designed correctly, i.e., a free-fall overflow V-notch weir is installed with sufficient static head available, the formation of air pockets does not influence the (average) flow distribution over the individual pipes. The flow that was prevented from entering into the pipe will accumulate above it, increasing static head until the obstruction is cleared and thus will still enter through that specific pipe. However, if the FDB is designed without weirs or without sufficient static head, the formation of these air pockets will affect the flow to individual pipes. Figure 4.9 shows examples of gas pockets that have formed in an inlet pipe escaping back up the pipe.

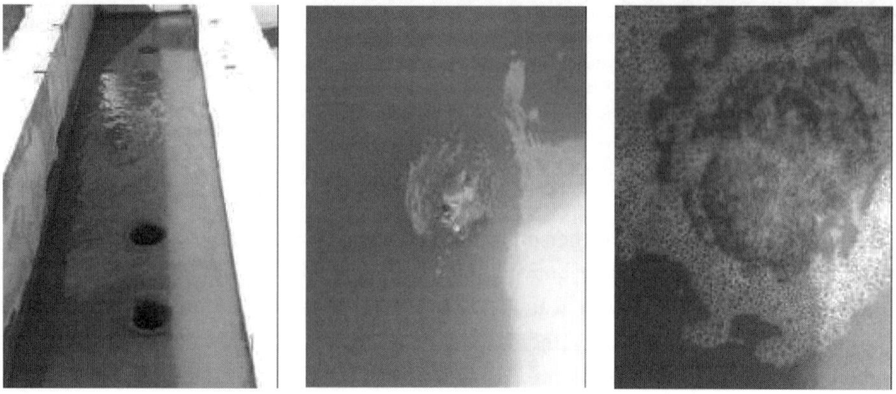

Fig. 4.9. Large air bubbles escaping from an inlet pipe resulting in the temporary blockage of the pipe.

Pipe diameters must be wide enough for gas bubbles to escape upwards. Following the calculations according Talaia (2007), a gas bubble with a diameter exceeding 2 mm has a rise velocity of about $0.2\,m \cdot s^{-1}$. For most full-scale UASB STPs, a diameter of 100 mm will be sufficient to achieve this aim. As to the second design requirement, van Haandel and Lettinga (1994) have suggested using a design grit settling velocity of $0.3\,m \cdot s^{-1}$ for grit removal units. Therefore, it is required to have a velocity at the outlet of the influent pipe that is greater than this to prevent the deposition of grit in the immediate vicinity of the outlet. To resolve the conflict between these two design requirements, installment, of inlet pipes with different inlet and outlet diameters is proposed (van Haandel and Lettinga 1994; Chernicharo 2007). The latter, obviously, will add to the costs. Very few full-scale reactors are equipped with inlets where the diameter of the pipe is reduced from top to bottom.

4. Gas–Liquid–Solid (GLS) Separator

The two main requirements for a well-performing UASB reactor are: (1) to ensure good contact between the incoming substrate and the anaerobic biomass; and (2) to maintain a large anaerobic sludge mass in the system. A well-designed influent distribution system guarantees the former, while a well-designed GLS separator is essential for the latter. By definition, the GLS separator separates the three phases present in the UASB reactor, i.e., the biogas (G), the liquid (L) and the suspended solids (S). Sewage is fed to the bottom of the reactor through the influent distribution system. Following an upflow path, the sewage flows through the thick anaerobic sludge blanket. It then passes through the GLS separator and finally the treated effluent is withdrawn at the top of the reactor. The GLS separator performs two important functions in this process: (1) the biogas formed in the sludge bed is prevented from entering into the settling compartment, but is instead collected in the gas hood for subsequent usage; and (2) the solids that enter into the settling compartment are allowed to settle in the quiescent environment that develops in the absence of gas bubbles and can therefore be retained for a large part in the reactor, thereby increasing the reactor sludge concentration while simultaneously improving the effluent quality considerably.

To prevent biogas bubbles from entering into the settler compartment an overlap is created between the deflector and the gas hood. In the literature an overlap of at least 10–20 cm is recommended, which is necessary because rising gas bubbles can otherwise be entrained with the effluent into the settling compartment.

With regard to the suspended solids retention efficiency, the design guidelines for the GLS separator are less clear. Van Haandel and Lettinga (1994) considered three main design criteria for the liquid velocity for a given reactor feed flow rate Q_i:

- the reactor upflow velocity in the digestion compartment v_1, which is defined by the reactor surface area A_1 and the influent flow rate Q_i;
- the aperture velocity v_2 at the level of the apertures between the deflector plates with an area A_2, which is the smallest area in the reactor; and
- the overflow velocity v_3 at the level of the discharge weirs or effluent pipes, defined by the area A_3.

For the average reactor feed flow, Chernicharo (2007) has recommended maximum values of 0.7, 2.5 and $0.8\,m \cdot h^{-1}$ for v_1, v_2 and v_3, respectively. Cavalcanti (2003) demonstrated at pilot scale that the installation of parallel plates above the conventional settler significantly improved suspended solids retention and thus also the maximum sludge concentration in the digestion zone. In these

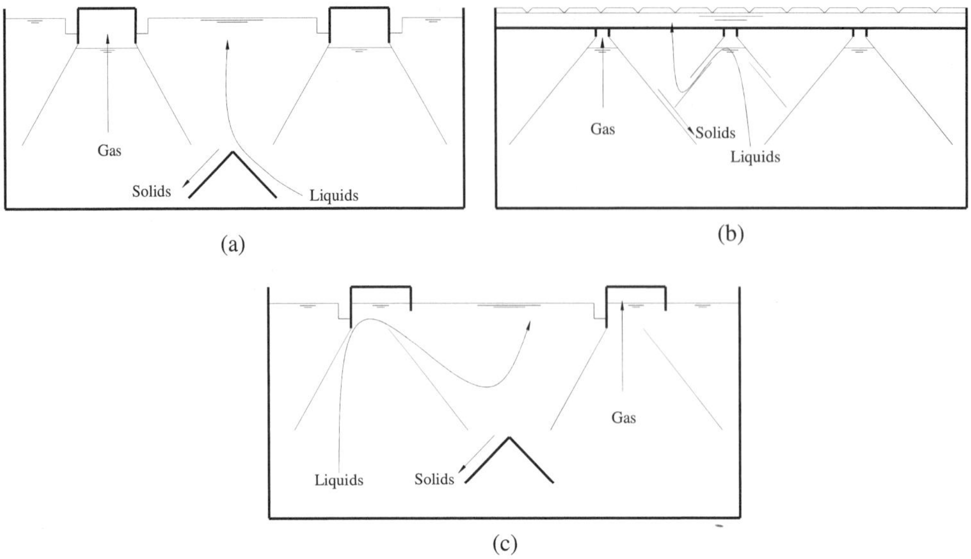

Fig. 4.10. GLS separator designs: (a) the conventional design as installed in most full-scale UASB STPs, (b) the alternative gas lift design with perpendicularly mounted effluent collection gutters and (c) the hybrid design installed in several Brazilian reactors.

experiments average sludge concentrations (based on the total liquid reactor volume) of 25–35 g-TSS·l^{-1} were achieved, which is more or less in the same range as the 64 m^3 pilot plant in Cali which retained 30–40 g-TSS·l^{-1}. In Brazil, typical (full-scale STPs) values of 15–18 g-TSS·l^{-1} are observed for conventional GLS designs.

4.1. *GLS separator designs*

A number of different GLS designs are implemented in full-scale systems. Some common examples are depicted in Fig. 4.10, in which the design of Fig. 4.10a is regarded as more or less standardised design.

In this design the gas box is pressurised, which allows transport of the biogas without additional equipment, such as compressors, to its point of flaring or utilisation. Ensuring that the gas hood is completely gas tight can be difficult and expensive. If the gas hood is made of concrete, often epoxy coatings are used, which has the added advantage of providing protection against the corrosive conditions. A disadvantage of using epoxy coatings is that improper application, e.g., when the concrete is not correctly pre-treated, may considerably reduce the effectiveness and durability of the coating. An alternative is to construct the gas hood entirely from a gas tight material, such as glass-reinforced polyesters (GRP). In the standardised conventional configuration, the GLS separator is not submerged which reduces the surface area available for settling at the overflow level. Elevated GLS separators sometimes lead to surface loads that are over 50% higher than the applied cross-sectional liquid upflow velocity in the reactor. However, there are also certain advantages associated with a GLS separator (partially) located above the reactor liquid level, as it acts as a support structure for the effluent collection devices and the FDBs. Additionally, the top of the GLS separators can be used as walkways to provide access to the influent distribution and effluent collection systems as these require regular cleaning. Care should be taken to the applied angle of the sides of the GLS separator. Angles exceeding 50° may lead to very deep GLS separator constructions, losing a significant part of the reactor volume for sludge accumulation. In some reactors even up to 60% of the height is used for the GLS separator device.

Although the alternative in Fig. 4.10b has interesting features, the perpendicular flow in the settling compartment reduces the effective settling time of the particles. Also the supposed advantages of the created gas lift effect enhancing solids retention have not yet been proven (Fig. 4.10c).

4.2. *Materials of construction*

Several materials of construction are used for GLS separators. Often the individual components that make up the GLS separator are constructed from different materials. For instance, the top gas box and the deflector can be constructed from concrete with epoxy coating to prevent corrosion and gas leakage, whereas the gas collection plates are constructed from materials such as concrete, canvas (heavy-duty woven cotton with plastic coating), corrugated iron, rigid PVC and GRP (Fig. 4.11).

(a) (b) (c)

(d) (e)

Fig. 4.11. Materials of constructions used for GLS separators: (a) corrugated iron, (b) concrete, (c) GRP, (d) canvas (PVC-coated woven cotton), and (e) rigid PVC.

All materials are found to be more or less suitable with the exception of the corrugated iron gas collection plates, which will fully corrode in several years of operation. Each material has its own pros and cons. Concrete gas collection plates require extensive form work and therefore result in high labour requirements. GRP plates can be pre-fabricated and are installed relatively easily, but on the other hand they are more expensive. PVC plates are prone to distortion or even rupture, but are relatively inexpensive to manufacture and install. The PVC-coated polyester sheets are relatively new on the UASB market (although frequently used for inflatable gas holders) and were pioneered in Brazil. The material selection for the GLS separator will be based on the local procurement and construction costs, as well as on the skill level of the civil contractor.

4.3. *Deflector type*

The main function of the deflector is, as the name indicates, to deflect the upflowing gas and solids away from the aperture to the settling zone and into the gas hood, where the gas will detach and the solids will settle back on the sludge bed. Different types of deflectors are constructed that mainly differ in the use of special devices for possible gas collection and conveyance (Fig. 4.12). The first deflector design, shown in Fig. 4.12a, has an inverted V-shape, allowing the biogas–liquid–solids mixture to enter into the deflector. The installation of pipes approximately 20 cm from the top of the deflector allows the transport of the biogas–liquid–solids mixture into the gas hood. Because of the position of the pipes below the deflector top a gas–liquid interface is present in the top part of the deflector. This makes the formation of a scum layer inevitable. Scum build-up may eventually result in the pipes becoming clogged, which can result in the uncontrolled discharge of gas over the bottom of the deflector, resulting in periodic escape of gas through the apertures into the settling compartment. The use of pipes to direct the biogas–liquid–solids mixture into the gas hoods may result in locally increased turbulence near the aperture in those sections of the deflector where the pipes are installed, which should be minimised.

The second design (Fig. 4.12b) is very similar to the first except that the pipes are located at the very top of the deflector, which prevents a gas–liquid interface from developing and therefore avoids the problems associated with scum formation.

The third design (Fig. 4.12c) differs from the previous two designs in that the bottom of the deflector is now closed. Hence there is no risk of a gas–liquid interface with consequential scum formation. However, as the bottom of the deflector is closed but not inclined, there is a distinct possibility that biogas will be entrained with the effluent flow into the settler section. This is especially likely because the deflector is never perfectly level and/or symmetrical, which will result localised gas escape over the length of the deflector.

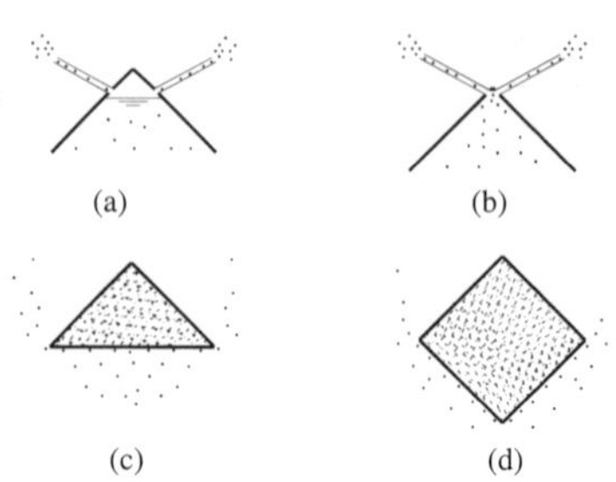

Fig. 4.12. Schematic representation of several deflector designs.

The last design (Fig. 4.12d) is closed as well but an inclined bottom section is added. The biogas is therefore actively deflected into the gas collection compartment while there is no risk of a gas–liquid interface developing. On the other hand, the addition of a bottom part to the deflector results in an increased total deflector height compared to the first three designs. However, it must be noted that the sludge bed should be maintained below the widest point of the deflector, in order to prevent biogas production above the deflector where it can enter directly into the settling compartment. As it is difficult to control the exact height of the sludge bed, it is recommended to keep the maximum sludge bed height at least 30 cm below this widest point. It is therefore concluded that the additional height of the deflector with an inclined bottom does not in fact reduce the volume of digestion zone; without any problem the bottom site can be submerged in the sludge blanket zone.

5. Effluent Collection

Effluent should be collected as uniformly as possible from the top of the UASB reactor to ensure optimum performance of the GLS separator. The two main alternatives for the collection of effluent from the top of the UASB reactor are effluent gutters fitted with V-notch overflow weirs and submerged perforated pipes. Mostly, V-notch overflow weirs are preferred as they can be controlled and cleaned easily, guaranteeing an even upflow in the settling compartment. In order to ensure that effluent is drawn off uniformly from the surface of the reactor there should be sufficient head available at the V-notch weirs to allow the weir to be aligned. Van Haandel and Lettinga (1994) suggest a minimum liquid height of 25 mm, below which it becomes difficult to adjust the overflow weirs. The flow rate over a V-notch weir can be calculated using the equation for a triangular weir:

$$Q = 1.465 \cdot h^{5/2} \cdot \tan \frac{1}{2}\alpha \tag{4.2}$$

where Q represents flow rate ($m^3 \cdot h^{-1}$), h liquid height at the weir (m) and α the angle. Assuming an angle of 90° is used, a head of 25 mm will result in a flow rate of $0.52\,m^{-3} \cdot h^{-1}$ per V-notch. Knowing that the liquid velocity in the UASB reactor is normally $0.5\text{–}1.2\,m \cdot h^{-1}$, it can be calculated that there should be a V-notch for approximately every 1–2 m^2 of overflow surface area. However, visual observations at full-scale UASB STPs showed an excessive number of V-notches in the installed effluent weirs. On the other hand, some full-scale reactors do not apply V-notches at all. Uneven withdrawal of effluent results in preferential flows in the settler section, which can result in an excessive loss of solids and the formation of dead zones in the reactor, these zones are subsequently not available for digestion.

The alternative of using submerged perforated pipes for the collection of effluent is only sporadically applied. The main advantage of this system is the less stringent levelling requirements. The flow rate through a circular opening can be calculated using the following equation:

$$Q = C_c \cdot A \cdot \sqrt{2 \cdot g \cdot h} \tag{4.3}$$

where C_c represents the contraction coefficient (0.66 for a circular opening), A represents the area of the opening (m^2), g gravitational acceleration coefficient ($m \cdot s^{-2}$), and h liquid height at the weir (m). In order to prevent clogging of the circular openings, a minimum diameter of 30 mm is suggested. Using Eqs. (4.2) and (4.3) it can be demonstrated that flow through a V-notch at a liquid head of 25 mm is 75% higher than at a head of 20 mm, while for a similar increase in liquid head the increase in flow rate is only 12% when a perforated pipe is used. Hence a misalignment of only 5 mm over the length of a V-notch gutter, which for a typical gutter length of 20 m amounts to an alignment error of only 0.025%, may lead to a difference in local effluent overflow rates of 75%, while there will be only a 12% difference in flow rate when perforated pipes are used. Perforated pipes also reduce turbulence at the overflow weir and, thereby, reduce malodorous emissions at the overflow point.

It is often suggested that the need to periodically clean the openings due to clogging by solids is a disadvantage of the perforated pipe system. However, V-notch weirs are also prone to clogging especially when the number of V-notches installed is excessive, resulting in low flow rates per individual V-notch. Therefore, both systems require provisions to allow for their periodic cleaning. Arguably the main disadvantage of using perforated tubes is the build-up of a scum layer on the surface of the settler. Therefore, provisions should be put in place for the removal of this scum layer especially if the reactor is covered which is increasingly the case. It should be noted that a scum layer will also build up on V-notch systems, whenever scum baffles are installed or regular access for cleaning is not possible due to the presence of a fixed reactor cover. Furthermore, it is quite possible that the frequency of cleaning in the case of submerged pipes is low, provided that the aperture diameter and flow rate are sufficiently high, as scum will then be removed ('sucked in') through the holes. A flow rate of at least 1 $m^3 \cdot h^{-1}$ should make the apertures reasonably self-cleaning. On the other hand, this relatively high flow will slightly affect the UASB effluent suspended solids content owing to the carry over of floating scum (see Section 7.1).

6. Sludge Withdrawal and Sludge Sampling

6.1. *Sludge withdrawal*

One of the main conditions for a satisfactory treatment performance of a UASB STP is to maximise the anaerobic sludge mass in the system, while simultaneously preventing sludge losses with the effluent. UASB reactors can retain a maximum sludge mass that is dependent on the size of the digestion zone and the maximum sludge concentration, which in turn depends on the phase separator efficiency in retaining suspended solids, the nature of the retained sludge, the degree of digestibility, preferential liquid flow streams, sludge filtering capacity, etc. Once the maximum storage capacity of the UASB reactor is reached, excess sludge will have to be discharged, otherwise unintentional sludge washout will occur. Without intentional sludge discharge, the rate of suspended solids washout in the effluent will be equal to the net sludge production and entrapment rate and this will considerably increase effluent COD and TSS values. Therefore, an efficient system to control the sludge blanket height is essential. This includes a reliable and accurate system for sludge withdrawal.

It is recommended that the excess sludge withdrawal system is designed for sludge to be removed from at least three levels in the reactor. One discharge location should be situated approximately 15–20 cm from the reactor floor, another should be located approximately at half way between the reactor floor and the lowest part of the GLS separator device, and the third one 50–75 cm beneath the widest point of the deflector, typically 1.5–2.0 m from the reactor floor. In addition, a drain must be installed, as a provision to completely empty the reactor for maintenance purposes. If such a drain is indeed installed, the lowest discharge point could be eliminated from the design. Instead of a bottom drain, 900 mm diameter manholes can be provided in an UASB reactor for regular cleaning of settled bottom sludge during operation and maintenance period. According to the authors, lack of this facility in all existing full-scale UASB-based STPs in India largely contributes to the observed failure of the system performance (Table 4.3, Sato *et al.* 2006).

For optimal performance of an UASB reactor, it is important to discharge sludge with a low volumetric sludge activity, which in the case of a municipal UASB is located in the less dense flocculent upper layer. On the other hand, due to sedimentation of inert particulate material in the UASB reactor, the inorganic sludge fraction of the bottom sludge will gradually increase. This will have two consequences: (1) the volumetric activity will decrease and (2) the density of the sludge may increase to a point where the inlet pipes become blocked and/or removal of the sludge from the reactor becomes impossible. Therefore, the normal routine should be to discharge sludge from the upper levels and only occasionally from the lower level or drain to prevent the build-up of inerts.

In a number of full-scale UASB STPs, the installed sludge withdrawal system was inadequate, e.g., only the bottom withdrawal system was installed, which results in a continuous discharge of the most active part of the sludge bed. Another downside of this strategy is that the sludge density will eventually reduce to very low levels reaching only 20 kg-TSS·m^{-3}, in contrast to values up to 60–100 kg-TSS·m^{-3} observed elsewhere. Also, a thick sludge bed layer has much better filtration capacities resulting in a much higher TSS removal efficiency.

As to the layout of the sludge withdrawal system, 2–3 withdrawal pipes per discharge level are required, each with a single opening per pipe. Considering that the surface area of UASB reactors tends to be about 500 m^2, the density of sludge withdrawal points is approximately 1 per every 100 m^2.

The withdrawn excess sludge from an anaerobic sewage treatment reactor is well stabilised owing to the long SRTs and can be dried by applying sludge drying beds generally within a week (van Haandel and Lettinga 1994). No odours arise from the sludge drying beds.

6.2. *Sludge sampling*

In order to control the sludge blanket height in a municipal UASB, which is a prerequisite for any optimised sludge inventory management system, it is absolutely necessary that the sludge concentration profile within the digestion zone can be accurately determined. Based on the difference between the measured and the target height of the sludge blanket level, it can be decided whether or not sludge withdrawal should be initiated. Furthermore, from this concentration profile the total sludge mass present in the reactor can be calculated. There are currently two systems used for sampling a UASB reactor (Fig. 4.13).

The most common system consisted of a series of valves distributed over the height of the wall of the digestion compartment (Fig. 4.13a). Chernicharo (2007) recommended that the sample ports are located at 50 cm intervals over the height of the digestion section. However, the proposed system is prone to wall effects and care should be taken that sampling is indeed representative. In addition, too quick sampling will create a vortex in the reactor body leading to misjudgement of the reactor's sludge quantity. The alternative system consists of a sample opening installed in the top part of the GLS separator device; a hydraulic seal should be used to

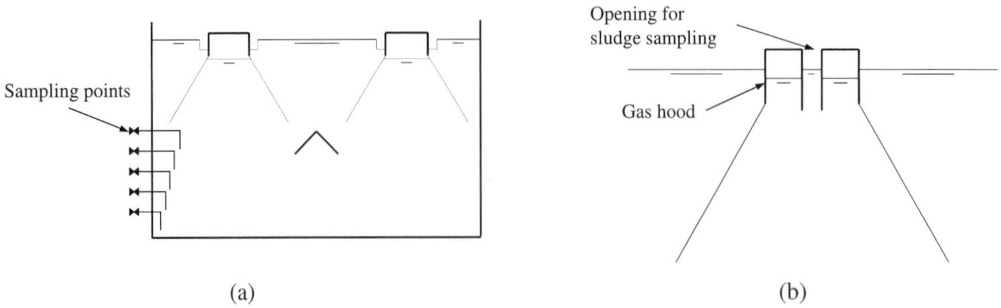

Fig. 4.13. Sludge sampling systems: (a) The most commonly employed system consisting of a series of sampling connections distributed over the height of the wall of the digester; and (b) the recommended procedure of taking samples from the top through the gas hood.

prevent the escape of biogas during sampling. A simple sampling device can be introduced via this opening, and samples can be withdrawn from any level.

7. Scum Removal

Scum is generally defined as a layer of floating material that develops at gas–liquid interfaces of bioreactors. The exact properties of the scum layer are highly dependent on the characteristics of the raw sewage. However, in general, the scum layer tends to be thin and somewhat fluid during the early stages of the development, but it gradually becomes very thick and viscous. The problems associated to scum formation are the following:

- blockage of part of the effluent weirs, resulting in imbalanced effluent withdrawal;
- potential for anaerobic decay with associated odour problems, procreation of flies and other unwanted vermin, and in general a very unpleasant appearance; and
- accumulation in the gas hood thereby disrupting the collection and removal of the produced biogas, and potentially leading to uncontrolled loss of biogas to the settling section or even ruptures in the GLS system.

During pilot plant studies Halalsheh *et al.* (2005, 2010) observed scum layers as thick as 1 m during the anaerobic treatment of concentrated sewage in Amman, Jordan. As a scum layer can develop wherever a gas–liquid interface is present, there are two possible locations in a UASB reactor where scum formation is a severe risk: (1) on top of the settler section and (2) on the gas–liquid interface in the gas hood. The removal of scum from both locations is arguably the most troublesome operational procedure currently associated with municipal UASB reactors. Therefore, before discussing systems and procedures for scum removal, it should be pointed out that the prevention of scum formation is much preferred over remedial measures. As scum formation is associated with, amongst others, the presence of free oil, grease and fats, hairs and fibres, it should be attempted to remove these constituents upfront at the headworks of the system. The use of fat traps, coarse and fine screening (6 mm), and aerated grit removal systems are highly recommended.

7.1. *Scum in the settler section*

Scum baffles were installed at many of the STPs with gutter systems for effluent withdrawal, as it has been suggested that the retention of scum results in a slight improvement in the effluent quality (van Haandel and Lettinga 1994). However, Souza *et al.* (2006) have shown that the production of scum under Latin American conditions is actually quite low at approximately 0.01 $l \cdot kg\text{-}COD^{-1}_{applied}$. Thus the installation of scum baffles is only necessary if the UASB reactor is the only means of treatment, or when it is followed by a series of ponds, where scum may accumulate, cause odour problems, and is difficult to remove. If scum baffles or perforated pipes are installed, provisions should also be put in place to allow for the removal of the scum layer that will be retained on the surface of the settler section. One approach is to allow a scum layer to develop into a thick viscous layer for approximately 3–4 months at which point it can be manually removed from the surface of the settler section using shovels and wheel barrows. This system was effective although clearly a very unpleasant job and very labour intensive.

An alternative method is to install scum removal bowls every 2–3 m in the settler section. The bowls are connected to a gate valve, which, when opened, allows the surface of the settler section, with its associated scum layer, to be drained and sent to the post-treatment unit. Return of this scum to the UASB reactor is not recommended as it will immediately result in a novel scum build-up.

In principle, scum layer build-up can be prevented if scum baffles are not installed (Chernicharo *et al.* 2009). However, in several full-scale plants a thick scum layer developed irrespective the free outflow of the effluent (Fig. 4.14). This is probably due to the excessive number of V-notches, far exceeding the recommendations discussed previously. If the flow per V-notch is too low, there is a high risk of blockage. In principle this can be remedied easily by cleaning the effluent weirs with a brush where and when needed, provided the reactor's surface is accessible. In the example of Fig. 4.14 the reactor was entirely covered with poured concrete and the settling section was only accessible by manholes. Therefore, in each design, provisions should be put in place to allow the removal of a scum layer.

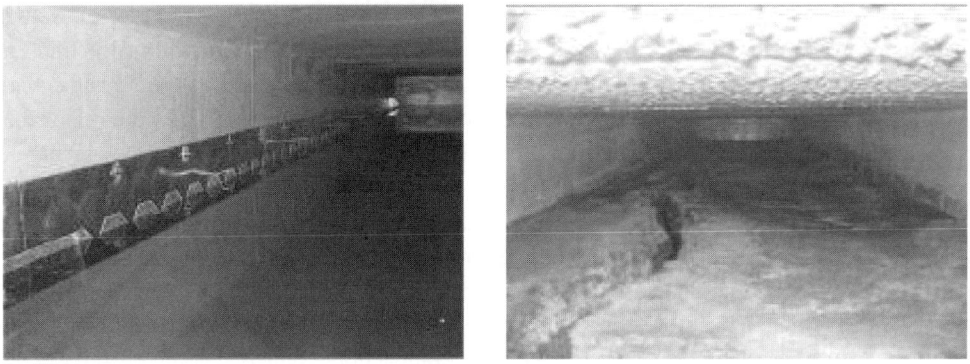

(a) (b)

Fig. 4.14. Images taken inside the settler of a fully closed UASB system without scum layer baffles: (a) the clean effluent gutter clearly visible during commissioning of the STP; and (b) a 40 cm scum layer developed after one year of operation.

7.2. *Scum in the gas hood*

The removal of scum from the gas hood is arguably an even more difficult task. Several methods have been implemented for this purpose, each with varying degrees of success. A commonly used method is the periodic removal of accumulated scum through access points (manholes) installed at regular intervals along the length of the GLS separator device. There are a number of disadvantages associated with this method:

- The access holes are so small that scum removal is very difficult, labour intensive and with only partial results at best;
- Due to the risk of explosion and/or sulfide toxicity, it is necessary to inert the gas hood prior to opening the covers as well as during the cleaning procedure. Although this safety provision is frequently disobeyed in practice;
- It is difficult to hermetically seal the access points after use. This is an absolute necessity, as otherwise biogas will escape to the environment resulting in environmental and safety issues (e.g., odour, H_2S toxicity, explosion risks); and
- Finally if the gas hood is not hermetically sealed it is impossible to operate the gas hoods at sufficient overpressure (15–40 mbar) to allow for the transport of the produced biogas to the gas treatment and utilisation facilities without the use of ventilators or compressors.

Another method that is commonly applied is spraying of the gas hood surface with water in an attempt to break the scum layer and disperse the scum. Unfortunately, this system was not effective at any of the known UASB reactors where the system was installed.

Recently, Chernicharo *et al.* (2009) proposed a novel system using the auto-generative gas pressure under the gas hood. The basic working mechanism relies on the sequential reduction and increase of pressure inside the gas hood to facilitate the removal of scum through weirs installed in the gas box. Excellent results under pilot conditions call for full-scale trials. The drained scum can be diverted to the post-treatment step or to the sludge discharge line.

8. Reactor Covers and Emission Prevention

The early full-scale UASB STPs were designed and built without covers on top of the settler sections. However, due to concerns about odorous emissions from UASB reactors, the trend in more recent full-scale designs is to include covers. From the STPs investigated, 60% had covers installed. There are essentially two options for covering the settler section, i.e., either removable covers made from materials such as aluminium, GRP and thin concrete slabs, or having a fixed cover of cast in place concrete.

As discussed above, it is necessary to have access to the settler section for the regular cleaning of V-notch weirs or the openings in the effluent pipes, whichever are installed and for periodic scum removal if required. Therefore, from an operation and maintenance perspective, the fixed cover is not recommended as it seriously limits access to the settler compartment. Considering that the primary reason for covering the settler section is odour control, it is interesting to note that practically all covered UASBs were not connected to a vent gas treatment system. Therefore, if the cover is not perfectly gas tight there will still be odour emissions, although at a reduced rate. In fact, in most STPs it is observed that the source of most odour problems is coming from the raw sewage inlet chamber upstream of the coarse screens. To a lesser extent odour problems are observed from the effluent overflow channel including the free fall overflows to either the post-treatment unit or effluent outfall. Reactor coverage becomes increasingly important for addressing the greenhouse gas emissions that generally occur at the top part of anaerobic reactors where the anaerobic environment is exposed to air. Here the P_{CH4} drops from about 0.8 bar to about nil, releasing all dissolved CH_4 at the effluent overflow weirs. In fact, up to 50% of the produced CH_4 could be released to the atmosphere unless provisions are put in place to prevent this. The high percentage of solubilised CH_4 can be attributed to the low COD load of the sewage and the relatively low temperature most of the plants are operated at. At 20 °C and 1 bar of CH_4 partial pressure, about 33 l of CH_4 is solubilised in 1 m^3 of effluent (Wilhelm *et al.* 1977). Assuming a flow of 50,000 $m^3 \cdot d^{-1}$, a COD concentration of 500 $mg \cdot l^{-1}$, a COD removal efficiency of 75%, a P_{CH4} of 0.8 bar, and a specific CH_4 production rate of 0.15 $m^3 \cdot kg\text{-}COD_{rem}^{-1}$, the amount of CH_4 produced is estimated as 2,813 $m^3 \cdot d^{-1}$ (i.e., 50,000 × 0.5 × 0.75 × 0.15), of which the solubilised CH_4 is estimated as 1,320 $m^3 \cdot d^{-1}$ (i.e., 50,000 × 0.033 × 0.8), equivalent to 47% of the total CH_4 production. This will get even worse when more diluted sewage is treated, or the sewage temperature is lower, or less COD is removed. The specific CH_4 production of 0.15 $m^3 \cdot kg\text{-}COD_{rem}^{-1}$ obviously depends on the biodegradability of the sewage and the amount of entrapped and converted primary sludge. Lettinga *et al.* (1993) and Uemura and Harada (2000) reported somewhat higher values, viz. 0.19 and 0.16 $m^3 \cdot kg\text{-}COD_{rom}^{-1}$, resulting in a higher recovery percentage.

CH_4 emission prevention can be obtained by covering the reactor and using a low pressure to collect the accumulating gases, which are subsequently directed towards the flare or gas engine for combustion. Alternatives to prevent atmospheric CH_4 emission include the oxidation of CH_4 in an aerated unit located on the top of the UASB reactor. This could be achieved by, for example, placing bio-rotors at the top of the UASB or combining a UASB with a post-treatment using activated sludge in a single tank (Fig. 4.15).

During the early developments of anaerobic sewage treatment, CH_4 emission was simply ignored or not taken into consideration in the full-scale design because of ignorance and/or financial limitation. Nowadays, uncontrolled greenhouse gas emissions should be avoided and

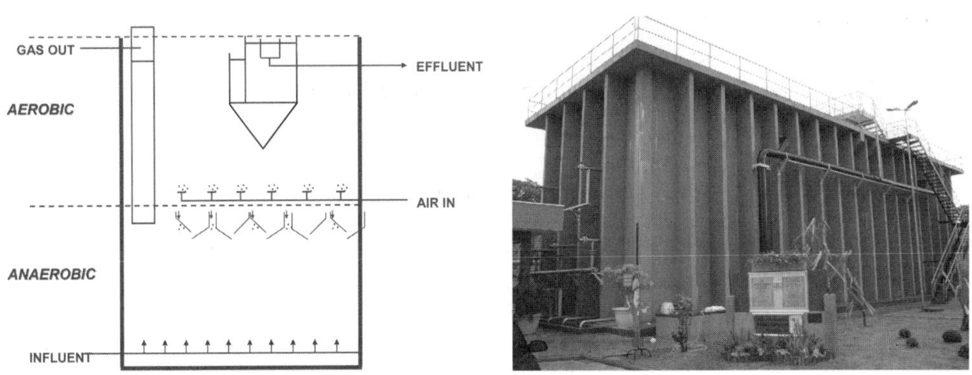

(a) (b)

Fig. 4.15. BIOPAQ® UBox sewage treatment plant: (a) The schematic design showing a combined anaerobic treatment with activated sludge, and (b) a full-scale plant in Igaraçu do Tietê, Brazil, treating wastewater of 25,000 population equivalents (printed with permission from Paques).

non-flaring of captured CH_4 should be prohibited. If instead, all the energy is used, with the increasing energy prices and tradable CO_2 credits, anaerobic sewage treatment will become an affordable investment for many developing countries.

At present, the produced biogas is hardly recovered and very few full-scale plants are equipped with a biogas motor. In most full-scale reactors the biogas is (or intended to be) flared. In general the biogas collection and disposal system is rather poor, with significant corrosion issues due to inappropriate material selection. In some cases, flares did not work and biogas was vented directly to the atmosphere. Furthermore, in most known STPs the installed gas meter fails and/or is not operational.

9. Future Outlook: Treatment of Concentrated Sewage

Domestic sewage is a dilute wastewater of complex characteristics, with a relatively high suspended solids content (i.e., a low $COD_{soluble}/COD_{total}$ ratio) and a low temperature. The suspended solids may constitute 50–65% of the total COD. Therefore, total COD conversion is largely limited by hydrolysis of particulate matter. Particularly when the sewage temperature drops to below 20 °C, the biological conversion capacity will determine the overall COD removal rather than the prevailing hydrodynamic conditions. In fact, because of the low temperature and the high TSS/COD ratio, the COD concentration range in which the HRT determines the volume of the UASB reactor (as shown in Eq. (4.1)) is relatively small. When temperature drops and non-digested sludge starts to accumulate in the sludge bed, the hydrolytic and methanogenic capacity of the sludge bed will gradually decrease, deteriorating both particulate and soluble COD removal, and eventually leading to reactor failure. Apparently, the primary design criterion, even with dilute domestic sewage, is the reactor solids retention time (SRT), which should be above a minimum value in order to maintain the methanogenic conversion capacity of the sludge. With dilute domestic sewage (COD below 1,000 $mg{\cdot}l^{-1}$) under tropical conditions (temperature above 20 °C), this condition will always be met. The prevailing SRT depends on various sewage characteristics, such as:

- sewage temperature,
- influent suspended solids concentration,
- rate of solids digestion in the reactor,
- filtering capacity of the sludge bed, which is determined by the applied upflow velocities and sludge characteristics,
- growth and decay of new sludge,
- sludge retention in the settler, determined by the applied liquid velocities and
- withdrawal of excess sludge.

The SRT can be calculated using Eq. (4.4):

$$SRT = \frac{X_{reactor} \cdot V_{reactor}}{Q_{effl} \cdot X_{effl} + Q_{excess\text{-}sludge} \cdot X_{excess\text{-}sludge}} \tag{4.4}$$

where X represents concentration of viable biomass ($kg{\cdot}m^{-3}$), V reactor volume (m^3) and Q flow rate ($m^3 \cdot d^{-1}$). As a rule of thumb, the minimum SRT should always be more than three times the doubling time (T_d) of the biomass responsible for the rate limiting step. With dilute domestic sewage under tropical conditions, these are the methanogens, with an estimated T_d at

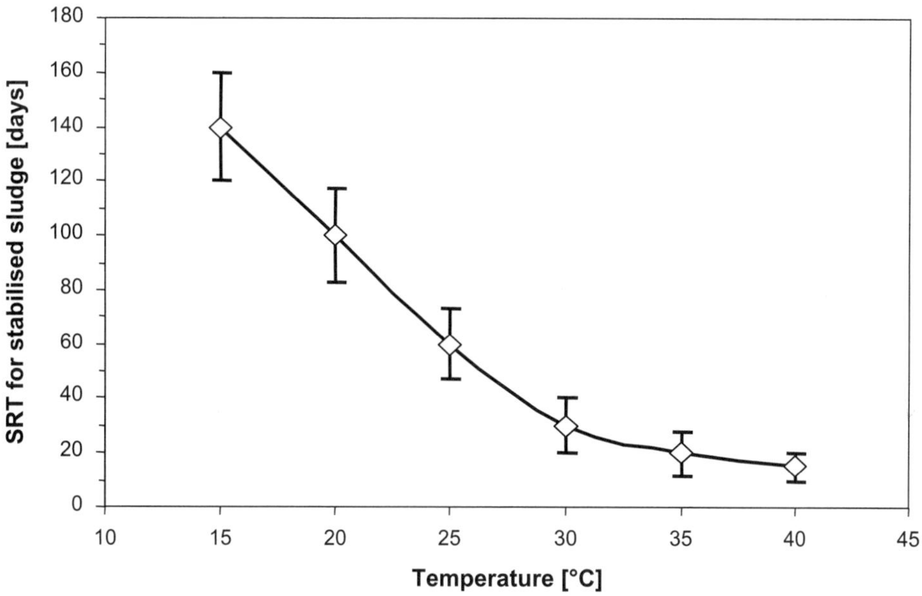

Fig. 4.16. Required SRT for domestic sewage treatment as a function of temperature.

25 °C of about 10 days. Therefore, SRTs of existing full-scale sewage treatment systems will never be below 30 days. The impact of temperature on the required SRT in the UASB reactor is illustrated in Fig. 4.16. Realising the importance of the SRT, it becomes clear that the conventional UASB reactor design for municipal wastewater needs reconsideration when temperature drops and COD exceeds 1,000 $mg{\cdot}l^{-1}$. In many arid climate countries with limited water supply, such as. Middle East, Northern Africa and Arabian peninsula, sewage COD concentrations range 1,000–2,500 $mg{\cdot}l^{-1}$. Furthermore, the temperate climates in the Middle East and Northern Africa are characterised by cold winters, particularly in mountainous areas.

Recent experiences in Jordan and Palestine show municipal sewage COD concentrations reaching 2,500 $mg\text{-}COD{\cdot}l^{-1}$ at a TSS/COD ratio of 0.6 (Mahmoud *et al.* 2008), whereas winter temperatures may drop to 15 °C. Applying the conventional UASB reactor design, the HRT needs to be increased to 20–24 h (Halalsheh *et al.* 2005, 2010). This, obviously, will affect the hydrodynamics of the system requiring changes in influent distribution to prevent short circuiting. Alternatively, the large suspended solids load can be addressed in separate reactor units such as a primary clarifier or enhanced solids removal in upflow filter systems, coupled to a sludge digester (Elmitwalli *et al.* 2002a,b). A novel approach is to link the UASB reactor to a coupled digester with sludge exchange (Mahmoud *et al.* 2004). With the latter system, accumulating solids will be digested at higher temperatures, whereas the methanogenic activity in the reactor will be increased by a return digested sludge flow.

Recently, the first full-scale reactor in the Middle East region has been started up (Nada *et al.* 2006) in Fayoum, south of Cairo, Egypt. The design is based on the conventional approach taking into account the relatively high strength of the sewage, resulting in a somewhat higher HRT with an average of 12 h. Pilot trials in Amman showed the feasibility of the system as an ideal pre-treatment method for a low-cost reduction in the COD load, while generating energy

Table 4.5. UASB pilot trials at Amman–Zarqa, waste stabilisation pond site 'Khirbet As Samra', Jordan

Average influent characteristics		Treatment performance (including post-clarification)	
Flow	170,000 $m^3 \cdot d^{-1}$	COD removal	up to 80%
BOD	750 $mg \cdot l^{-1}$	BOD removal	up to 85%
COD	1,900 $mg \cdot l^{-1}$	SS removal	up to 80%
TSS	400–600 $mg \cdot l^{-1}$	Pathogens	negligible
$NH_4 + $-N	25–90 $mg \cdot l^{-1}$	CH_4 production	0.15 $m^3 \cdot COD_{rem}^{-1}$
TKN	70–120 $mg \cdot l^{-1}$	Potential CH_4 production	27,000 $m^3 \cdot d^{-1}$
P-total	10–30 $mg \cdot l^{-1}$	(equivalent to ~5MW, assuming 40% efficiency)	
Temperature	16 °C (winter) 28 °C (summer)		

Note: Flow estimates concern the year 2000 (Ammary 2007; Halalsheh *et al.* 2005). Flow projections for 2025 are exceeding 425.000 $m^3 \cdot d^{-1}$ (Ammary 2007).

for post-treatment. Table 4.5 summarises the key influent characteristics and the most important results (Halalsheh *et al.* 2005, 2010).

Although prospects for full-scale application in Amman looked very promising (Table 4.5), it was decided to change the former pond system into a modern activated sludge plant. With regard to sustainability in domestic sewage treatment this decision is considered *a wasted opportunity*, particularly as more concentrated municipal wastewaters are in fact ideal for anaerobic pre-treatment. The recovered energy can then be beneficially used at the treatment site for extensive treatment up to discharge or reuse standards. Any excess energy may serve as power supply for irrigation pumps or for settlements in the vicinity of the plant.

10. Concluding Remarks

The UASB system has developed into a mature technology that has been applied successfully for the treatment of municipal sewage in regions with a warm climate. However, the performance of the large STPs built in the last ten years is inferior to that of the original pilot- and full-scale STPs built in the late 1980s and early 1990s. In all cases the inferior performance can be explained by poor design and construction, insufficient or unqualified operator attendance or to the presence of high SO_4^{2-} concentrations in the sewage. Despite publication of the recommended design guidelines by van Haandel and Lettinga (1994), many of the commissioned full-scale UASB systems have structural shortcomings. The hydraulic profiles of some plants were insufficient to achieve uniform distribution of flow and flow division devices, such as the FDB, are poorly constructed. In general, the GLS separators are better designed, although the efficacy of the gas lift designs is questionable when a low strength wastewater such as sewage is treated, as the gas production rate may be too low to induce the required mammoth flow. Effluent collection remains problematic, particularly in UASB reactors with a fixed cover preventing easy access to the effluent weirs for cleaning. Scum removal from the gas hood is without doubt the most troublesome design issue. Only a very few full-scale reactors manage to operate without aggravating scum removal problems.

As to the subject of insufficient/unqualified operator attendance, this might be due to earlier publications where the UASB was portrayed as a low-cost and low-maintenance solution. Whereas the UASB is indeed a low maintenance system, low maintenance does not equal the absence of maintenance, which seems to be the meaning taken all too often. The best operating full-scale reactors that were designed in the past two decades for sewage temperatures between 17 °C and 27 °C achieve removal efficiencies of nearly 67%, 70% and 75% for COD, BOD and TSS, respectively, which are a little bit below the projected design values in Table 4.4. A further increase by 6–10% in removal efficiency seems possible when all critical design features are taken into account and operator attendance is maximised.

Considering the present concern on fossil fuel consumption, anaerobic sewage treatment offers a feasible alternative for treating the huge flow of domestic and municipal wastewaters in many parts of the world. In light of the current greenhouse gas discussion, recovery of all produced CH_4 should be an intrinsic part of the treatment plant design. Owing to its compactness, high-rate anaerobic sewage treatment can be applied in the urban areas as well. The latter will lead to huge costs reduction in constructing sewerage networks, pumping stations and conveyance networks. It must be realised that in 2000 only 35% of the produced municipal wastewaters in Asia were treated, whereas in Latin America this value was only 15% (WHO/UNICEF 2000). Since then, advancements are limited and largely vary (Chile reaching 80% and Haiti remaining at 10% or less). In Africa, the generated wastewaters are hardly collected and sewage treatment, with the exception the Mediterranean part and South Africa, is nearly absent. With the increase in basic understanding of the anaerobic process and increase in the number of full-scale experiences at any scale, anaerobic treatment will undoubtedly become one of the prime methods for treating organically polluted wastewater streams.

References

Ammary, B. Y. (2007) Wastewater reuse in Jordan: Present status and future plans. *Desalination* **211**, 164–176.

Cavalcanti, P. F. F. (2003) *Integrated Application of the UASB Reactor and Ponds for Domestic Sewage Treatment in Tropical Regions.* Thesis, Wageningen University, the Netherlands.

Chernicharo, C. A. L. (2006) Post-treatment options for the anaerobic treatment of domestic wastewater. *Rev. Environ. Sci. Biotechnol.* **5**(1), 73–92.

Chernicharo, C. A. L. (2007) *Anaerobic Reactors*, Biological Wastewater Treatment Series (Vol. 4), IWA Publishing, London.

Chernicharo, C. A., Almeida, P. G., Lobato, L. C., Couto, T. C., Borges, J. M., and Lacerda Y. S. (2009) Experience with the design and start up of two full-scale UASB plants in Brazil: Enhancements and drawbacks. *Wat. Sci. Technol.* **60**(2), 507–515.

CPCB (2003) *Status of Sewage Treatment Plants in Ganga Basin. Central Pollution* Control Board (CPCB), Ministry of Environments & Forests (MoEF), Delhi, India.

CPCB (2007) *Evaluation of Operation and Maintenance of Sewage Treatment Plants in India.* Central Pollution Control Board (CPCB), Ministry of Environments & Forests (MoEF), Delhi, India.

Draaijer, H., Maas, J. A. W., Schaapman, J. E., and Khan, A. (1992) Performance of the 5 MLD UASB reactor for sewage treatment in Kanpur, India. *Wat. Sci. Technol.* **25**(7), 123–133.

Draaijer, H., Pereboom, J. H. F., and Sontakke, V. N. (1994) Four years experience with the 5 MLD UASB reactor for sewage treatment at Kanpur, India. *Proceedings of 7th International Symposium on Anaerobic Digestion*, January 27–31, Cape Town, South Africa.

Elmitwalli, T. A., Sklyar, V., Zeeman, G., and Lettinga G., (2002a) Low temperature pre-treatment of domestic sewage in an anaerobic hybrid or an anaerobic filter reactor. *Bioresour. Technol.* **82**, 233–239.

Elmitwalli, T. A., Oanh, K. L. T., Zeeman, G., and Lettinga, G. (2002b) Treatment of domestic sewage in a two-step anaerobic filter/anaerobic hybrid system at low temperature. *Wat. Res.* **36**(9), 2225–2232.

Halalsheh, M., Sawajneh, Z., Zu'bi, M., Zeeman, G., van Lier J. B., Fayyad, M., and Lettinga G. (2005) Treatment of strong domestic sewage in a 96 m^3 UASB reactor operated at ambient temperatures: Two-stage versus single-stage reactor. *Bioresour. Technol.* **96**, 577–585.

Halalsheh, M., Zeeman, G., Lettinga, G., and van Lier, J. B. (2010) Anaerobic Pre-treatment of Strong Sewage. VDM Publishing House Ltd, Mauritius, *In Press*.

Lettinga, G., de Man, A., van der Last, A. R. M., Wiegant, W., Van Knippenberg, K., Frijns, J., and van Buuren, J. C. L. (1993) Anaerobic treatment of domestic sewage and wastewater. *Wat. Sci. Technol.* **27**(9), 67–93.

Mahmoud, N., Zeeman, G., Gijzen, H., and Lettinga, G. (2004) Anaerobic sewage treatment in a one-stage UASB reactor and a combined UASB-digester system. *Wat. Res.* **38**(9), 2348–2358.

Mahmoud, N., Zeeman, G., and van Lier J. B. (2008) Adapting UASB technology for sewage treatment in Palestine and Jordan. *Wat. Sci. Technol.* **57**(3), 361–366.

Nada, T., Ali, H. I., Farid, M. N., El-Gohary, F. A., and Moawad, A. (2006) A full-scale application for anaerobic wastewater treatment technology (up-flow anaerobic sludge blanket UASB) in Egypt. *Conference National Research Centre*, March 13–15, Cairo, Egypt.

Oliveira, S. C. and von Sperling, M. (2009) Performance evaluation of UASB reactor systems with and without post-treatment. *Wat. Sci. Technol.* **59**(7), 1299–1306.

Royal Haskoning Consulting Engineers (1989) *Anaerobic Treatment of Domestic Wastewater under Tropical Conditions — Final Report.* Royal Dutch Consulting Engineers and Architects.

Sato, N., Okubo, T., Onodera T., Ohashi, A., and Harada, H. (2006) Prospects for a self-sustainable sewage treatment system: A case study on full-scale UASB system in India's Yamuna River Basin, *J. Environ. Manage.* **80**(3), 198–207.

Schellinkhout, A. and Collazos, C. J. (1992) Full-scale application of the UASB technology for sewage treatment. *Wat. Sci. Technol.* **25**(7), 159–166.

Schellinkhout A. and Osorio E. (1994) Long-term experience with the UASB technology for sewage treatment on large scale. *Proceedings of 7th International Symposium on Anaerobic Digestion*, January 27–31, Cape Town, South Africa.

Schellinkhout, A., Lettinga, G., van Velsen, A. F. M., Louwe Kooijmans, J., and Rodriguez, G. (1985) The application of the UASB-reactor for the direct anaerobic treatment of domestic waste water under tropical conditions. *Proceedings of Seminar on Anaerobic Treatment of Sewage*, M. S. Switzenbaum (ed.), pp. 259–276, June 27–28, Amherst, USA.

Sharma, S. (2002) Wouldn't you like a nice anaerobic sludge blanket? www.cleanganga.com, visited 2007.

Souza, C. L., Silva, S. Q., Aquino, S. F., and Chernicharo, C. A. L. (2006) Production and characterization of scum and its role in odour control in UASB reactors treating domestic wastewater. *Wat. Sci. Technol.* (54), 201–208.

Talaia, M. A. R. (2007) Terminal velocity of a bubble rise in a liquid column. *World Acad. Sci., Eng. Technol.* **28**, 264–268.

Tare, V. and Nema, A. (2006) Sewage treatment through UASB technology — expectations and reality. *Proceedings of 22nd National Convention on Environmental Engineering*. Institution of Public Health Engineering and IT-BHU, Varanasi, April 7, 2006.

Uemura, S. and Harada, H. (2000) Treatment of sewage by a UASB reactor under moderate to low temperature conditions. *Bioresour. Technol.* **72**(3), 275–283.

van Haandel, A. C. and Lettinga, G. (1994) *Anaerobic Sewage Treatment. A Practical Guide for Regions with a Hot Climate.* John Wiley and Sons. Chichester, pp. 84.

van Lier, J. B., Mahmoud, N., and Zeeman G. (2008) Anaerobic wastewater treatment. In: Henze, M., van Loosdrecht, M. C. M., Ekama, G. A., and Brdjanovic D. (eds.), *Biological Wastewater Treatment, Principles, Modelling and Design, Chapter 16*, ISBN: 9781843391883, IWA Publishing, London, UK, pp. 415–456.

Vieira, S. M. M. and Souza, M. E. (1986) Development of technology for the use of the UASB reactor in domestic sewage treatment. *Wat. Sci. Technol.* **18**(12), 109–121.

von Sperling, M. and Chemicharo, C. A. L. (2005) *Biological Wastewater Treatment in Warm Climate Regions (Vol. 1).* IWA Publishing, London. pp. 750–751.

WHO/UNICEF (2000) *Global Water Supply and Sanitation Assessment 2000 Report.* World Health Organisation, United Nations Children's Fund, Geneva, p. 80.
Wiegant, W. M. T., Kalker J. J., Sontakke, V. N., and Zwaag R. R. (1999) Full scale experience with tannery water management: An integrated approach. *Wat. Sci. Technol.* **39**(5), 169–176.
Wilhelm, E., Battino, R., and Wilcock, R. J. (1977) Low-pressure solubility of gases in liquid water. *Chem. Rev.* **77**, 219–262.

Chapter 5

Application of UASB Technology for Sewage Treatment with a Novel Post-treatment Process

Shigeki Uemura[1,*] and Hideki Harada[2]

[1]*Department of Civil and Environmental Engineering, Kisarazu National College of Technology, 2-11-1 Kiyomidai-higashi, Kisarazu, Chiba 292-0041, Japan*

[2]*Department of Civil Engineering, Tohoku University, 06 Aoba, Sendai, Miyagi 980-8579, Japan*

**uemura@wangan.c.kisarazu.ac.jp*

The upflow anaerobic sludge blanket (UASB) process has been shown to be effective for removal of organic substances from sewage, especially under warm climate conditions. However, treatment of sewage by UASB alone is not sufficient to meet the effluent discharge standards set by many developing countries, meaning that UASB effluent needs post-treatment process. In combination with properly selected processes, a sustainable and appropriate wastewater treatment system can be established. In this chapter, the down-flow hanging sponge (DHS) process is introduced as an appropriate post-treatment process for UASB-treated sewage, with emphasis on its basic concept, mechanism of treatment, history of development, and recent successful results of a demonstration study at a full-scale DHS plant in India.

1. Introduction

At present, the most popular anaerobic sewage treatment process is the upflow anaerobic sludge blanket (UASB) process, which was developed by a research group at the Wageningen Agricultural University in the Netherlands in the 1970s (Lettinga *et al.* 1980). The most important feature of the UASB is that sludge granules of 1–5 mm diameter can be developed inside the reactor without any support medium. The granules have a high density, excellent mechanical strength, and a high settling velocity in combination with a high specific methanogenic activity (Lettinga *et al.* 1980; Hulshoff Pol and Lettinga 1987). The granules form a blanket through which the influent wastewater flows. Organic substances in the wastewater are digested by anaerobic microbes, while the wastewater flows through this sludge blanket. As a result of anaerobic digestion of the organic substances, biogas consisting of methane, carbon dioxide, hydrogen, nitrogen, hydrogen sulfide, etc. is generated. Due to its high biomass concentrations, the conversion rate in UASB is several times higher than that in conventional anaerobic processes. Neither mechanical mixing within the reactor nor recirculation of sludge and effluent is needed, resulting in little external energy requirement (Lettinga *et al.* 1983, 1993, 1997).

The UASB process was originally developed for the treatment of high strength industrial wastewaters through the recovery of energy in the form of methane. In the years since UASB was introduced, many researchers have proved that the process works are very effective for the treatment of a wide range of organic wastewater (Hulshoff Pol and Lettinga 1987; Lettinga and Hulshoff Pol 1991; Lettinga *et al.* 1983, 1993, 1997). The construction, startup, operation, and maintenance of the UASB process are comparatively simple and inexpensive. Because of these benefits, UASB has become the most popular of all available anaerobic treatment processes. Numerous studies on small- and large-scale UASB processes have been carried out on the treatment of low strength domestic/municipal wastewater, and UASB has been rapidly recognized as a good option in the treatment of sewage especially in developing countries (Lettinga *et al.* 1997).

Uncontrolled urbanization coupled with the lack of environmental and sanitary infrastructure in developing countries has led to severe pollution of local aquatic bodies. Although there is an urgent need for improved water supply and sanitation service levels, the high cost of the infrastructure and the scarcity of development funds are major problems for these countries. Under these circumstances, introduction of expensive and energy consuming conventional wastewater treatment processes like the activated sludge process (ASP) is unrealistic. The process selection objectives that should be applied in most developing countries can be summarized as follows:

(1) The cost of the process should be as low as possible. It is preferable to treat sewage at the 1/10 cost to achieve, say, 60–70% COD removal than to polish it to the highest standards, such as 99% removal.
(2) The process should be simple and as easy to maintain as possible. This is because manpower in developing countries may not have adequate training to operate sophisticated processes, such as ASP.
(3) The size of the treatment facility should be as compact as possible. Large areas of land are no longer available even in the emerging megalopolises of developing regions. Thus, processes using lagoon and stabilization pond are not suitable.
(4) Sewage treatment plants should be constructed in a decentralized fashion in developing countries. Unlike in developed countries, small-sized plants should be constructed rather than a few large centralized plants.

Among existing anaerobic treatment processes, the UASB process has to a large extent been proven to satisfy the factors listed above, due to the following factors:

(1) UASB is technologically simple to operate, and there is no need of highly skilled personnel for its operation.
(2) The UASB equipment requires only a few mechanical parts, making it less vulnerable to breakdowns.
(3) The process is more tolerant of flow variations or shock loads.
(4) High strength wastewater can be treated efficiently with no energy penalty.
(5) Excess sludge production is substantially lower compared to conventional aerobic treatment processes.

These positive factors have made UASB an attractive option for the treatment of municipal sewage in developing countries. However, it is usually difficult to meet local discharge standards using the UASB treatment alone. Therefore, an appropriate post-treatment process is a further prerequisite to UASB adoption.

In this chapter, the application of the UASB process to the direct treatment of sewage is first summarized. Then the conventional post-treatment processes for UASB-treated sewage are reviewed. Among the post-treatment processes, the down-flow hanging sponge (DHS) process has been recognized as an appropriate post-treatment process for UASB-treated sewage. Lastly, we introduce the basic concept and treatment mechanism of DHS, review its history of development, and summarize the recent successful performance of a demonstration study with a full-scale DHS plant in India.

2. Treatment of Sewage by the UASB Process

2.1. *Performance of full-scale UASB reactors under moderate to tropical conditions*

Many full-scale UASB reactors have already been applied to sewage treatment, especially in developing countries, and have shown successful performance. The countries introducing UASB for sewage treatment are almost all located in tropical or subtropical regions such as Jordan (Halalsheh *et al.* 2005), Columbia (Schellinkhout and Collazos 1992), India (Draaijer *et al.* 1992), Brazil (Vieira and Garcia 1992; Vieira *et al.* 1994; Dixo *et al.* 1995; Chernicharo and Nascimento 2001; Florencio *et al.* 2001), and Mexico (Monroy *et al.* 2000). Some representative data recently obtained from these reactors are presented in Table 5.1.

In full-scale UASB sewage treatment plants, there must be a sand trap, screens, and a grid chamber for removal of coarse material, and drying beds for excess sludge drawn from the reactor. It is important to notice that a UASB reactor can replace the primary settler of a conventional aerobic treatment plant.

Application of the UASB processes to full-scale sewage treatment has been successful under moderate to tropical conditions with temperatures above 18–20 °C (Table 5.1). It was recently reported that granular sludge with good settleability having a sludge volume index (SVI) <50 ml·g-SS^{-1} was developed in a pilot-scale UASB reactor that treated municipal sewage (Ohya *et al.* 2009). However, to our best knowledge, granulation has thus far not been observed in any of the existing full-scale UASB plants treating raw sewage. Although only flocculent sludge developed in all the cases, almost all of the UASB plants achieved acceptable values of BOD, COD, and SS reduction. Thus, sludge granulation is not always necessary for successful anaerobic sewage treatment by UASB.

The UASB process can be started up with or without inoculation. In fact some full-scale UASB reactors have been started up without being inoculated at all (Schellinkhout and Collazos 1992; Draaijer *et al.* 1992; Halalsheh *et al.* 2005). The inoculation can also be done with digested sludge or sludge from any other anaerobic treatment processes. It is obvious that the use of well-developed granular sludge gives some specific benefits, such as high settleability and high specific methanogenic activity, etc. (van Haandel and Lattinga 1994).

Excess sludge production from UASB reactors is far less than that from conventional aerobic treatment process; e.g., the sludge yield for UASB reactors reported so far varies from 0.03 to 0.2 kg-SS per kg-COD removed (Cavalcanti *et al.* 1999; Lettinga *et al.* 1983), which compares with 0.42–0.46 kg-VSS·(kg-COD)$^{-1}$ removed in conventional ASP (Henze *et al.* 2000). This excellent feature of UASB is due to its long sludge retention time (SRT), which contributes to mineralization of the sludge and SS accumulated in the sludge bed. For example, in the case of a pilot-scale UASB reactor that was well controlled and operated at a hydraulic retention time (HRT) of 8 h under temperatures ranging from 10 to 28 °C over 700 days, the sludge yield was only 0.020 kg-VSS·(kg-COD)$^{-1}$ removed while the calculated SRT was

Table 5.1. Recent representative results of full-scale UASB plants for sewage treatment.

Country	Vol. (m^3)	T (°C)	Influent conc. ($mg\ l^{-1}$) COD	BOD	SS	Inoculum	HRT (h)	Removal (%) COD	BOD	SS	PE	References
Jordan	60	18–25	1,419–1,612	—	352–431	None	23–27	51–62	—	55–62	—	Halalseh *et al.* (2005)
Jordan	96	18–26	1,505–1,650	—	376–417	None	5–10	57–60	—	50–87	—	Halalseh *et al.* 2005
Colombia	35	amb.	—	—	—	—	5–19	66–72	79–80	69–70	—	Schellinkhout and Collazos (1992)
Colombia	2 × 3,300	amb.	—	—	—	None	5.2	18–44	17–48	—	160,000	Schellinkhout and Collazos (1992)
India	1,200	20–30	563	214	418	None	6	74	75	75		Draaijer *et al.* (1992)
Brazil	122	18–28	188–459	104–255	67–236	—	4.4–14.5	54–65	60–72	64–76	—	Vieira and Garcia (1992)
Brazil	67.5	16–23	700–1,000	515	379	Digested sludge	7	74	80	87	1,410	Vieira *et al.* (1994)
Brazil	160	26.5	—	771	681	—	7	—	89	90	—	Dixo *et al.* (1995)
Brazil	9	—	592	376	—	—	5.5	45	48	—	—	Chernicharo *et al.* (2001)
Brazil	810	30.2–31	290–563	—	139–204	Digested sludge anaerobic sludge activated sludge	8.8–9.7	59–75	—	—	13,000	Florencio *et al.* (2001)
Mexico	2,200	20	600	—	—	—	20.3	—	75–80	—	—	Monroy *et al.* (2000)
Mexico	5 × 16,740	22.5	2,400	—	—	—	18.6	80	—	—	—	Monroy *et al.* (2000)

approximately 250 days, which was sufficient for almost complete mineralization of the sludge (Ohya *et al.* 2009).

2.2. *Application of UASB to sewage treatment under psychrophilic conditions*

As mentioned above, UASB process has been successfully operated for sewage treatment in tropical temperatures above 18–20 °C. The performance of UASB reactors for sewage treatment under psychrophilic conditions (<20 °C) was well summarized by Lettinga *et al.* (2001). Those results show that anaerobic treatment of sewage by UASB might also be an attractive option even under moderate temperature conditions. However, degradation of the SS entrapped in the sludge bed is markedly reduced under psychrophilic conditions, resulting in a significant decrease in the hydrolysis rate of the SS. Although a SRT of 15 days is sufficient for providing hydrolysis and methanogenesis at a process temperature of 25 °C, a SRT of 75 days is required for a process temperature of 15 °C. Hydrolysis of proteins and lipids strongly declines with decreasing temperature, especially approaching 15 °C (Lettinga *et al.* 2001). Moreover, the internal mixing by the evolved biogas was not optimal in the UASB reactor treating sewage, especially under psychrophilic conditions, producing dead space in the reactor, and as a result, leading to a reduction in the treatment efficiency (de Man *et al.* 1988).

Inoculation of granular sludge may be necessary at lower temperatures. However, Uemura and Harada (2000) reported that when treating raw domestic sewage at low temperatures in the range between 13 and 25 °C by a UASB reactor inoculated with granular sludge, accumulation of SS occurred in the granular sludge bed, which led to the deterioration of the overall methanogenic activity and reactor performance. The hydrolysis rate of the entrapped organics was significantly affected by temperature, that is, 58% of the entrapped particulate organics were liquefied at 25 °C, decreasing to 33% at 13 °C. Moreover, the seeded granules tended to be partly broken due to autolysis, decreasing its methanogenic activity during sustained operation under psychrophilic conditions.

An expanded granular sludge bed (EGSB) reactor was developed by the use of effluent recirculation combined with taller reactors in order to improve the sludge-wastewater contact (van der Last and Lettinga 1992). The EGSB reactor is characterized by the fact that the granular sludge bed is expanded due to high upflow velocity more than 4 $m \cdot h^{-1}$ (versus less than 1–2 $m \cdot h^{-1}$ in UASB), eliminating dead zones.

The EGSB has been shown to be fairly efficient for the removal of soluble organic substances even at low temperatures (Table 5.2). When the biogas production rate is very low and, consequently the mixing intensity is also poor, the expanded sludge bed contributes to better contact between the incoming soluble organic substances and the seeded granular sludge. Compared to UASB reactors, higher organic loading rates can be accommodated in EGSB reactors (Table 5.2). Consequently, the gas production is also higher, further improving the mixing inside the reactor.

However, the EGSB reactor is unfortunately inadequate for sewage treatment as shown by van der Last and Lettinga (1992). This is due to the high upflow velocity applied in the EGSB, by which the influent suspended solids are blown through the sludge bed and leave the reactor (van Haandel and Lettinga 1994). Thus, more sophisticated reactors or systems are needed for anaerobic treatment of sewage at the lower temperatures that prevail in winter or in cold regions.

Table 5.2. Performance of EGSB for low strength wastewater treatment at low temperatures.

Feed	T (°C)	Reactor volume (l)	HRT (h)	Upflow velocity ($m\,h^{-1}$)	Influent COD ($mg\,l^{-1}$)	Loading (kg-COD $m^{-3}d^{-1}$)	Effluent COD ($mg\,l^{-1}$)	COD removal (%)	References
Presettled domestic sewage	6–12	120	2–7	5–7	391 ± 120	<4.7	—	16–34	van der Last and Lettinga (1992)
Synthetic wastewater (acetate, propionate, butyrate)	10–12	4.3	1.6–2.5	6–10	500–800	7–12	—	>90	Rebac *et al.* (1995)
Dilute brewary wastewater	15–30	225.5	1.3–2.4	—	666–772	9–14.4	—	70–91	Kato *et al.* (1997)
Synthetic wastewater (acetate, propionate, butyrate)	8	8.6 (4.3L × 2)	2.9–4.1	10	550–1,100	4.9–6.1	—	87–98	van Lier *et al.* (1997)
Melting wastewater	6–15	140 (70 × 2)	2.4–4.9	4–6	713–1,053 (COD sol.)	4.3–7.1 (COD sol.)	202–464 (COD sol.)	47–78 (COD sol.)	Rebac *et al.* (1998)
Synthetic wastewater (acetate, propionate, butyrate)	3–9	8.6 (4.3 × 2)	1.5–5.3	—	500–900	3.5–15.5	—	80–>90	Lettinga *et al.* (1999)
Synthetic wastewater (sugar, potato starch, peptone, meat extract, urea)	11–25	4.7	3.5–5.7	3.0–8.0	383–849	1.6–4.5	10–96	78–96	Chu *et al.* (2005)
Synthetic wastewater (scurose, acetate, propionate, yeast extract)	20	16.8	1.5–12	5	600–800	1.6–12	91–235	60–90	Kawasaki *et al.* (2005)
Synthetic wastewater (scurose, acetate, propionate, yeast extract)	20	2	1.0–1.5	5	300–400	4.8–6.4	—	57–63	Yoochatchavall *et al.* (2008)
Synthetic wastewater (scurose, acetate, propionate, yeast extract)	5–15	2	3–6	5	600–800	3–6	—	65–80	Syutsubo *et al.* (2008)

—: not reported.

2.3. *State-of-the-art of sewage treatment by UASB in India*

In this section, the history and current circumstances of Indian sewage treatment plants (STPs) using UASB are presented. The Indian government launched the Ganga Action Plan to improve water quality of the Ganges River in 1983, constructing 35 sewage treatment plants, of which three plants employed the UASB process. Under a bilateral cooperation agreement between India and the Netherlands for the design and construction of full-scale UASB reactors for sewage treatment, the first STP employing the UASB process was built in 1989. The COD, BOD, and SS removals were 74, 75, and 75%, respectively, at an HRT of 6 h (Draaijer *et al.* 1992). Following this successful application, the Indian government installed two other UASB reactors for sewage treatment at Kanpur (36,000 $m^3 \cdot d^{-1}$) and at Mirzapur (14,000 $m^3 \cdot d^{-1}$). A polishing pond called the final polishing unit (FPU) was utilized as a post-treatment for those STPs. The popularity of such polishing ponds in India can be attributed to their simple design and operation, and low investment costs, which are prerequisites in developing countries.

Based on the experience with Ganga Action Plan, the Indian government decided to launch a project to implement 16 full-scale UASB reactors (with a total capacity of 598,000 $m^3 \cdot d^{-1}$) in the Yamuna River basin under its Yamuna Action Plan (YAP) in 1990. This indeed indicates that UASB was recognized as a standard method for sewage treatment in India. System configurations of full-scale STPs at Kanpur and Mirzapur, employing a combination of UASB and FPU processes, were used as the STP design models under YAP (Fig. 5.1). Raw sewage flows through

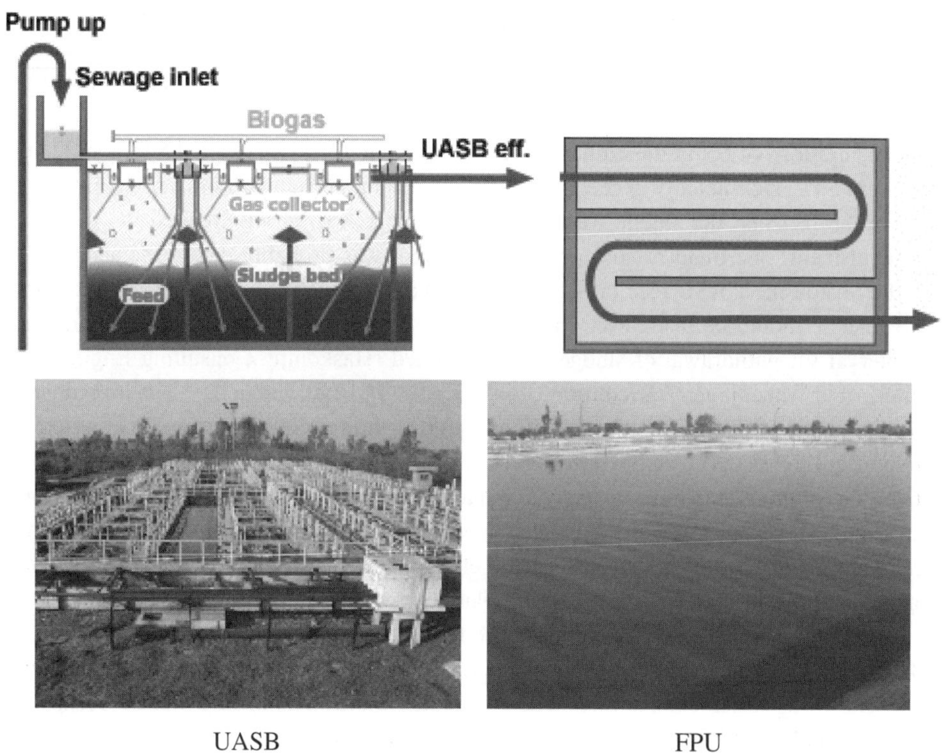

Fig. 5.1. Typical Indian sewage treatment plant consisting of UASB and FPU.

a screen and grit chamber to UASB, and then to the FPU. Feed inlets of the UASB reactor were installed at selected distances to achieve an equal distribution of sewage. Biogas from the UASB was collected in a gas dome. On the other hand, the FPU was designed by installing two baffles to form three lanes.

All STPs using UASB under the YAP program were nearly identical in design and construction, e.g., the HRT was uniformly designed to be 8.4–10.7 h for UASB and 1.0–1.6 days for FPU (Cavalcanti *et al.* 2001). The FPUs were designed with a significantly shorter HRT than the 7–9 days typical of conventional polishing ponds, because it was considered that the removal of residual pollutants from the UASB reactors was much easier (Starkenburg *et al.* 2001; Hammad 1996; Cavalcanti *et al.* 2001).

The combined systems were designed to handle 150–200 $mg \cdot l^{-1}$ of influent BOD and 275–400 $mg \cdot l^{-1}$ of influent SS for meeting the required standard of 50 $mg\text{-}SS \cdot l^{-1}$ in the final effluent. In May 2000, Indian government added fecal coliform as an additional parameter to the discharge standard, along with a maximum permissible limit of 1,000 $MPN \cdot (100\text{-}ml)^{-1}$ (Khan *et al.* 2001).

We originally carried out an investigation on the sewage treatment efficiency of 15 STPs employing the combined system constructed under the YAP program located in ten Indian towns (Sato *et al.* 2006). The results of the investigation are shown in Table 5.3. The numerical values express the treatment capacity of the respective STPs, i.e., the actual flow rate varies from 10,000 $m^3 \cdot d^{-1}$ at the Faridabad STP to 81,000 $m^3 \cdot d^{-1}$ at the Ghaziabad STP.

The concentrations of COD in effluent from the combined UASB-FPU systems ranged from 120 $mg \cdot l^{-1}$ at the Saharanpur STP to 446 $mg \cdot l^{-1}$ at the Panipat STP. Meanwhile, BOD ranged from 42 $mg \cdot l^{-1}$ at the Karnal STP to 154 $mg \cdot l^{-1}$ at the Faridabad STP. None of the STPs was found to meet the standard of 30 $mg \cdot l^{-1}$ for BOD. A possible reason for this failure is the higher concentration of pollutants in the influent due to the incorporation of industrial wastewater into municipal sewage (Sato *et al.* 2006).

Concentrations of SS in the effluent varied from 63 $mg \cdot l^{-1}$ at the Karnal STP to 622 $mg \cdot l^{-1}$ at the Agra STP, significantly exceeding the Indian discharge standard limit of 50 $mg \cdot l^{-1}$. This is due to the fact that the sludge washed out from the UASB short-circuited via the FPU to the final effluent. The main reason for sludge washout was probably due to the accumulation of sludge inside the UASB reactor. According to Indian guidelines, a profile of sludge in the UASB reactor should be made every month. Based on this profile, the amount of sludge and time interval for withdrawal of sludge are determined (Haskoning Consulting Engineers and Architects and Wageningen Agricultural University 1994). However, in the STPs investigated, it was revealed that no monthly sludge profiles were made and excess sludge was not periodically removed from the UASB reactors. Furthermore, it was found that 11 of the 15 STPs had FPUs in operation for more than three years, which may explain the accumulation of too much sludge in the ponds. However, most of these FPUs have yet to be cleaned up even though an Indian guideline for maintenance of STP prescribes a clean-up operation every 1–2 years, or whenever the accumulation of sludge exceeds 40 cm (Public Health Engineering Department 1995a,b; Foundation for Greentech Environmental Systems 2004).

The concentration of fecal coliform bacteria is one of the most important parameters to monitor for sewage treatment facilities. These bacteria should be removed by the FPU, especially when the treated water is to be used for irrigation (Cavalcanti *et al.* 2001). The combined systems of UASB and FPU provided a poor-quality effluent with 1.1×10^5–2.2×10^6 $MPN \cdot (100\text{-}ml)^{-1}$ of fecal coliform in the final effluent. Those STPs were constructed without consideration

Table 5.3. Performance of existing Indian sewage treatment plant employing UASB and FPU.

Sampling point	Actual flowrate (MLD)	T (°C)	BOD ($mg\,l^{-1}$) Sewage	BOD ($mg\,l^{-1}$) UASB	BOD ($mg\,l^{-1}$) FPU	COD ($mg\,l^{-1}$) Sewage	COD ($mg\,l^{-1}$) UASB	COD ($mg\,l^{-1}$) FPU	SS ($mg\,l^{-1}$) Sewage	SS ($mg\,l^{-1}$) UASB	SS ($mg\,l^{-1}$) FPU	Fecal coliform (MPN/100 ml) Sewage	Fecal coliform (MPN/100 ml) UASB	Fecal coliform (MPN/100 ml) FPU
Yamuna Nagar 25MLD	25	18.1	250	108	65	702	324	135	372	423	197	4.0×10^5	2.0×10^5	1.1×10^5
Yamuna Nagar 10MLD	10	18.8	318	184	—	939	585	—	374	270	—	9.2×10^6	3.3×10^6	—
Panipat 35MLD	20	16.3	196	131	116	487	377	322	320	531	311	1.0×10^6	2.5×10^5	3.2×10^5
Panipat 10MLD	6	18	—	—	—	985	680	446	411	513	306	7.9×10^6	3.3×10^6	—
Sonepat 30MLD	20	22.8	160	121	93	481	333	243	189	294	235	—	—	—
Faridabad 20MLD	9	17.6	—	—	—	1,194	261	160	2,500	468	311	—	—	—
Faridabad 45MLD	48	20.8	318	111	98	1,055	354	248	920	222	107	7.8×10^5	7.8×10^5	3.3×10^5
Faridabad 50MLD	45	22.2	365	179	154	1,113	950	363	593	860	443	2.0×10^5	2.0×10^5	2.2×10^5
Noida 27MLD	35	18.8	247	79	85	674	336	194	558	230	342	2.3×10^6	2.3×10^6	4.9×10^5
Gurgaon 30MLD	29	16.6	318	154	104	870	447	320	435	288	214	7.9×10^6	1.1×10^6	4.9×10^5
Ghaziabad 56MLD	56	25.7	—	—	—	418	245	134	185	362	259	—	—	—
Ghaziabad 70MLD	81	23.5	293	151	85	829	359	288	458	412	206	2.3×10^6	7.8×10^5	4.9×10^5
Agra 78MLD	65	17.5	264	126	64	762	469	222	514	677	622	3.3×10^6	2.7×10^6	2.2×10^6
Saharanpur 38ML	30	25.0	—	—	—	363	152	120	169	137	102	—	1.3×10^6	7.9×10^5

—: Not determined.

given to fecal coliform removal because there was no standard regarding fecal coliform in wastewater discharge when the existing STPs were designed. This parameter was only recently added to the discharge standard (10^4 MPN·(100-ml)$^{-1}$). Cavalcanti *et al.* (2001) reported that it takes 7–9 days to fulfill the removal of fecal coliform from wastewater in order to achieve the 10^4 MPN·(100-ml)$^{-1}$ in a plug flow regime such as that found in the FPU. Thus, an effective means of removing fecal coliform by using disinfecting facilities is required to comply with the standard.

Testing results have shown that the effluent from the STPs investigated failed to comply with applicable discharge standards in terms of BOD, SS, and fecal coliform concentrations (Table 5.3). Rather the disappointing performance of the UASB–FPU system could be accounted for not only by the unsatisfactory potential of FPU but also by inappropriate operation and maintenance as mentioned above. For example, investigations should be carried out to determine whether strong industrial wastewater is being discharged. If unacceptable wastewater is being discharged, the authorities should force the industries to stop this discharge or treat the wastewater before discharging it. Moreover, the periodic removal of excess sludge from the UASB and FPU should be done to improve SS removal in the UASB–FPU system.

Although FPUs were designed with a significantly shorter HRT, each FPU requires much more land area than that of a UASB. For example, the FPU for the 40 million-liter per day (MLD) Karnal STP requires 11.5 times as much land area as that of the UASB at the same plant (35,525 m^2 for the FPU and 3,072 m^2 for the UASB). Nonetheless, existing FPUs are unable to satisfy the effluent standard, especially for fecal coliform. Conversely, the physical improvement of FPUs such as adding an aerator and extending the retention time might be required. In conclusion, the post-treatment process should be re-evaluated with other technologies.

2.4. *Post-treatment for UASB-treated sewage*

As shown in the previous section, UASB is an effective process for removal of organic substances from sewage in regions with a warm climate. However, treatment of sewage by UASB alone is insufficient to meet the effluent discharge standards set by many countries, meaning that UASB effluent needs further treatment. Several post-treatment processes have been tested for UASB, such as, UASB itself (two-stage UASB), anaerobic ponds and lagoons, anaerobic hybrid reactors, trickling filter, aerobic ponds, dissolved air floatation, rotating biological contactor, sequencing batch reactors, and ASP. Selection of post-treatment depends upon the targeted effluent quality, the quality of UASB effluent, land availability, and so forth. Recent representative results are summarized in terms of pond processes (Table 5.4) and other sophisticated processes (Table 5.5), respectively. Tables 5.4 and 5.5 give some detail on the specification and operational parameters of the combined treatment systems and their overall performance. The target of post-treatment is mostly to remove residual organics. Removal of pathogens may require a tertiary treatment facility which provides a specific treatment unit like chemical or physical disinfection.

Aerobic treatment processes can more readily remove residual organic substances, nutrients, and pathogens. Among conventional aerobic processes, ASP is the most popular and effective sewage treatment process found in developed countries. However, the use of ASP may not be suitable for developing countries due to its high cost of construction, operation, and maintenance (due mainly to aeration cost), and its excessive land requirements.

Table 5.4. Performance of pond processes as a post-treatment of UASB treating sewage.

	Effluent of UASB				Process			Effluent of post treatment				
T (°C)	COD ($mg{\cdot}l^{-1}$)	BOD ($mg{\cdot}l^{-1}$)	SS ($mg{\cdot}l^{-1}$)	F.coli (No/100 ml)	Type	Vol. (m^3)	HRT (day)	COD ($mg{\cdot}l^{-1}$)	BOD ($mg{\cdot}l^{-1}$)	SS ($mg{\cdot}l^{-1}$)	Fecal coliform (No/100 ml)	References
26.5	—	83	66	9.2×10^6	1FP + 3MP	4	20	—	43	72	6.1×10^2	Dixo *et al.* (1995)
20–35	126	23	35	1.3×10^6	3MP + 7DP	6.3	4.2	49	8	11	5.0×10^3	van der Steen *et al.* (1999)
18.3–21	203	—	—	2.8×10^7 (*E.coli*)	SP + FishP	28 + 23	10.5 + 9.7	121	—	—	2.5×10^5 (*E.coli*)	Ghosh *et al.* (1999)
26	254	169	190	2.3×10^7	PP	32.5	5	188	59	68	7.5×10^4	Cavalcanti *et al.* (2001)
	192	150	99	2.4×10^6			15	108	24	18	5.8×10^2	
25	71	12.5	—	—	7DP	1.19	21	12.8	1	<10	—	Caicedo *et al.* (2002)

DP: Duckweed pond, FP: facultative pond, MP: maturation pond, PP: polishing pond, SP: stabilization pond, —: not reported.

Table 5.5. Performance of post-treatment processes for UASB treating sewage.

	Effluent of UASB			Process			Effluent of post-treatment			
T (°C)	COD (mg·l⁻¹)	BOD (mg·l⁻¹)	SS (mg·l⁻¹)	Type	Vol. (m^3)	HRT (day)	COD (mg·l⁻¹)	BOD (mg·l⁻¹)	SS (mg·l⁻¹)	References
—	114	—	32	ASP	—	2.8	50	—	13	Von Sperling *et al.* (2001)
—	105	29	35	AeF	6.3	0.5	37.8	9.4	9.8	Goncalves *et al.* (1998, 1999)
—	186	—	77	AeF	12×10^3	0.30–0.38	80	—	26	Goncalves *et al.* (2002)
25	215	—	98	SBR	7	2.625	43	12	6	Guimaraes *et al.* (2003)
—	128	52	33	TF	60	1.20	64	17	14	Chernicharo and Nascimento (2003)
19	151	76	—	RBC (two-stage)	55 × 2	0.75 × 2	25	7	—	Castillo *et al.* (1997, 1999)
12	300	120					57	10		
14	276	—	—	RBC (three-stage)	—	4 + 4 + 2	43	—	—	Tawfik *et al.* (2002)
12–15	300	—	—	RBC (two-stage)	60 × 2	5	61	—	—	Tawfik *et al.* (2004)
30.3	180	—	115	EGSB	157.5	4	87	—	11	Kato *et al.* (2003)
—	142	48	36	AnF	102	1.5	118	27	21	Chernicharo and Machado (1998)
	142	48	36	AnF	102	1.5	114	28	32	

ASP: Activated sludge process, AeF: submerged aerated filter (upflow), AnF: anaerobic filter, EGSB: expanded granular sludge bed, RBC: rotating biological contactor, SBR: sequencing batch reactor, TF: trickling filter, —: not reported.

3. Down-Flow Hanging Sponge Process

3.1. *Basic concept of DHS*

To overcome the drawbacks of ASP mentioned above, the DHS reactor has been proposed and investigated as an appropriate post-treatment process for UASB-treated sewage, which was found to be very efficient not only in the removal of residual COD but also in nitrogen and pathogen removal. The DHS process was originally conceived by the research group of Prof. Harada in Tohoku University (and before in Nagaoka University of Technology), Japan. The process is an attached growth-type treatment process employing polyurethane sponges as support media for the growth of bacteria (Fig. 5.2).

A concept of DHS is based on the conventional trickling filter. Therefore, air dissolves into the wastewater as it flows down the DHS reactor, and thus there is no need for external aeration. However, unlike trickling filters using rocks as a packing material, DHS employs polyurethane sponge as a support material. Polyurethane sponge is insoluble, not biodegradable, low cost, and highly mechanically stable. Furthermore, sponge material has a void ratio of more than 95%, which provides an excellent site for growth and attachment of active biomass, resulting in a significant increase in SRT. Sludge is retained both inside and outside the sponge carrier. The enormous amount of biomass and very long SRT expedite the treatment process. Moreover, the DHS method provides adequate time for self-degradation of any attached biomass, reducing the production of excess sludge from the process. In the five years of continuous operation of the DHS G2 pilot plant, sludge production was found to be almost negligible (Machdar *et al.* 1997, 2000).

When designing the reactor, a gap between consecutive sponge materials must be provided: the gap works as a zone for the direct contact of air and wastewater that flows down the reactor. This facilitates the diffusion of air into the wastewater, increasing DO in the wastewater and thus

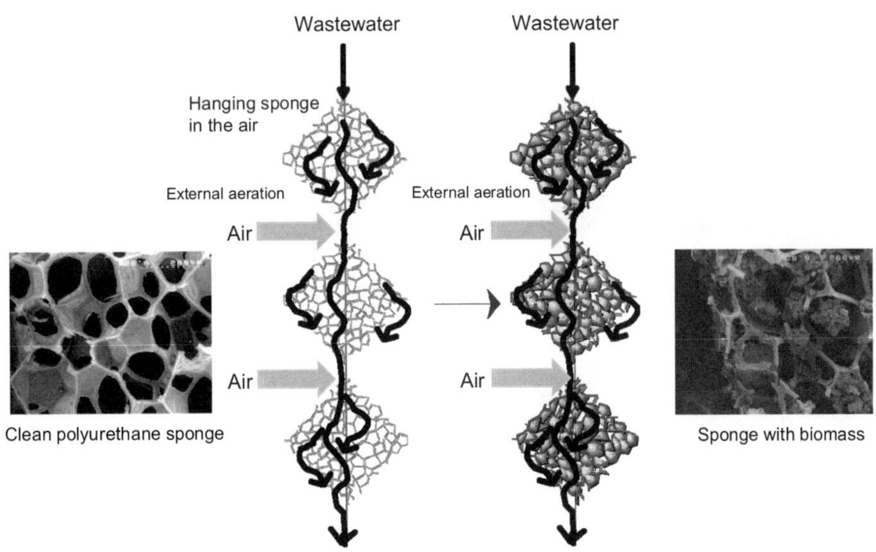

Fig. 5.2. Concept of DHS.

curtailing the need of any external aerating device. It has also been proven that DHS can work at an extremely low HRT (almost 2 h based on sponge volume) with good efficiency toward the removal of organics and ammonium oxidation. Thus, DHS requires less area compared to other conventional processes. Moreover, the startup of DHS is easy: the process attains stable-state operation within 4–6 weeks after the startup. Because of these advantages, DHS overcomes the drawbacks of the conventional post-treatment processes.

3.2. *History of DHS evolution*

The first prototype of DHS was developed in 1995, and is referred to as the first generation DHS (DHS G1 or cube type DHS) (Fig. 5.3). It was assembled with sponge cubes 1.5 cm on each side, and connected to each other diagonally in series with the help of a nylon string (Agrawal *et al.* 1997; Araki *et al.* 1999).

Though the performance of DHS G1 was outstanding, the scale-up of the system with similar design was somewhat unrealistic. The second-generation DHS (DHS G2 or curtain-type DHS) was developed to overcome the few drawbacks of DHS G1 (Fig. 5.3). The DHS G2 was

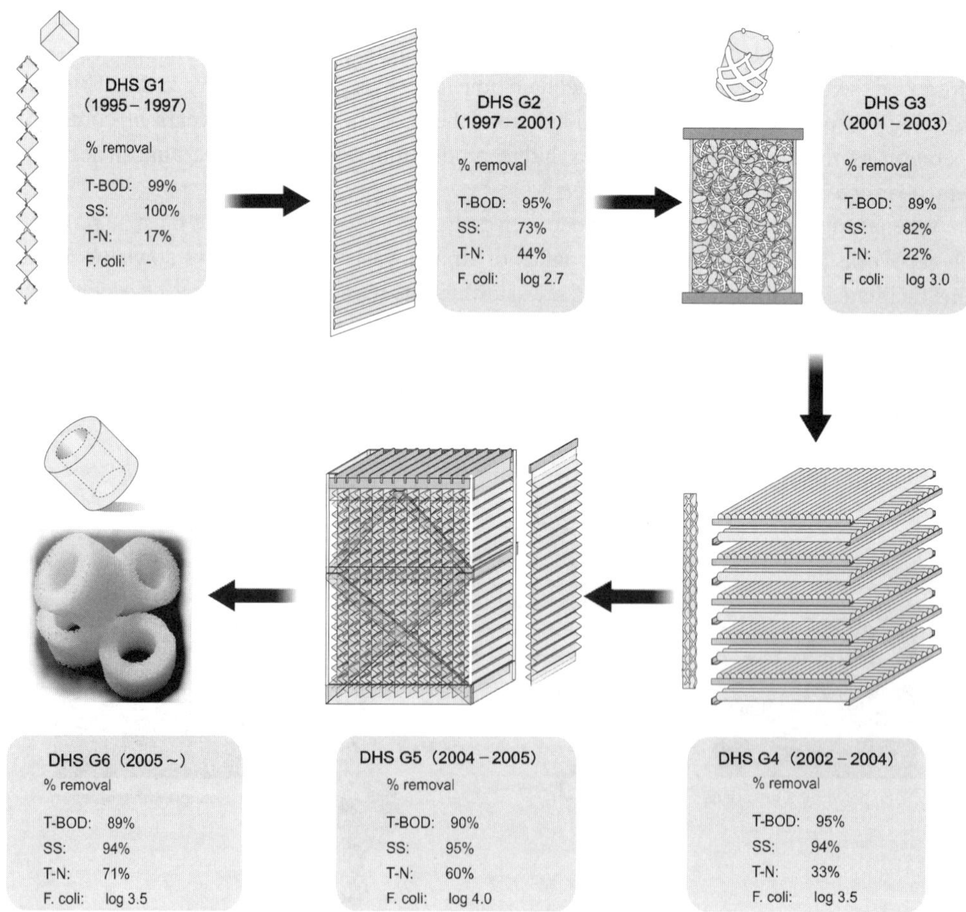

Fig. 5.3. An overview of different configurations of DHS reactor.

constructed with long triangular polyurethane sponge strips (75 cm in length and triangular sides of 3 cm) tiled on both sides of a plastic sheet (2 m height) with 0.9 cm of intervals between each consecutive sponge strip. Two such curtains with 38 strips were constructed for the main body of the pilot-scale DHS G2 (Machdar *et al.* 1997, 2000; Uemura *et al.* 2002).

After experimental testing of DHS G2, scale-up difficulties were again evident; in this case the hanger and the plastic sheet of the curtain appeared to make the construction of a full-scale plant more expensive. The third-generation DHS (DHS G3) was developed to overcome this drawback. The basic design concept was similar to that of the trickling filter. However, instead of the solid packing materials used in a trickling filter, small sponge pieces with an outer supporting material were used. Construction of the process was easier as it involved random packing of the sponge units (Fig. 5.3). Percentage occupancy of the sponge unit in the reactor was 36–38%. (Tawfik *et al.* 2006a,b).

The fourth-generation DHS (DHS G4) was developed the following year. The reactor was constructed with long sponge strips (2.5 cm × 2.5 cm × 50 cm), which were placed inside a net-like cylindrical plastic cover to provide rigidity (Fig. 5.3). This assembly was considered as a single sponge unit for the DHS reactor. Fifteen such sponge units were arranged in a row, which were then stacked one above another but oriented 90° to each other to make a 20 level stack (the gap between consecutive rows was 0.7–1.0 cm). This assembly was considered as a single module with 300 sponge units and 39% occupancy of the sponge by volume. Four such modules were stacked vertically with a predetermined gap between each pair for the construction of the whole reactor (Tandukar *et al.* 2005, 2006a,b).

After the study on DHS G4, the fifth-generation DHS (DHS G5) was developed. It was a direct improvement of the design of the second-generation DHS. Several DHS curtains were assembled to construct a DHS module. A single-curtain unit was constructed by pasting a sponge sheet with an undulating (wavy) surface on both sides of a thin plastic sheet. The sponge module was then stacked one above the other for the construction of a pilot-scale DHS reactor. With this design, the sponge occupancy could be raised to 55–57%, reducing the required volume and making the overall reactor more compact.

A new DHS design was developed simultaneously along with the fifth generation. It was based on the concept of DHS G3 and was named the sixth generation DHS (DHS G6). The only difference was the type of sponge material. Unlike the soft sponge used in the third generation design, which was placed inside a plastic ring to provide rigidity (Fig. 5.3), the sixth generation DHS employed a sponge that was hardened by introducing epoxy resin into the polyurethane network structure. The presence of the epoxy resin reduced the void volume of the sponge to 70% from more than 95% in the ordinary polyurethane sponge. This approach removed the need for the supporting plastic ring because the sponge was fully rigid by itself. In this way the cost for the sponge component could be reduced.

3.3. *Characteristics of DHS sludge yield*

The levels of excess sludge production from DHS evaluated in the previous studies were as follows: 0.09 kg-TSS·(kg-COD)$^{-1}$ removed at HRT = 2.7 h at around 15 °C (DHS G3, Tawfik *et al.* 2006a), 0.12 kg-VSS·(kg-BOD)$^{-1}$ removed at HRT = 2 h at 20–25 °C (DHS G4, Tandukar *et al.* 2006a), 0.02 kg-TSS·(kg-COD)$^{-1}$ removed at HRT = 2.5 h at 9–32 °C (DHS G5, Tandukar *et al.* 2007) and 0.05 kg-TSS·(kg-COD)$^{-1}$ removed at HRT = 1.5 h at 14–26 °C (practical scale G2 plant in India, Okubo *et al.* 2008). Those observations show that the excess sludge production from the DHS process is quite negligible. Moreover, typical production of excess sludge from UASB treating domestic sewage varied from 0.03 to 0.2 kg-SS·(kg-COD)$^{-1}$ removed (Cavalcanti *et al.* 1999; Lettinga *et al.* 1983). For example, Tandukar *et al.* (2007) stated that the gross excess sludge production from a UASB-DHS G5 system was only 0.06 kg-SS·(kg-COD)$^{-1}$ removed, which was 1/15 of that from an ASP operated at the same conditions (0.88 kg-SS·(kg-COD)$^{-1}$ removed).

Lower sludge production in the DHS process might be due to a higher degree of sludge mineralization, which is further enhanced by longer SRT and high temperatures. The SRT values reported for DHS to date have been as follows; 168 days (G3, Tawfik *et al.* 2006a), 108–329 days (G3, Tawfik *et al.* 2008), and 55 days (practical scale G2 plant in India, Okubo *et al.* 2008). We now consider the results of an experiment in a pilot-scale DHS G3 conducted to investigate the factors affecting the repression of excess sludge production (Matsunaga *et al.* 2009). The excess sludge production in the DHS G3 reactor was 0.32 kg-SS·(kg-COD)$^{-1}$ removed at temperatures below 15 °C (winter time) while 0.02 kg-SS·(kg-COD)$^{-1}$ removed at above 25 °C (summer time). The specific oxygen uptake rate (OUR) of the sludge harvested from the 0.4m portion from the top of the DHS, for example, decreased with decreasing temperature from 0.179 at 25 °C to 0.027 g-O_2·g-VSS^{-1}·d^{-1} at 15 °C, implying that the sludge production in DHS is affected by the OUR which changed in conjunction with temperature (Fig. 5.4).

The COD removal rates based upon sludge VSS along with the DHS height are also shown in Fig. 5.4. It should be noticed that, at 25 °C, the potential OUR of the sludge always exceeded the sludge COD removal rate at each height. However, at 15 °C, the OUR was less than the sludge COD removal rate. Thus, the combination of the low sludge COD removal rate with the high OUR is suggested to be one of the possible reasons for lower sludge production from DHS. The sludge OUR originated from respiration of not only bacteria but also protozoa and micrometazoa. Protozoa and micrometazoa are well known to play an important role for sludge reduction in biofilm reactors such as TF and RBC. By predation of those microorganisms, excess sludge production in TF or RBC is 20–30% less than that in ASP (Henze *et al.* 2000). Therefore, a variety of microfauna in the DHS G3 sludge was investigated (at the 0.4 m height, Table 5.6). In Table 5.6, the microfauna of the sludge in an ASP that treated the same sewage is also presented

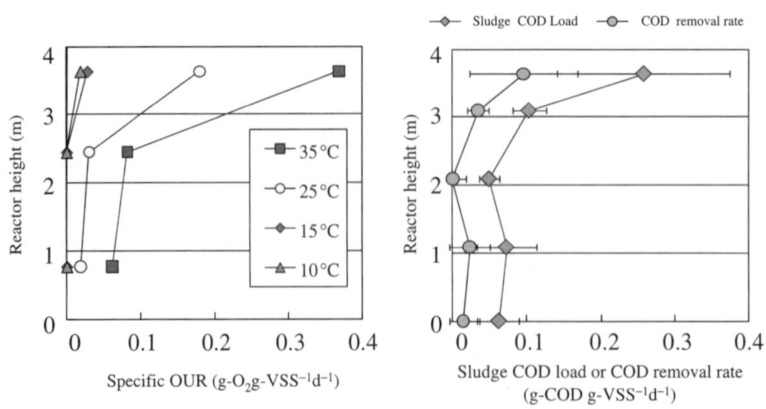

Fig. 5.4. Specific OUR and sludge COD removal rate in DHS sludge.

Table 5.6. Microfauna of DHS sludge.

	DHS sludge		ASP sludge	
	No/mg-SS	No/ml	No/mg-SS	No/ml
Protozoa				
Sarcodina	2,200	83,700	920	1,200
Ciliophora	1,050	40,100	3,400	4,400
Micrometazoa				
Nematoda	50	1,740	ND	ND
Rotifera	40	1,590	190	250
Oligotrichia	20	880	20	25
Daphnia	10	400	ND	ND

ND: not detected.

as a reference. The ASP sludge contained no nematoda and daphnia while those were found in the DHS sludge, indicating that the DHS sludge contains a wider variety of microfauna compared to ASP sludge. Furthermore, although each number concentration per sludge weight was similar in both the ASP and DHS, the number concentration per unit sludge volume in the DHS sludge was one to two orders of magnitude larger than that in ASP sludge. This is due to the fact that the DHS retained a large amount of sludge which was almost tenfold larger than that in ASP. In conclusion, the repression of excess sludge production in DHS is due to the combined effect of large amounts of retained sludge and high sludge respiration which was caused by a wide variety of microfauna including a large number of protozoa and micrometazoa.

Regarding the physical sludge characteristics, excess sludge from a DHS system was found to have better settling properties than the ASP sludge. The SVI for excess sludge from ASP was 140 $ml \cdot g\text{-}SS^{-1}$, whereas the value was as low as 25 $ml \cdot g\text{-}SS^{-1}$ for excess sludge from DHS (Tandukar *et al.* 2007). Better settleability assures shorter settling time and removes the need for any mechanical settling devices.

3.4. *Performance of a full-scale DHS G2 plant in India*

Based on a long-term investigation on the performance of DHS G2 (curtain-type DHS), a demonstration experiment was carried out in India by installing a full-scale plant based on the DHS G2 design (Okubo *et al.* 2008; Onodera *et al.* 2009). The DHS G2 plant was installed at an actual STP site in Karnal, Haryana state, India in 2002. The STP consists of a UASB and FPU. The average flow rate of sewage at the plant was 26,300 m^3 per day during the study period (designed flow rate = 40,000 $m^3 \cdot d^{-1}$). Four UASB reactors were installed at the STP, with a total effective volume of 14,100 m^3 and a designed HRT of 8.5 h; the actual HRT was measured to be 13 h. For post-treatment, the UASB effluent was forwarded to an FPU in which baffles had been installed to create three lanes for establishing plug flow conditions (effective capacity = 40,000 m^3, designed HRT = 1 day, actual HRT = 1.5 days).

The DHS plant employed curtain-type sponge modules which were made by tiling polyurethane sponge bars on curtain-like plastic sheets (Fig. 5.5). Each hanging sponge module consisted of 38–43 right-triangular polyurethane sponges (25 mm on each side) in a tiled configuration attached to both sides of the plastic sheets. The total volume of sponge was 31.1 m^3, which occupied 24.7% of the tower volume. The wastewater was supplied to the top of the

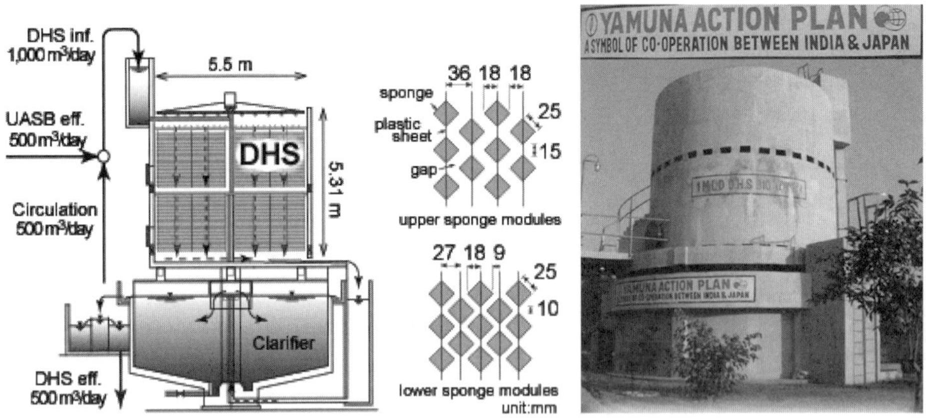

Fig. 5.5. Schematic diagram of DHS reactor installed at Indian sewage treatment site.

Table 5.7. Summary of long-term experiment on UASB-DHS and UASB-FPU at Karnal, India.

Parameter	Raw sewage	UASB eff.	DHS eff.	FPU eff.
pH	7.15 (0.16)	7.01 (0.13)	7.86 (0.13)	7.27 (0.14)
DO, $mg \cdot l^{-1}$	0	0	5.55 (0.99)	0.10 (0.71)
ORP, mV	-173 (54)	-179 (85)	222 (36)	-63 (145)
BOD-total, $mg \cdot l^{-1}$	154 (50)	53 (17)	6 (4)	41 (14)
BOD-soluble, $mg \cdot l^{-1}$	60 (21)	28 (6.6)	3 (2)	23 (10)
COD_{Cr}-total, $mg \cdot l^{-1}$	440 (140)	166 (47)	33 (14)	131 (136)
COD_{Cr}-soluble, $mg \cdot l^{-1}$	139 (46)	79 (21)	22 (11)	64 (19)
Fecal coliform, MPN/100 ml	5.35×10^6	3.09×10^6	6.24×10^4	8.41×10^5
NH_4^+-N, $mg \cdot l^{-1}$	25 (8)	26 (8)	5 (5)	25 (9)
NO_3^--N, $mg \cdot l^{-1}$	—	—	5 (2)	—
SS, $mg \cdot l^{-1}$	226 (96)	66 (20)	8 (4)	57 (18)
Turbidity, FAU	255 (67)	126 (24)	17 (6)	110 (24)
Removal		**by UASB**	**by DHS**	**by UASB-FPU**
BOD-total, %		64 (13)	96 (3)	71 (12)
BOD-soluble, %		47 (25)	95 (4)	58 (19)
COD_{Cr}-total, %		60 (13)	92 (4)	67 (13)
COD_{Cr}-soluble, %		39 (22)	83 (9)	50 (21)
Fecal coliform, $\log_{10}$		0.24	1.93	0.80
NH_4^+-N, %		—	82 (15)	6 (23)
SS, %		61 (13)	95 (3)	65 (15)
Turbidity, %		48 (14)	93 (3)	54 (14)

—: not detected. Figures in parenthesis represent standard deviation.

column using a rotating distributor arm powered by the difference in water head between the inlet chamber and the distribution arm, from where it would trickle down through the sponge modules. The DHS effluent was returned to the inlet chamber at a circulation ratio of 100%.

The experimental results taken during the whole experimental period of over 900 days are summarized in Table 5.7 with the results on the performance of the FPU. These results showed that DHS exhibited an excellent performance over the FPU, and was almost equivalent to that of conventional ASP. The calculated SRT of the DHS G2 was extremely long, i.e., approximately

69 days, which is almost ten times longer than that of conventional ASP. The excess sludge production of the DHS was calculated as 0.06 kg-SS·(kg-COD)$^{-1}$ removed (0.18 kg-SS·(kg-BOD)$^{-1}$ removed). Moreover, the excess sludge production from the UASB-DHS system can be assumed to be 0.09–0.29 kg-SS·(kg-COD)$^{-1}$ removed. Thus, the UASB-DHS combined system requires much less external energy input and sludge production than conventional ASP.

4. Conclusion

The UASB process is demonstrably effective for removal of organic substances from sewage under moderate to warm climate conditions. Although anaerobic treatment of sewage by UASB might also be an attractive option even under moderate temperature conditions, more sophisticated reactors might be necessary at the lower temperatures that prevail during winter or in cold regions.

The use of UASB for sewage treatment has recently been increasing in many developing countries, especially those located in subtropical and tropical regions. For example, in India, UASB has been recognized as a standard method for sewage treatment. However, treatment of sewage by UASB alone is insufficient to achieve the effluent discharge standards of many countries. So UASB effluent needs post-treatment. In combination with other proper processes, a sustainable wastewater treatment system can be established. In India, FPU is utilized for post-treatment process for almost all of the existing UASB sewage treatment plants. The popularity of the FPU process can be attributed to the facility's simple design and operation and low investment costs, which are prerequisites in developing countries. However, according to our recent study, the effluent from the existing Indian sewage treatment plants consisting of UASB and FPU processes failed to comply with applicable discharge standards. This was mainly due to the designed short HRT and the lack of proper operation and maintenance for removing excess sludge and scum. Taking into account the fact that FPU facilities require much larger land areas and are unable to satisfy the effluent standard, the post-treatment process should be re-evaluated with other technologies.

The DHS process has been recently developed as an appropriate post-treatment process for UASB that treats sewage. DHS employs polyurethane sponge as a support material, which has a void ratio of more than 95%. The most important feature of DHS is that no external aeration is required because air dissolves into the wastewater as it flows down the reactor. Moreover, the sponge media provide excellent sites for growth and attachment of active biomass that has quite high OUR and a variety of microfauna, repressing the production of excess sludge from the process. The demonstration study conducted with an existing UASB and a full-scale DHS G2 plant verified that the UASB-DHS system exhibits excellent performance that is almost equivalent to that of conventional ASP. Thus, it can be concluded that DHS is an attractive alternative as a post-treatment process for UASB that treats sewage in developing countries, and can be used in place of other conventional technologies.

References

Agrawal, L. K., Ohashi, Y., Mochida, E., Okui, H., Ueki, Y., Harada, H., and Ohashi, A. (1997) Treatment of raw sewage in a temperate climate using a UASB and the hanging sponge cubes process. *Wat. Sci. Technol.* **36**, 433–440.

Araki, N., Ohashi, A., Machdar, I., and Harada, H. (1999) Behaviors of nitrifiers in a novel biofilm reactor employing hanging sponge-cube as attachment site. *Wat. Sci. Technol.* **39**(7), 23–31.

Cavalcanti, P. F. F., Medeiros, E. J. S., Silva, J. K. M., and van Haandel, A. (1999) Excess sludge discharge frequency for UASB reactors. *Wat. Sci. Technol.* **40**(8), 211–220.

Cavalcanti, P. F. F., Haandel, A. V., and Lettinga, G. (2001) Polishing ponds for post treatment of digested sewage, Part 1: Flow-through ponds. *Wat. Sci. Technol.* **44**, 237–245.

Chernicharo, C. A. L., and Nascimento, M. C. P. (2001) Feasibility of a pilot-scale UASB/trickling filter system for domestic sewage treatment. *Wat. Sci. Technol.* **44**(4), 221–228.

de Man, A. W. A., van der Last, A. R. M., and Lettinga, G. (1988) The use of EGSB and UASB anaerobic systems for low strength soluble and complex wastewaters at temperatures ranging from 8 to 30 °C. In: *Proceedings of the 5th International Symposium on Anaerobic Digestion* (Hall, E. R. and Hobson, P. N. eds.), pp.197–209.

Dixo, N. G. H., Gambrill, M. P., Catunda, P. F. C., and Haandel, A. C. V. (1995) Removal of pathogenic organisms from the effluent of an upflow anaerobic digester using waste stabilization ponds. *Wat. Sci. Technol.* **31**(12), 275–284.

Draaijer, H., Maas, J. A. W., Schaapman, J. E., and Khan, A. (1992) Performance of the 5-MLD UASB reactor for sewage treatment at Kanpur. *India. Wat. Sci. Technol.* **25**(7), 123–133.

Florencio, L., Kato, M. T., and Morais, J. C. D. (2001) Domestic sewage treatment in full-scale UASBB plant at Mangueira, Recife, Pernambuco. *Wat. Sci. Technol.* **44**(4), 71–77.

Foundation for Greentech Environmental Systems (2004) Case study on sewage treatment plants and low-cost sanitation under river action plans. New Delhi, India.

Halalsheh, M., Sawajneh, Z., Zu'bi, M., Zeeman, G., J. Lierb, M. F., and Lettinga, G. (2005) Treatment of strong domestic sewage in a 96 m^3 UASB reactor operated at ambient temperatures: Two-stage versus single-stage reactor. *Biores. Technol.* **96**(5), 577–585.

Hammad, S. M. (1996) Performance of a full scale UASB domestic waste water treatment plant. *J. Instit. Public Health Eng.* India **1**, 11–19.

Haskoning Consulting Engineers, Architects and Wageningen Agricultural University (1994) *Design, Construction, Operation, and Maintenance of UASB-Reactors for Domestic Wastewater*. Nijmegen, Netherlands.

Henze, M., Harremoës, P., Jansen, J. l. C., and Arvin, E. (2000) *Wastewater Treatment: Biological and Chemical Processes*. 3rd edition, Springer, Berlin.

Hulshoff Pol, L. and Lettinga, G. (1987) New technologies for anaerobic wastewater treatment. *Wat. Sci. Technol.* **18**(12), 41–53.

Khan, A., Khan, P., Wiegant, W., Schaapman, J. E., and Sikka, B. (2001) Implementation of UASB technology in river conservation projects in India — policy development for wastewater treatment. In: *Proceedings of Ninth International Conference on Anaerobic Digestion*, pp. 151–156. Belgium.

Lettinga, G., Roersma, R., and Grin, P. (1983) Anaerobic treatment of raw domestic sewage at ambient temperatures using a granular bed UASB reactor. *Biotechnol. Bioeng.* **25**, 1701–1723.

Lettinga, G. and Hulshoff Pol, L. W. (1991) UASB-process design for various types of wastewaters. *Wat. Sci. Technol.* **24**(8), 87–107.

Lettinga, G., van Velsen, A. F. M., Homba, S. W., de Zeeuw, W., and Klapwijk, A. (1980) Use of the upflow sludge blanket (USB) reactor concept for biological wastewater treatment, especially for anaerobic treatment. *Biotechnol. Bioeng.* **22**, 699–734.

Lettinga, G., de Man, A., van der Last, A. R. M., Wiegant, W., van Knippenberg, K., Frijns, J., and van Buuren, J. C. L. (1993) Anaerobic treatment of domestic sewage and wastewater. *Wat. Sci. Technol.* **27**(9), 67–73.

Lettinga, G., Field, J., van Lier, J., Zeeman, G. G., and Hulshoff Pol, L. W. (1997) Advanced anaerobic wastewater treatment in the near future. *Wat. Sci. Technol.* **35**(10), 5–12.

Lettinga, G., Rebac, S., and Zeeman, G. (2001) Challenge of psychrophilic anaerobic wastewater treatment. *TRENDS in Biotechnol.* **19**(9), 363–370.

Machdar, I., Harada, H., Ohashi, A., Sekiguchi, Y., Okui, H., and Ueki, K. (1997) A novel and cost effective sewage treatment system consisting of UASB pretreatment and aerobic post-treatment units for developing countries. *Wat. Sci. Technol.* **36**(12), 189–197.

Machdar, I., Sekiguchi, Y., Sumino, H., Ohashi, A., and Harada, H. (2000) Combination of a UASB reactor and a curtain-type DHS (downflow hanging sponge) reactor as a cost-effective sewage treatment system for developing countries. *Wat. Sci. Technol.* **42**(3/4), 83–88.

Matsunaga, K., Onodera, T., Uemura, S., Yamada, M., Yamauchi, M., Takahashi, M., Kubota, K., Harada, H., and Yamaguchi, T. (2009) Evaluation of the factors affecting reduction of excess sludge generation in a down-flow hanging sponge reactor applied as post-treatment of UASB treating municipal wastewater. *Environ. Eng. Res.* **46**, 623–628 (in Japanese).

Monroy, O., Fama, G., Meraz, M., Montoya, L., and Macarie, H. (2000) Anaerobic digestion for wastewater treatment in Mexico: State of the technology. *Wat. Res.* **34**, 1803–1816.

Okubo, T., Harada, H., Onodera, T., Uemura, S., Yamaguchi, T., and Ohashi, A. (2008) Development of a novel sewage treatment technology with minimum energy requirement-performance of organic matters removal by practical-scale DHS reactor-. *J. Environ. Syst. Eng. G JSCE* **64**(2), 187–195 (in Japanese).

Onodera, T., Harada, H., Okubo, T., Uemura, S., Yamaguchi, T., and Ohashi, A. (2009) Performance of full-scale down-flow hanging sponge (DHS) reactor for nitrification and fecal coliform removals. *J. Japan Sewage Works Assoc.* **46**, 106–116 (in Japanese).

Ohya, A., Onodra, T., Yamaguchi, T., Syutubo, K., Araki, N., Tagawa, T., Yamazaki, S., Yoneyama, Y., Ohashi, A., and Harada, H. (2009) Evaluation of the treatment characteristics and the retained sludge properties in an UASB reactor treating municipal sewage. *Environ. Eng. Res.* **46**, 629–635 (in Japanese).

Public Health Engineering Department (1995a) *Detailed Project Report: 10 MLD UASB Sewage Treatment Plant, Panipat Zone I.* Public Health Engineering Department, Government of Haryana, Chandigarh, India.

Public Health Engineering Department (1995b) *Detailed Project Report: 30 MLD UASB Sewage Treatment Plant, Gurgaon.* Public Health Engineering Department, Government of Haryana, Chandigarh, India.

Sato, N., Okubo, T., Onodera, T., Ohashi, A., and Harada, H. (2006) Prospects for a self-sustainable sewage treatment system: A case study on full-scale UASB system in India's Yamuna River basin. *J. Environ. Manage.* **80**, 198–207.

Schellinkhout, A. and Collazos, C. J. (1992) Full-scale application of the UASB technology for sewage treatment. *Wat. Sci. Technol.* **25**(7), 159–166.

Starkenburg, W. V., Maas, J. A., Kazmi, K. A., and Wiegant, W. M. (2001) The Mirzapur domestic wastewater UASB treatment plant: Six years of operation. In: *Proceedings of the 9th International Conference on Anaerobic Digestion Part 2*, pp. 567–569. Belgium.

Tandukar, M., Uemura, S., Machdar, I., Ohashi, A., and Harada, H. (2005) A low cost municipal sewage treatment system with a combination of UASB and the "fourth generation" Downflow Hanging Sponge (DHS) reactor. *Wat. Sci. Technol.* **52**(1–2), 323–329.

Tandukar, M., Uemura, S., Ohashi, A., and Harada, H. (2006a) Combining UASB and the 'Fourth Generation' DHS (Down-Flow Hanging Sponge) reactor for municipal wastewater treatment. *Wat. Sci. Technol.* **53**(3), 209–218.

Tandukar, M., Machdar, I., Uemura, S., Ohashi, A., and Harada, H. (2006b) Potential of a combination of UASB and DHS (Downflow Hanging Sponge) reactors as a novel sewage treatment system for developing countries: A long-term evaluation. *J. Environ. Eng. ASCE* **132**(2), 166–172.

Tandukar, M., Ohashi, A., and Harada, H. (2007) Performance comparison of a pilot-scale UASB and DHS system and activated sludge process for the treatment of municipal wastewater reactor for municipal wastewater. *Wat. Res.* **41**, 2697–2705.

Tawfik, A., Ohashi, A., and Harada, H. (2006a) Sewage treatment in a combined up-flow anaerobic sludge blanket (UASB)-down-flow hanging sponge (DHS) system. *Biochem. Eng. J.* **29**, 210–219.

Tawfik, A., El-Gohary, F., Ohashi, A., and Harada, H. (2006b) The influence of physical–chemical and biological factors on the removal of faecal coliform through down-flow hanging sponge (DHS) system treating UASB reactor effluent. *Wat. Res.* **40**, 1877–1883.

Tawfik, A., El-Gohary, F., Ohashi, A., and Harada, H. (2008) Optimization of the performance of an integrated anaerobic–aerobic system for domestic wastewater treatment. *Wat. Sci. Technol.* **58**(1), 185–194.

Uemura, S. and Harada, H. (2000) Treatment of sewage by a UASB reactor under moderate to low temperature conditions. *Biores. Technol.* **72**, 275–282.

Uemura, S., Takahashi, K., Takaishi, A., Machdar, I., Ohashi, A., and Harada, H. (2002) Removal of indigenous coliphages and fecal coliforms by a novel sewage treatment system consisting of UASB and DHS units. *Wat. Sci. Technol.* 46(11–12), 303–309.

Van der Last, A. R. and Lettinga, G. (1992) Anaerobic treatment of domestic sewage under moderate climatic (Dutch) condition using upflow reactors at increased superficial velocities. *Wat. Sci. Technol.* **25**(7), 167–178.

Van Haandel, A. C. and Lettinga, G. (1994) *Anaerobic Sewage Treatment: A Practical Guide for Regions with Hot Climate.* John Willey & Sons, Chichester, England.

Vieira, S. M. M. and Garcia, Jr. A. D. (1992) Sewage treatment by UASB reactor. operation results and recommendations for design and utilization. *Wat. Sci. Technol.* **25**(7), 143–157.

Vieira, S. M. M., Carvalho, J. L., Barijan, F. P. O., and Rech, C. M. (1994) Application of the UASB technology for sewage treatment in a small community at Sumare, Sao Paulo State. *Wat. Sci. Technol.* **30**(12), 203–210.

Chapter 6

Anaerobic Granulation and Granular Sludge Reactor Systems

Joo-Hwa Tay[1], Kuan-Yeow Show[2,*], Duu-Jong Lee[3] and Zhen-Peng Zhang[4]

[1]*Department of Environmental Science and Engineering, Fudan University, Shanghai, China*

[2]*Faculty of Engineering and Green Technology, Universiti Tunku Abdul Rahman, Kampar, Perak, Malaysia*

[3]*Department of Chemical Engineering, National Taiwan University, Taipei 106, Taiwan*

[4]*Beijing Enterprises Water Group Limited, Beijing 100195, China*

**kyshow@gmail.com*

Advanced anaerobic processes integrating granular sludge are emerging technologies for efficient wastewater treatment. As a result of increasing acceptance toward the anaerobic granular processes, a growing interest in the technology has resulted in a remarkable increase of full-scale plants installed worldwide over the last three decades. While the technology is well accepted and much research work had been conducted on granule formation, insight unfolding the granulation remains unclear. This chapter presents a review of literature documenting theoretical elucidations on anaerobic granulation and recent advances in granulation research with hydrogen production. Applications of anaerobic granulation in various reactor systems and the future trend are also outlined.

1. Introduction

Microbial aggregation in which cells are organized into dense and fast settling granules has been extensively studied due to its practical significance in anaerobic biological wastewater treatment. Anaerobic granular sludge is composed of dense microbial communities striving upon a syntrophic relationship. Complete degradation of industrial waste involves complex interactions between the syntrophic species populated within the granule matrix. Fitzpatrick *et al.* (1989) conjectured that there are several possible advantages for microorganisms to form granules or aggregates. These include (1) aggregation (or granulation) leading to heterogeneous ordered populations of syntrophic microorganisms in the form of multicellular association under favorable physiological conditions, (2) syntrophic interactions between adjacent organisms that facilitate genetic exchange, (3) granulation may protect cells from predators such as anaerobic ciliates, (4) the diffusion distance for fermentation intermediates is minimized within granules, which is a very efficient means to conserve energy available in the degradation system, and (5) under unfavorable conditions for growth (e.g., extreme pH), a more favorable microenvironment can be maintained within the aggregates so that metabolism can be sustained.

In granular sludge reactors, the large size and high density of individual granules enable them to settle rapidly, which enhance the separation of purified liquid from the biomass (Noyola and Mereno 1994; Batstone and Keller 2001). Considering the present applications of self-aggregation technology, reactors based on the anaerobic granular sludge appear to be the most promising. These reactor systems provide efficient and stable operational performance, attributable to its effective retention of the fast settling granules associated with high biomass concentration and rich microbial diversity. Two popular granular reactor systems, namely the sludge-based upflow anaerobic sludge blanket (UASB) and expanded granular sludge blanket (EGSB) technologies, comprise a predominant share of industrial applications. They have been used in treating high organic strength industrial wastewaters from brewery, food processing, distillations and fermentation, sugar processing, and so forth (Frankin 2001).

Global warming arising from unprecedented emissions of greenhouse gases presents the greatest test humans have ever faced. With global concerns over energy shortages and greenhouse gas formation through combustion of fossil fuels, use of anaerobic processes for biofuel clean energy other than methane clearly fits this need. Application of anaerobic granular sludge processes for biohydrogen production emerged only in recent years. The biohydrogen process is still developing and has shown its potential in enhancing hydrogen production as compared to the suspended sludge systems.

A major problem associated with granular sludge systems is the prolonged startup required for successful development of granules. In cases where inoculation is seeded with flocculent sludge, it usually takes several months or longer before the sludge develops into granules. In order to reduce the lengthy start-up of granular sludge systems, technologies for enhanced and rapid production of anaerobic granules are highly desirable and sought after. This chapter reviews current research work on anaerobic granulation and the factors influencing granulation. Current applications of anaerobic granulation technology and the future trends are also outlined.

2. Granulation

The mechanisms leading to the formation of granules are yet to be completely understood, albeit extensive documentation of research works on anaerobic granulation is going on in the last three decades. It appears that anaerobic granulation is purely a spontaneous biochemical evolvement. Many theories and models have been proposed attempting to shed light on granule formation. This section compiles current hypotheses in granulation documented in the literature.

2.1. *Granulation with methane production*

Most studies on anaerobic granulation were conducted whereby methane is produced as the major gaseous product. Microbial adhesion or self-immobilization is generally regarded as the onset of anaerobic granulation process. It can be defined in terms of the energy involved in the interaction of bacterium-to-bacterium or bacterium-to-solid surface. In a thermodynamic sense, when one bacterium approaches another, the interactions involve repulsive electrostatic force, attractive van de Waals force, and repulsive hydration interaction. Based on the thermodynamics, some physicochemical hypotheses for anaerobic granulation have been developed.

Lettinga *et al.* (1980) postulated that, in the presence of nuclei or micro-carrier, anaerobic bacteria could attach onto the particle surfaces to form embryonic granules. Subsequently, mature granules can be further developed from the growth of these attached bacteria under appropriate operating conditions. The inert nuclei model was supported by experimental evidence such that addition of zeolite or hydro-anthracite particles with a diameter of 100 μm into inoculated sludge seemed to be effective in promoting the formation of anaerobic granules (Hulshoff Pol 1989).

Hulshoff Pol *et al.* (1988) suggested that microbial aggregation is a protective microbial response against high selection pressures. In UASB reactors, upflow liquid acts as a selection pressure of biomass by washing out light and dispersed particles while retaining heavier biomass. Rapid development of anaerobic granules could be accomplished through a purely physical aggregation from the hydraulic stress applied on the anaerobic flocculant sludge (Noyola and Mereno 1994). The results showed that flocculant anaerobic sludge could be converted into a relatively active granular sludge by enhancing agglomeration through only short hydraulic stress of less than 8 hours. It had been reported that under very weak hydraulic selection pressure operating conditions, no anaerobic granulation was observed (Alphenaar *et al.* 1993; O'Flaherty *et al.* 1997). It appears that relatively high selection pressure in terms of upflow liquid velocity is favorable to granule formation.

Pereboom and Vereijken (1994) proposed that granules originated from fines formed by attrition and from colonization of influent suspended solids. The shear forces due to liquid and gas turbulence are responsible for attrition of small particles from biomass. Increase in granule size is only due to microbial growth. It was reported that wastewaters of high concentrations of suspended solids would result in narrow granule size distributions, while influent of little or no suspended solids would lead to good distribution of size.

As bacteria have negatively charged surfaces under normal pH conditions, a basic idea to expedite anaerobic granulation is to reduce the electrostatic repulsion between negatively charged bacteria by introducing multi-valence positive ion into the seed sludge. The positive ion hence initiates cell-to-cell interaction which is a crucial initiation toward granulation. It had been reported that addition of Ca, Mg, and Al ions would increase the rate of anaerobic granulation in UASB reactors (Schmidt and Ahring 1993; Teo *et al.* 2000; Yu *et al.* 2001; Liu *et al.* 2002). On the other hand, high calcium concentrations of above 500 $mg \cdot l^{-1}$ (Guiot *et al.* 1988; Thiele *et al.* 1990; Yu *et al.* 2001) were found detrimental to granulation.

Moreover, the multi-valence positive ion may promote sludge granulation by bonding with extracellular polymers (ECPs) (Forster and Lewin 1972; Rudd *et al.* 1984). Calcium ion may bridge ECPs to ECPs and link cells to ECPs to form an initial three dimensional structure of microbial community, in which bacteria could grow further. ECPs themselves can mediate both cohesion and adhesion of cells, and play a vital role in maintaining structural integrity of microbial matrix. It had been reported that ECP could change the surface negative charge of the bacteria, and thereby bridge two neighboring cells physically to each other, as well as to bridge cells with other inert particulate matters (Schmidt and Ahring 1996, 1994; Shen *et al.* 1993).

Show *et al.* (2004a,b) and Wang *et al.* (2004) investigated the influence of a coagulant polymer on start-up, sludge granulation, and the associated reactor performance in laboratory-scale UASB reactors. The results indicated that adding the polymer at a concentration of 20 $mg \cdot l^{-1}$ markedly reduced the start-up time by as much as 46%. The organic loading capacities of the reactors were also increased by more than 33%. The researchers hypothesized that positively charged polymer form bridges among the negatively charged bacterial cells through electrostatic charge attraction. The bridging effect would enable greater interaction between biosolids resulting in preferential development and enhancement of granulation.

It has been hypothesized that under normal physiological conditions, strong hydration repulsion at the outer surfaces of bacteria is the main force keeping the cells apart. Thus local dehydration of the short-distance-apart surfaces would be a prerequisite for cell-to-cell aggregation. The local dehydration and hydrophobic interaction model as proposed by Wilschut and Hoekstra (1984) shows that when bacterial surfaces are strongly hydrophobic, irreversible adhesion will occur. Theoretically, increasing the hydrophobicity of cell surfaces would cause a corresponding decrease in the excess Gibbs energy of the surface, which in turn promotes cell-to-cell interaction and further serves as driving force for cell self-separation from liquid phase (Rouxhet and Mozes 1990; van Loosdrecht *et al.* 1987).

The discussion so far seems to suggest that each physicochemical model accounts for contribution of only one or two factors to the initial granulation process in UASB reactor. As these factors exert their influences under specific environmental conditions and in specific stages during the entire granulation process, the physicochemical models provide somewhat limited elucidation on anaerobic granulation. It seems impossible to develop a pure thermodynamic model with a satisfactory confidence level. The fact is that bacteria cannot be simply treated as physically-defined dead colloidal particles, and bacteria indeed have no well-defined surface boundary, simple geometry, or uniform molecular surface composition. It follows that the physicochemical forces alone are not able to completely explain the entire microbial granulation process. Hence the physicochemical phenomena involved in microbial granulation ought to be related to the biological triggers controlling the granulation.

Anaerobic granulation is a complex process in which biological factors are involved other than physicochemical forces. In the past two decades, significant research progress had been made in understanding microbiological characteristics of anaerobic granules and interactions among different microbial species in the granules. In view of the development, a series of structural models for anaerobic granulation has been developed to interpret the observed phenomena.

Like the polymer-bonding model as discussed earlier, a Capetown model suggests that ECPs are produced by *Methanobacterium* strain AZ, a hydrogen-utilizing methanogen (Palns *et al.* 1987; Sam-Soon *et al.* 1988). Under the conditions of high hydrogen partial pressure and limited cysteine, the amino acids (except cysteine) would be over-secreted. Excessive amino acids could induce ECP formation, and consequently *Methanobacterium* strain AZ and other genera are enmeshed in the ECP matrix, which in turn leads to the initiation of anaerobic granulation.

Wiegant (1988) proposed a spaghetti model postulating that the development of UASB granules is initiated by attachment of filamentous *Methanosaeta* (formerly known as *Methanothrix*) on precursors, followed by a formation of a three-dimensional network through a branched-growth process. Other bacteria, such as *Methanosarcina*, could be easily entrapped in this network (Sanchez *et al.* 1994; Wu *et al.* 1996). The structured aggregates further develop through cellular multiplication of the entrapped bacteria, and become denser and spherical by the action of hydrodynamic shear force attributed to upflow liquid and biogas. It must be emphasized that in the spaghetti model, formation of the structured aggregate is a crucial stage of the overall granulation process.

The microbial species in an anaerobic community form a syntrophic relationship in which each bacteria group constitutes a significant link in a complex chain of bioconversion. The syntrophic microcolony model suggests that the syntrophic relationship eventually leads to the formation of stable microcolonies or consortia, viz. initial granules. Anaerobic granule indeed can be regarded as the congregation of cells to form fairly stable, contiguous, multicellular associations under physiological conditions in a defined biological system. In order to maintain high metabolic efficiency, the granule-associated cells would assemble in an organized structure, and signaling mechanisms in organizing the syntrophic species can be predicted (Shapiro 1998). Therefore, it appears from the syntrophic microcolony model that the driving force for sludge granulation should be a result of the needs for bacterial survival or balance and for optimal combination of different biochemical functions of multiple species under the prevailing culture conditions.

Based on microscopic observations, a multi-layer model for anaerobic granulation was initially proposed by MacLeod *et al.* (1990) and Guiot *et al.* (1992). According to this model, the microbiological composition of granules is different in each layer. The inner layer mainly consists of methanogens that may act as nucleation centers necessary for the initiation of granule development. H_2-producing and H_2-utilizing bacteria are dominant species in the middle layer, and a mixed species including rods, cocci, and filamentous bacteria takes predominant position in the outermost layer. To convert a target organic to methane, the spatial organizations of methanogens and other species in the granules are essential. The layered structure of granules is supported by the works of Ahring *et al.* (1993) and Lens *et al.* (1995) with immunological and histologic methods, with a dynamic model (Arcand *et al.* 1994), with microelectrodes (Santegoeds *et al.* 1999), and with fluorescence in situ hybridization using 16S rRNA-targeted oligonucleotides (Sekiguchi *et al.* 1998, 1999; Tagawa *et al.* 2000). A distinct layered structure was also found in the methanogenic–sulfidogenic aggregates, with sulfate-reducing bacteria in the outer 50–100 μm and methanogens in the inner layers (Santegoeds *et al.* 1999).

From microscopic examination and activity measurements, Dubourgier *et al.* (1987) suggested that granulation starts with the covering of filamentous *Methanosaeta* by colonies of cocci or rods (acidogenic bacteria), forming microflocs of 10–50 μm. Subsequently, *Methanosaeta* filaments, due to its filamentous morphology and surface properties, might establish bridges between several microflocs forming larger granules of size greater than 200 μm. Further development of acidogenic and syntrophic bacteria favors the granules growth. The authors supported the idea that *Methanosaeta* plays a vital role in enhancing granule strength by forming a network that stabilizes the overall structure. Subsequent work by Morgan *et al.* (1991a,b) also supported the suggestions on the importance of *Methanosaeta* and bacterial polymers in the granulation.

2.2. *Granulation with hydrogen production*

Potential of hydrogen as an ideal alternative energy has attracted worldwide attention in recent years. This is largely due to the concerns on energy sustainability and environmental impact such as global warming arising from the use of depleting fossil fuels. Because of sustainable energy production with practically pollution-free features, biological processes such as anaerobic fermentation, biophotosynthesis, and photodecomposition to harness hydrogen from renewable sources like water or organic compounds have received considerable attention (Das and Veziroglu 2001). Biohydrogen production using fermentative, photosynthetic bacteria, or algae is an environmentally friendly and energy saving process, which appears to be a promising alternative to the conventional fossil fuel-based hydrogen production processes. Among these processes, anaerobic hydrogen fermentation appears to be favorable, since hydrogen can be generated at higher rates with a concomitant reduction in environmental organic pollutants. Feasibility of anaerobic hydrogen production from organic wastes has been demonstrated by some research groups in different countries all over the world (Chang and Lin 2004; Fang and Liu 2002; Wu *et al.* 2006; Zhang *et al.* 2006). Nevertheless, application of granular sludge processes for anaerobic hydrogen production has begun only in the recent years. The formation mechanism of hydrogen-producing granules has yet to be established. While the process is still being developed, it has shown a greater potential in hydrogen production as compared to the suspended sludge systems (Kotsopoulos *et al.* 2006; Liu and Fang 2003).

The physical, physicochemical, and hydraulic characteristics of hydrogen-producing granules, such as bioparticle morphology, ECP contents, and settling velocity, have been investigated by several research groups (Fang *et al.* 2002a; Mu and Yu 2006; Zhang *et al.* 2004). Some researchers pointed out that polysaccharides might be the key ECP constituents for hydrogen-producing granules and hence might play a more important role in the formation of hydrogen-producing granules (Fang *et al.* 2002a; Mu and Yu 2006). However, in recent studies on rapid formation of hydrogen-producing granules, Zhang *et al.* (2008a) found that proteins might be more crucial than ECP polysaccharides to the initiation of granulation. Knowledge with respect to microbial species and spatial distribution of microbial components of the hydrogen-producing granules or biofilms remains limited, even though some advanced molecular techniques have been extensively applied to evaluate microbial communities and their micro-environment of various granules and biofilms (Chiu *et al.* 2006; Fang *et al.* 2002a,b; Tay *et al.* 2001a).

A novel approach to rapidly initiate granulation of hydrogen-producing sludge was developed in an anaerobic continuously stirred tank reactor (Zhang *et al.* 2007a). Applying acid incubation on the culture demonstrated an ability to initiate microbial granulation of sludge within 5 days of reactor operation (Fig. 6.1). The granulation was much shorter than that normally required of 2–4.5 months reported in other studies (Fang *et al.* 2002a; Yu and Mu 2006). Lowering the system pH through the acid incubation could result in improvement in surface physicochemical properties of the culture favoring microbial granulation. The improved zeta potential, hydrophobicity,

Fig. 6.1. Photographs of hydrogen-producing CSTR and granules.

and extracellular protein facilitated the interactions between microbial cells which accelerated granule development. Formation of granular sludge significantly increased the reactor biomass which in turn improved the hydrogen production rate.

It has been established that microbial aggregation begins with the decrease in surface charges of cells. This may lead to a notion that microbial adsorption behavior can be explained by the DLVO theory. This theory, as presented by Derjaguin, Landau, Verwey, and Overbeek, with an original aim of explaining colloid stability is well accepted in describing the adhesion behavior of microbial cells (Hulshoff Pol *et al.* 2004; Schmidt and Ahring 1996; Tay *et al.* 2006). The DLVO theory considers the overall Gibbs free energy as a function of the sum of attractive and repulsive forces over a distance between the cells. Since bacteria have negatively charged surface under normal pH conditions, the electrostatic repulsion dominates or is balanced by a weak attraction so that microbial cells are located at certain distance from each other (secondary minimum). The abrupt increase in zeta potential at a pH of 2.0 could whittle down the electrostatic repulsion between microbial cells. As a result, the van der Waals forces are predominant. Consequently, irreversible adhesion of microbial cells might occur with the reduction in electrostatic repulsion when bacterium reaches the primary minimum.

It has been reported that microbial aggregation could be initiated and accelerated via different approaches to altering bacterial surface charges. Perez *et al.* (1998) reported that the auto-agglutination of the pure cells increased as the zeta potential changed from -39 mV at pH 7 to -7 mV at pH 2. Reducing electrical repulsion between cell and solid surface could improve the microbial adhesion performance (Bonet *et al.* 1993; Tsuneda *et al.* 2003). Moreover, microbial granulation is accelerated by introducing multi-valence positive ions and cationic polymers to reduce electrostatic repulsion between negatively charged bacteria (Mahoney *et al.* 1987; Show *et al.* 2004a,b; Wang *et al.* 2004).

When a cell approaches another one at a certain distance (primary minimum), short-range interaction forces like hydrogen bonds or hydrophobic interaction would become crucial. Previous studies indicated that there is a strong correlation between microbial auto-aggregation and cell surface hydrophobicity (Baldi *et al.* 1999; Del Re *et al.* 2000; Kos *et al.* 2003; Liu *et al.* 2003, 2004; Tay *et al.* 2000). Improved hydrophobicity of bacteria is conducive to creating a favorable environment for the contact of bacterium-to-bacterium or impelling bacteria out of aqueous phase environments to aggregate. Consequently, improved hydrophobicity of cell surface observed in the present study was likely to contribute to microbial aggregation, along with the reduced surface charges, while the ECP could serve as the construction materials for microbial aggregation by polymeric interactions with microbial cells (Erdincler *et al.* 2001; Schmidt and Ahring 1996; Tsuneda *et al.* 2003).

Although the role of ECP bonding in initiating microbial aggregation remains unclear, associated increases in ECP production with the development of granules indicate that the function of ECP is more likely to maintain structural integrity of microbial granules. The electrostatic repulsion between the cells increased to original level once the acid incubation ceased. At the same time, hydrophobic interactions between the microbial cells became weak. Both of these factors would drive aggregated cells apart again. Nevertheless, disintegration of granules was not observed throughout the study, suggesting that the irreversible adhesion previously established could be maintained by ECP bonding. In fact, it was found that hydrogen-producing bacteria in the granules were surrounded with a large amount of ECP. Notably, besides the microbial surface physicochemical properties, the hydrodynamic, biological, and other forces may also

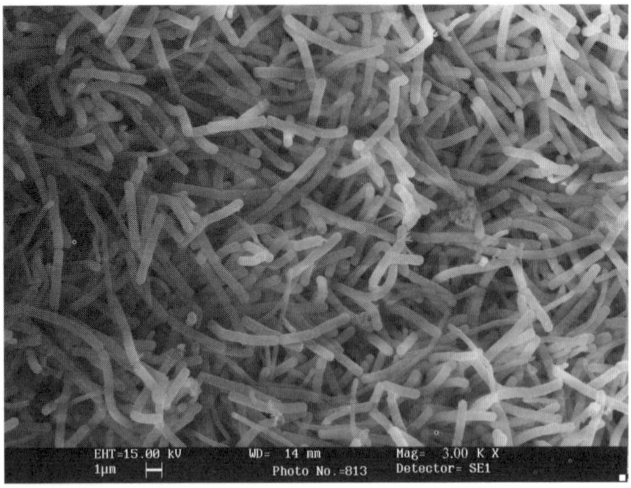

Fig. 6.2. SEM image of hydrogen-producing bacteria *Clostridium pasteurianum*.

influence or contribute to microbial adhesion and aggregation. For example, appropriate operating hydraulic retention time provided the hydraulic selection pressures favorable for aerobic granulation (Pan *et al.* 2004).

Investigation on physiological changes of microbial culture would contribute to a better understanding of the mechanisms involved in the granulation. Microbiological factors affecting cell immobilization were investigated through examining species composition, bacteria, and their extracellular polymers distribution in the hydrogen-producing granules (Zhang *et al.* 2008b). Phylogenetic analysis showed that the microbial community of the granules was uniquely constituted by the rod-shaped hydrogen-producing bacteria, which were identified to be *Clostridium pasteurianum* by analyzing their nucleotide sequences (Fig. 6.2). Besides rod-like bacteria, coccus and filamentous bacteria (Mu and Yu 2006) or fusiform bacteria (Fang *et al.* 2002a; Wu *et al.* 2006) were also found in the sucrose-fed hydrogen-producing granules. This is most likely due to different feedstocks and seed sludge used. It is interesting to note that *C. pasteurianum*-like bacterium (similarity, 98%) was also found to be unique microbial species in the self-flocculated hydrogen-producing granules cultivated by Wu *et al.* (2006) in a CSTR operated at an HRT of 0.5 h and a sucrose concentration of 30 g-COD$\cdot$l^{-1}, which had a fusiform shape (3–5 μm in length and 1–2 μm in width), rather than a rod shape as observed in the present study. These findings seem to imply that *C. pasteurianum* or the strains close to *C. pasteurianum* might be the crucial microbial species participating in the formation of hydrogen-producing granules.

A common notion postulates that filamentous bacteria are favorable anaerobes for microbial immobilization (Macleod *et al.* 1990; Morgan *et al.* 1991b). For example, Wiegant (1988) observed that steady granules could be developed from the precursors containing mainly filamentous *Methanosaeta* in UASB reactors. Initially, *Methanosaeta* form very small aggregates via bridging between microbial cells or microflocs, or attaching to finely dispersed matter. Cellular multiplication and entrapment of non-attached bacteria led to the growth of the precursors that eventually developed into granules. An irony finding derived from hydrogen-producing granules indicated that there were no filamentous bacteria found in the granules. Contrary to the complex microbial groups present in methanogenic granules, the hydrogen-producing granules consisted merely single species of *Clostridia* arising from acid incubation observed by Zhang *et al.* (2008b). *C. pasteurianum* is a non-homoacetogenic hydrogen-producing species present either singly or in pairs which excrete intracellular polysaccharide reserve (Fang *et al.* 2002a). Because of its ability to secrete much more ECP, *C. pasteurianum*-like bacterium has a high affinity to attach onto all kinds of surface, microbial cells, or support materials, thus forming granules or biofilm. ECP is thought to serve as a bioglue to facilitate *Clostridium* interaction and further strengthening of the aggregate *pasteurianum* appears to be crucial for microbial species to microbial granulation because of its structure through formation of a polymeric network. It can be concluded that, *Clostridium* species has strong ability to secrete ECP.

The role of ECP in the microbial granulation had been noted previously. *Methanosarcina*, typical methanogens, are responsible for the nucleus formation as the anaerobic cultures are fed with acetate, especially at the high organic loading rates (Bochem *et al.* 1982; Wu *et al.* 1996). The formation of *Methanosarcina* granules is believed to be due to their capacity to secrete much more ECPs and form clumps of bacteria (Chen and Lun 1993). The granules are developed as the clumps reach macroscopic dimensions and other species, such as *Methanosaeta*, tend to inhabit in the *Methanosarcina* clumps.

The ECP composition affects the surface properties of the bacterial flocs, such as surface charges and hydrophobicity. It should be stressed that the key ECP contents of *Clostridia*-predominated granules are distinctly different from those of methanogenic granules, and the roles of ECP contents might also be different in the process of granulation. The ECP content of methanogenic granules has been reported to range between 0.6 and 20% of the volatile suspended solids (VSS) content, depending on the granular sludge examined, the extraction method, and the analytical methods for ECP (Schmidt and Ahring 1996). The ECP in methanogenic granules consists mainly of proteins and polysaccharides, typically in a ratio of 2:6. Although the same ECP composition was found for hydrogen-producing granules, much lower ratios of proteins to polysaccharides ranging from 0.2 to 0.6 were observed by Zhang *et al.* (2007a,b) and other researchers (Fang *et al.* 2002a; Mu and Yu 2006). Because of such a significant difference, some researchers pointed out that unlike the association of proteins with methanogenic granulation, polysaccharides might be the key ECP constituent for hydrogen-producing granules and hence might play a more important role in the formation of hydrogen-producing granules (Fang *et al.* 2002a; Mu and Yu 2006).

Using a multiple fluorochrome staining scheme and confocal laser scanning microscopy (CLSM) technique to study the internal structure of the granules, Zhang *et al.* (2008b) observed that proteins were distributed uniformly in healthy hydrogen-producing granules, while other ECP contents (polysaccharides and lipids) and cells were mostly accumulated on the outer layer of the granules (Fig. 6.3). Deeper inside the granule was a core made mainly of proteins. On account of the relative abundance of proteins in the granule center, Zhang *et al.* (2008b) hypothesized a possible mechanism involved in the evolution of hydrogen-producing granules. Microbial aggregation begins as soon as a protein-rich matrix or nucleus has been established. The protein matrix might contain microbial cells and proteins could be enriched by secretion of the attached cells or by entrapping extraneous proteins in the form of macromolecules, colloids, and slimes. Based on this hypothesis, the granulation phenomena attributable to acid incubation discussed earlier can be explained. The acid incubation stimulated the production of ECP proteins, and the concentration of ECP proteins reached a higher level within a short period, which promoted the

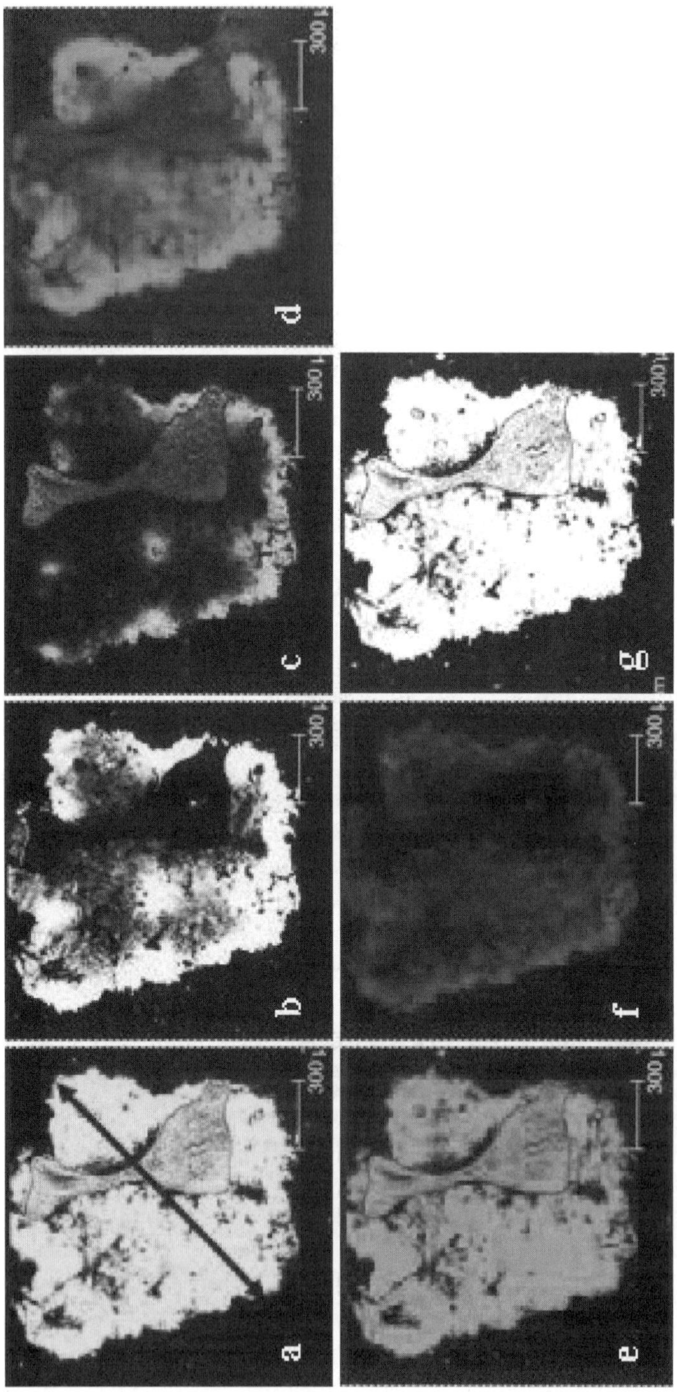

Fig. 6.3. Confocal images of hydrogen-producing granules. (a) Green: proteins; (b) yellow: lipids; (c) light blue: α-polysaccharides; (d) red: living cells; (e) pink: dead cells; (f) blue: β-polysaccharides; and (g) merging of all images.

process of granulation. On the other hand, a lower accumulation rate of ECP proteins without acid incubation might limit the formation of hydrogen-producing granules.

The strong agglutination property of *C. pasteurianum* might be attributed to its high productivity of extracellular polymers which would serve as a bioglue to facilitate cell-to-cell or cell-to-substratum interaction. As abundant proteins were noted in the granule cores indicated by the confocal laser scanning microscope images, it can be derived that rapid formation of the hydrogen-producing granules could be due to the establishment of precursor protein-rich microbial nuclei.

3. Granular Sludge Reactors

Much emphasis has been placed in the last three decades on applications of anaerobic granulation technology for sustainable development. Substantial effort is now being installed in exploring broader applications of anaerobic granulation technology for removal of high strength and nasty organic wastewaters by converting them into a renewable energy source. Apparent advantages derived are from low operating costs, compact reactor construction, production of energy in the form of biogas, low surplus sludge production, which result in overall favorable economics. For anaerobic treatment to compete with alternative technologies such as aerobic or physicochemical treatment, it has to be cost-effective in terms of investment and operating costs, reliable, and durable. System designs have focused on increased process control to secure optimal operating conditions and system "compactness" in order to reduce investment costs.

The key to successful application of anaerobic treatment is to uncouple the liquid and the solids (biomass) retention times in the reactor system. In order to achieve high system loading rates, short hydraulic retention times should be applied, while at the same time maintaining positive net solids (biomass) retention. Thus, various reactor designs were developed over the past two decades that are based on different ways of retaining biomass within the reactor system. Anaerobic reactor systems integrating granular sludge for waste degradation are discussed in this section.

3.1. *Upflow anaerobic sludge blanket (UASB) reactor*

An embryonic form of the UASB system was first described by Coulter *et al.* (1957). Lettinga *et al.* (1979) reported the first feasibility study on UASB in an attempt to combine the contact reactor and the clarifier of the anaerobic contact process in one chamber. The UASB is typically constructed with influent pumped to the bottom of the unit and flow upward through a blanket of sludge where pollutants are degraded by biomass. The UASB has a unique gas–solid–liquid separator for separation of biogas from sludge granules and treated effluent. The remarkable design of the three-phase separator serves two key functions using just one single reactor, namely maximizing biomass retention without the need of an external clarifier. The UASB technology is well accepted because of its high organic removal efficiency, simplicity, low capital and operating costs, and low footprint requirement. Moreover, they are characterized by low sludge production and low energy input.

Early development of UASB in the 1980s had been used in treating food industry wastewaters, such as beet sugar, corn, and potato starch processing. The application has since been broadened to treat a variety of wastewaters including those from piggery, alcohols, fatty acids, slaughterhouse, milk fat, distillery and fermentation, pulp and paper wastes, brewery, food processing, chemicals, pharmaceuticals, and so on. Recent studies showed that UASB can be applied in treating wastewaters containing concentrated proteins (Fang 1994) and aromatic compounds such as phenol (Wen *et al.* 1995). Effluent recirculation is widely used to avoid toxic impact to the microorganisms.

Supplement glucose improves the anaerobic degradation of phenol (Tay *et al.* 2001b). The maximum concentration of phenol could go as high as 6,000 $mg \cdot l^{-1}$ and this is far too toxic to living aquatic organisms. Phenol can be degraded to methane and carbon dioxide through phenol metabolizers and hydrogen utilizing and acetotrophic methanogens. The phenol removal efficiency was also the best at 98%, compared with 88% without glucose supplement. Moreover, it also exhibited greater resistance to those adverse conditions and the system recovered faster than the other system without the glucose supplement.

Removal of chlorinated phenols (CPs) is possible in UASB reactors (Droste *et al.* 1998). Halogenated organic pollutants are labeled as toxic and recalcitrant in the environment. Effluents containing CPs and related compounds are especially problematic to treat due to their persistence and their high solubility in fat. CP compounds were able to be metabolized to mineral end products to a large extent at loading rates where the reactor's performance was not hindered. There was no accumulation of phenol in any of the reactors in the experimental conditions.

Treatment of polyethylene terephthalate (PET) wastewater with UASB is proven feasible in full-scale application (Polanco *et al.* 1999). The wastewater from the esterification process consists of mainly un-reacted raw material, largely ethylene glycol, and products of the secondary or degradation reactions, such as terephthalic acid esters, methanol, acetaldehyde and crotonaldehyde playing the major part. There is also another wastewater stream, called the second stream, from the melt spinning process where a bath of chemicals is showered to improve the physical characteristics of the fiber. It was reported that the anaerobic biodegradability was 90% and 75% for esterification wastewater and second stream wastewater, respectively.

A main problem associated with the granular sludge UASB systems is the inherent long start-up required for the development of anaerobic granules. In cases where inoculation is seeded with flocculant sludge obtained from municipal wastewater sludge digesters, it usually takes several months or even a much longer period before the system can be fully functional. In order to reduce the prolonged start-up of granular sludge-based UASB systems, technologies for enhanced and rapid production of anaerobic granules are highly desirable and sought after. Much research work has been carried out to address these issues.

An unconventional approach to accelerate start-up and granulation processes in UASB reactors by stressing the organic loading rate (OLR) had been developed (Show *et al.* 2004a). Three UASB reactors treating a synthetic feed with chemical oxygen demand (COD) of 2,500 $mg \cdot l^{-1}$, at a mesophilic temperature of 35 °C were studied. The results indicate that the start-up of reactors could be significantly accelerated under stressed loading conditions. Start-up times of the moderately and severely stressed reactors for operating at OLRs of 1 to 16 $g\text{-}COD \cdot l^{-1} \cdot d^{-1}$ ranged from 10 to 80 days and 13 to 90 days, respectively. Comparing with 17–120 days needed in the control reactor to reach the same OLRs, the start-up times were shortened by 25–41%. The extent of acceleration depends on the level at which the reactor is stressed. Applying stress and the extent of stress level in starting up the reactors did not reduce the reactor loading capacity, as all the reactors reached a similar maximum OLR of 16 $g \cdot l^{-1} \cdot d^{-1}$ at the end of operation.

Show *et al.* (2004a) explored further into the possibility of accelerating development of granulation with the unconventional approach of stressed loadings. The researchers found that under stressed loading conditions, the sludge particles began to form granules earlier in both the stressed reactors after 24 and 30 days of start-up operation. Comparing with the control reactor without applying stress, the times taken to form granule were reduced by 45% and 32% in the severely and moderately stressed units, respectively. The granule formation occurred earlier in the severely stressed reactor than the moderately stressed unit.

While the results obtained had established significant acceleration in start-up and granulation processes, the characteristics of granules developed were greatly influenced by the level of stress exerted. Characterization of bioparticles revealed that the granules developed in the moderately stressed reactor exhibited superior characteristics in terms of settleability, strength, microbial activity and morphology, and granular sludge growth, as compared with both the control reactor operated without stress and the unit which was over-stressed (Show *et al.* 2004a). The results presented indicate that the unconventional start-up approach could offer a practical solution for the inherent long start-up in UASB systems with concomitant saving in time and cost.

A different approach to rapid start-up is to adjust the Microbial Load Index values (Tay and Yan 1997). Research findings show that under high Microbial Load Indices of 0.8 and 0.6, granulation developed well in 3–4 months of operation, making a fast start-up. However, with low Microbial Load Indices of 0.3 and 0.2, there was still no granulation after 6 months. It was observed that there are three phases during the process of granulation, namely acclimation, granulation, and maturation. A stepwise and gradual increment in the sludge loading rate must be followed too to avoid the scenario of overloading or starving at different stages.

Use of polymers enhanced the anaerobic granulation process (El-Mamouni *et al.* 1998). Chitosan, natural polymer, out-performed Percol 763, a synthetic polymer in terms of granules formation rate. Chitosan yielded a granulation rate as high as $56\,m \cdot d^{-1}$, compared to $35\,m \cdot d^{-1}$ with Percol 763 in acidic pH. The high granulation yield of chitosan was most probably attributed to its polysaccharides structure, acting similar to the extracellular polymeric substances (EPS) in aggregating anaerobic sludge.

Show *et al.* (2004b) investigated the effects of cationic polymers on reactor start-up and granule development. The experimental results demonstrated that adding the polymer in the seeding stage markedly accelerated the start-up time by as much as 50% and the granule formation by 30% through the use of an appropriate dosage of polymer. In addition, granules developed in polymer-assisted reactor (Fig. 6.4) exhibited better settleability, strength, and methanogenic activity at all organic loading rates tested. Organic loading capacities of polymer-assisted reactor were also increased significantly from 24 to $40\,g\text{-}COD \cdot l^{-1} \cdot d^{-1}$. It was hypothesized that the cationic polymer is able to form bridges among the negatively charged bacterial cells. The bridging enables greater interaction between biosolids resulting in preferential development and enhancement of biogranulation in UASB reactors.

Competition between methanogenesis and sulfidogenesis in anaerobic wastewater treatment exists (Zhou and Fang 1998). High concentrations of sulfate in wastewater can adversely affect the methane production in anaerobic treatment processes. Sulfidogens degrade substrate into bicarbonates and intermediates in the reduction process of sulfate to sulfide. Sulfidogens and methanogens coexist in many anaerobic ecosystems as they have similar physiologies. They are strictly anaerobes, and in favor of similar optimum temperature and pH. Results showed

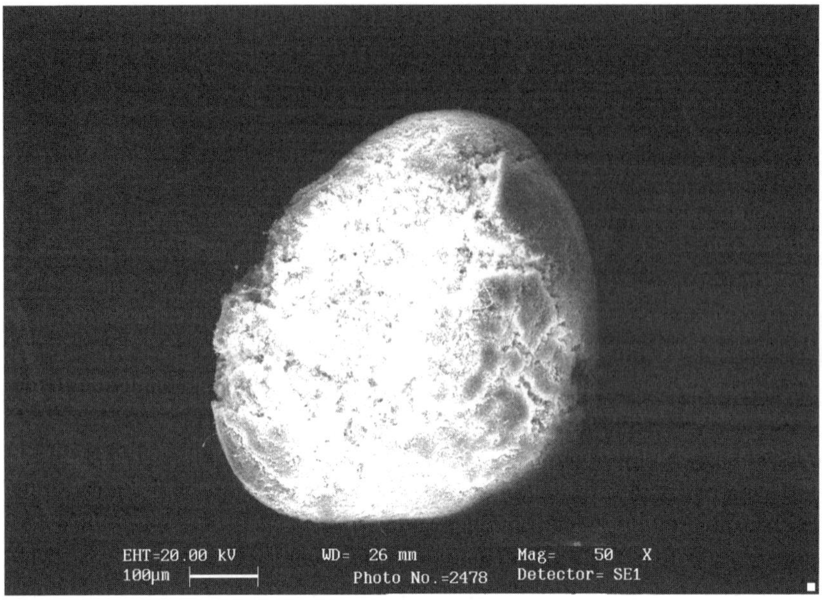

Fig. 6.4. Polymer-assisted granule developed in a UASB reactor.

that after acclimation, a benzoate removal efficiency of 99.5% was consistent regardless of the sulfate concentrations. Sulfidogenesis slowly out-competed methanogenesis during the acclimation phase. This was indicated by the increased sulfate-reducing efficiency of from 48% up to 99% while it was accompanied by the decrease in methane production from 1.02 to 0.39 l-methane$\cdot l^{-1} \cdot d^{-1}$.

3.2. *Expanded granular sludge bed (EGSB) reactor*

It was reported that the expanded granular sludge bed reactor (EGSB) system is able to overcome shortcomings of the upflow anaerobic sludge blanket reactor in the chemical industry (Zoutberg and de Been 1997). The most striking feature is the growth of biomass in a granular form, similar to the UASB granules. The process is especially suitable to treat wastewater containing compounds that are toxic in high concentrations and that only can be degraded in low concentrations (chemical industry). It is also possible to operate the reactor as an ultra-high loaded anaerobic reactor (to 30 g-COD$\cdot l^{-1} d^{-1}$) for applications in other sectors of industry (e.g., brewery, yeast, sugar, corn ethanol production, etc.).

Mesophilic conditions at 35 °C to treat slaughterhouse wastewater in an EGSB system appear to be a feasible option (Nunez and Martinez 1999). The average COD removal efficiency was 67%. Total suspended solid (TSS) removal was 90%. Fats were 85% removed and no accumulation of fats on the sludge was observed. The specific methanogenic activity of the sludge was about three times higher than that of the sludge inoculated into the reactor. The sludge activity did not change significantly after one year of operation.

Thermophilic sulfite and sulfate reduction offers good prospects as part of an alternative technology to conventional off-gas desulfurization technologies (Weijma *et al.* 2000). Methanol can be efficiently used as electron and carbon source to obtain high sulfite and sulfate elimination rates in thermophilic bioreactors.

A comparison of the behavior of EGSB and UASB reactors in diluted (e.g., ethanol, diluted beer) and concentrated (e.g., coffee) wastewater treatment has been made. There were no major differences in the removal rates during the operation with coffee wastewater. It was likely that for this effluent the process is limited by the reaction kinetics instead of by the mass transfer, due to the complex nature of the waste. With diluted beer, EGSB reactor indicated a better performance than the UASB (Jeison and Chamy 1999). The granular sludge-based UASB and expanded granular sludge bed (EGSB) processes gradually take a large portion of applications. Although UASB remains as the predominant technology in use, ESGB type processes are increasingly gaining more popularity (Frankin 2001).

3.3. *Hybrid anaerobic reactors*

The design of anaerobic hybrid reactor is basically combining that of a UASB and a biofilter. The designs include flexi-rings or other media in the upper portion of the reactor to trap biomass and support fixed film growth. This results in a hybrid sludge blanket-fixed film reactor. Simultaneous removal of trichlorfon, with glucose added as carbon source for degradation requirement of trichlorfon in a hybrid bioreactor was modeled (Chen *et al.* 1998). The hybrid bioreactor has both suspended granular biomass and immobilized biofilm. Evaluation of the respective roles of these two types of biomass with a mathematical model was developed and the model was also verified well with experimental results. It has been found that the granular biomass plays a key role in removing both substances in the system. This was due to complete coexistence of both trichlorfon-degrading and glucose-removing bacteria in each type of the granules. Such a system is applicable to the treatment of complex industrial wastewaters that contain easily biodegradable organics as well as refractory pollutants.

Anaerobic thermophilic (55 °C) treatment of thermomechanical pulping whitewater in reactors based on biomass attachment (biofilm) and entrapment (granules) was studied (Jahren *et al.* 1999). Up to 70% COD removals were achieved in all reactors. The hybrid reactor exhibited degradation rates up to 10 g-COD$\cdot l^{-1} \cdot d^{-1}$. The anaerobic multi-stage reactor, consisting of three compartments, each packed with granular sludge and carrier elements, showed degradation rates up to 9 g-COD$\cdot l^{-1} \cdot d^{-1}$. Clogging and short-circuiting eventually became a problem in the multi-stage reactor, probably caused by too high packing of the carriers.

Pentachlorophenol (PCP) is remarkably and efficiently degraded in a hybrid reactor supplied with a mixture of fatty acids (propionic, butyric, acetic, and lactic acids) and methanol (Montenegro *et al.* 2001). The reduction of COD was around 97% and methane was found to be 86% in the biogas production. The efficiency of volatile fatty acids breakdown was 93%, 64%, and 74%, respectively, for butyric, propionic, and acetic acids. PCP total removal of more than 99% was reached by granular sludge activities formed during 21 months of reactor operation. The removal rate was 1.07 mg-PCP$\cdot$g-VSS$^{-1} \cdot d^{-1}$ during the highest PCP concentration addition.

3.4. *Anaerobic continuous stirred tank reactor (CSTR)*

A common bioreactor type used for anaerobic digestion is the continuously stirred tank reactor (CSTR). Vanderhaegen *et al.* (1992) cultivated anaerobic granules in CSTRs, and found that granular sludge disappeared within three weeks when the reactors were incubated statically instead of being shaken. The hydrodynamic shear force appeared to be necessary for maintaining the integrity of granular sludge, and anaerobic granulation may be not reactor type-dependent, but related to the way by which the reactor is operated. It should be pointed out that hydrodynamic flow patterns in UASB and CSTR are quite different. In CSTR, the bacteria move in a stochastic way, i.e., there is no regular liquid upflow pattern. The experiments by Vanderhaegen *et al.* (1992) challenged the general belief that liquid upflow pattern is essential for the development of granular methanogenic sludge.

It is generally believed that spontaneous granulation of hydrogen-producing bacteria can occur with reduced HRT in CSTRs (Fang *et al.* 2002a; Oh *et al.* 2004; Yu and Mu 2006). A CSTR is frequently used for continuous hydrogen production (Chen and Lin 2003; Yu *et al.* 2003). In such a conventional system, hydrogen-producing bacteria are well suspended in the mixed liquor and less suffered from the mass transfer resistance, but biomass retention is substantially influenced by the culture HRT and washout of biomass may occur at a shorter HRT. Hydrogen production rates are thus restricted considerably by a low CSTR biomass retention and low hydraulic loading (Lay *et al.* 1999; Yu *et al.* 2002).

Zhang *et al.* (2007a) reported that the formation of granules in an anaerobic CSTR was accomplished within a very short period of 3–5 days after the acclimated sludge was subject to an acid incubation for 24 h by reducing the culture pH from 5.5 to 2.0 (Fig. 6.5). In the control tests without acid incubation, formation of hydrogen-producing granules was unsuccessful. The formation of granule under acid incubation was much shorter than those reported in other studies which took several months under normal conditions (Fang *et al.* 2002a; Yu and Mu 2006).

Based on further investigations, the researchers (Zhang *et al.* 2007a,b, 2008a,b) found that changing the culture pH would result in improvement in surface physicochemical properties such as reduced electrostatic repulsion, improved hydrophobic surface, more ECP secretion and a higher ratio of proteins/polysaccharides. It was suggested that the improvement in surface physicochemical properties would favor microbial agglutination and immobilization.

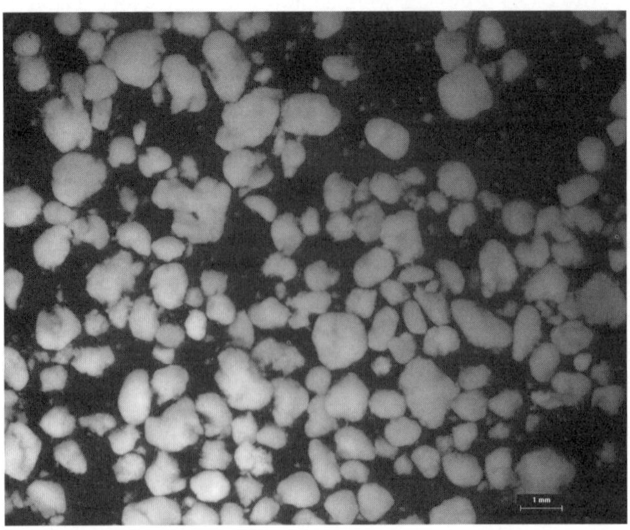

Fig. 6.5. Images granules developed in the anaerobic fluidized-bed reactor, 10×; average diameter: 1.5 ± 0.3 mm.

Show *et al.* (2007) and Zhang *et al.* (2007b) investigated the influence of HRT, ISC, and hence OLR on hydrogen production performance of granular sludge in a CSTR, which was conducted by examining different HRTs and substrate concentrations and associated OLRs. The study found that reducing OLR at a given amount of biomass, i.e., reducing SLR might result in the culture underfed and hence metabolic shift. It was deduced that relatively low hydrogen yield obtained at an OLR of 2.5 g-glucose$\cdot l^{-1}\cdot h^{-1}$ was possibly attributed to the metabolic shift induced by the reduced SLR. Formation of granular sludge significantly increased the overall reactor biomass concentration to as much as 16.0 g-VSS$\cdot l^{-1}$, which enabled the CSTR to operate at an OLR of up to 20 g-glucose$\cdot l^{-1}\cdot h^{-1}$ and hence enhanced the performance in hydrogen production. The granule-based CSTR was demonstrated to be a promising biosystem for hydrogen production.

3.5. *Anaerobic baffled reactor (ABR)*

With the latest development, split feed anaerobic baffled reactor (ABR) was introduced and it shortened the start-up period and gave higher process performance (Uyanik *et al.* 2002), when using improved seed material, even for the treatment of particularly problematic wastewater, viz. ice-cream wastewater. An SEM analysis revealed that granulation process occurred rapidly in the ABR compared with other reactor configurations, and that the reactor contained a highly mixed population of methanogens in all compartments. The use of polymer-conditioned anaerobic sludge and granular sludge as seed proved advantageous over the use of suspended growth anaerobic sludge, and the "improved" SFABR consequently performed more efficiently and also showed greater stability than the conventional ABR.

3.6. *Internal circulation (IC) reactor*

In order to overcome the washout-related operation problem in conventional UASB system, Pereboom and Vereijken (1994) presented a modified design of two-stage UASB reactor, namely the Internal Circulation (IC) reactor. IC reactor consists of two inter-connected UASB compartments on top of each other. It seems that the high turbulence and adequate mixing characteristics make the IC reactor more attractive for reducing clogging and handling high-strength organic wastewater. The influence of the added external circulation on the development of granular sludge was evaluated (Ding and Wang 2005). The results show that the added external circulation can enhance biomass granulation, accelerate granule development, and improve sludge characteristics in the IC reactor.

3.7. *Anaerobic sequencing batch reactor (ASBR)*

Anaerobic SBR (ASBR), which has a much simpler configuration than UASB, is a modified design of batch fed process. The major differences between ASBR and USAB are: (a) ASBR does not require feed distribution; (b) there is no three-phase separator in ASBR; and (c) an upflow hydraulic pattern is absent in ASBR system. As compared to UASB system, the non-continuous operation mode of ASBR is the major disadvantage. Wirtz and Dague (1996) successfully cultivated a granular sludge blanket with an ASBR in five months after seeding the reactor with non-granular primary digester sludge. Angenent and Sung (2001) found that MLVSS retained in ASBR is 2.5-fold higher than that in UASB, and the performance in term of COD removal in ASBR is comparable to that of UASB. The research conducted by Kennedy and Lentz (2000) also showed that at low OLRs, the performances of continuous UASB and anaerobic SBR were quite similar, but continuous UASB reactors performed more favorably than the anaerobic SBR at high OLRs.

3.8. *Anaerobic migrating blanket reactor (AMBR)*

Angenent and Sung (2001) developed a novel anaerobic wastewater treatment system, namely anaerobic migrating blanket reactor (AMBR), which is a continuously fed, compartmentalized reactor without the requirement of elaborate gas–solid–separation and feed-distribution systems. It was found that granules in an AMBR tended to be darker in color, smaller and denser than granules in an UASB reactor operated under conditions similar to those in the AMBR. The internal structure of AMBR is more complex than that of UASB, especially the multiple-point mechanic mixing to overcome sludge clogging as well as to improve feed distribution. As compared to UASB system, the possible advantages of an AMBR with more than three compartments include smaller biomass migration rates, less chance of short-circuiting, operation in a step feed mode for high-strength wastewater during shock loads, and difficult-to-degrade compounds would be degraded more efficiently.

4. Future Trends

The anaerobic granular systems have been known for their unique ability to treat high-strength wastewaters. With global concerns over energy shortages and greenhouse gas formation through combustion of fossil fuels, more efforts toward conversion of wastes into renewable energy is needed. Greater efforts are now sought after for broader applications of anaerobic granulation system for ridding the environment of unwanted organic materials by converting them into biogas — a renewable energy source. The anaerobic granulation process leading toward efficient methane and hydrogen production from wastewaters clearly fits this need. Research toward even broader application is clearly of importance. Problems that need addressing are process reliability, toxicity causes and effects, odor production and control, and better understanding of refractory organic degradation.

It becomes clear that anaerobic granular processes are established technology for a wide variety of industrial applications. The technology is accepted in the industrialized western world as well as in less developed countries. The granular sludge-based UASB and EGSB processes gradually take a large portion of these applications. Although UASB still is the predominant technology in use, at present ESGB type processes are gaining more popularity driven by economics (Frankin 2001). With the recent development of EGSB and 'Staged Multi-Phase Anaerobic' (SMPA) reactor systems, this may lead to a very promising new generations of anaerobic treatment system (Lettinga *et al.* 1997). New insights into the anaerobic degradation of very different categories of compounds, and into process and reactor technology will lead to very promising new generations of anaerobic treatment systems (Lettinga *et al.* 1997). These concepts will provide a higher efficiency at higher loading rates, and are applicable for extreme environmental conditions (e.g., low and high temperatures) and to inhibitory compounds. Moreover, by integrating the anaerobic process with other biological methods (sulfate reduction, micro-aerophilic organisms) and with physical–chemical methods, a complete treatment of the wastewater can be accomplished at very low costs, while at the same time valuable resources can be recovered for reuse.

In the fields of psychrophilic and thermophilic anaerobic treatment, specific reactor development may contribute to further enhance volumetric conversion capacities. Due to reduced water usage, both COD and salt concentrations tend to increase for industrial effluents. As a consequence, there is a need for the development of anaerobic reactors retaining flocculant or granular biomass. The membrane bioreactors (MBRs) offer a solution for certain niches in wastewater treatment. However, poor oxygen transfer economy and biomass fouling are major problems of MBR to be overcome. Membrane bioreactors coupled with granular-based anaerobic processes are worth exploring into.

A promising application of anaerobic granular systems is associated with Certified Emission Reduction or Carbon Credit set under the Kyoto Protocol. In addition to contribution to sustainable development with energy recovery in the form of methane, carbon credits can be claimed by application of advanced anaerobic processes in wastewater treatment for reducing emissions of greenhouse gases. As anaerobic granular systems are capable of handling high organic loadings concomitant with high-strength wastewater and short hydraulic retention time, they could render much more carbon credits than other conventional anaerobic systems.

Looking at the prospects of carbon trading, it may not be an unreasonable expectation that, in the future, wastewater treatment will experience a global shift toward employment of highly efficient granular sludge-based anaerobic processes in maximizing energy production and greenhouse gas emission reductions.

References

Ahring, B. K., Schmidt, J. E., Winther-Nielsen, M., Macario, A. J. L., and de Macario, E. C. (1993) Effect of medium composition and sludge removal on the production, composition and architecture of thermophilic (55 °C) acetate-utilizing granules from an upflow anaerobic sludge blanket reactor. *Appl. Environ. Microbiol.* **59**, 2538–2545.

Alphenaar, P. A., Visser, A., and Lettinga, G. (1993) The effect of liquid upflow velocity and hydraulic retention time on granulation in UASB reactors treating wastewater with a high-sulfate content. *Biores. Technol.* **43**, 249–258.

Angenent, L. T. and Sung, S. (2001) Development of anaerobic migrating blanket reactor (AMBR), a novel anaerobic treatment system. *Wat. Res.* **35**, 1739–1747.

Arcand, Y., Guitot, S. R., Desrochers, M., and Chavarie, C. (1994) Impact of the reactor hydrodynamics and organic loading on the size and activity of anaerobic granules. *Chem. Eng. Biochem. Eng. J.* **56**, 23–35.

Baldi, F., Ivosevic, N., Minacci, A., Pepi, M., Fani, R., Svetlicic, V., and Zutic, V. (1999) Adhesion of *Acinetobacter venetianus* to diesel fuel droplets studied with in situ electrochemical and molecular probes. *Appl. Environ. Microbiol.* **65**(5), 2041–2048.

Batstone, D. J. and Keller, J. (2001) Variation of bulk properties of anaerobic granules with wastewater type. *Wat. Res.* **35**, 1723–1729.

Bochem, H. P., Schoberth, S. M., Sprey, B., and Wengler, P. (1982) Thermophilic biomethanation of acetic-acid — morphology and ultrastructure of a granular consortium. *Can. J. Microbiol.* **28**(5), 500–510.

Bonet, R., Simon-Pujol, M., and Congregado, F. (1993) Effects of nutrients on exopolysaccharide production and surface properties of *Aeromonas salmonicida*. *Appl. Environ. Microbiol.* **59**(8), 2437–2441.

Chang, F. Y. and Lin, C. Y. (2004) Biohydrogen production using an up-flow anaerobic sludge blanket reactor. *Int. J. Hydrogen Energy* **29**(1), 33–39.

Chen, J. and Lun, S. Y. (1993) Study on mechanism of anaerobic sludge granulation in UASB reactors. *Wat. Sci. Technol.* **28**(7), 171–178.

Chen, G. H., Ozaki, H., and Terashima, Y. (1998) Modeling containing simultaneous removal of trichlorfon and glucose in a hybrid bioreactor magnetically immobilized biomass. *Wat. Sci. Technol.* 38(4–5), 179–186.

Chen, C. C. and Lin, C. Y. (2003) Using sucrose as a substrate in an anaerobic hydrogen-producing reactor. *Adv. Environ. Res.* **7**(3), 695–699.

Chiu, Z. C., Chen, M. Y., Lee, D. J., Tay, S. T. L., Tay, J. H., and Show, K. Y. (2006) Diffusivity of oxygen in aerobic granules. *Biotechnol. Bioeng.* **94**(3), 505–513.

Coulter, J. B., Soneda, S., and Ettinger, M. B. (1957) Anaerobic contact process for sewage disposal. *Sew. Ind. Wastes* **29**, 468.

Das, D. and Veziroglu, T. N. (2001) Hydrogen production by biological processes: A survey of literature. *Int. J. Hydrogen Energy* **26**(1), 13–28.

Del Re, B., Sgorbati, B., Miglioli, M., and Palenzona, D. (2000) Adhesion, autoaggregation and hydrophobicity of 13 strains of *Bifidobacterium longum*. *Lttr. Appl. Microbiol.* **31**(6), 438–442.

Ding, J. and Wang, D. (2005) Influence of external circulation on sludge characteristics during start-up of internal circulation reactor. *J. Central South Univ. Technol.* **12**(4), 425–429.

Droste, R. L., Kennedy, K. J., Lu, J. G., and Lentz, M. (1998) Removal of chlorinated phenols in upflow anaerobic sludge blanket reactors. *Wat. Sci. Technol.* **38**(8–9), 359–367.

Dubourgier, H. C., Prensier, G., and Albagnac, G. (1987) Structure and microbial activities of granular anaerobic sludge. In: Lettinga, G., Zehnder, A. J. B., Grotenhuis, J. T. C., and Hulshoff Pol, L. W. (eds.), *Granular Anaerobic Sludge: Microbiology and Technology*, pp. 18–33. The Netherlands: Pudoc Wageningen.

El-Mamouni, R., Leduc, R., and Guiot, S. R. (1998) Influence of synthetic and natural polymers on the anaerobic granulation process. *Wat. Sci. Technol.* **38**, 341–347.

Erdincler, A., Koseogly, S., and Onay, T. (2001) The role of blending in polymer conditioning of waste activated sludge. *Wat. Sci. Technol.* **44**(6), 63–66.

Fang, H. H. P. (1994) Performance and granule characteristics of UASB process treating wastewater with hydrolyzed proteins. *Wat. Sci. Technol.* **30**(8), 55–63.

Fang, H. H. P. and Liu, H. (2002) Effect of pH on hydrogen production from glucose by a mixed culture. *Biores. Technol.* **82**(1), 87–93.

Fang, H. H. P., Liu, H., and Zhang, T. (2002a) Characterization of a hydrogen-producing granular sludge. *Biotechnol. Bioeng.* **78**(1), 44–52.

Fang, H. H. P., Zhang, T., and Liu, H. (2002b) Microbial diversity of a mesophilic hydrogen-producing sludge. *Appl. Microbiol. Biotechnol.* **58**(1), 112–118.

Fitzpatrick, J. A., Roam, G. D., and You, B. T. (1989) Anaerobic microbial aggregation and model development of a UASBR. Paper presented at a *Symposium on Anaerobic Digestion*, 15–18 May, 1989, Taipei, Taiwan.

Forster, C. F. and Lewin, D. C. (1972) Polymer interaction at activated sludge surfaces. *Effl. Wat. Treat. J.* **12**, 520–525.

Frankin, R. J. (2001) Full-scale experiences with anaerobic treatment of industrial wastewater. *Wat. Sci. Technol.* **44**(8), 1–6.

Guiot, S. R., Gorur, S. S., Bourque, D., and Samson, R. (1988) Metal effect on microbial aggregation during upflow anaerobic sludge bed-filter (UBF) reactor start-up. In: Lettinga, G., Zehnder, A. J. B., Grotenhuis, J. T. C., and Hulshoff Pol, L. W. (eds.), *Granular Ananerobic Sludge: Microbiology and Technology*, pp. 187–194. The Netherlands: Wageningen.

Guiot, S. R., Pauss, A., and Costerton, J. W. (1992) A structured model of the anaerobic granules consortium. *Wat. Sci. Technol.* **25**, 1–10.

Hulshoff Pol, L. W., Heijnekamp, K., and Lettinga, G. (1988) The selection pressure as a driving force behind the granulation of anaerobic sludge. In: Lettinga, G., Zehnder, A. J. B., Grotenhuis, J. T. C., and Hulshoff Pol, L. W. (eds.), *Granular Ananerobic Sludge: Microbiology and Technology*, pp. 153–161. The Netherlands: Wageningen.

Hulshoff Pol, L. W. (1989) *The Phenomenon of Granulation of Anaerobic Sludge*. PhD thesis, Agricultural University of Wageningen, Wageningen, The Netherlands.

Hulshoff Pol, L. W., de Castro Lopes, S. I., Lettinga, G., and Lens, P. N. L. (2004) Anaerobic sludge granulation. *Wat. Res.* **38**(6), 1376–1389.

Jahren, S. J., Rintala, J. A., and Ødegaard, H. (1999) Anaerobic thermophilic (55 °C) treatment of TMP whitewater in reactors based on biomass attachment and entrapment. *Wat. Sci. Technol.* **40**(11–12), 67–76.

Jeison, D. and Chamy, R. (1999) Comparison of the behaviour of expanded granular sludge bed (EGSB) and upflow anaerobic sludge blanket (UASB) reactors in dilute and concentrated wastewater treatment. *Wat. Sci. Technol.* **40**(8), 91–98.

Kennedy, K. J. and Lentz, E. M. (2000) Treatment of landfill leachate using sequencing batch and continuous upflow anaerobic sludge blanket (UASB) reactors. *Wat. Res.* **34**, 3640–3656.

Kos, B., Suskovic, J., Vukovic, S., Simpraga, M., Frece, J., and Matosic, S. (2003) Adhesion and aggregation ability of probiotic strain *Lactobacillus acidophilus* M92. *J. Appl. Microbiol.* **94**(6), 981–987.

Kotsopoulos, A., Zeng, J., and Angelidaki, I. (2006) Biohydrogen production in granular up-flow anaerobic sludge blanket (UASB) reactors with mixed cultures under hyper-thermophilic temperature (70 °C). *Biotechnol. Bioeng.* **94**(2), 296–302.

Lay, J. J., Lee, Y. J., and Noike, T. (1999) Feasibility of biological hydrogen production from organic fraction of municipal solid waste. *Wat. Res.* **33**(11), 2579–2586.

Lens, P. N. L., De Beer, D., Cronenberg, C., Ottengraf, S., and Verstraete, W. (1995) The use of microsensors to determine distributions in UASB aggregates. *Wat. Sci. Technol.* **31**, 273–280.

Lettinga, G., van Velsen, A. F. M., de Zeeuw, W., and Hobma, S. W. (1979) Feasibility of the upflow anaerobic sludge blanket process. In: *Proceedings of National Conference On Environmental Engineering*, In American Society of Civil Engineers (pp. 35), San Francisco, July.

Lettinga, G., van Velsen, A. F. M., Hobma, S. W., de Zeeuw, W., and Klapwijk, A. (1980) Use of the upflow sludge blanket (USB) reactor concept for biological waste water treatment especially for anaerobic treatment. *Biotechnol. Bioeng.* **22**, 699–734.

Lettinga, G., van Lier, J., Zeeman, G., and Hulshoff Pol, L. W. (1997) Advanced anaerobic wastewater treatment in the near future. *Wat. Sci. Technol.* **35**(10), 5–12.

Liu, Y., Xu, H. L., Show, K. Y., and Tay, J. H. (2002) Anaerobic granulation technology for wastewater treatment. *World J. Microbiol. Biotechnol.* **18**(2), 99–113.

Liu, H., and Fang, H. H. P. (2003) Hydrogen production from wastewater by acidogenic granular sludge. *Wat. Sci. Technol.* **47**(1), 153–158.

Liu, Y., Yang, S. F., and Tay, J. H. (2003) Elemental compositions and characteristics of aerobic granules cultivated at different substrate N/C ratios. *Appl. Microbiol. Biotechnol.* **61**(5–6), 556–561.

Liu, Y., Yang, S. F., Tay, J. H., Liu, Q. S., Qin, L., and Li, Y. (2004) Cell hydrophobicity is a triggering force of biogranulation. *Enzyme Microbiol. Technol.* **34**(5), 371–379.

MacLeod, F. A., Guiot, S. R., and Costerton, J. W. (1990) Layered structure of bacterial aggregates produced in an upflow anaerobic sludge bed and filter reactor. *Appl. Environ. Microbiol.* **56**, 1598–1607.

Mahoney, E. M., Varangu, L. K., Cairns, W. L., Kosaric, N., and Murray, R. G. E. (1987) The effect of calcium on microbial aggregation during UASB reactor start-up. *Wat. Sci. Technol.* **19**(1–2), 249–260.

Montenegro, M. D. A. P., Moraes, E. D. M., Soares, H. M., and Vazoller, R. F. (2001) Hybrid reactor performance in pentachlorophenol (PCP) removal by anaerobic granules. *Wat. Sci. Technol.* **44**(4), 137–144.

Morgan, J. W., Evison, L. M., and Forster, C. F. (1991a) Internal architecture of anaerobic sludge granules. *J. Chem. Technol. Biotechnol.* **50**, 211–226.

Morgan, J. W., Evison, L. M., and Forster, C. F. (1991b) Upflow sludge blanket reactors: The effect of bio-supplements on performance and granulation. *J. Chem. Technol. Biotechnol.* **52**, 243–255.

Mu, Y. and Yu, H. Q. (2006) Biological hydrogen production in a UASB reactor with granules. I: Physicochemical characteristics of hydrogen-producing granules. *Biotechnol. Bioeng.* **94**(5), 980–987.

Noyola, A. and Mereno, G. (1994) Granulation production from raw waste activated sludge. *Wat. Sci. Technol.* **30**, 339–346.

Nunez, L. A. and Martinez, B. (1999) Anaerobic treatment of slaughterhouse wastewater in an expanded granular sludge bed (EGSB) reactor. *Wat. Sci. Technol.* **40**(8), 99–106.

O'Flaherty, V., Lens, P. N. L., De Beer, D., and Colleran, E. (1997) Effect of feed composition and upflow velocity on aggregate characteristics in anaerobic upflow reactors. *Appl. Microbiol. Biotechnol.* **47**, 102–107.

Oh, Y. K., Kim, S. H., Kim, M. S., and Park, S. (2004) Thermophilic biohydrogen production from glucose with trickling biofilter. *Biotechnol. Bioeng.* **88**(6), 690–698.

Palns, S. S., Loewenthal, R. E., Dold, P. L., and Marais, G. v. R. (1987) Hypothesis for pelletisation in upflow anaerobic sludge blanket reactor. *Water SA* **13**, 69–80.

Pan, S., Tay, J. H., He, Y. X., and Tay, S. T. L. (2004) The effect of hydraulic retention time on the stability of aerobically grown microbial granules. *Lett. Appl. Microbiol.* **38**(2), 158–163.

Pereboom, J. H. F. and Vereijken, T. L. F. M. (1994) Methanogenic granule development in full scale internal circulation reactors. *Wat. Sci. Technol.* **30**, 9–21.

Perez, P. F., Minnaard, Y., Disalvo, E. A., and De Antoni, G. L. (1998) Surface properties of bifidobacterial strains of human origin. *Wat. Sci. Technol.* **64**(1), 21–26.

Polanco, F. F., Hidalgo, M. D., Polanco, M. F., and Encina, P. A. G. (1999) Anaerobic treatment of polyethylene terephthalate (PET) wastewater from lab to full-scale. *Wat. Sci. Technol.* **40**(8), 229–236.

Rouxhet, P. G. and Mozes, N. (1990) Physical chemistry of the interaction between attached microorganisms and their support. *Wat. Sci. Technol.* **22**, 1–16.

Rudd, T., Sterritt, R. M., and Lester, J. N. (1984) Complexation of heavy metals by extracellular polymers in the activated sludge process. *J. Wat. Pollut. Control Fed.* **56**, 1260–1268.

Sam-Soon, P. A., Looewenthal, R. E., Dold, P. L., and Marais, G. v. R. (1988) Pelletization in upflow anaerobic sludge bed reactors. In: Hall, E. R. and Hobson P. N. (eds.), *Anaerobic Digestion*, 55–60. Pergamon Press, Oxford, U. K.

Sanchez, J. M., Arijo, S., Munoz, M. A., Morinigo, M. A., and Borrego, J. J. (1994) Microbial colonization of different support materials used to enhance the methanogenic process. *Appl. Microbiol. Biotechnol.* **41**, 480–486.

Santegoeds, C. M., Damagaad, L. R., Hesselink, C., Zopfi, J., Lens, P. N. L., Muyzer, G., and de Beer, D. (1999) Distribution of sulfate-reducing and methanogenic bacteria in anaerobic aggregates determined by microsensor and molecular analysis. *Appl. Environ. Microbiol.* **65**, 4618–4629.

Schmidt, J. E. and Ahring, B. K. (1993) Effects of magnesium on thermophilic acetate-degrading granules in upflow anaerobic sludge blanket (UASB) reactor. *Enzyme Microbiol. Technol.* **15**, 304–310.

Schmidt, J. E. and Ahring, B. K. (1994) Extracellular polymers in granular sludge from different upflow anaerobic sludge blanket (UASB) reactors. *Appl. Microbiol. Biotechnol.* **42**, 457–462.

Schmidt, J. E. and Ahring, B. K. (1996) Granular sludge formation in upflow anaerobic sludge blanket (UASB) reactors. *Biotechnol. Bioeng.* **49**, 229–246.

Sekiguchi, Y., Kamagata, Y., Syutsubo, K., Ohashi, A., Harada, H., and Nakamura, K. (1998) Diversity of mesophilic and thermophilic granular sludge determined by 16S rRNA gene analysis. *Microbiol.* **22**, 2655–2665.

Sekiguchi, Y., Kamagata, Y., Nakamura, K., Ohashi, A., and Harada, H. (1999) Fluorescence *in situ* hybridization using 16S rRNA-targeted oligonucleotides reveals localization of methanogens and selected uncultured bacteria in mesophilic and thermophilic sludge granules. *Appl. Environ. Microbiol.* **65**, 1280–1288.

Shapiro, J. A. (1998) Thing about bacterial populations as multicellular organisms. *Annu. Rev. Microbiol.* **52**, 81–104.

Shen, C. F., Kosaric, N., and Blaszczyk, R. (1993) The effect of selected heavy metals (Ni, Co and Fe) on anaerobic granules and their extracellular polymeric substance (EPS). *Wat. Res.* **27**, 25–33.

Show, K. Y., Tay J. H., Yang, L., Wang, Y., and Lua, C. H. (2004a) Effects of stressed loading on startup and granulation in upflow anaerobic sludge blanket reactors. *J. Environ. Eng. ASCE* **130**(7), 743–750.

Show, K. Y., Wang, Y, Foong, S. F., and Tay, J. H. (2004b) Accelerated start-up and enhanced granulation in upflow anaerobic sludge blanket reactors. *Wat. Res.* **38**(9), 2293–2304.

Show, K. Y., Zhang, Z., Tay, J. H., Liang, T. D., Lee, D. J., and Jiang, W. J. (2007) Production of hydrogen in a granular sludge-based anaerobic continuous stirred tank reactor. *Int. J. Hydrogen Energy*, **32**(18), 4744–4753.

Tagawa, T., Syutsubo, K., Sekiguchil, Y., Ohashi, A., and Harada, H. (2000) Quantification of methanogen cell density in anaerobic granular sludge consortia by fluorescence *in-situ* hybridization. *Wat. Sci. Technol.* **42**, 77–82.

Tay, J. H. and Yan, Y. G. (1997) Anaerobic biogranulation as microbial response to substrate adequacy. *J. Environ. Eng.* **123**(10), 1002–1010.

Tay, J. H., Xu, H. L., and Teo, K. C. (2000) Molecular mechanism of granulation. I: H+ translocation-dehydration theory. *J. Environ. Eng. ASCE* **126**(5), 403–410.

Tay, T. L. S., Ivanov, V., Kim, I. S., Feng, L., and Tay, J. H. (2001a) Quantification of ratios of *Bacteria* and *Archaea* in methanogenic microbial community by fluorescence *in situ* hybridization and fluorescence spectrometry. *World J. Microbiol. Biotechnol.* **17**(6), 583–589.

Tay, J. H., He, Y. X., and Yan, Y. G. (2001b) Improved anaerobic degradation of phenol with supplemental glucose. *J. Environ. Eng.* **127**(1), 38–45.

Tay, J. H., Tay, S. T. L., Liu, Y., Show, K. Y., and Ivanov, V. (2006) *Biogranulation Technologies for Wastewater Treatment.* Oxford: Elsivier Science.

Teo, K. C., Xu, H. L., and Tay, J. H. (2000) Molecular mechanism of granulation. II: Proton translocating activity. *J. Environ. Eng.* **126**, 411–418.

Thiele, J. H., Wu, W. M., Jain, M. K., and Zeikus, J. G. (1990) Ecoengineering high rate biomethanation system: design of improved syntrophic biomethanation catalysis. *Biotechnol. Bioeng.* **35**, 990–999.

Tsuneda, S., Jung, J., Hayashi, H., Aikawa, H., Hirata, A., and Sasaki, H. (2003) Influence of extracellular polymers on electrokinetic properties of heterotrophic bacterial cells examined by soft particle electrophoresis theory. *Colloid Surf. B.* **29**(2–3), 181–188.

Uyanik, S., Sallis, P. J., and Anderson, G. K. (2002) The effect of polymer addition on granulation in an anaerobic baffled reactor (ABR): Part I. Process performance. *Wat. Res.* **36**, 933–942.

Vanderhaegen, B., Ysebaert, E., Favere, K., Van Wambeke, M., Peeters, T., Panic, V., Vandenlangenbergh, V., and Verstraete, W. (1992) Acidogenesis in relation to in-reactor granule yield. *Wat. Sci. Technol.* **25**, 21–30.

van Loosdrecht, M. C. M., Lyklema, J., Norde, W., Schraa, G., and Zehnder, A. J. B. (1987) Electrophoretic mobility and hydrophobicity as a measure to predict the initial steps of bacterial adhesion. *Appl. Environ. Microbiol.* **53**, 1898–1901.

Wang, Y., Show, K. Y., Tay, J. H., and Sim, K. H. (2004) Effects of cationic polymer on start-up and granulation in UASB reactors. *Chem. Technol. Biotechnol.* **79**(3), 219–228.

Weijma, J., Stams, A. J. M., Hulshoff Pol, L. W., and Lettinga, G. (2000) Thermophilic sulfate reduction and methanogenesis with methanol in a high rate anaerobic reactor. *Biotechnol. Bioeng.* **67**(3), 354–363.

Wen, T. C., Cheng, S. S., and Lay, J. J. (1995) A kinetic model of a recirculated upflow anaerobic sludge blanket treating phenolic wastewater. *Wat. Environ. Res.* **67**(6), 1005–1006.

Wiegant, W. M. (1988) The spaghetti theory on anaerobic granular sludge fermentation, or the inevitability of granulation. In: Lettinga, G., Zehnder, A. J. B., Grotenhuis, J. T. C., and Hulshoff Pol, L. W. (eds.), *Granular Anaerobic Sludge: Microbiology and Technology*, pp. 146–152. The Netherlands: Wageningen.

Wilschut, J. and Hoekstra, D. (1984) Membrane fusion: From liposome to biological membrane. *Trend Biochem. Sci.* **9**, 479–483.

Wirtz, R. A. and Dague, R. R. (1996) Enhancement of granulation and start-up in anaerobic sequencing batch reactor. *Wat. Environ. Res.* **68**, 883–892.

Wu, W. M., Jain, M. K., and Zeikus, J. G. (1996) Formation of fatty acid-degrading anaerobic granules by defined species. *Appl. Environ. Microbiol.* **62**, 2037–2044.

Wu, S. Y., Hung, C. H., Lin, C. N., Chen, H. W., Lee, A. S., and Chang, J. S. (2006) Fermentative hydrogen production and bacterial community structure in high-rate anaerobic bioreactors containing silicone-immobilized and self-flocculated sludge. *Biotechnol. Bioeng.* **93**(5), 934–946.

Yu, H. Q., Tay, J. H., and Fang, H. H. P. (2001) The role of calcium in sludge granulation during UASB reactor start-up. *Wat. Res.* **35**, 1052–1060.

Yu, H., Zhu, Z., Hu, W., and Zhang, H. (2002) Hydrogen production from rice winery wastewater in an upflow anaerobic reactor by using mixed anaerobic cultures. *Int. J. Hydrogen Energy*, **27**(11–12), 1359–1365.

Yu, H. Q., Hu, Z. H., and Hong, T. Q. (2003) Hydrogen production from rice winery wastewater by using a continuously-stirred reactor. *J. Chem. Eng. Japan* **36**(10), 1147–1151.

Yu, H. Q. and Mu, Y. (2006) Biological hydrogen production in a UASB reactor with granules. II: Reactor performance in 3-year operation. *Biotechnol. Bioeng.* **94**(5), 988–995.

Zhang, J. J., Li, X. Y., Oh, S. E., and Logan, B. E. (2004) Physical and hydrodynamic properties of flocs produced during biological hydrogen production. *Biotechnol. Bioeng.* **88**(7), 854–860.

Zhang, Z. P., Show, K. Y., Tay, J. H., Liang, D. T., Lee, D. J., and Jiang, W. J. (2006) Effect of hydraulic retention time on biohydrogen production and anaerobic microbial community. *Process Biochem.* **41**(10), 2118–2123.

Zhang, Z. P., Show, K. Y., Tay, J. H., Liang, D. T., Lee, D. J., and Jiang, W. J. (2007a) Rapid formation of hydrogen-producing granules in an anaerobic continuous stirred tank reactor induced by acid incubation. *Biotechnol. Bioeng.* **96**(6), 1040–1050.

Zhang, Z. P., Tay, J. H., Show, K. Y., Yan, R., Liang, D. T., Lee, D. J., and Jiang, W. J. (2007b) Biohydrogen production in a granular activated carbon anaerobic fluidized bed reactor. *Int. J. Hydrogen Energy* **32**(2), 185–191.

Zhang, Z., Show, K. Y., Tay, J. H., Liang, T., and Lee, D. J. (2008a) Biohydrogen production with anaerobic fluidized bed reactors- a comparison of biofilm-based and granule-based systems. *Int. J. Hydrogen Energy* **33**(5), 1559–1564.

Zhang, Z., Show, K. Y., Tay, J. H., Liang, T., and Lee, D. J. (2008b) Enhanced continuous biohydrogen production by immobilized anaerobic microflora. *J. Energy Fuels* **22**(1), 87–92.

Zhou, G. M. and Fang, H. H. P. (1998) Competition between methanogenesis and sulfidogenesis in anaerobic wastewater treatment. *Wat. Sci. Technol.* **38**(8–9), 317–324.

Zoutberg, G. R. and de Been, P. (1997) The Biobed® EGSB (expanded granular sludge blanket) system covers shortcomings of the UASB reactor in the chemical industry. *Wat. Sci. Technol.* **35**(10), 183–188.

Chapter 7

Anaerobic Membrane Reactors

David C. Stuckey
Department of Chemical Engineering and Chemical Technology,
Imperial College of Science, Technology and Medicine,
Prince Consort Road, London SW7 2BY, UK
d.stuckey@ic.ac.uk

In the last decade anaerobic membrane reactors (AnMBRs) have evolved from aerobic MBRs, with the membrane either external or submerged within the reactor. These reactors can achieve high chemical oxygen demand (COD) removals (~98%) at low hydraulic retention times (HRTs) of 3 hours. In addition, since the membrane stops biomass being washed out of the reactor, they are capable of enhancing performance with inhibitory substrates, and at psychrophilic and thermophilic temperatures, and enable nitrogen removal via Anammox. As with MBRs, fouling is an issue, but addition of activated carbon or resins/precipitants can remove soluble organic products (SMPs) and enhance flux. Due to their low energy use and solids production, and a solids-free effluent, they also have considerable potential to enhance nutrient and water recycling. Nevertheless, more work is needed to compare fouling in aerobic and anaerobic systems, to determine how much knowledge in aerobic systems can be transferred, to determine how reactor operation influences fouling, to evaluate the effect of different additives to the reactor on membrane fouling, to determine whether nitrogen removal can be incorporated into AnMBRs, to determine methane solubility at low temperatures and recover it from the effluent, and to establish sound mass and energy balances on pilot scale plants to evaluate the economics of AnMBRs.

1. Introduction

Over the past 50 years process engineers have struggled with fact that anaerobic bacteria grow very slowly due to their low energy yields per gram of substrate, while coping with pressure to separate hydraulic retention time (HRT) from the solids retention time (SRT). If this can be achieved, then anaerobic digestion (AD) reactors can be shrunk to small sizes while being able to cope with high organic loading rates (4–30 kg-COD$\cdot$m$^{-3}\cdot$d^{-1}). In recent years there has been increasing pressure on designing anaerobic reactors that not only have very short HRTs, but also reduce the footprint of the process by intensifying or combining unit operations — this has led to the development of the anaerobic membrane reactor (AnMBR) which incorporates solids removal and chemical oxygen demand (COD) reduction in one reactor. This design has evolved from aerobic membrane bioreactors (AMBRs), which are more developed but in this current climate have obvious drawbacks, such as high energy use and solids yields, and emission of greenhouse gases, such as carbon dioxide and nitrous oxide (if nitrifying/denitrifying). Hence, there is increasing interest in AD due to these concerns, and that AnMBRs enhance water and nutrient (nitrogen and phosphorus) recycling possibilities. The AnMBR can now be operated at very low HRTs (3 hours — Hu and Stuckey 2006) and produce a solids-free effluent with high COD removals due to the retention of slowly degradable organics within the reactor. In addition, it can retain slow growing bacteria which would be normally washed out of many reactors, and enable these bacteria to grow and persist under unfavourable conditions, e.g., under high salinity (Vyrides and Stuckey 2009a).

This chapter will review the current state of the art, although this has been done recently with good and comprehensive reviews on AnMBRs by Liao *et al.* (2006), and permeate flux and fouling in AnMBRs by Berube *et al.* (2006) and Liao *et al.* (2004). In addition, Meng *et al.* (2009) carried out a recent insightful review of the literature on membrane fouling and membrane materials in both AMBRs and AnMBRs. Hence this chapter will focus on recent advances, and on specific areas and questions such as the role of SMPs in membrane fouling, and fouling amelioration using additives such as powdered activated carbon (PAC) and polymers/metal salts/biopolymers, while drawing on work on both AMBRs and AnMBRs in these past reviews.

2. History

The association of membranes and anaerobic digestion seems to be obvious, and was first used in 1978 by Grethlein to improve the quality of septic tank effluent. In the early 1980s Dorr-Oliver developed a system containing a CSTR reactor with external cross-flow membrane unit (Sutton *et al.* 1983; Li *et al.* 1985), however, this never reached full scale, undoubtedly due to membrane costs. In 1985 Japan started a program to develop anaerobic membrane reactors called the Aqua Renaissance '90 where a variety of configurations and wastewaters were evaluated (Kimura 1991; Kataoka *et al.* 1992; Minami 1994), and the COD removals were generally over 90%. In the UK, Anderson and co-workers (Anderson *et al.* 1986a, b) also developed anaerobic reactors with external membrane modules.

The first time this design made a significant impact was when the ADUF (anaerobic digestion ultrafiltration) system was developed in South Africa by Ross and co-workers (Ross *et al.* 1990, 1992; Strohwald and Ross 1992). The ultrafiltration (UF) membrane used was a tubular asymmetrical polyethersulphone (PES) low transmembrane pressure (TMP) one manufactured locally with a molecular weight (MW) cut-off of between 20 and 80 kDa. The unit was operated at high cross-flow velocities of 2–3.5 $m \cdot s^{-1}$ to maintain a high flux of 1,100 $l \cdot m^{-2} \cdot d^{-1}$ at solids levels below 40 $g \cdot l^{-1}$, resulting in OLRs as high as 10 $kg\text{-}COD \cdot m^{-3} \cdot d^{-1}$.

In the last two decades work had progressed steadily with researchers examining the role of both external and internal membrane modules, the composition of fouling materials, how to reduce fouling by adsorption and precipitation in the reactor, the role of membranes in reduced fouling, operation at reduced and elevated temperatures, and the microbial ecology in the reactor. The reactor has also been used to treat a variety of both domestic and industrial wastewaters, and these topics will be addressed in more detail below.

3. Membrane Configuration

3.1. *Pressure-driven external cross-flow membrane*

The combination of anaerobic reactors and membranes can be engineered in primarily three different ways, as shown in Fig. 7.1 (Liao *et al.* 2006). In the first variation the membrane is

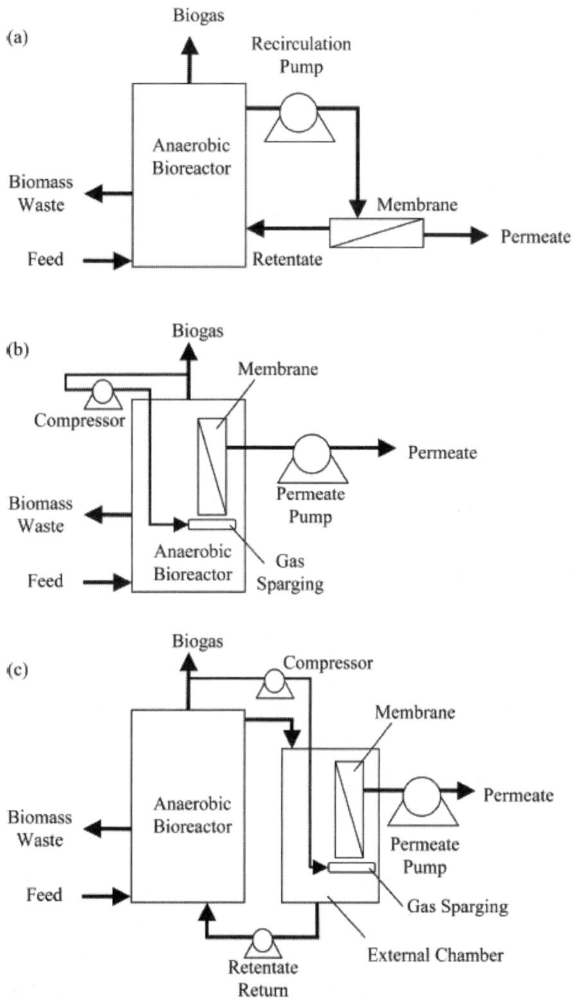

Fig. 7.1. Schematic of anaerobic membrane bioreactor configurations. (a) Pressure-driven external cross-flow membrane. (b) Vacuum-driven submerged membrane with the membrane immersed directly into the reactor. (c) Vacuum-driven submerged membrane with the membrane immersed in an external chamber (Liao *et al.* 2006).

situated outside the anaerobic reactor which makes membrane cleaning and replacement simple, but involves an external pump which circulates the biomass at quite high velocities (2–4 $m \cdot s^{-1}$), thereby scouring the membrane surface to reduce membrane fouling.

This also provides quite a high pressure to force the liquid through the membrane. While high fluxes result (60 $l \cdot m^{-2} \cdot h^{-1}$; LMH), the power/energy costs are high, and it is clear that some pump types (e.g., rotary) lead to floc and cell shear with a decrease in overall particle size, and an increase in soluble organics in anaerobic systems (Kim *et al.* 2001), and this has also been shown to happen in aerobic systems (Wisniewski and Grasmick 1998). The correlation between aerobic and anaerobic membrane behaviour is not clear at this point, but they seem to behave in similar ways, at least at the macroscale. Interestingly, the use of centrifugal pumps (Kim *et al.* 2001) does not seem to result in as much shear and size reduction. This reduction in particle size and increase in soluble organics in turn leads to rapidly reducing fluxes. However, until recently there has been confusion in the literature about the effect of sidestream pumping and shear on methanogenic activity.

The initial work on the question of pump shear on methanogenic activity was by Brockmann and Seyfried (1996). These authors showed that increasing recycle resulted in reducing substrate-specific sludge activities assays (in terms of ml-CH_4·g-$MLSS^{-1}$·h^{-1}), and a 50% loss of activity was observed after only recirculating the biomass 20 times through the external membrane unit, while 100 cycles resulted in 90% loss of activity. They postulated that this loss in activity was due to the reduction in floc size and an impairment of the syntrophic association between acidogenic and methanogenic bacteria. In terms of floc size distribution, Wisniewski and Grasmick (1998) showed with aerobic sludge that increasing velocities in the membrane unit resulted in substantial reductions in floc size, and this was confirmed by Choo and Lee (1998) who showed that after only 12 days of recirculation the average floc size dropped from 16 to 3 μm, and that there was an increase in the amount of colloidal particles; over this time there was also a fourfold decrease in flux to 10 LMH. While not concentrating on COD removal in this paper, it was clear that there was little decrease in performance. Hence, it appears that a reduction in floc size to very small diameters is not a priori evidence that the symbiotic association is still not functioning. Indirect evidence for this also comes from work by Voolapalli and Stuckey (1998) who used hydrophobic gas membranes to extract hydrogen from methanogenic cultures. These authors found that the hydrophobic surface attracted methanogens, and that floc sizes decreased to less than 5 μm, but the COD removal was better than the control under shock loads. Kim *et al.* (2001) also found that the type of pump on the recycle of an aerobic reactor made a big difference with a rotary pump resulting in a far higher shear on the flocs than a centrifugal one, and a reduction in the SOUR of 22% compared to none with the centrifugal pump. Whether these results can be extrapolated to anaerobic systems is not known, but clearly pump type could have a large impact on anaerobic systems.

Finally, in a recent paper by Padmasiri *et al.* (2007) explored this question in some depth using swine manure with a cross-flow external membrane module with velocities of up to 2 m·s^{-1}. The reactor started up successfully at the initial loading rate of 1 g-VS·l^{-1}·d^{-1}. After a two-fold increase in loading rate and flow velocity on day 75, performance deteriorated. The methanogenic population dynamics in the reactor were monitored with terminal restriction fragment length polymorphism (T-RFLP). Changes in the relative levels of *Methanosarcinaceae* and *Methanosaetaceae* were consistent with changes in VFA concentrations, i.e., high and low levels of acetate corresponded to a high abundance of *Metharios arcinaceae* and *Methariosaetaceae*, respectively. The levels of hydrogenotrophic methanogens (*Methanomicrobiales*) increased during decreased reactor performance suggesting that syntrophic interactions involving hydrogenotrophic methanogens remained intact regardless of the degree of shear in the AnMBR. The authors concluded that the decrease in reactor performance can be explained by an increase in the rate of hydrolysis caused by increased shear and mixing, and that this lead the methanogens being overwhelmed with an increase in VFAs. While this seems plausible, it is not clear whether the presence of insoluble and non-biodegradable fibres in the pig slurry may ameliorate the effect of shear, and preserve the microbial consortia. It appears that shear itself may inhibit methanogens by increasing cell lysis and enhance both extracellular polysaccharide (ECP) and soluble microbial products (SMPs) release. However, it is clear that external membrane modules have some drawbacks despite their high fluxes and convenience in terms of cleaning and replacement (Jeison *et al.* 2009).

3.2. *Vacuum-driven submerged membrane immersed directly into the reactor*

Another way of operating these reactors is to use a vacuum (or even hydrostatic head) on the membrane to draw the effluent through. Despite this being a common method in aerobic treatment, e.g., Kubota plant, there have not been many examples of this type of setup in use. In addition, the membrane can either be immersed in the reactor tank itself (Fig. 7.1b) (submerged anaerobic membrane bioreactor — SAMBR), or in a separate reactor (Fig. 7.1c), which requires a pump, but the flow is not through a module. This type of setup is used in many aerobic plants, e.g., GE-Zenon, since it is easier to clean the module.

The advantages of having the membrane submerged in the reactor are that the energy required for pumping is eliminated, although biogas needs to be recycled from the headspace to underneath the membranes to provide gas bubble shear to keep the membranes relatively clear from fouling (Vyrides and Stuckey 2009a). In addition, the biomass in the reactor is subjected to less severe shear than in a side-stream membrane unit, and hence based on the previous discussion the biomass should be less stressed. Finally, the capital cost would be lower since external modules and pumps are not required. The first papers in this area were by Anderson *et al.* (1986a) and Kayawake *et al.* (1991), however, it is only recently that more literature on the submerged configuration has been published, and this will be discussed in more detail later under different subheadings (Hu and Stuckey 2006; Van Zyl *et al.* 2008; Akram and Stuckey 2008a; Bohdziewicz *et al.* 2008; Jeison and van Lier 2008a; Walker *et al.* 2009a).

The final configuration which has been developed recently is that of sequential membrane reactors where effluents from one reactor (often coarse with large pore sizes) is treated by another membrane reactor with a smaller pore size. This flowsheet has been used in the treatment of the organic fraction of municipal solid waste (OFMSW) where the (acid) hydrolysis reactor contains a coarse steel membrane (30–140 μm) and the leachate containing fine colloidal particles is treated in either an anaerobic filter (Walker *et al.* 2009a, b), or in a submerged anaerobic reactor (Trzcinski and Stuckey 2009a, b). Another configuration is where a membrane is used in the initial hydrolysis tank, e.g., with pig waste (Lee *et al.* 2001), to stop solids leaving and enhancing their SRT, or after an anaerobic reactor to eliminate solids in the effluent (Kimura 1991).

4. Membrane Pore Size and Composition

The pore size of most commercial aerobic MBRs tend to be in the coarse ultrafiltration (UF) to fine microfiltration range (MF) (Judd 2006), and virtually all work on anaerobic systems has used aerobic membranes as the tacit assumption is that they are the most suitable ones to use. Intuitively, membrane pore size should have an effect on flux, with larger pores resulting in greater fluxes. Hernandez *et al.* (2002) evaluated immersed membranes made of polypropylene (10 μm) and fibreglass (100 μm) in an anaerobic granular reactor. While fluxes dropped rapidly with both membranes, after 90 days "steady state" was reached where the 100 μm membrane stabilised at 70 LMH, while the 10 μm one achieved a flux of only 10 LMH. However, in the more established aerobic MBR literature, Judd (2006) states that "the effects of pore size on membrane fouling are strongly related to the feed solution characteristics, and in particular, the particle size distribution. This has led to conflicting trends reported in the literature, with no consistent general trend noted between pore size and hydraulic performance. This can, in part, be attributed to the complex and changing nature of the biological suspension in MBR systems and the comparatively large pore size distribution (in) the membranes used, along with operational facets such as the system hydrodynamics and the duration of the test". At this point, there is no reason to believe that this observation is not applicable to anaerobic MBRs as well.

Another factor that may have a big effect on the flux and molecular weight (MW) cut-off of specific membranes in addition to pore size is the formation of a gel polarisation layer on the membrane surface caused by an increase in concentration of rejected solutes at the surface. This results in the interfacial concentration exceeding the solubility of the solute and a "gel layer" forming which controls both the flux and MW of solutes able to pass through the membrane (Elmaleh and Abdelmoumni 1997). In addition, some bacteria from suspension will be located on this gel layer, either because they are metabolising it, or that they are physically forced there by the flow through the membrane. While this could be termed a "biofilm", it is not really what has been commonly understood as a biofilm due to both its composition and how it is formed. This may also result in COD removal as the solutes pass through it. Once this layer builds up, it's thickness depends on the hydrodynamic shear in the reactor, e.g., gas volume and bubble size, and when this is low then this gel layer can reject quite low MW solutes. Chang *et al.* (2001) investigated the effect of membrane fouling on solute rejection in aerobic systems, and found that membranes with the same MW cut-off but different hydrophobicities resulted in different effluent CODs. They explained this by the cake layer deposited on the membrane surface playing an important role in solute rejection. Very recently, Akram and Stuckey (2010) investigated the effect of varying gassing rates on the biological performance, flux and volatile fatty acids (VFAs) retention in a 3 l submerged anaerobic membrane bioreactor (SAMBR). They found that the gel/cake layer present under low gassing conditions enhanced the retention of high molecular weight (MW) residual COD and the electrostatic exclusion of acetic acid resulting in a higher COD inside reactor than with high gassing rates and higher COD removal. The fluxes were similar at high and low gas rates as the same type of dynamic equilibrium was established on the membrane surface at a constant transmembrane pressure (TMP) of 0.2 bars.

Most commercial membranes are made from hydrophobic polymers like polysulfone, polyethersulfone, polypropylene, polyethylene, and polyvinylidenefluoride (PVDF) due to their excellent chemical resistance and thermal and mechanical properties. Hydrophobic membranes, however, are easily susceptible to fouling, while membranes with hydrophilic surfaces are less susceptible to fouling which is often reversible. Thus, the ideal membrane combines superior bulk properties of hydrophobic polymers with the surface chemistry of hydrophilic materials. Over the last few years researchers have devised various strategies to reduce fouling of membrane surfaces. These methods include chemical, low-temperature plasma, and photochemical techniques (Hilal *et al.* 2005). However, most of the modifications have been assessed using aerobic rather than anaerobic cultures. Nevertheless, a lot of this knowledge can be extrapolated to anaerobic systems as in the main they can be assumed to be similar, although no work has yet been done on comparing the similarities and differences between the behaviour of aerobic and anaerobic cultures in MBRs.

Surface modification of membranes can change the properties of the membrane while preserving its macroporous structure, and there are two main methods: coating and grafting. Coating involves dipping the membrane into a solution containing the polymer(s) bearing the antifouling property, thereby coating its surface. Grafting covalently immobilises hydrophilic species onto the membranes from solutions, and an in-depth review of this area can be seen in Hilal *et al.* (2005). Sainbayar *et al.* (2001) modified a polypropylene membrane with ozone and graft polymerisation, and used these membranes in an anaerobic MBR. This treatment resulted in only a 13.5% increase in flux over a virgin membrane, and flux enhancement depended on the degree of grafting. Liu *et al.* (2010) modified a membrane made of chitosan/cellulose acetate blend with heparin or quaternary ammonium, or immobilised silver ions on it. The modifications either increased the surface hydrophilicity, altered the surface charge, or endowed the surface with an anti-bacterial function. They showed that the hydrophilicity of a membrane, although important, is not an effective indicator of the tendency of membrane biofouling with aerobic organisms, while the nature of the surface charges plays a very important role. It was found that the anti-adhesion modifications were more effective than the anti-bacterial one which tries to kill bacteria already attached to the membrane surface. However, the best approach to minimising membrane biofouling was to modify membrane surfaces so that they have effective anti-adhesion property plus additional anti-bacteria function, with the latter as a safe guard in case some bacteria do attach to the surface of a membrane. Li *et al.* (2010) surface-modified polypropylene microporous membranes (PPHFMMs) by the sequential photoinduced graft polymerisation of acrylic acid and acrylamide. The grafting density and the grafting chain length, which played important roles in the antifouling characteristics, were controlled in the first step and the second step, respectively. Antifouling characteristics of the modified membranes were evaluated in an aerobic submerged membrane bioreactor (SMBR), and showed better filtration performances than the unmodified membrane, and the acrylic acid grafted membranes were better than the acrylamide modified membranes. Finally, Gu *et al.* (2009) modified the surface of polypropylene membranes by the sequential photoinduced graft polymerisation of D-gluconamidoethyl methacrylate (GAMA), and the effect of the grafting chain length on the antifouling characteristics in an aerobic submerged membrane bioreactor was investigated. After continuous operation in the bioreactor for about 70 h, reduction from pure water flux was 90.7% for the virgin PPHFMM, and ranged from 80.8 to 87.2% for the modified membranes, increasing with increasing chain length. The results demonstrated that the antifouling characteristics for the glucopolymer-modified membranes were improved with an increase in GAMA chain length. Hence, in summary it appears that surface modification can enhance fluxes slightly, however, whether this gain in flux is economic is unclear.

5. Membrane Fouling and Flux

One of the most critical areas of operation, fouling reduces membrane flux and hence increases the area of membranes per volume of reactor, and therefore the capital cost. Hence there has been a large amount of work in this area over the years, most of it again related to aerobic MBRs, and the following articles have reviewed this work (Chang *et al.* 2002; Liao *et al.* 2004; Le-Clech *et al.* 2006; Drews *et al.* 2006; Meng *et al.* 2009). Fouling is an extremely complex phenomenon, and while probably very similar to aerobic reactors, a lot less is known about anaerobic fouling. Fouling is caused by a combination of components in the reactor (soluble organics, colloidal particles from the feed and cell lysis, and inorganic precipitates), and these in turn are influenced by a range of parameters such as the composition of the biological system, membrane type, hydrodynamic conditions, reactor operating conditions, process performance targets, and the chemical system. These parameters are summarised in Fig. 7.2, and the fouling components as well as some of these parameters will be discussed in more detail below. Figure 7.3 (Meng *et al.* 2009) is a pictorial summary of fouling mechanisms in biological membrane reactors.

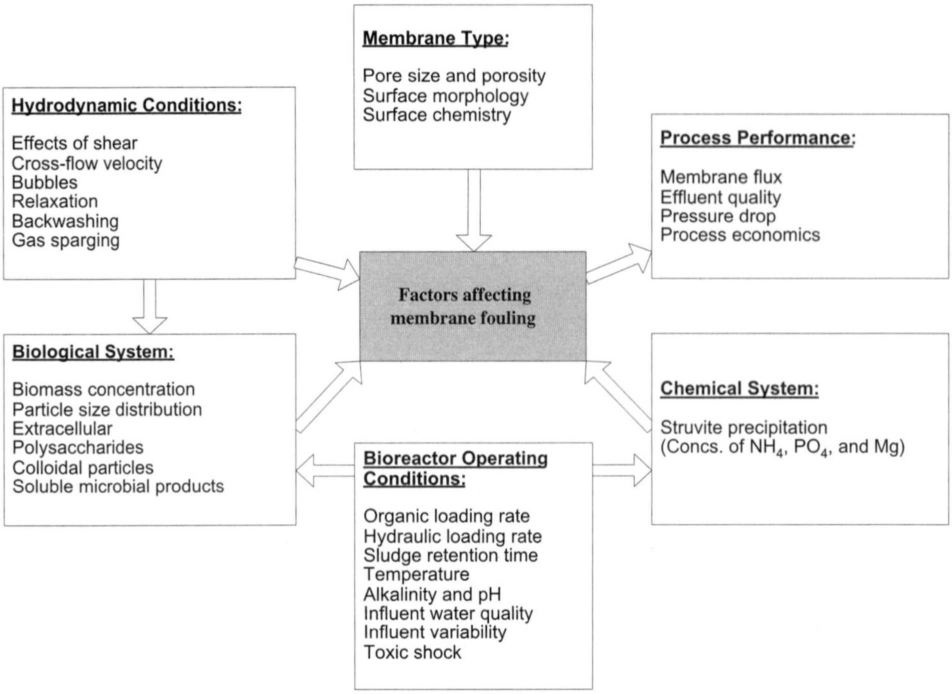

Fig. 7.2. The parameters and interactions that influence process performance in an anaerobic membrane bioreactor (Semmens 2003).

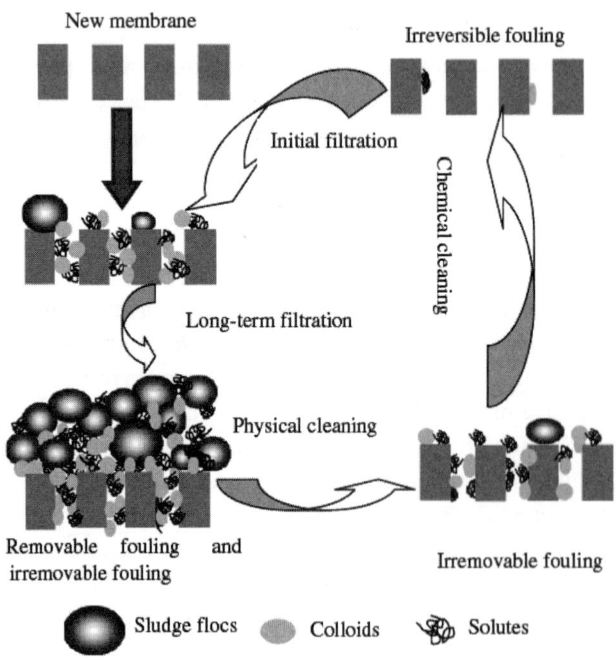

Fig. 7.3. Schematic illustration of the formation and removal of removable and irremovable fouling in MBRs (from Meng *et al.* 2009).

5.1. *Soluble organics*

These are primarily from the high MW fraction which are generated as either soluble microbial products (SMPs) from cell metabolism and lysis (Barker and Stuckey 1999), or as extracellular polysaccharides (ECP) which in anaerobic systems only constitute around 20% of the SMPs (Aquino and Stuckey 2004a). SMPs are critical in membrane fouling but until recently not much was known about their composition or what parameters influence their production in the reactor. In anaerobic MBRs with high solids retention times (HRTs) the production of SMPs from cell lysis (biomass-associated products — BAPs) is high, and much of the soluble COD in the reactor is SMPs (including ECPs — Vyrides and Stuckey 2009b). In addition, the way the reactor is operated influences SMP production considerably. Aquino and Stuckey (2003) showed that nutrient deficiencies in a CSTR increase SMP production, while the presence of toxic compounds in the waste being treated also enhances their concentration (Aquino and Stuckey 2004a). Finally, if the reactor is subjected to organic or shock loads then SMP production increases (Aquino and Stuckey 2004b). It has also been shown in anaerobic CSTRs that even the type of feed can alter the amount of SMPs produced per mole of carbon, and at SRTs of 20 days from 2% to 17% of the feed COD can be converted into SMPs (Barker and Stuckey 2001). Hence it appears that under any type of stress SMP production can increase dramatically in an anaerobic culture which in turn can lead to increases in fouling. The composition of these SMPs is extremely complex (Barker *et al.* 1999), but Aquino and Stuckey (2006a, b) have analysed some of the low MW SMPs and these are often substituted ring compounds, even when the culture is fed simple carbohydrate substrates. Modelling of the production of SMPs under a range of conditions, even during transient shocks, is now possible and recent models allow us to predict SMP concentration under a range of conditions, and hence their possible influence on fouling (Barker and Stuckey 2001; Aquino and Stuckey 2008). However, in most cases the high MW solutes are the ones causing membrane fouling, and the composition of these, other than the generic class of compounds called ECPs, is still unclear but would be partly cell lysis products which are not easily degradable. Cho and Fane (2002) and Okamura *et al.* (2009) also found that increasing membrane resistance was linked to higher levels of extracellular polymeric substance (EPS).

In contrast to the view that all ECP/EPSs and SMPs cause fouling in membrane reactors, Drews *et al.* (2006, 2008) concluded in lab-, pilot-, and full-scale aerobic reactors that no correlation between polysaccharides, measured as sugars, and membrane fouling was observed in many cases. Nevertheless, they did find fouling was related to SMP using membranes with larger pore sizes and at short sludge ages (SRTs). However, it is clear from the paragraph above that SMPs/ECPs are not just polysaccharides, and these authors use these terms loosely and interchangeably. In addition, Kimura *et al.* (2005) also measured membrane flux, but the relationship between aerobic and anaerobic membrane behaviour, as discussed before, is still unclear. Hence basing membrane fouling on measuring just polysaccharides will not give an accurate picture of the role of soluble organics in fouling membranes, and the lack of information about anaerobic rather than aerobic systems is more acute.

5.2. *Colloidal particles*

The effect of fine colloids (0.05–0.4 μm) on MF and UF membrane fouling has been understood for quite a long time. Choo and Lee (1996) were the first to show their effect in pumped external module anaerobic systems where the high shear led to an increase in the number of colloids and pore clogging. Colloids close to the size of membrane pores seem to cause the most problems since they block the pores of the UF/MF membrane. However, there has been very little information on submerged anaerobic membrane reactors until recently due to the lack of experience with this mode of operation. Akram and Stuckey (2008a) investigated the start-up of a submerged anaerobic membrane system and found that the low flux (2 LMH) during high strength treatment was attributed to fine colloids, and high amounts of SMP. Teychene *et al.* (2008) showed that in an aerobic system the fouling layer was controlled by the deposition of SMPs while colloids played a less significant role.

5.3. *Inorganic fouling*

The role of inorganic fouling in AnMBRs depends strongly on the chemical composition of the wastewater treated. The common precipitate is struvite ($MgNH_4PO_4{\cdot}6H_2O$ — Doyle and Parsons 2002), but also other phosphate and calcium salts can precipitate, especially on inorganic membranes (Ognier *et al.* 2002; Kang *et al.* 2002). However, fouling via inorganic or organic entities is not totally independent, Wang *et al.* (2008) using scanning electron microscopy (SEM) and energy-dispersive X-ray (EDX) analysis found that membrane surfaces were covered with a compact gel layer which was formed by organic substances and inorganic elements such as Mg, Al, Fe, Ca, and Si. The organic foulants coupled with inorganic precipitation and enhanced the formation of a gel layer and thus caused membrane fouling in the MBR, and this was shown by Costa *et al.* (2006) in ultrafiltration. Hence the separation between organic and inorganic fouling is somewhat simplistic, and concentration polarisation will lead to higher concentrations of both organic and inorganic ions at the membrane surface. In anaerobic as opposed to aerobic systems, the concentration of both ammonia and carbonate ions is also so much higher due to higher loads and protein hydrolysis, and the chemistry of carbon dioxide equilibrium, so precipitation with these ions is more likely.

Trzcinski (2009) used SEM and EDX analysis to investigate the morphology and structure of the inorganic precipitate found on a submerged membrane in an AnMBR used to treat landfill leachate. The crystals contained calcium, oxygen, and phosphorus, and the intersecting plates were characterized by significantly higher amounts of phosphorus compared to the background, and calcite was not considered due to the high phosphorus content (Fig. 7.4). The high amount of phosphorus suggests that it could either be dicalcium phosphate dihydrate ($CaHPO_4{\cdot}2H_2O$) or amorphous calcium phosphate ($Ca_3(PO_4)_2$). Doyle and Parsons (2002) observed that calcium ion can affect struvite crystal growth and leads to the formation of an amorphous calcium phosphate. Therefore, it is very likely that struvite precipitation did not occur because the available phosphates precipitated with calcium on the membrane keeping the solution undersaturated with respect to struvite.

5.4. *Measures to manage membrane fouling*

There are three main ways of managing fouling in an AnMBR: (a) operating at high fluxes for short periods of time and then relaxing/backflushing and cleaning the membranes with aggressive acids/bases/oxidants; (b) operating at below "critical flux" levels and only relaxing/backflushing and cleaning occasionally; or (c) trying to reduce the degree of fouling by operating the reactor in

Fig. 7.4. Intersecting plate-shaped precipitate on the membrane in the AnMBR (left), and its EDX analysis (right) (Trzcinski 2009).

such a way to minimise SMP/colloid production, using hydrodynamics in the reactor to minimise the fouling layer, or treating the reactor contents to remove the primary foulants. The first two ways have been covered extensively by a number of recent reviews (Gander *et al.* 2000; Liao *et al.* 2004, 2006; Berube *et al.* 2006; Meng *et al.* 2009), so the third way of managing flux will be focussed upon here.

Theoretically, it should be possible to operate an AnMBR in such a way as to reduce the production of colloids and SMPs to minimise fouling since there are some correlations in the literature between these parameters and operating conditions such as HRT, SRT, shock loads, nutrient deficiencies, and the presence of toxins; this evidence has been detailed above. Ng *et al.* (2006) operated four aerobic/anoxic reactors with the same feed but different mean cell retention times (MCRT-3, 5, 10, and 20 days), and found that fouling was not controlled by either MLSS or the zeta potential of the biomass, but by increasing soluble microbial products and ECPs, which both increased with decreasing MCRTs. Hence, the strategy of operating with high MCRTs appears to be sensible provided this does not lead to poorer effluent quality due to the increase in SMPs (Barker and Stuckey 1999). However, despite this paper few researchers have tried to explore this area, probably because there are many constraints on operating parameters that override the importance of membrane fouling, e.g., meeting effluent standards, running at low HRTs. Hence this strategy only remains a potential method of fouling management.

Hydrodynamics can be used in a number of ways to reduce fouling. Firstly in side-stream membranes the flowrate (cross-flow velocity) can be increased to reduce the fouling layer and hence increase the flux; however, Gander *et al.* (2000) amongst others have shown that the energy cost of side-stream membrane reactors is considerably higher than submerged units. The second method is by increasing the gas flow (biogas) under the membranes, thereby increasing bubble shear across the membrane and enhancing the membrane flux. This has been shown by many workers in aerobic systems (Psoch and Schiewer 2006; Phattaranawik *et al.* 2007) both in submerged and side-stream membranes, however, there has been little work in anaerobic reactors. Akram (2007) found that lowering the gassing rate can lead to enhanced removal of soluble COD by increasing the gel layer on the surface through reduced turbulence; this was also found to result in the significant rejection of low MW solutes such as acetic acid. Vyrides and Stuckey (2009a) found that in order to reduce energy consumption the gas sparging could be cycled between 5 minutes off and 10 minutes on — this in turn resulted in enhanced COD removal due to the increased thickness of the fouling layer, but slightly lower fluxes. Hence there appears to be a trade-off between membrane flux and effluent quality.

Another way of enhancing membrane flux is by trying to adsorb/precipitate some of the key foulants from solution using either activated carbon, or cationic polymers, biopolymers, EDTA, or metal salts. Park *et al.* (1999) were the first to evaluate the addition of powdered activated carbon (PAC) to anaerobic MBRs, and found that addition of up to $5\,g{\cdot}l^{-1}$ enhanced both flux and COD removal. However, some of their experiments were in stirred cells, and when added to the side-stream membrane reactor, they only followed flux decline over 10 days — the key question is whether PAC would simply saturate and then become ineffective, or whether it would be bioregenerated and continue to be effective over a longer time period. Hu and Stuckey (2007) were the first to address both of these questions, operating a submerged membrane MBR (SAMBR) for a period of three months on a synthetic sewage ($460\,mg{\cdot}l^{-1}$) with both PAC, granular activated carbon (GAC), and a control. The average COD removal with PAC increased by 22.4% compared with the control, while with GAC it was not significantly better; this effect was presumed to be due to PAC having a significantly greater surface area per mass than GAC, and hence absorbing more fine colloidal particles and high molecular weight organics. These effects manifested themselves by the PAC reactor having considerably lower TMPs and higher fluxes than both the GAC and control reactor. This effect persisted for over three months suggesting that the absorbed solutes were being degraded over time regenerating the surface for new solutes to adsorb. Akram and Stuckey (2008b) used PAC with a higher synthetic feed strength ($4\,g{\cdot}l^{-1}$) and an HRT as low as 6 hours to achieve COD removals as high as 98%. Size exclusion chromatography showed that the membrane acted as an ultrafilter ($<30\,kDa$) as the cake or gel layer significantly reduced the amount of organics passing through the membrane. Particle size distributions demonstrated that PAC also adsorbed fine colloids, and PAC addition resulted in significant flux improvement from 2 to 9 LMH in the presence of $1.67\,g{\cdot}l^{-1}$ PAC. However, addition of $3.4\,g{\cdot}l^{-1}$ reduced the flux to 5 LMH due partly to enhanced viscosity. Interestingly, Trzcinski (2009) found using identical reactors but feeding OFMSW leachate that PAC made a significant difference in the first 3 days, but then flux and COD removal returned to earlier levels, presumably because the organics adsorbed onto the PAC were not easily biologically degradable, e.g., humic acids.

Koseoglu *et al.* (2008) tried a different approach by evaluating the effect of adding various chemicals on filterability and fouling reduction in aerobic MBR mixed liquors; three cationic polymers, a biopolymer, a starch, and two metal salts were assessed. All tested additives were able to remove SMP, but to varying extents; 33%, 45%, 51%, 36%, 38%, 54%, and 56% for the three cationic polymers, $FeCl_3$, PACl (alum polymer), chitosan, and starch (Mylbond 168), respectively. All additives significantly reduced fouling rates and increased permeability values; at their optimum dosages, the cationic polymers MPE50, MPL30, and KD452 provided 96%, 80% and 74% reductions in fouling rates, respectively. Critical flux enhancements achieved by MPL30, MPE50, KD452, FeC13, PACl, chitosan, and starch were 38%, 46%, 38%, 14%, 14%, 0%, and 22% in comparison with raw mixed liquor. Cationic polymers increased critical flux values to levels above 50 LMH, however, SMP removal from MBR mixed liquors and further improvement in filtration performance did not always correlate. Wu *et al.* (2009) also evaluated PAC, zeolite, or coagulants (polyamide, polyaluminium chloride) on an anaerobic effluent to reduce fouling before a microfiltration step. Polyaluminium chloride ($10\,mg{\cdot}l^{-1}$) was found to be the most effective among the adsorbents and coagulants tested.

Okamura *et al.* (2009) found in aerobic submerged MBRs that the primary biofoulant was SMP which included soluble EPS, and that filtration resistance directly correlated with the EPS concentration. When the microfiltration liquid contained over 80 mg·l^{-1} of EPS, a gel accumulated on the membrane containing EPS and uronic acid concentrations 50 times higher than the original liquid, with only traces in the filtrate. The gel contained high molecular weight solutes (106–108 Da), suggesting the presence of polysaccharides which were much smaller than the pore size of the membrane. Interestingly, a strong complexing agent, ethylenediaminetetraacetic acid (EDTA), decreased filtration resistance suggesting that polysaccharides containing uronic acid units could undergo intermolecular or intramolecular ionic cross-linking by polyvalent cations and form the gel, thus clogging the membrane pores.

Specific inorganic contaminants can be removed in the reactor itself. Kim *et al.* (2007) controlled the formation of struvite by passing the reactor contents through a dialyzer, and the permeate was then passed through a fluidised zeolite column to remove the ammonia. Interestingly, using both organic and inorganic membranes the authors noted that the flux with the inorganic one improved by 15–20%, while the organic one showed no significant difference. However, whether the additional capital cost offsets the improved flux over time can only be determined using an economic analysis. In addition, Chesters (2009) reported on the use of a new anti-scalant (Genesys PHO: 2–5 mg·l^{-1}) to control calcium phosphate scale by inhibition so that the solubility of calcium phosphate is raised 150 times. This may be more cost-effective than the previous method.

Finally, fouling could be partially controlled by the addition of non-degradable particles with low specific gravity which physically scour the membrane surface. Akram (2007) added ion exchange resins (IX) to both buffer VFA peaks during organic shock loads, and enhance flux in a submerged anaerobic MBR (SAMBR). The effect of higher organic shock loads (20 kg-COD·m^{-3}·d^{-1}) in the presence of various quantities (1.67 g·l^{-1} and 5 g·l^{-1}) of IX resins were studied at 20 hours HRT. The VFAs level increased significantly during an organic shock in the presence of 1.67 g·l^{-1} of Amberlite IRC-50 due to consecutive stresses, however, the presence of IX resins arrested pH fluctuations and improved the stability of the SAMBR. Long acclimatisation time and sufficient quantities of IX resins (5 g·l^{-1}) improved stability due to higher buffer capacity while producing less VFAs and SMP during shock period. The rigid Amberlite IRC-50 particles scoured the cake layer and the flux improved from 1.5 (for 1.67 g·l^{-1} IX resins) to 4 LMH (3.4 g·l^{-1} of resins). The phenomenon of physical scouring was verified using rigid glass beads (3.4 g·l^{-1}) that doubled the flux to 3 LMH in three weeks.

6. Effect of Operating Parameters

How the AnMBR is operated influences its performance, and the key parameters are temperature, hydrodynamics (scouring gas flow-covered above), HRT, SRT, and organic loading rate which in turn influences the microbial ecology.

6.1. *Operating temperature*

Normally, anaerobic processes are operated at around their optimum mesophilic temperature (35 °C) due to their slow growth rate. However, there are good reasons why it would be helpful to be able to operate at higher temperatures, i.e., thermophilic (55 °C) with hot industrial wastewaters, e.g., palm oil mill effluent (POME), due to the higher rates of reaction, and the viscosity of the biomass suspension being lower leading to higher fluxes (Ross *et al.* 1990). In addition, at lower temperatures, i.e., psychrophilic (10 °C), the energy required for domestic wastewater treatment in temperate climates would be a lot lower, and anaerobic digestion could be operated at ambient temperatures all year round. Nevertheless, there may be issues about the loss of methane in the effluent at low temperatures due to its enhanced solubility. At both temperature extremes membranes can stop the washout of slow growing species and enhance performance and stability of anaerobic digestion. Until recently there was very little information available about any temperatures other than mesophilic. Murata *et al.* (1994) treating sewage using an external membrane unit coupled with a CSTR obtained a VSS reduction rate at 55 °C of 67–78%, but the difference between 35 °C and 55 °C was not large in relation to the heating energy. At 35 °C, VSS reductions were 51–79% at a VSS volumetric load of 0.93–1.94 $kg{\cdot}m^{-3}{\cdot}d^{-1}$, an HRT of 4.2–8.4 d and a SRT of 42–335 d; this compares with conventional AD at an HRT of 15–30 d. Decreasing the temperature from 30 °C to 25 °C did not greatly reduce the VSS reduction rate; a rate from 67% to 70% was maintained with a VSS volumetric load of about 1 $kg{\cdot}m^{-3}{\cdot}d^{-1}$.

Jeison and van Lier (2007, 2008a) operated a single submerged tubular polysulphone membrane in two thermophilic reactors fed on a VFA feed or a glucose/VFA feed with OLRs up to 15 and 20 $kg{\cdot}m^{-3}{\cdot}d^{-1}$, respectively. The degree of acidification significantly affected the physical properties of the sludge, greatly affecting the attainable flux (~7 LMH), and due to high TMPs when the critical flux was exceeded, restricted the flux to close to the critical flux. Since reversible cake formation was identified as the limiting factor for the operational flux, increasing surface shear should result in higher fluxes. Preliminary tests over a short time showed that cross-flow operation may be a feasible alternative to reduce particle deposition, with fluxes increasing by an order of magnitude, although with concordant energy costs. This will be the case only if the high shear stress does not negatively affect the physical properties of the sludge, i.e., inducing a decrease in the particle size. Finally, they carried out a crude economic analysis costing membranes from €50–150$\cdot m^{-2}$, but assumed a membrane lifetime of only three years — based on current information this is very pessimistic, and lifetimes of 5–7 years could now be expected. This analysis showed that membrane costs were still the largest component compared to energy costs. Recently, Kanai *et al.* (2010) briefly discussed the application of flat sheet Kubota membranes to thermophilic industrial wastewaters with high ammonia levels (3720 $mg\text{-}TKN{\cdot}l^{-1}$) with COD removals of between 75% and 92%. Unfortunately, the paper contained no details on HRT, OLR, or even fluxes.

Working at low temperatures has been the goal of many workers in the area of AD, and there have been a number of reviews (Kashyap *et al.* 2003; Lettinga *et al.* 2001). The issue at low temperatures is that the kinetic parameters such as reaction and hydrolysis rate decrease (Veeken and Hamelers 1999), and the half-rate constant for methanogens seems to increase (Speece 1996); this leads to lower COD removals, and higher effluent VFA levels. Despite attempts to select for psychrophilic methanogens in reactors, even after many months the organisms present are still mesophilic (Akram 2007), although the existence of psychrophilic methanogens is now well documented (Cavicchioli 2006). Akram (2007) operated lab-scale 3 l SAMBRs at 35 °C, 25 °C and 15 °C on a synthetic feed of 4 $kg\text{-}COD{\cdot}m^{-3}$ and 20 h HRT (OLR-4.8 $kg\text{-}COD{\cdot}m^{-3}{\cdot}d^{-1}$). At 25 °C under a high sparging rate (5 LPM) COD removal was 98%. The methanogenic activity of the 15 °C inoculum at 15 °C and 35 °C showed an increased contribution of hydrogenotrophic methanogenesis, and although the biomass was still mesophilic it became psychrotolerant. Three weeks after the change to 15 °C, the persistence of acetate and propionate at high levels at steady state showed an increase in Ks for methanogens and acetogens, and reflected a new steady-state dynamics within the microbial population. This phenomenon of high levels of unmetabolised VFAs at low temperatures, even after 300 days, has been observed before for anaerobic filters (Matsushige *et al.* 1990). It seems that the propionate-degrading bacteria are more affected by temperature than the acetate-degrading archea in the long term (Langenhoff and Stuckey 2000). Propionate degradation has also been identified as the rate-limiting step in psychrophilic anaerobic digestion (Collins *et al.* 2003), particularly during acetate build-up. As effluent COD was primarily VFAs, their accumulation led to a drop in performance at 15 °C (COD removal 65%); however, a low sparging rate (2 LPM) increased VFA retention due to the membrane gel layer, and increased COD removal to 91%.

The group of O'Flaherty has worked extensively on psychrophilic methanogenesis, although mainly using UASBs. McKeown *et al.* (2009) evaluated the feasibility of long-term (>3 years), low-temperature (4–15 °C) operation of an expanded granular sludge bed-anaerobic filter treating a 3.75–10 kg-COD·m^{-3} VFA-based wastewater. Despite transient perturbations, COD removal and methane concentrations exceeded 80% and 65%, respectively, at an OLR of 10 kg-COD·m^{-3}·d^{-1}. Over time 50% of the granules disintegrated leading to an OLR reduction to 3.75–5.0 kg-COD·m^{-3}·d^{-1}. From then on, stable and efficient performance was observed at 4–10 °C, indicating the adaptation of the mesophilic inoculum to psychrophilic operating conditions. Physiological activity assays indicated the development of psychroactive syntrophic and methanogenic populations, including psychrophilic propionate- oxidising and hydrogenotrophic methanogenic activity. The data suggest that mesophilic inocula can physiologically adapt to sub-optimal operational temperatures: treatment efficiencies and sludge loading rates at 4 °C (day 1,243) were comparable to those achieved at 15 °C (day 0). Furthermore, long-term, low-temperature bioreactor operation may act as a selective enrichment for psychrophilic methanogenic activity from mesophilic inocula. Madden *et al.* (2010) investigated the stability of low-temperature UASBs undergoing transient perturbations using COD removal, while microbial ecology was monitored using DGGE. Gene fingerprinting data indicated that *Methanosaeta* sp. were the persistent, keystone members of the archaeal community, and likely were pivotal for the physical stability and maintenance of the granular biofilms. Cluster analyses of denaturing gradient gel electrophoresis (DGGE) data suggested that temporal shifts in microbial community structure were predominantly independent of the applied perturbations. The observed efficient and stable bioprocess performance at low temperatures highlights the potential for long-term, low-temperature bioreactor operation, and the use of membranes can enhance organism retention.

6.2. *Hydraulic retention time and solids retention time*

These two parameters are key in optimising the performance of an AnMBR — low HRTs mean smaller reactors and lower costs, while the long SRTs potentially available for MBRs mean very low sludge yields for disposal. In an extensive summary of the literature to date, Liao *et al.* (2006) found that the lowest HRTs for a wide range of wastewaters was 8–12 hours. However, Wen *et al.* (1999) found that with sewage and a submerged membrane that an HRT as low as 4 hours was possible with 97% COD removal, even at temperatures down to 13 °C, with almost complete methane recovery in the gas phase, contrary to some expectations. Hu and Stuckey (2006) used laboratory-scale SAMBRs to evaluate their ability to treat dilute wastewaters

(460 mg-COD·l^{-1}), and found COD removals of >90% soluble COD at HRTs as low as 3 h. MW analysis using HPLC showed that the membrane filtered out most of the high molecular weight soluble organics, resulting in high COD removals. This appears to be the lowest recorded HRT using AnMBRs, and demonstrates that good removals are possible at low HRTs. SRTs in AnMBRs can be as low as 25 d and as high as 335 d (Liao *et al.* 2006; Murata *et al.* 1994), and this can be controlled fairly easily with sludge wastage rates. Clearly at long SRTs the sludge will be minimal and very non-biodegradable, however, one factor which may reduce the attractiveness of long SRTs is the fact that biomass-associated products (BAPs), which form part of SMPs in the digester, tend to increase with SRTs, and this tends to increase effluent COD (Barker and Stuckey 1999).

6.3. *Microbial ecology of anaerobic membrane reactors*

Despite interesting questions about the influence of infinite SRT on microbial ecology due to membrane retention, there are very few studies in this area. In addition, it is not clear whether the microbial ecology has any influence on flux through SMP production and membrane fouling, and none of the recent comprehensive reviews mention microbial ecology, although Briones and Raskin (2003) have reviewed the ecology of anaerobic systems. Kozariszczuk *et al.* (2000) in a brief paper used fluorescent *in situ* hybridisation (FISH) probes to examine the relationship between reactor performance and microbial ecology. Padmasiri *et al.* (2007) examined the microbial ecology of a 6 l external membrane reactor treating swine manure over 300 days. The reactor was successful started up at an OLR of 1 kg-VS·l^{-1}·d^{-1}. After a twofold increase in OLR followed by twofold increase in flow velocity through the membrane module on day 75, the performance of the AnMBR. The methanogenic population dynamics in the reactor were monitored with terminal restriction fragment length polymorphism (T-RFLP). Changes in the relative levels of *Methanosarcinaceae* and *Methanosaetaceae* were consistent with changes in VFA concentrations, i.e., high and low levels of acetate corresponded to a high abundance of *Methanosarcinaceae* and *Methanosaetaceae*, respectively. The levels of hydrogenotrophic methanogens (*Methanomicrobiales*) increased during decreased reactor performance suggesting that syntrophic interactions involving hydrogenotrophic methanogens remained intact regardless of the degree of shear in the AnMBR.

Yeon *et al.* (2009) examined the role of bacterial quorum sensing (QS) in aerobic membrane fouling. Bacteria regulate specific group behaviours such as biofilm formation in response to population density using autoinducers. In this study, the concept of bacterial QS was applied to MBRs as a new biofouling control paradigm. The research was conducted in three phases: (1) measuring autoinducer signals in MBRs, (2) correlate QS activity and membrane biofouling, and (3) apply QS-based membrane biofouling control. A bioassay with *Agrobacterium tumefaciens* reporter strain proved that *N*-acyl homoserine lactone (AHL) autoinducers were produced in the MBR. Furthermore, thin layer chromatographic analysis identified at least three different AHLs in the biocake, of which *N*-octanoyl-homoserine lactone was the most abundant. During continuous MBR operation, the fouling layer showed strong AHL activity simultaneously with an abrupt increase in the transmembrane pressure implying that QS is closely associated with membrane biofouling. Porcine kidney acylase I, which can inactivate the AHL molecule, was shown to prevent biofouling by quenching AHL autoinducers. Hence it appears that QS could be a novel target for biofouling control in anaerobic MBRs.

Finally, Trzcinski *et al.* (2010) used a novel three-stage membrane process to achieve 75% VS removal at 35 °C of the organic fraction of municipal solid waste (OFMSW) at an OLR of 4 kg-VS·m^{-3}·d^{-1}, an SRT of 66 days, and an HRT of 20 days. The process consisted of an anaerobic hydrolytic reactor (HR) where the solids and liquid fractions were separated with a mesh. The leachate was pumped to a submerged anaerobic membrane bioreactor (SAMBR) and the treated permeate was polished in an aerobic membrane bioreactor (AMBR). denaturing gradient gel electrophoresis (DGGE) and DNA sequencing analyses indicated that the increase in methane content in the HR caused by the excess sludge recycle from the SAMBR was associated with an increase in the number of hydrogenotrophic species, mainly *Methanobrevibacter* sp., *Methanobacterium formicicum*, and *Methanosarcina* sp. At 20 °C VS removal dropped to 50% in the HR, and some DGGE bands disappeared compared to 35 °C samples, while some (*Ruminococcus flavefaciens*) reduced in intensity. The species associated with the COD-polishing properties of the AMBR correspond to the genera *Pseudomonas*, *Hyphomonas* and *Hyphomicrobiaceae*. These results highlight the positive effect of recycling the excess sludge from the SAMBR to re-inoculate the HR with hydrogenotrophic species.

7. Application to Synthetic and Industrial Wastewaters

It is beyond the scope of this chapter to review all the papers covering the use of AnMBRs with both synthetic and industrial wastewaters as there is extensive coverage of the literature up to 2006 in Liao *et al.* (2006). However, one of the most interesting recent developments made possible by the use of membranes is the potential incorporation of nitrogen removal in AnMBRs using the Anammox process. This process takes nitrite and ammonia in wastewater, reducing and oxidising them, respectively, to nitrogen gas (Jetten *et al.* 2009). While detected in many treatment processes, their energy yields and hence growth rates are so low they cannot be maintained at high enough concentrations to be effective. With the use of AnMBRs their potential is now available, and there have been a number of recent publications. van der Star *et al.* (2008) cultivated high concentrations of these organisms in an MBR successfully, while Wang *et al.* (2009) starting with an activated sludge seed also obtained a climax community with Anammox bacteria as the dominant population. Feng *et al.* (2008) partially nitrified a synthetic inorganic ammonia wastewater in an aerated biofilm reactor (NH_4^+/NO_2^- ratios of 1:1 to 1:1.3) to act as a feed to an Anammox process. However, this approach may be problematic with high COD wastewaters due to their oxygen demand, and the competition from heterotrophic organisms. Kappell *et al.* (2005) achieved >90% COD removal and >95% nitrogen removal with an aerated membrane nitrifying biofilm in an anaerobic CSTR, Hu *et al.* (2009) installed aerated membranes into an anaerobic baffled reactor (ABR), to achieve simultaneous removal of nitrogen and COD. Total VFA and COD concentrations in the effluent decreased by 68.1% and 59.5% respectively, with increased nitrogenous pollutant remove efficiency by 83.5%, at influent COD concentration of 1,600 mg·l^{-1} and NH_4^+-N concentration of 80 mg·l^{-1}.

Since the review of Liao *et al.* in 2006 there have been a number of other interesting studies on AnMBRs, including: the treatment of high strength Fischer–Tropsch process wastewater (van Zyl *et al.* 2008); Saddoud and Sayadi (2007) pre-acidified a difficult slaughterhouse WW before treating with an external module AnMBR; Zhang *et al.* (2008) treated palm oil mill effluent (POME) using an EGSB and aerobic biodegradation as the first stage, and UF and RO units for the second stage removing almost all the COD, solids and colour; and Lee *et al.* (2009)

produced hydrogen in a submerged AnMBR and compared it with a CSTR — the MBR's H_2 production rate was 2.43–2.56 $l\text{-}H_2 \cdot l^{-1} \cdot d^{-1}$, which was about 2.6 times higher than that of the CSTR, reflecting the MBR's higher H_2 productivity. Finally, Bijmans *et al.* (2009) used formate in a high-rate AnMBR to reduce sulfate to hydrogen sulfide.

In the area of sewage treatment, Ho *et al.* (2007) investigated a non-woven fabric filter and a poly-tetrafluoroethylene (PTFE) composite membrane to determine their applicability to treat low strength wastewater in an AnMBR, while Ho and Sung (2009) investigated a cross-flow AnMBR to treat synthetic municipal wastewater at different HRTs. The AnMBR was operated at COD loading rates of 1–2 $kg\text{-}COD \cdot m^{-3} \cdot d^{-1}$ for 280 days. The permeate COD was always lower than 40 $mg \cdot l^{-1}$, but approximately 30% of COD was not available for methane recovery, irrespective of applied HRTs, as a result of the COD loss by dissolved methane, sulfate reduction, and untreated COD in the permeate. The fraction of methane recovered from the synthetic municipal wastewater decreased from 48% to 35% with a decrease in HRT from 12 to 6 hours. Therefore, AnMBR operation with relatively long HRTs and SRTs may be favourable, to enhance methane recovery and reduce or eliminate sludge production. Lew *et al.* (2009) and An *et al.* (2009) also examined the use of external membrane modules with anaerobic digestion of sewage.

8. Potential Application for Anaerobic Membrane Reactors

It is clear from the review above that anaerobic membrane reactors are receiving increasing attention, although there is considerably less work done than on aerobic MBRs. This in the main has been driven by considerations of energy use, solids disposal, greenhouse gas emissions, and resource recovery. Is the application of AnMBRs limited to specific types of wastewater? Liao *et al.* (2006) in their comprehensive review discuss the role of AnMBRs with wastes varying in both COD and solids concentrations. They conclude that there is considerable opportunity for high strength, high solids wastewaters, while with low strength wastewaters with both high and low solids opportunities should grow (Fig. 7.5). The only wastewaters ruled out at present are the high strength low solids ones where current designs such as the UASB and EGSB provide efficient and economic forms of treatment; these conclusions are also supported by Jeison and Van Lier (2008b). Nevertheless, AnMBRs result in solids-free effluents and with pressures increasing on tighter standards and resource recovery, it may be in the future that even high strength low solids wastewaters will provide opportunities for AnMBRs. In addition, soluble wastes with high ammonia contents, low temperatures, or with inhibitory compounds present may benefit from membranes to enable adapted populations to develop.

9. Research Needs and Conclusions

Based on the selective review of the literature above, it appears that there are a number of key areas that need more work:

While there has been a lot of work carried out on aerobic membrane reactors, and a deepening understanding of the factors controlling membrane fouling, there is considerably less work in the area of anaerobic membrane reactors. How do the key areas relate — can the understanding in aerobic systems, including fouling, be transferred directly to anaerobic systems? Work is needed to do an in-depth analysis of the similarities and differences between aerobic and anaerobic MBR fouling.

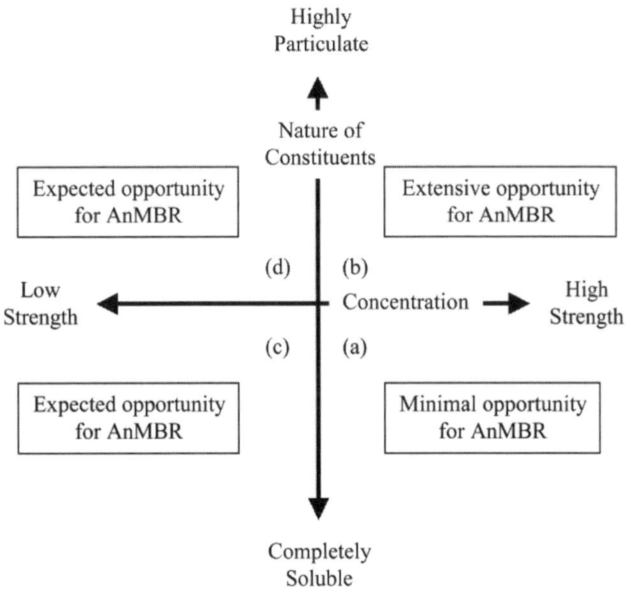

Fig. 7.5. Applicability of AnMBRs to different types of wastewaters (from Liao *et al.* 2006).

While there has now been substantial work on membrane fouling in AnMBRs, there is little understanding of the effect of system operating parameters, HRT, SRT MLVSS, feed composition and strength, shock loads, toxins, shear on the production of colloids, SMPs, and ECPs which are the main causes of fouling in AnMBRs.

While membrane fouling is a key to the economic performance of AnMBRs, its prevention via system additives to ameliorate fouling is still poorly understood. This is a key area where fluxes could be improved at little cost, but more understanding and research is necessary.

One factor holding back the low-temperature treatment of sewage in AnMBRs is the putative loss of methane in the effluent due to its increased solubility. In order to improve the energy balance of the process, more work is needed on factors controlling methane solubility at low temperatures, and whether it is possible to recover the soluble methane using gas membranes either inside the reactor, or on the effluent from the reactor.

Explore the "shear rate dilemma" in more depth. Do side-stream membrane systems hold more promise due to their higher fluxes, or are submerged membranes more cost-effective? Allied to this question is the one about floc size and anaerobic symbiosis — is there a minimum floc size for stable methanogenesis?

Including nitrogen removal in AnMBRs using the Anammox process with little additional cost, e.g., membrane-aerated biofilms, or nitrate addition, and no production of greenhouse gases is potentially promising, but more work is needed to explore the limitations of the process and its costs.

Finally, more work is needed at both laboratory and pilot scales to establish sound numbers for the energy use and solids production of AnMBRs operating on real-world wastewaters, e.g., sewage, so that the benefits of this new technology can be substantiated compared to more established technologies such as activated sludge or even aerobic MBRs.

In conclusion, the last decade has brought increased interest in aerobic MBRs due to concerns about energy, greenhouse gas emissions, and resource recovery. This review has shown considerable research effort directed towards increasing the use on AnMBRs, but many questions remain to be answered to increase their uptake.

References

Akram, A. (2007) *Stability and Performance Improvement of a Submerged Anaerobic Membrane Bioreactor (SAMBR) for Wastewater Treatment.* PhD thesis, Imperial College London.

Akram, A. and Stuckey, D. C. (2008a) Biomass acclimatisation and adaptation during start-up of a submerged anaerobic membrane bioreactor (SAMBR). *Environ. Technol.* **29**(10), 1053–1065.

Akram, A. and Stuckey, D. C. (2008b) Flux and performance improvement in a submerged anaerobic membrane bioreactor (SAMBR) using powdered activated carbon (PAC). *Process Biochem.* **43**(1), 93–102.

Akram, A. and Stuckey, D. C. (2010) Influence of gassing rates on the performance and solute rejection of a submerged anaerobic membrane bioreactor (SAMBR). submitted to *Wat. Res.*

An, Y. Y., Yang, F. L., Bucciali, B., and Wong, F. (2009) Municipal wastewater treatment using a UASB coupled with cross-flow membrane filtration. *J. Environ. Eng. ASCE,* **135**(2), 86–91.

Anderson, G. K., Saw, C. B., and Fernandes, M. I. A. P. (1986a) Applications of porous membranes for biomass retention in biological wastewater treatment process. *Process Biochem.* **21**(6), 174–182.

Anderson, G. K., James, A., Saw, C. B., and Le, M. S. (1986b). Crossflow micro-filtration — A membrane process for biomass retention in anaerobic-digestion. In: *Proceedings Part II of 4th World Filtration Congress*, Ostend, Belgium. 22–25 April, pp. 11–17.

Aquino, S. F. and Stuckey, D. C. (2003) Production of soluble microbial products (SMP) in anaerobic chemostats under nutrient deficiency. *J. Environ Eng. ASCE* **129**(11), 1007–1014.

Aquino, S. F. and Stuckey, D. C. (2004a) Soluble microbial products formation in anaerobic chemostats in the presence of toxic compounds. *Wat. Res.* **38**(2), 255–266.

Aquino, S. F. and Stuckey, D. C. (2004b) The effect of organic and hydraulic shock loads on the production of soluble microbial products in anaerobic digesters. *Wat. Environ. Res.* **76**(7), 2628–2636.

Aquino, S. F. and Stuckey, D. C. (2006a) Chromatographic characterization of dissolved organics in effluents from two anaerobic reactors treating synthetic wastewater. *Wat. Sci. Technol.* **54**(2), 193–198.

Aquino, S. F., Hu, A. Y., Akram, A., and Stuckey, D. C. (2006b) Characterization of dissolved compounds in submerged anaerobic membrane bioreactors (SAMBRs). *J. Chem. Technol. Biotechnol.* **81**(12), 1894–1904.

Aquino, S. F. and Stuckey, D. C. (2008) Integrated model of the production of soluble microbial products (SMP) and extracellular polymeric substances (EPS) in anaerobic chemostats during transient conditions. *Biochem. Eng. J.* **38**(2), 138–146.

Barker, D. J. and Stuckey, D. C. (1999) A review of soluble microbial products (SMP) in wastewater treatment systems. *Wat. Res.* **33**(14), 3063–3082.

Barker, D. J., Mannucchi, G. A., Salvi, S. M. L., and Stuckey, D. C. (1999) Characterisation of soluble residual chemical oxygen demand (COD) in anaerobic wastewater treatment effluents. *Wat. Res.* **33**(11), 2499–2510.

Barker, D. J. and Stuckey, D. C. (2001) Modeling of soluble microbial products in anaerobic digestion: The effect of feed strength and composition. *Wat. Environ. Res.* **73**(2), 173–184.

Berube, P. R., Hall, E. R., and Sutton, P. M. (2006) Parameters governing permeate flux in an anaerobic membrane bioreactor treating low-strength municipal wastewaters: A literature review. *Wat. Environ. Res.* **78**(8), 887–896.

Bijmans, M. F. M., Dopson, M., Peeters, T. W. T., Lens, P. N. L., and Buisman, C. J. N. (2009) Sulfate reduction at pH 5 in a high-rate membrane bioreactor: Reactor performance and microbial community analyses. *J. Microbiol. Biotechnol.* **19**(7), 698–708.

Bohdziewicz, J., Neczaj, E., and Kwarciak, A. (2008) A Landfill leachate treatment by means of anaerobic membrane bioreactor. *Desalination* **221**(1–3), 559–565.

Briones, A. and Raskin, L. (2003) Diversity and dynamics of microbial communities in engineered environments and their implications for process stability. *Curr. Opin. Biotechnol.* **14**(3), 270–276.

Brockmann, M. and Seyfried, C. F. (1996) Sludge activity and cross-flow microfiltration — a non-beneficial relationship. *Wat. Sci. Technol.* **34**(9), 205–213.

Cavicchioli, R. (2006) Cold-adapted archaea. *Nat. Rev. Microbiol.* **4**, 331–343.

Chang, I. S., Bag, S. O., and Lee, C. H. (2001) Effects of membrane fouling on solute rejection during membrane filtration of activated sludge. *Process Biochem.* **36**(8–9), 855–860.

Chang, I. S., Clech, P. L., Jefferson, B., and Judd, S. (2002). Membrane fouling in membrane bioreactors for wastewater treatment. *J. Environ. Eng.* **128**(11), 1018–1029.

Chesters, S. P. (2009) Innovations in the inhibition and cleaning of reverse osmosis membrane scaling and fouling. *Desalination* **238**(1–3), 22–29.

Cho, B. D. and Fane, A. G. (2002) Fouling transients in nominally sub-critical flux operation of a membrane bioreactor. *J. Membr. Sci.* **209**(2), 391–403.

Choo, K. H. and Lee, C. H. (1996) Effect of anaerobic digestion broth composition on membrane permeability. *Wat. Sci. Technol.* **34**(9), 173.

Choo, K. H. and Lee, C. H. (1998) Hydrodynamic behaviour of anaerobic biosolids during crossflow filtration in the membrane anaerobic bioreactor, *Wat. Res.* **32**, 3387–3397.

Collins, G., Woods, A., McHugh, S., Carton, M. W., and O'Flaherty, V. (2003) Microbial community structure and methanogenic activity during start-up of psychrophilic anaerobic digesters treating synthetic industrial wastewaters. *FEMS Microb. Ecol.* **46**, 159–170.

Costa, A. R., de Pinho, M. N., and Elimelech, M. (2006) Mechanisms of colloidal natural organic matter fouling in ultrafiltration. *J. Membr. Sci.* **281**(1–2), 716–725.

Doyle, J. D. and Parsons, S. A. (2002) Struvite formation, control and recovery. *Wat. Res.* **36**(16), 3925–3940.

Drews, A., Lee, C. H., and Kraume, M. (2006) Membrane fouling — a review on the role of EPS. *Desalination* **200**(1–3), 186–188.

Drews, A., Vocks, M., Bracklow, U., Iverson, V., and Kraume, M. (2008) Does fouling in MBRs depend on SMP? *Desalination* **231**(1–3), 141–149.

Elmaleh, S. and Abdelmoumni, L. (1997) Cross-flow filtration of an anaerobic methanogenic suspension. *J. Membr. Sci.* **131**, 261–274.

Feng, Y. J., Tseng, S. K., Hsia, T. H., Ho, C. M., and Chou, W. P. (2008) Aerated membrane-attached biofilm reactor as an effective tool for partial nitrification in pretreatment of anaerobic ammonium oxidation (ANAMMOX) process. *J. Chem. Technol. Biotechnol.* **83**(1), 6–11.

Gander, M., Jefferson, B., and Judd, S. (2000) Aerobic MBRs for domestic wastewater treatment: A review with cost considerations. *Sep. Purif. Technol.* **18**(2), 119–130.

Grethlein, H. E. (1978) Anaerobic digestion and membrane separation of domestic wastewater, *J. Wat. Pollut. Control Fed.* **50**, 754.

Gu, J. S., Yu, H. Y., Huang, L., Tang, Z. Q., Li, W., Zhou, J., Yan, M. G., and Wei, X. W. (2009) Chain-length dependence of the antifouling characteristics of the glycopolymer-modified polypropylene membrane in an SMBR. *J. Membr. Sci.* **326**(1), 145–152.

Hernandez, A. E., Belalcazar, L. C., Rodriguez, M. S., and Giraldo, E. (2002) Retention of granular sludge at high hydraulic loading rates in an anaerobic membrane bioreactor with immersed filtration. *Wat. Sci. Technol.* **45**(10), 169–174.

Hilal, N., Ogunbiyi, O. O., Miles, N. J., and Nigmatullin, R. (2005) Methods employed for control of fouling in MF and UF membranes: A comprehensive review. *Sep. Sci. Technol.,* **40**(10), 1957–2005.

Ho, J. K., Khanal, S. K., and Sung, S. (2007) Anaerobic membrane bioreactor for treatment of synthetic municipal wastewater at ambient temperature. *Wat. Sci. Technol.* **55**(7), 79–86.

Ho, J. H. and Sung, S. W. (2009) Anaerobic membrane bioreactor treatment of synthetic municipal wastewater at ambient temperature. *Wat. Environ. Res.* **81**(9), 922–928.

Hu, A. Y. and Stuckey, D. C. (2006) Treatment of dilute wastewaters using a novel submerged anaerobic membrane bioreactor. *J. Environ Eng. ASCE* **132**(2), 190–198.

Hu, A. Y. and Stuckey, D. C. (2007) Activated carbon addition to a submerged anaerobic membrane bioreactor: Effect on performance, transmembrane pressure, and flux. *J. Environ Eng. ASCE* **133**(1), 73–80.

Hu, S. W., Yang, F., Liu, S. T., and Yu, L. Q. (2009) The development of a novel hybrid aerating membrane-anaerobic baffled reactor for the simultaneous nitrogen and organic carbon removal from wastewater. *Wat. Res.* **43**(2), 381–388.

Jeison, D. and van Lier, J. B. (2007) Thermophilic treatment of acidified and partially acidified wastewater using an anaerobic submerged MBR: Factors affecting long-term operational flux. *Wat. Res.* **41**(17), 3868–3879.

Jeison, D. and van Lier, J. B. (2008a) Feasibility of thermophilic anaerobic submerged membrane bioreactors (AnSMBR) for wastewater treatment. *Desalination* **231**(1–3), 227–235.

Jeison, D. and Van Lier, J. B. (2008b) Anaerobic wastewater treatment and membrane filtration: A one night stand or a sustainable relationship? *Wat. Sci. Technol.* **57**(4), 527–532.

Jeison, D., Telkamp, P., and van Lier, J. B. (2009) Thermophilic sidestream anaerobic membrane bioreactors: The shear rate dilemma. *Wat. Environ. Res.* **81**(11), 2372–2380.

Jetten, M. S. M., van Niftrik, L., Strous, M., Kartal, B., Keltjens, J. T., and Op den Camp, H. J. M. (2009) Biochemistry and molecular biology of anammox bacteria. *Crit. Rev. Biochem. Mol. Biol.* **44**(2), 65–84.

Judd, S. (2006) *The MBR Book: Principles and Application of Membrane Bioreactors in Water and Wastewater Treatment*. Elsevier (Oxford, UK).

Kang, I. J., Yoon, S. H., and Lee, C. H. (2002) Comparison of the filtration characteristics of organic and inorganic membranes in a membrane-coupled anaerobic bioreactor. *Wat. Res.* **36**(7), 1803–1813.

Kanai, M., Ferre, V., Wakahara, S., Yamamoto, T., and Moro, M. (2010) A novel combination of methane fermentation and MBR — Kubota Submerged anaerobic membrane bioreactor process. *Desalination* **250**(3), 964–967.

Kappell, A. S., Semmens, M. J., Novak, P. J., and LaPara, T. M. (2005) Novel application of oxygen-transferring membranes to improve anaerobic wastewater treatment. *Biotechnol. Bioeng.* **89**(4), 373–380.

Kashyap, D. R., Dadhich, K. S., and Sharma, S. K. (2003) Biomethanation under psychrophilic conditions: A review. *Bioresour. Technol.* **87**(2), 147–153.

Kataoka, N., Tokiwa, Y., Tanaka, Y., Fujiki, K., Taroda, H., and Takeda, K. (1992) Examination of bacterial characteristics of anaerobic membrane bioreactors in three pilot-scale plants for treating low strength wastewater by application of colony forming curve analysis method. *Appl. Environ. Microbiol.* **58**, 2751.

Kayawake, E., Narukami, Y., and Yamagata, M. (1991) Anaerobic digestion by a ceramic membrane enclosed reactor. *J. Ferment. Bioeng.* **71**, 122.

Kim, J. S., Lee, C. H., and Chang, I. S. (2001) Effect of pump shear on the performance of a crossflow membrane bioreactor. *Wat. Res.* **35**, 2137–2144.

Kim, J., Lee, C. H., and Choo, K. H. (2007) Control of struvite precipitation by selective removal of NH_4^+ with dialyzer/zeolite in an anaerobic membrane bioreactor. *Appl. Microbiol. Biotechnol.* **75**(1), 187–193.

Kimura, S. (1991) Japan's aqua renaissance '90 project, *Wat. Sci. Technol.* **23**(7–9), 1573.

Kimura, K., Yamato, N., Yamamura, H., and Watanabe, Y. (2005) Membrane fouling in pilot-scale membrane Bioreactors (MBRs) treating municipal wastewater. *Environ. Sci. Technol.* **39**(16), 6293–6299.

Koseoglu, H., Yigit, N. O., Iversen, V., Drews, A., Kitis, M., Lesjean, B., and Kraume, M. (2008) Effects of several different flux enhancing chemicals on filterability and fouling reduction of membrane bioreactor (MBR) mixed liquors. *J. Membr. Sci.* **320**(1–2), 57–64.

Kozariszczuk, M., Wenzel, W., Kraume, M., and Szewzyk, U. (2000) Microbiology and chemical engineering — new possibilities for interdisciplinary cooperation in the development of small anaerobic wastewater treatment plants. *Wat. Sci. Technol.* **41**(1), 17–20.

Langenhoff, A. M. A. and Stuckey, D. C. (2000) Treatment of dilute wastewater using an anaerobic baffled reactor: Effect of low temperature. *Wat. Res.* **34**, 3867–3875.

Le-Clech, P., Chen, V., and Fane, T. A. G. (2006) Fouling in membrane bioreactors used in wastewater treatment. *J. Membr. Sci.* **284**(1–2), 17–53.

Lee, D. Y., Li, Y. Y., and Noike, T. (2009) Continuous H_2 production by anaerobic mixed microflora in membrane bioreactor. *Bioresour. Technol.* **100**(2), 690–695.

Lee, S. M., Jung, J. Y., and Chung, Y. C. (2001) Novel method for enhancing permeate flux of submerged membrane system in two-phase anaerobic reactor. *Wat. Res.* **35**(2), 471–477.

Lettinga, G., Rebac, S., and Zeeman, G. (2001) Challenge of psychrophilic anaerobic wastewater treatment. *Trends Biotechnol.* **19**(9), 363–370.

Lew, B., Tarre, S., Beliavski, M., Dosoretz, C., and Green, M. (2009) Anaerobic membrane bioreactor (AnMBR) for domestic wastewater treatment. *Desalination* **243**(1–3), 251–257.

Li, A., Kothari, D., and Corrado, J. J. (1985) Application of membrane anaerobic reactor system for the treatment of industrial wastewaters. In: *Proceedings of the 39th Industrial Waste Conference*, Purdue University, Ann Arbor Science, Ann Arbor, MI, 627.

Li, W., Zhou, J., Gu, J. S., and Yu, H. Y. (2010) Fouling control in a submerged membrane-bioreactor by the membrane surface modification. *J. Appl. Polym. Sci.* **115**, 2302–2309.

Liao, B. Q., Bagley, D. M., Kraemer, H. E., Leppard, G. G., and Liss, S. N. (2004) A review of biofouling and its control in membrane separation bioreactors. *Wat. Environ. Res.* **76**(5), 425–436.

Liao, B. Q., Kraemer, J. T., and Bagley, D. M. (2006) Anaerobic membrane bioreactors: Applications and research directions. *Crit. Rev. Environ. Sci. Technol.* **36**(6), 489–530.

Liu, C. X., Zhang, D. R., He, Y., Zhao, X. S., and Bai, R. B. (2010) Modification of membrane surface for anti-biofouling performance: Effect of anti-adhesion and anti-bacteria approaches. *J. Membr. Sci.* **346**(1) 121–130.

Madden, P., Chinalia, F. A., Enright, A. M., Collins G., and O'Flaherty, V. (2010) Perturbation-independent community development in low-temperature anaerobic biological wastewater treatment bioreactors. *Biotechnol. Bioeng.* **105**(1) 79–87.

Matsushige, K., Inamori, Y., Mozuochi, M., Hosomi, M., and Sudo, R. (1990) The effects of temperature on anaerobic filter treatment for low-strength organic wastewater. *Environ. Technol.* **11**, 899–910.

McKeown, R. M., Scully, C., Mahony, T., Collins, G., and O'Flaherty, V. (2009) Long-term (1243 days), low-temperature (4–15 °C), anaerobic biotreatment of acidified wastewaters: Bioprocess performance and physiological characteristics. *Wat. Res.* **43**(6), 1611–1620.

Meng, F. G., Chae, S. R., Drews, A., Kraume, M., Shin, H. S., and Yang, F. L. (2009) Recent advances in membrane bioreactors (MBRs): Membrane fouling and membrane material. *Wat. Res.* **43**(6), 1489–1512.

Minami, K. (1994) A trial of high performance anaerobic treatment on wastewater from Kraft pulp and mill. *Desalination* **98**, 273.

Murata, M., Kimuro, H., Kanekuni, N., Ohkuma, N., Ogasawara, H., and Fujioka, T. (1994) Small-scale sewage plant experiment by pretreatment and methanization of suspended-solids. *Desalination* **98**(1–3), 217–224.

Ng, H. Y., Tan, T. W., and Ong, S. L. (2006) Membrane fouling of submerged membrane bioreactors: Impact of mean cell residence time and the contributing factors. *Environ. Sci. Technol.* **40**(8), 2706–2713.

Ognier, S., Wisniewski, C., and Grasmick, A. (2002) Characterisation and modelling of fouling in membrane bioreactors. *Desalination* **146**(1–3), 141–147.

Okamura, D., Mori, Y., Hashimoto, T., and Hori, K. (2009) Identification of biofoulant of membrane bioreactors in soluble microbial products. *Wat. Res.* **43**(17), 4356–4362.

Padmasiri, S. I., Zhang, J. Z., Fitch, M., Norddahl, B., Morgenroth, E., and Raskin, L. (2007) Methanogenic population dynamics and performance of an anaerobic membrane bioreactor (AnMBR) treating swine manure under high shear conditions. *Wat. Res.* **41**(1), 134–144.

Park, H., Choo, K. H., and Lee, C. H. (1999) Flux enhancement with powdered activated carbon addition in the membrane anaerobic bioreactor. *Sep. Sci. Technol.* **34**(14), 2781–2792.

Phattaranawik, J., Fane, A. G., Pasquier, A. C. S., and Bing, W. (2007) Membrane bioreactor with bubble-size transformer: Design and fouling control. *AICHE J.* **53**(1), 243–248.

Psoch, C. and Schiewer, S. (2006) Anti-fouling application of air sparging and backflushing for MBR. *J. Membr. Sci.* **283**(1–2), 273–280.

Ross, W. R., Barnard, J. P., Le Roux, J., and de Villiers, H. A. (1990) Application of ultrafiltration membranes for solids-liquid separation in anaerobic digestion systems: The ADUF process. *Wat. SA* **16**, 85.

Ross, W. R., Barnard, J. P., Strohwald, N. K., Grobler, C. J., and Sanetra, J. (1992) Practical application of the ADUF process to the full-scale treatment of maize-processing effluent. *Wat. Sci. Technol.* **25**(10), 27.

Saddoud, A. and Sayadi, S. (2007) Application of acidogenic fixed-bed reactor prior to anaerobic membrane bioreactor for sustainable slaughterhouse wastewater treatment. *J. Hazard. Mater.* **149**(3), 700–706.

Sainbayar, A., Kim, J. S., Jung, W. J., Lee, Y. S., and Lee, C. H. (2001) Application of surface modified polypropylene membranes to anaerobic membrane bioreactor. *Environ. Technol.* **22**, 1035.

Semmens, M. J. (2003) Personal Communication.

Speece, R. E. (1996) *Anaerobic Biotechnology for Industrial Wastewaters*. Tennessee, USA. Archae Press.

Strohwald, N. K. H. and Ross, W. R. (1992) Application of the ADUF process to brewery effluent on a laboratory scale, *Wat. Sci. Technol.* **25**(10), 95.

Sutton, P. M., Li, R. R., and Korchin, S. R. (1983) Dorr-Oliver's fixed film suspended growth anaerobic systems for industrial wastewater treatment and energy recovery. In: *Proceedings of the 37th Industrial Waste Conference*, Purdue University, Lafayette, IN, 667.

Teychene, B., Guigui, C., Cabassud, C., and Amy, G. (2008) Toward a better identification of foulant species in MBR processes. *Desalination* **231**(1–3), 27–34.

Trzcinski, A. P. (2009) *Anaerobic Membrane Bioreactor Technology for Solid Waste Stabilization.* PhD thesis, Imperial College London.

Trzcinski, A. P. and Stuckey, D. C. (2009a) Anaerobic digestion of the organic fraction of municipal solid waste in a two-stage membrane process. *Wat. Sci. Technol.* **60**(8), 1965–1978.

Trzcinski, A. P. and Stuckey, D. C. (2009b) Continuous treatment of the organic fraction of municipal solid waste in an anaerobic two-stage membrane process with liquid recycle. *Wat. Res.* **43**(9), 2449–2462.

Trzcinski, A. P., Ray, M. J., and Stuckey, D. C. (2010) Performance of a three-stage membrane bioprocess treating the organic fraction of municipal solid waste and evolution of its archaeal and bacterial ecology. *Bioresour. Technol.* **101**(6), 1652–1661.

van der Star, W. R. L., Miclea, A. I., van Dongen, U. G. J. M., Udo, G. J. M., Muyzer, G., Picioreanu, C., and van Loosdrecht, M. C. M. (2008) The membrane bioreactor: A novel tool to grow anammox bacteria as free cells. *Biotechnol. Bioeng.* **101**(2), 286–294.

Van Zyl, P. J., Wentzel, M. C., Ekama, G. A., and Riedel, K. J. (2008) Design and start-up of a high rate anaerobic membrane bioreactor for the treatment of a low pH, high strength, dissolved organic waste water. *Wat. Sci. Technol.* **57**(2), 291–295.

Veeken, A. and Hamelers, B. (1999) Effect of temperature on hydrolysis rates of selected biowaste components. *Bioresour. Technol.* **69**(3), 249–254.

Voolapalli, R. K. and Stuckey, D. C. (1998) Stability enhancement of anaerobic digestion through membrane gas extraction under organic shock loads. *J. Chem. Technol. Biotechnol.* **73**(2), 153–161.

Vyrides, I. and Stuckey, D. C. (2009a) Saline sewage treatment using a submerged anaerobic membrane reactor (SAMBR): Effects of activated carbon addition and biogas-sparging time. *Wat. Res.* **43**(4), 933–942.

Vyrides, I. and Stuckey, D. C. (2009b) Effect of fluctuations in salinity on anaerobic biomass and production of soluble microbial products (SMPs). *Biodegradation* **20**(2), 165–175.

Walker, M., Banks, C. J., and Heaven, S. (2009a) Development of a coarse membrane bioreactor for two-stage anaerobic digestion of biodegradable municipal solid waste. *Wat. Sci. Technol.* **59**(4), 729–735.

Walker, M., Banks, C. J., and Heaven, S. (2009b) Two-stage anaerobic digestion of biodegradable municipal solid waste using a rotating drum mesh filter bioreactor and anaerobic filter. *Bioresour. Technol.* **100**(18), 4121–4126.

Wang, T., Zhang, H. M., Yang, F. L., Liu, S. T., Fu, Z. M., and Chen, H. H. (2009) Start-up of the Anammox process from the conventional activated sludge in a membrane bioreactor. *Bioresour. Technol.* **100**(9), 2501–2506.

Wang, Z. W., Wu, Z. C., and Yin, X. (2008) Membrane fouling in a submerged membrane bioreactor (MBR) under sub-critical flux operation: Membrane foulant and gel layer characterization. *J. Membr. Sci.* **325**(1), 238–244.

Wen, C., Huang, X., and Qian, Y. (1999) Domestic wastewater treatment using an anaerobic bioreactor coupled with membrane filtration. *Process Biochem.* **35**, 335–340.

Wisniewski, C. and Grasmick, A. (1998) Floc size distribution in a membrane bioreactor and consequences for membrane fouling. *Colloid. Surfaces A Physchem. Eng. Aspects* **138**(2–3), 403–411.

Wu, B., An, Y. Y., Li, Y. Z., and Wong, F. S. (2009) Effect of adsorption/coagulation on membrane fouling in microfiltration process post-treating anaerobic digestion effluent. *Desalination* **242**(1–3), 183–192.

Yeon, K. M., Cheong, W. S., Oh, H. S., Lee, W. N., Hwang, B. K., Lee, C. H., Beyenal, H., and Lewandowski, Z. (2009) Quorum sensing: A new biofouling control paradigm in a membrane bioreactor for advanced wastewater treatment. *Environ. Sci. Technol.* **43**(2), 380–385.

Zhang, Y. J., Yan, L., Qiao, X. L., Chi, L., Niu, X. J., Mei, Z. J., and Zhang, Z. J. (2008) Integration of biological method and membrane technology in treating palm oil mill effluent. *J. Environ. Sci. China* **20**(5), 558–564.

Chapter 8

Anaerobic Baffled Reactor (ABR) for Wastewater Treatment

David C. Stuckey

Department of Chemical Engineering and Chemical Technology,
Imperial College of Science, Technology and Medicine,
Prince Consort Road, London SW7 2BY, UK
d.stuckey@ic.ac.uk

The anaerobic baffled reactor (ABR) was developed in the early 1980s, and consists of a series of compartments (up to 8) in one reactor which are baffled to force the incoming wastewater up through a series of sludge blankets, thereby reducing the loss of biomass. It can also be operated with granules or internal media which enhance its stability. Hence the sludge retention time (SRT) can be separated from the hydraulic retention time (HRT), leading to good chemical oxygen demand (COD) and solids removal, low sludge production, and a small footprint. Tracer studies have shown that the reactor approximates to a series of completely stirred tank reactors (CSTRs), and this promotes the separation of bacterial trophic groups down the length of the ABR enhancing performance and stability. Due to its design, the reactor can tolerate both severe hydraulic and organic shock loads without failing, and is now starting to be used at full scale in many developing countries as a cheap and efficient method of low cost sanitation, and industrial wastewater treatment, often with the added benefit of energy production. This chapter reviews the current state of knowledge with ABRs focussing on: reactor development, hydrodynamics, performance, biomass characteristics and retention, soluble microbial products (SMPs), modelling, and full-scale operation.

1. Introduction

The application of anaerobic technology to the treatment of wastewaters is critically dependent on the successful development of high-rate anaerobic bioreactors. These reactors achieve a high reaction rate per unit reactor volume (in terms of kg-COD$\cdot$m$^{-3}\cdot$d^{-1}) by retaining the biomass (solids retention time — SRT) in the reactor independently of the incoming wastewater (hydraulic residence time — HRT), in contrast to continually stirred tank reactors (CSTRs), thus reducing reactor volume and ultimately allowing high volumetric loading rates, e.g., 10–40 kg-COD$\cdot$m$^{-3}\cdot$d^{-1} (Iza *et al.* 1991). High-rate reactors may be classified into four broad groups depending on biomass detention: fixed-film, suspended growth, membrane and hybrid. The highest loading rates achieved to date were with the anaerobic attached film expanded bed (AAFEB) reactor (120 kg-COD$\cdot$m$^{-3}\cdot$d^{-1} — Switzenbaum and Jewell 1980), but its inherent complexity and high operating costs limit its practical use on a wide scale.

Around the same time as Lettinga developed the UASB (Lettinga *et al.* 1980), McCarty and co-workers at Stanford noticed that most of the biomass present within an anaerobic rotating biological contactor (RBC — Tait and Friedman 1980) was actually suspended, and when they removed the rotating discs they developed the anaerobic baffled reactor (ABR) (McCarty 1981). However, baffled reactor units had previously been developed slightly earlier by Chynoweth *et al.* (1980). The ABR is starting to be used on a wider scale, and has several advantages over other well-established systems which have been summarised in a past review (Barber and Stuckey 1999).

The most significant advantage of the ABR is its ability to separate suspended growth acidogenesis and methanogenesis down the reactor, allowing the reactor to behave as a two-phase system without the associated control problems and high costs (Weiland and Rozzi 1991). In addition, due to its construction the ABR can operate on hydrostatic head without the use of electricity, and hence can be used in developing countries (Pillay *et al.* 2008). Two-phase operation can increase acidogenic and methanogenic activity by a factor of up to four as acidogenic bacteria accumulate within the first stage (Cohen *et al.* 1980, 1982), and different bacterial groups can develop under more favourable conditions. The advantages of two-phase operation have been extensively documented (Pohland and Ghosh 1971; Ghosh *et al.* 1975; Cohen *et al.* 1980, 1982). These benefits have catalysed the development of other staged reactor configurations such as the multiplate anaerobic reactor (El-Mamouni *et al.* 1992; Guiot *et al.* 1995), upflow staged sludge bed (USSB) (van Lier *et al.* 1994, 1996), staged anaerobic filter (Alves *et al.* 1997), and recently the carrier anaerobic baffled reactor (Feng *et al.* 2009), all of which have showed considerable potential for wastewater treatment. Disadvantages of the baffled reactor design at pilot/full scale include the requirement to build shallow reactors to maintain acceptable liquid and gas upflow velocities, and problems with maintaining an even distribution of the influent (Tilche and Vieira 1991).

The objective of this chapter is to review the currently available literature on the anaerobic baffled reactor, focussing on reactor development, hydrodynamics, performance, biomass characteristics and retention, modelling, and full-scale operation. Finally, a concluding section will summarise the state of the art for the ABR.

2. Reactor Development

The first ABR was developed by Fannin *et al.* (1981, 1982) who added vertical baffles to a plug flow reactor treating a high solids kelp slurry to enhance the reactor's ability to maintain slowly growing methanogens, which were being replaced by the influent solids. With a constant loading rate of $1.6\,\text{kg-COD}\cdot\text{m}^{-3}\cdot\text{d}^{-1}$ methane levels increased from 30% to over 55% with a methane yield of $0.34\,\text{m}^3\cdot\text{kg-VSS}^{-1}$ after the baffles were added. In a later study, Bachmann *et al.* (1983, 1985) compared the performance of two baffled reactors before and after narrowing the downflow chambers and slanting the baffle edges. Although methane production rates and reactor efficiency were improved in the modified design, a decrease in the methane content of the biogas was also noted. A full description of the evolution of the design of the ABR can be seen in Barber and Stuckey (1999), and its structure is shown in Fig. 8.1.

To remove nitrogen in the ABR, Barber and Stuckey (2000a,b) incorporated nitrifiers immobilised in polyvinyl alcohol (PVA) beads in the aerated seventh compartment of an eight-compartment reactor, with Kaldnes particles (Rostron *et al.* 2001) in the last compartment to reduce solids loss caused by aeration. Due to the bead structure and mass transfer problems, only 50% of the incoming ammonia was nitrified. With recycled effluent almost all the nitrate/nitrite (~96%) was reduced to nitrogen gas, and there was a small increase (~5%) in chemical oxygen demand (COD) removal over the ABR. Similar to this strategy is the development of the carrier ABR (CABR), where a medium is added (e.g., bamboo rings, coke particles) to some or all of

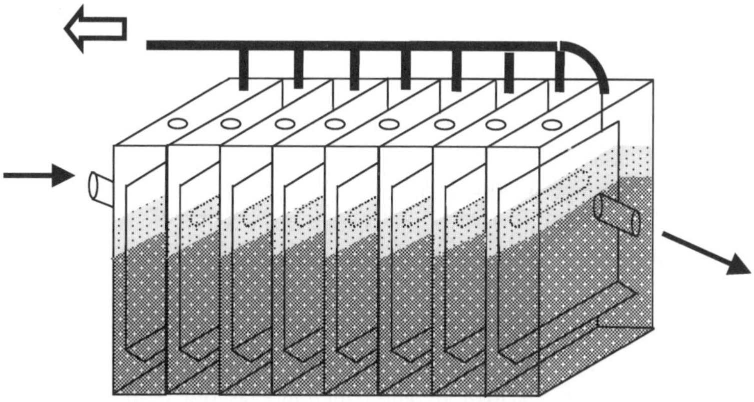

Fig. 8.1. Schematic figure of an ABR with sludge blanket.

the ABR compartments to provide a surface for bacteria to grow on, and to prevent the washout of biomass — in essence a series of upflow anaerobic filters (Feng *et al.* 2008a,b, 2009; Wang *et al.* 2009). This modification seems to help during shock loads and results in good performance under most conditions.

Another modification of the ABR is to seed the reactor with granular UASB sludge during start-up. Grobicki (1989) had noticed granules forming under certain conditions in the ABR, but it was not predictable. Freese and Stuckey (2000) studied the effect of starting up an ABR with granules and suspended biomass. It was found that using granules resulted in a far faster start-up, but that nutrient deficiencies resulted in poor granules, and that if the feed was changed from carbohydrate to protein/carbohydrate then a "fuzz" developed on the granules preventing gas escape and they were washed down and out of the reactor. Akunna, Baloch and their co-workers (Akunna and Clark 2000; Baloch and Akunna 2003; Baloch *et al.* 2008) also developed the concept of a granular-bed ABR (GRRABR) where with whiskey wastes the acidogens tended to be suspended while the methanogens were immobilised in the granules.

The final evolution of the ABR resulted in similar manifestations of the design — the periodic ABR (PABR) developed by Skiadas, Lyberatos and their co-workers (Skiadas and Lyberatos 1998; Skiadas *et al.* 2000; Stamatelatou *et al.* 2002). This design used a concentric circular reactor where the chambers are similar to an ABR with baffles, but it can be fed into any chamber from 1 to 8. The authors claim that this mode of operation allows the PABR to be operated as a UASB, ABR or a combination in the middle, and hence it is best suited to handle time-varying loads at maximal conversion rates. Similar in design is the anaerobic migrating blanket reactor ABR (AMBR) where three CSTR compartments are fed so that the entry and exit points can be varied over time (Angenent *et al.* 2000, 2001). While marginally better than a UASB or standard ABR under some conditions, these types of reactors lose some of the advantages of the ABR, e.g., separation of the various anaerobic trophic groups.

3. Reactor Hydrodynamics

3.1. *Flow patterns*

The hydrodynamics and degree of mixing that occur within a reactor of this design strongly influences the extent of contact between substrate and bacteria, thus controlling mass transfer and reactor performance. Grobicki and Stuckey (1992) used Li^+ to conduct residence time distribution studies in a number of baffled reactors (4–8 compartments), both with and without biomass, at various HRTs, and used Levenspiel (1972) "dispersion" and "tanks in series" models to analyse their data. The models provided the degree of mixing and the amount of unused volume (dead space) within the reactor. They found low levels of dead space ($<8\%$ hydraulic dead space in an empty reactor) in comparison with other anaerobic reactor designs, e.g., 50–93% in an anaerobic filter (Young and Young 1988), and $\sim 50\%$ in a CSTR (Stuckey 1983). Dead space increased to 18% on the addition of 8 g-VSS·l^{-1}, however, no direct correlation between hydraulic dead space and HRT could be drawn. At low HRT, the presence of biomass had no significant effect on hydraulic dead space, which was found to be a function of flowrate and number of baffles. This contrasted with biological dead space, which was found to be a function of biomass concentration, gas production and flowrate, and which increased with increasing flowrates.

At high loading rates caused by low HRT, gas production as well as increased flowrates kept sludge beds partly fluidised. Therefore, the contradictory effects of hydraulic and biological dead space prevented a correlation being derived between HRT and overall dead space. Biological dead space was established as the major contributor to overall dead space at high HRT, but its effect decreased at lower HRT since gas production disrupted channelling within the biomass bed. Severe channelling, caused by large hydraulic shocks, was found to be beneficial since most of the biomass was not entrained in the flow, and this resulted in low washout and a fast recovery in performance (Nachaiyasit and Stuckey 1997b; Grobicki and Stuckey 1992). Langenhoff and co-workers (Langenhoff *et al.* 2000; Langenhoff and Stuckey 2000) evaluated dead space with low strength feeds and at low temperatures at different HRTs. Little difference in mixing or dead space was observed under the different conditions tested. The average dead space measured was 20–37%, and the flow pattern within the reactors showed an intermediate behaviour between plug flow and ideally mixed. Krishna *et al.* (2008) confirmed these observations recently. Unfortunately, no work has been done on the rate at which solid particles/biomass moves down the reactor.

3.2. *Effect of effluent recycle*

Recycling effluent tends to reduce removal efficiency because the reactor approaches a completely mixed system. However, recycle can alleviate the problems of low pH in the feed, and discourage gelatinous bacterial growth at the reactor inlet for the treatment of a complex protein/carbohydrate wastewater (Bachmann *et al.* 1985). Nevertheless, recently Bayrakdar *et al.* (2009) investigated the treatment of acidic (pH 4.5–7.0) wastewater containing sulfate (1–2 g·l^{-1}) and Zn (65–200 mg·l^{-1}) at 35 °C with lactate ($COD/SO_4^{2-} = 0.67$) as a carbon and energy source for sulfate reducing bacteria (SRB). Although the feed pH was steadily decreased from 7.0 to 4.5, effluent pH was always within 6.8–7.5, and 99% Zn removal was attained due to the formation of Zn-sulfide precipitate. Hence it appears that the ABR can withstand very low feed pHs in the first compartment without failing, and may not need pH adjustment, and this has been confirmed at full scale with Tapioca starch wastewaters in Thailand and in our laboratory (unpublished work).

Another benefit of recycle is the dilution of toxicants and reduction of substrate inhibition in the influent (Grobicki and Stuckey 1991). Kennedy and Barriault (2007) treated aircraft deicing fluid and found with a recycle ratio of 6:1 that the minimum HRT decreased from 27 to 17 hours, the organic loading rate increased from 6.2 to 9.9 $g \cdot l^{-1} \cdot d^{-1}$, and COD removal increased from 89 to 93%. They explained this increase due to the dilution of ethylene glycol toxicity in the front of the ABR. Nevertheless, recycle should have a negative effect on reactor hydrodynamics by causing increased mixing (which encourages solids loss, and enhancing the amount of dead space — Grobicki and Stuckey 1992). Nachaiyasit (1995) showed that dead space doubled to approximately 40% when the recycle ratio was increased from zero to 2, and discovered a fall in both gas production and methane composition down the reactor when the recycle ratio was increased. Saritpongteeraka and Chaiprapat (2008) treated concentrated rubber latex wastewater using different pH adjustment chemicals and recycling ratios (R). Two ABRs, one receiving wastewater pretreated with NaOH and the other with ash, were operated at 35 °C under identical HRTs from 10 to 1.25 d. Both ABRs had high COD and sulfate removals at an HRT of 10 d (82.7%/96.2% — ABR NaOH, 80.77%/96.60% ABR Ash). Increasing R (0.3 and 0.5) raised the hydraulic loading on the system and resulted in a drop of organic removal efficiency and methane yield. In contrast, Ghaniyari-Benis *et al.* (2009) found that recycle had no effect on COD removal. Hence the benefits of recycle will depend on the type of wastewater being treated.

4. Reactor Performance

4.1. *Start-up*

The objective of start-up is to develop the most stable microbial culture (granular or flocculent) in each compartment for the waste stream in question as quickly as possible. Important factors governing the start-up of the ABR have been summarised (Barber and Stuckey 1999), and will not be discussed here. Initial loading rates should be low so that slow growing microorganisms are not over loaded, and both gas and liquid upflow velocities should be low ($<2\, m \cdot h^{-1}$) so that flocculent and granular growth is encouraged. The recommended initial loading rate is approximately 1.2 $kg\text{-}COD \cdot m^{-3} \cdot d^{-1}$ (Henze and Harremoës 1983); however, while Nachaiyasit (1995) originally noted adequate performance with an initial loading rate of 13 $kg\text{-}COD \cdot m^{-3} \cdot d^{-1}$, an accumulation of intermediate products caused reactor souring and eventual failure after two weeks of operation. This experience of overloading was replicated by Freese and Stuckey (2000) when OLRs of 13 and 4.35 $kg\text{-}COD \cdot m^{-3} \cdot d^{-1}$ were used with flocculent sludge, although with granular sludge a good start-up was achieved in 30 days with an OLR of 4.8 $kg\text{-}COD \cdot m^{-3} \cdot d^{-1}$, and 8 after 45 days: hence the use of granules enables fast start-ups. Bodkhe (2009) used no seed with domestic sewage and managed to successfully start-up in only 90 days at 6 days HRT, however, the OLR was only 0.067 $kg\text{-}COD \cdot m^{-3} \cdot d^{-1}$.

In terms of strategy, Barber and Stuckey (1998) showed that an initial long HRT (80 hours) at 4 $g\text{-}COD \cdot l^{-1}$ (OLR = 1.2 $kg\text{-}COD \cdot m^{-3} \cdot d^{-1}$) which is reduced stepwise to 20 hours at constant feed concentration over 80 days (OLR = 4.8 $kg\text{-}COD \cdot m^{-3} \cdot d^{-1}$) provides greater reactor stability and superior performance than a reactor started up with a constant and low HRT (20 hours) coupled to a stepwise increase in substrate concentration. These findings were linked to better solids accumulation, promotion of methanogenic populations and faster recovery to hydraulic shocks in the reactor started at the longer retention time. However, with a salty complex nutrient-deficient oil well "produced water", Ji *et al.* (2009) took around 170 days to achieve a stable start-up at an OLR of only 0.2 $kg\text{-}COD \cdot m^{-3} \cdot d^{-1}$.

4.2. *Treatment applications*

This section reviews the performance of the ABR treating a variety of wastewaters, in particular, low and high strength, low temperature, sulfate containing waste and industrial wastewaters.

4.2.1. *Low strength treatment*

Various authors have treated low-strength wastewaters effectively in the ABR, as shown in Barber and Stuckey (1999). Dilute wastewaters result in a low mass transfer driving force between biomass and substrate, and subsequently biomass activities will be greatly reduced according to Monod kinetics. As a result, treatment of low-strength wastewaters has been found to encourage the dominance of scavenging bacteria such as *Methanosaeta* in the ABR (Polprasert *et al.* 1992). It appears that biomass retention is enhanced significantly due to lower gas production rates suggesting that low hydraulic retention times (6 $\rightarrow$ 2 hours) are feasible during low strength treatment. Another positive consequence of low retention times when treating dilute wastewaters is an increase in hydraulic turbulence, which can lower apparent K_S values (Kato *et al.* 1997) thus enhancing treatment efficiency.

Langenhoff *et al.* (2000) evaluated the treatment of dilute wastewaters (500 mg-COD$\cdot$l^{-1}): one a soluble milk waste, the other a synthetic colloidal waste containing particles of dog food and rice. They reduced the HRT in the duplicate ABRs from 40 hours down to 6 hours over 160 days, and the one being fed milk was reduced further to 5, 4, 2.85 and 1.3 hours. COD removals were around 90% at 20 hours, 80–95% at 10 hours, and 80% at 6 hours. Even at 2.85 hours COD removal was 60%, while finally at 1.3 hours only 40% removal occurred, and there was little difference between removals of the soluble and colloidal feed, although the latter was slightly less degradable. Langenhoff and Stuckey (2000), using the same soluble milk waste, then evaluated the effect of low temperatures on both COD removal and hydrodynamics at 10 hours HRT. At 35 °C COD removal was 95%, 70% at 20 °C and 60% at 10 °C. Interestingly, almost all the COD in the effluent at these lower temperatures was not unmetabolised feed or volatile fatty acids (VFAs), but soluble microbial products (SMPs). Also, despite the higher viscosity at the lower temperatures the residence time distributions (RTD) did not change, and the reactor behaved like eight CSTRs in series. Finally, it was found that even after 60 days at 10 °C the methanogens in the ABR were still mesophilic, and there was no selection for psychrophiles resulting in lower metabolic rates and lower COD removals.

Feng *et al.* (2008a) treated a dilute synthetic feed (300 mg-COD$\cdot$l^{-1}) in a 6-compartment Carrier ABR containing bamboo rings at 18, 12 and 9 hours HRT at 28, 18 and 10 °C. The results were very similar to Langenhoff *et al.* (2000) and at their highest load (9 h) and lowest temperature (10 °C) they achieved 42.6% COD removal.

4.2.2. *High strength treatment*

High strength wastewaters can be treated at low (<2 kg-COD$\cdot$m$^{-3}\cdot$d^{-1}) and high loading rates (20 kg-COD$\cdot$m$^{-3}\cdot$d^{-1}) depending on the HRT. Whereas low HRTs are possible and even necessary for dilute wastewaters, the opposite applies when treating concentrated waste due to the high gas mixing caused by high loads and COD removal. This can result in higher biomass wastage, and has led to modifications in reactor design in order to enhance solids retention, e.g., carriers or lamella separators within the ABR (He *et al.* 2009). When Boopathy and Tilche (1991) treated 115 g-COD·l^{-1} molasses alcohol stillage (OLR = 12.25 kg-COD·m^{-3}·d^{-1}) then switched to raw alcohol molasses (990 g-COD·l^{-1}, OLR = 28 kg-COD·m^{-3}·d^{-1}) they noticed an increase in overall gas production of over 65% within 3 weeks, a drop in COD removal of 20%, and an increase in volatile suspended solids of around 50% within 3 weeks. Higher levels of gas production increased sludge bed expansion, but the improved settling ability of the biomass may have reduced the effects of solids loss caused by the gas (Boopathy and Tilche 1991). In contrast, Baloch and Akunna (2003) only used a glucose-based feed strength of 5 g-COD·l^{-1}, but operated down to 6 hour HRTs (20 kg·m^{-3}d^{-1}) in a granular-bed anaerobic baffled reactor (GRABBR) and achieved SCOD removals of 97%, although the effluent solids varied from 40 to 240 mg·l^{-1} resulting in a true COD removal of only 92–96%. The solids were lower than a suspended solids system due to the granules in the reactor.

4.2.3. *Low temperature treatment*

The vast majority of work done so far on the ABR has been at mesophilic temperatures. Nachaiyasit and Stuckey (1995) found no significant reduction in COD removal when the ABR temperature was decreased from 35 to 25 °C, with steady state reached after only two weeks. Similar high treatment efficiencies at ambient temperature have also been noted for a medium strength phenolic wastewater (Holt *et al.* 1997). Nachaiyasit and Stuckey (1997a) further reduced the temperature to 15 °C, and COD removal fell from 93 to 83% after one month. However, despite the fact that the reactors were kept for long periods of time at reduced temperatures (12 weeks) their performance did not improve despite the increased intermediate acid concentrations. This may be due to the fact that K_S increases substantially as temperature falls (Lawrence and McCarty 1969), leaving low levels of VFAs that cannot be degraded. In conclusion, the work found that a combination of decreased catabolic rates, increased K_S, and higher levels of refractory material caused inferior performance at 15 °C, but that a drop in temperature from 35 to 25 °C had negligible effects on overall reactor performance.

Feng *et al.* (2008a) treated a dilute synthetic feed (300 mg-COD·l^{-1}) in a six compartment Carrier ABR containing bamboo rings at 18, 12 and 9 hours HRT at 28, 18 and 10 °C. Their results were very similar to Langenhoff and Stuckey (2000), and at their highest load (9 h) and lowest temperature (10 °C) they achieved 42.6% COD removal. Manariotis and Grigoropoulos (2002) used a three-chamber ABR to evaluate the treatment of low-strength synthetic wastewater (300–400 mg·l^{-1}). At 26 °C and HRTs of 24 and 12 h, COD removal averaged 87.2 and 91.0%, respectively. The ABR was then refed after two years without feeding; the response was rapid and removal averaged 85.3% even during the initial 10-day period. Lowering the temperature to 16 °C did not impair removal efficiency at HRTs of 24 and 12 hours. However, biogas release decreased by 30% and apparent COD conversion to methane dropped by 24 and 31%, respectively. However, the baffled reactor has been run as low as 13 °C (Orozco 1988).

A recent review on cold-archaea (Cavicchioli 2006) has shown that psychrophilic methanogens do exist, and can metabolise at temperatures as low as 1 °C, but some such as *Methanogenium frigidum* have optima around 15 °C but a maximum growth temperature of only 18 °C, while *Methanogenium burtonii* have optima around 23 °C and a maximum of 28 °C. Hence it should be possible to develop reactors capable of good to high COD removals at temperatures as low as 10 °C, a typical sewage temperature in Northern latitudes such as Europe and North America.

4.2.4. *Sulfate treatment*

Fox and Venkatasubbiah (1996) investigated the effects of sulfate reduction in the ABR by treating a sulfate containing pharmaceutical wastewater up to a final strength of 20 g-COD·l^{-1} with a COD:SO_4^{2-} ratio of 8:1. At steady state, 50% COD removal and 95% sulfate reduction was possible with a detention time of 1 day. After altering the COD:SO_4 ratio by adding glucose, isopropanol and sulfate, the authors noted a fall in potential sulfate reduction from >95% at COD:SO_4^{2-} = 150:1 to <50% at COD:SO_4^{2-} = 24:1. Increasing sulfate concentrations with glucose and isopropanol present showed inhibition of sulfate reduction caused by elevated sulfide concentrations. Increasing the inlet concentration from 2 g-COD·l^{-1} to 8 g-COD·l^{-1} (COD:SO_4 at 8:1) over 100 days caused an increase in the total effluent sulfide to toxic levels (200 mg·l^{-1}, or 80 mg·l^{-1} unionised H_2S assuming pH 7), with COD removal dropping to below 20%. The major contributor to the effluent VFA was acetate indicating inhibition of acetoclastic methanogenesis and a distinct lack of acetate cleaving sulphidogenesis.

Barber and Stuckey (2000c) investigated the effect of sulfate reduction on COD removal in an eight-compartment ABR at two COD-to-SO_4^{2-} ratios (40 [100 mg·l^{-1} of SO_4^{2-}] and 8 [500 mg·l^{-1} of SO_4^{2-}]) with an influent COD of 4,000 mg·l^{-1}. Sulfate reduction was primarily confined to the first four compartments, but was incomplete. The dissolved sulfide concentrations were less than 130 mg·l^{-1} (or 4% in the gaseous phase), and typically less than the levels expected to cause bacterial inhibition; however, various trophic groups, especially acetogenic bacteria, seemed to be acutely inhibited by hydrogen sulfide (H_2S). In contrast, low concentrations of reduced sulfide (15 mg·l^{-1} H_2S(aq)) generated from sulfate reduction stimulated reductive methanogenesis. Nevertheless, overall COD removal decreased from 98 to 90% with increased sulfate addition, and seemed to be influenced by residual COD production. The VFA results obtained during this work were consistent with literature findings, namely, acetate accumulation and lack of acetoclastic activity, improved propionate use and inert butyrate use. Vossoughi *et al.* (2003) carried out similar work with varying COD:SO_4^{2-} ratios and obtained very similar results.

Barber and Stuckey (2000d) extended their work to investigate trace metal bioavailability and chromium(III) removal in an ABR treating a synthetic waste identical to Barber and Stuckey (2000a). Sulfide precipitation reduced nutrient bioavailability as follows (most deficient first): Fe > Co > Ni. However, the metals had the following affinity for complexation: Ni > Co >> Fe, and complexation potential was greatest at the front and rear of the reactor. At the front, it was hypothesised that high concentrations of SMPs provided an excess of potential ligands, which may have sequestered the metals. However, in the last compartment, a higher pH would have altered the competition between protons and metal cations for complexing sites in favour of the metal cations. The complexing potential in the ABR increased with sulfide levels, and this coincided with higher residual COD production (SMPs). Although the addition of chromium (50 mg·l^{-1}) had no observable effect on reactor performance/stability, its removal was controlled by the following factors (most influential first): solubility > complexation > absorption > physical adsorption.

Patidar and Tare (2004) evaluated the effect of Fe, Ni, Zn, Co and Mo on the anaerobic degradation of sulfate laden organics using a bench-scale UASB, ABR and hybrid HABR, operating with OLRs of 1.9–5.75 kg-COD·m^{-3}·d^{-1}, sulfate loadings of 0.54–1.88 kg-SO_4^{2-}·m^{-3}·d^{-1}, and COD:SO_4^{2-} ratios of 2.0–8.6. Initially, no nutrient limitation was observed with COD removals of more than 94% in all three systems. Increasing sulfate loading resulted in Ni and Co limitations, but their supplementation restored COD removal in the UASB system. However, the baffled systems did not recover because of severe inhibition by sulfide, and the precipitation of nutrients can reduce process performance, leading to failure even before sulfide concentration reach toxic levels.

Bell *et al.* (2000) investigated the degradation of an industrial dye wastewater in an ABR containing over 50% tartrazine in a synthetic feed. Initially the dye waste was 5% of the feed, but this was increased to 10% over time — COD and colour removals were over 90% throughout the run. In a linked study, Plumb *et al.* (2001) used fluorescent in-situ hybridisation (FISH) and an archaeal 16S ribosomal DNA (rDNA) clone library and found that sulphidogenic bacteria, together with *Methanogenium hollandica* contribute considerably to the treatment process through metabolism of dye-associated sulphonate groups and subsequent conversion of sulfur compounds to carbon dioxide and methane.

Freese and Stuckey (2004) examined the effects of treating wastewaters with high levels of sulfate in a bench-scale ABR. A baseline feed of 4 g-COD·l^{-1} was supplemented with 0.4 g·l^{-1} of SO_4^{2-} (a COD:SO_4^{2-} of 10). At a 20 h HRT, the SO_4^{2-} content of the feed was increased stepwise, resulting in a decrease in the COD:SO_4^{2-} *t* ratio to a minimum of 1. Even at a ratio of 1 COD removal was 77%, but SO_4^{2-} reduction accounted for half of the COD removal. FISH was used to evaluate the effect of the extreme sulfate loading on the ABR. After 4 weeks of feeding S as much as 25% of the bacterial population was found to be *Desulfovibrio* which dominated the sulfate reducers, probably because the efficient acetate scavenger *Methanosaeta* were present in high numbers, thereby out-competing *Desulfobacter*. In the front of the ABR where sulfate reduction is expected, specific sulfate reduction rates (1.85 kg-SO_4^{2-}·kg-VSS^{-1}·day^{-1}) were significantly higher than the values seen in the literature. Bayrakdar *et al.* (2009) also found that even at a COD:SO_4^{-2} ratio of 0.67 COD and Zn removals could be over 95%.

Finally, one novel use of the ABR was investigated by Arnirfakhri *et al.* (2006) who used chemoautotrophic bacteria for the biodesulphurization of natural gas scrub liquors containing hydrogen sulfide. To increase the SRT the reactor was packed by pumice (0.5–2 cm diameter). A synthetic wastewater containing $S_2O_3^{2-}$ was used in start-up, and S^{2-} for running the ABR. Performance was evaluated at HRTs of 50, 28 and 9 h, with the equivalent loading rate of 0.11, 0.2 and 0.62 mmol S^{2-}·l^{-1}·h^{-1}. Sulfide was converted to elemental sulfur and sulfate, so that more than 61% of the sulfur in the end products was elemental sulfur, and the maximum sulfide removal rate was about 3.03 mmol-S^{2-}·l^{-1}·h^{-1}.

4.2.5. *Industrial wastewaters*

Early examples of using the ABR to treat industrial wastewaters have been summarised by Barber and Stuckey (1999). In the main these were food-processing wastewaters such as molasses and whey. In the past decade the ABR has demonstrated its versatility with a wider range of feeds, and due to its baffled and sequential stages, has tolerated a number of quite toxic and inhibitory wastes with good to high COD and toxin removals. Grover *et al.* (1999) treated Black Liquor and obtained a maximum COD removal of about 60% at an OLR of 5 kg-COD·m^{-3}·d^{-1} at an HRT of 2 d (pH 8.0 and 35 °C), while above 6 kg-COD·m^{-3}·d^{-1} was toxic and destabilised the reactor system. Kennedy *et al.* (2006) using similar waste achieved slightly lower COD removals (50%) at a slightly higher OLR (7 kg-COD·m^{-3}·d^{-1}).

Bell and co-workers (Bell *et al.* 2000; Bell and Buckley 2003) performed batch biodegradability and toxicity assays on a number of food dyes, and found they ranged from noninhibitory ($IC_{50} > 20\,g{\cdot}l^{-1}$) to inhibitory ($IC_{50}$ of $0.2\,mg{\cdot}l^{-1}$), while not being readily utilised by anaerobes as a sole substrate. The food dye tartrazine ($250\,mg{\cdot}l^{-1}$) was added to a sucrose–peptone feed at 20 h HRT and achieved a COD reduction of 50–60% and a colour reduction of about 95%. An industrial wastewater from a food dye manufacturer was then fed to a second reactor at a concentration of 10% (v/v). The organic content of the influent was reduced by about 70%, and colour was reduced by almost 90%.

5. Biomass Characteristics and Retention Capabilities

5.1. *Bacterial populations*

With the unique construction of the ABR as a series of CSTRs, it was postulated that various profiles of microbial communities will develop within the reactor. The microbial ecology within each compartment will depend on the type and amount of substrate present, as well as external parameters such as pH and temperature (Briones and Raskin 2003). In the acidification zone of the ABR (front compartments) bacteria capable of fast growth at high substrate levels and reduced pH will dominate. A shift to slower growing scavenging bacteria that grow better at higher pH will occur towards the end of the reactor. Various techniques have been applied to the ABR, and by far the most common observation involved a shift in population between the two acetoclastic methanogens, i.e., *Methanosarcina* sp. and *Methanosaeta* sp. At high acetate concentrations *Methanosarcina* outgrows *Methanosaeta* in the front of the reactor due to faster growth kinetics (doubling time 1.5 days compared with 4 days for *Methanosaeta*), however, at low concentrations in the rear *Methanosaeta* is dominant due to its scavenging capability (Yang and Moengangongo 1987; Yang *et al.* 1988; Polprasert *et al.* 1992) ($K_s = 30\,mg{\cdot}l^{-1}$ compared with $400\,mg{\cdot}l^{-1}$ for *Methanosarcina* — Gujer and Zehnder 1983). Hassouna (personal communication 1998) showed a shift in the activity of acid-producing bacteria down the length of an eight-compartment baffled reactor. Using the method of Owen *et al.* (1979), aliquots were removed from each compartment of ABRs treating a range of substrate concentrations. In the foremost compartments a glucose spike was readily converted to volatile acids within a few hours and this contrasted with the results from subsequent compartments which showed slower and slower degradation of the spike.

Uyanik *et al.* (2002) assessed the influence of polymer addition on granule formation within an ABR, and compared the ecology of the biomass in each compartment of two ABRs treating ice-cream wastewater. The average fraction of autofluorescent methanogens was lower in the initial compartments of the ABRs (1.5%), compared to the last compartment (15%), showing that each compartment of an ABR had a unique microbial composition. Scanning electron micrographs revealed that the dominant bacteria in the initial compartments of the ABR were those which could consume H_2/CO_2 and formate as substrate, i.e., *Methanobrevibacter*, *Methanococcus*, however, populations shifted over time to acetate utilisers, i.e., *Methanosaeta* and *Methanosarcina*, in the final compartments.

Roy *et al.* (2009) used an eight-compartment anaerobic reactor to study shifts in a bacterial community degrading swine manure at 25 °C. As found above, the reactor successfully separated the hydrolysis/acidogenesis stage from the methanogenesis stage. Bacterial

16S rDNA- and rRNA-based fingerprints obtained through amplicon length heterogeneity polymerase chain reaction (LH-PCR) were used to characterise the bacterial community structure and the metabolically active community, respectively. The rDNA-based fingerprints described a more temporal than compartmentalised distribution of similar bacterial communities. In contrast, the rRNA-based analyses described a distribution that was linked more to reactor performance parameters. Diversity indices calculated from fingerprint data were used to assess overall diversity shifts. The decrease in rRNA-based diversity observed through the reactor compartments was greater than the decrease in rDNA-based diversity, which indicated that the analysis of metabolically active bacterial diversity discriminated more than analysis based on the mere presence of bacteria. The observed decrease in diversity suggests that the bacterial community became specialised in degrading less diversified substrates through the compartments. These findings suggest that rRNA-based analyses are better for monitoring reactor performance.

5.2. *Granulation and floc sizes*

Although granulation is not necessary in the ABR for optimum performance, unlike suspended systems such as the UASB, various reports have noted the appearance of granules in the reactor, and some authors have started up ABRs using granules. Boopathy and Tilche (1991) started up HABRs with a low initial loading rate (0.97 kg-COD·kg-VSS^{-1}·d^{-1}) and upflow velocities below 0.46 m·h^{-1}, to encourage the growth of granular biomass. Granules of 0.5 mm appeared after one month in all compartments, and were comprised primarily of acetoclastic methanogens. Similarly, Tilche and Yang (1987) found *Methanosarcina* coated flocs were held together by fibrous bacteria resembling *Methanosaeta*. The flocs, which were formed after one month, were small with diameters less than 1.5 mm, and weak. Boopathy and Tilche (1992) noticed similar particles of both types described above, which grew from 0.5 mm after one month to 3.5 mm after three months in a hybrid reactor. Granules of *Methanosarcina* clusters were low density, full of gas cavities, and therefore lifted to the surface of the reactor due to high gas and liquid velocities during high loading. The particle size appeared to be partially dependent on substrate type, and there was little difference in particle size throughout the reactor when molasses alcohol stillage wastewater was treated. However, two weeks after switching to raw molasses with a 10-fold increase in inlet COD a profile emerged which showed a steady decrease in particle size down the reactor. Orozco (1988) reported a similar decrease in granule size from 5.4 mm in the first chamber down to 1.5 mm in the last chamber of a reactor treating dilute carbohydrate waste, and Ji *et al.* (2009) also noted the same profile in an ABR treating oilfield produced water.

Uyanik *et al.* (2002) studied the effect of polymer addition on granule formation in comparison to a control and found that the granules were denser. They also noted a bacterial stratification within the granules. Baloch *et al.* (2008) started up a GRABBR with granules and found that after steady state those in the methanogenic zone of the reactor were stable and densely packed with smooth regular surface. In contrast, granules in the acidogenic zone were less stable with an irregular fissured surface. The granules consisted of a vast diversity of species from the outer surface to the core and possessed a multi-layered structure, and viruses in the granules suggest the presence of bacteriophage. These could be responsible for destroying cells and weakening the internal structure of granules, and thus possibly causing the breaking of granules. Protozoa-like microorganisms on the exterior of the granules are believed to play an important role as bacterial predators and control the growth of bacterial cells.

On a laboratory scale (Barber and Stuckey 1998) floc size seemed to grow to a maximum near the centre of an eight-compartment reactor and then reduce towards the rear, and this was confirmed by Freese and Stuckey (2000). Typical floc sizes were 100, 230 and 175 μm in the front, middle and rear compartments, respectively. These authors postulated that the floc size was a function of both gas production and COD concentration, with the largest particles growing when COD concentrations were sufficiently high to support growth, and gas production low enough to avoid floc breakage.

6. Soluble Microbial Products (SMPs)

The precise definition of SMP is operational, but is currently regarded as "the pool of organic compounds that are released into solution from substrate metabolism (usually with biomass growth) and biomass decay" (Noguera *et al.* 1994), although for anaerobic systems VFAs are not included as SMP. Some of the SMPs have been identified as humic and fulvic acids, polysaccharides, proteins, nucleic acids, organic acids, amino acids, antibiotics, steroids, exocellular enzymes, siderophores, structural components of cells, e.g., cell walls, and products of energy metabolism. Crucially, SMPs have been found to comprise the majority of soluble organics in effluents, and hence are of particular interest in achieving consent levels for biochemical oxygen demand (BOD) and COD. SMP also exhibit several characteristics, such as toxicity and metal chelating properties, which may effect the performance of the treatment system, and their presence has also been shown to adversely affect the kinetic activity and the flocculating and settling properties of sludge (Barker and Stuckey 1999). It became clear that due to the staged structure of the ABR, SMPs were easy to measure and follow down the reactor, and hence the ABR was a good tool to study these compounds.

Schiener *et al.* (1998) were the first to examine the influence of operating parameters such as temperature, HRT and OLR on the production of SMPs in an ABR fed a simple sucrose-nutrients substrate (4 g-COD$\cdot$l^{-1}, 20 h, 35 °C). They found that decreasing operating temperature resulted in increases in SMP production (2.3% [90 mg$\cdot$l^{-1} — at baseline conditions] in the effluent at 35 °C $\rightarrow$ 9% at 15 °C), a decreased rate of degradation, and hence increases in effluent SMPs. In addition, decreasing HRTs (2% at 20 h $\rightarrow$ 16% at 5 h), and increasing OLRs (3% when the feed was 15 g-COD$\cdot$l^{-1}), also resulted in increased effluent SMPs. Quantitatively, between 26% and 48% of the incoming feed COD was converted to SMPs in the first compartment of the ABR. Molecular weight (MW) fractionation revealed a bimodal distribution in the effluent with 25% >100 kDa, and 30% < 1 kDa, while all MW fractions were between 62% and 82% anaerobically degradable after 11 days. Analysis revealed most of the high MW fraction to be a heteropolysaccharide, probably cell wall fragments.

Barker *et al.* (1999) characterised effluents from several different anaerobic processes (ABR, UASB, CSTR, return liquors, with different feeds — synthetic and industrial) by their molecular weight (MW) distributions (using the techniques of ultrafiltration and size exclusion chromatography), biodegradability (both aerobic and anaerobic) and activated carbon adsorption characteristics. The majority (up to 89% w/w) of the material present in the effluents was in the low MW range (<1 kDa), although there was a significant amount (up to 22% w/w) in the high MW range (>300 kDa). Differences in the distributions were observed according to reactor type, with effluents from ABRs containing a higher proportion of high MW material than effluents from other reactor types, probably due to the long SRTs (up to 150 days) and high proportion of biomass-associated products (BAPs). Aerobic degradation of the effluents was significantly more successful than anaerobic, despite the fact that the low MW material was found to be the most difficult to degrade aerobically. Activated carbon adsorption revealed that the low MW material was more difficult to adsorb than the high MW compounds, and that the Freundlich model was the most accurate for predicting removal efficiencies.

Barker *et al.* (2000) used similar tools as above to investigate SMPs in an ABR treating low-strength wastewater. Results indicated that high MW (>300 kDa) compounds were produced in the middle compartments of the reactor and formed 22% of the effluent COD. This fraction was found to be 86% aerobically degradable but only 4% anaerobically. Low MW (<1 kDa) organics represented the highest portion (36%) of the effluent COD, and were mainly found in the first compartment of the ABR and in the effluent. This fraction was more easily degraded under anaerobic conditions (33%) than aerobic conditions (17%). Nuclear magnetic resonance (NMR) analysis of the low MW fraction revealed the possible presence of alcohol, carboxylate and aromatic groups. SMP production increased with increasing HRT, probably due to enhanced biomass decay at high HRTs, and also increased with decreasing temperature, probably due to increased stress on the biomass and a reduced metabolism of the SMP at low temperatures. Finally, SMP production with higher levels of initial biomass was greater than for an ABR operating at the same conditions but with lower levels of initial biomass.

Preliminary data from Aquino and Stuckey (2002) on the identification of SMPs using GC/MS surprisingly revealed the presence of long chain alkenes (C-12–C-24) and alkanes (C-12–C-16), as well as some aromatic compounds such as terephthalates in significant concentrations. These compounds which are likely to come from cell lysis and endogenous decay may not be easily biodegradable, hence their presence in the effluent is likely to result in the residual COD. Aquino and Stuckey (2003) also investigated the effect of nutrient deficiency on SMP production in anaerobic chemostats fed on glucose as the sole carbon source. Under steady-state conditions the SMP/COD_{out} ratio was as high as 95% and the normalised SMP production (SMP/S_o) averaged 3%. During nutrient deficiency the SMP/COD_{out} ratio averaged 45% indicating that the importance of SMP in the effluent was reduced due to the enhanced production of VFAs. Nevertheless, under such stressed conditions the normalised production of SMP (SMP/S_o) increased reaching up to 37%. Analysis of extracellular polymers (ECPs) indicated that ECP_{carb} production was enhanced after nutrient cessation, and chemical analyses of the effluent suggested that most of the SMPs was not protein — or carbohydrate-like material. DNA analysis indicated that part of the SMP produced during nutrient deficiency might be due to enhanced cell lysis, although some organics might have been deliberately excreted to scavenge metal nutrients. BMP assays showed that the lack of N and P resulted in the highest SMP production which might have been released to dump electrons that could not be used in cell synthesis due to the lack of these macronutrients.

Aquino and Stuckey (2004) investigated the effects of toxicity on SMP formation in anaerobic CSTRs fed on glucose ($\sim$10 g-COD·l^{-1}). Addition of the toxic compounds chloroform and chromium increased the net accumulation of SMP, despite reducing the percentage of SMP in the effluent due to the overwhelming production of VFAs. With chloroform the SMP/S_0 increased from 2% to 8%, whereas in the reactor spiked with Cr(VI) the SMP/S_0 ratio reached as high as 20%, and in both cases net SMP accumulation was proportional to the concentration of toxicant. After the spikes biomass seemed to produce more extracellular polymeric substances (EPS) suggesting that this might have helped them to cope with the stress. Chromatography results indicate that some of the high MW compounds present in the SMP might have been due to EPS release into the bulk solution, and that other compounds, probably released as a result of cell lysis, were also present. Hydrolysis of EPS did not seem to contribute to SMP accumulation in the presence of toxic compounds, and DNA analysis suggested that in the presence of chromium cell lysis products were an important contributor to SMP accumulation.

Feng *et al.* (2008b) confirmed past work that SMPs increase with decreasing temperature and increasing OLR (decreasing HRTs) in a CABR. Patidar and Tare (2008) investigated the effect of sulfide concentrations on SMP production and effect of complexation on soluble Fe, Ni, Zn and Co concentrations in laboratory-scale UASB, ABR and hybrid ABR (HABR) degrading synthetic sulfate laden organics at 35 °C. Results indicate increased SMP concentration with increasing sulfide concentration, and due to their ligand properties significantly higher soluble Fe (0.43–0.66 $mg{\cdot}l^{-1}$), Ni (0.29–0.4 $mg{\cdot}l^{-1}$), Zn (0.15–0.19 $mg{\cdot}l^{-1}$) and Co (0.13–0.29 $mg{\cdot}l^{-1}$) concentrations in all three systems which appear adequate for stable process operation and performance.

7. Modelling

Bachmann *et al.* (1983) found similar treatment behaviour under identical conditions in an ABR, anaerobic filter and a rotating biological disc reactor. They considered the sludge particles in the ABR to be fluidised spheres with a surface area through which the solute must diffuse, and used a combination of a fixed film (Williamson and McCarty 1976) along with a variable order model (Rittmann and McCarty 1978) which incorporated the concepts of liquid-layer mass transfer, Monod kinetics and molecular diffusion. The specific surface area in each reactor chamber was estimated from a data set, and then used to simulate behaviour at different loadings. Initial predictions were good, but the model underestimated the level of COD removal at higher loading rates due to an unrealistic assumption of a constant diffusion layer depth which would decrease at higher loading rates due to increased gas production. Bachmann *et al.* (1985) used this model with a feed of 8 $g\text{-}COD{\cdot}l^{-1}$ and predicted: a decrease in treatment efficiency with: (a) decreasing influent substrate concentration at constant loading rates, (b) an increase in organic loading at constant influent substrate concentration, and (c) an increase in recycle ratio at constant HRT since the reactor approaches completely mixed behaviour. Reactor efficiency improved with reducing substrate concentration at constant HRT. Most of these findings were mirrored in the work of Xing *et al.* (1991) on the modelling of a HABR (150 l) treating 20 $kg\text{-}COD{\cdot}m^{-3}{\cdot}d^{-1}$ of molasses wastewater. The model assumed biomass weight and not concentration was the critical parameter, and that all substrate consumption occurred within a perfectly mixed granular sludge bed due to gas evolution.

Bachmann *et al.* (1983, 1985) assumed that the floc diameter was very large relative to the active biofilm depth. However, at high loading rates this is not the case in the ABR since sludge particles within the reactor act as fluidised spheres with a surface area through which the substrate must diffuse for consumption (Bachmann *et al.* 1985). These facts imply that a spherical model would provide a better fit than a simple planar one.

Nachaiyasit (1995) derived a spherical model using Monod kinetics combined with molecular diffusion of the substrate into the biomass aggregates based on the assumptions that substrate concentration could be described by COD, and biomass concentration by VSS. The calculation of diffusion layer thickness, and liquid phase mass transfer coefficient used the methods of Bachmann *et al.* (1985). In general, the model predicted better COD removal than was experimentally measured, and was most accurate for high loading rates (8 and 15 g-COD·l^{-1} at 20 hours) than at short retention times (10, 5 hours HRT with feed concentration of 4 g-COD·l^{-1}), but showed large deviations for the first couple of compartments for some of the simulations. However, the spherical model did provide a closer fit than the earlier planar fixed-film equations. A sensitivity analysis showed floc surface area and flowrate had the greatest influence on model predictions, so the model was modified using surface area as a fitting parameter. Nachaiyasit (1995) then compared the results obtained from different models based on the same assumptions and with the same experimental data, and found the closest fit with the spherical model.

Intrachandra (2001) followed up the work of Nachaiyasit by developing a hydrodynamic model to describe the ABR on the basis of reactor hydraulics, biological reaction kinetics and mass transfer within the biomass particle. A separate three-compartment ABR was used to enable accurate measurements to be made between compartments. A flow model of a CSTR followed by a dispersion plug flow reactor (PFR) was used to simulate the reactor hydrodynamics observed from tracer studies (Levenspiel 1972). The hydrodynamic model was evaluated by using parameters obtained from the flow model and then compared with the experimental data obtained from acetate impulse experiments. Simulation results revealed a significant effect of the specific area of the biomass, and showed that not all the surface area of the biomass present was used in substrate transport and consumption; hence the specific area of the biomass was used as a fitting parameter. After modifying the model it gave a better fit and suggested that none of the three factors investigated, i.e., kinetics, mass transfer and hydraulics, can be ignored in modelling an ABR.

Stamatelatou *et al.* (2003) modelled a PABR as two variable volume interacting sections, of constant total volume, one with high solids and the other with low solids, with the gas and liquid flows influencing the material flows between the two sections. The biomass was divided into acidogenic, acetogenic and methanogenic groups of microorganisms. The kinetic part of the model accounted for possible inhibition of acidogenesis, acetogenesis and methanogenesis by VFAs, and the model succeeded in predicting the reactor performance with step increases in the organic loading rate.

Kennedy and Barriault (2007) attempted to use a simple first-order compartment model for substrate removal, but it did not describe ABR performance well. The rate coefficients ($k_{(1)}$–$k_{(4)}$) or $k_{(avg)}$ were inconsistent, and no predictive correlation of k values with substrate concentration, HRT or OLR could be achieved in describing the naturally attenuated two-phase phenomenon.

In summary, it seems that the best model to use is similar to the one developed by Intrachandra (2001) based on reactor hydraulics, biological reaction kinetics and mass transfer within the biomass particle. While complex, and using surface area as a fitting parameter, it does manage to predict performance quite well, even under transient conditions. However, a number of other phenomena need to be incorporated into this model. Firstly, the structure of the reactor is known to cause a shift in the population dynamics of the two species (*Methanosarcina* and *Methanosaeta*) responsible for acetate consumption. Since both archae differ widely in both rate and K_S, acetate loadings and pH will have an effect on reactor performance in each compartment, and this needs to be incorporated. Secondly, channelling is an important phenomenon in the ABR (see *Reactor Hydrodynamics*), and will affect the accuracy of any model. Therefore a compromise will exist between optimal (required) compartment number, "safe" compartment number, and also upflow liquid velocity. Finally, it is also necessary to model reactor behaviour when hydrolysis is the rate-limiting step, as is the case with high solid influents and lipid containing wastewaters, since by assuming first-order rate kinetics it is possible to calculate the minimum solids retention time to achieve a given efficiency (Pavlostathis and Giraldo-Gomez 1991). With wastewaters containing a large amount of particulate material, it seems likely that COD removal will be low at the front end, and that the VFA profile formed will be shifted down the reactor; unless the reactor is modified, or extra compartments added, a drop of efficiency may result.

8. Full-scale Experience

The performance data of a full-scale plant, treating domestic waste from a small town in Colombia (Tenjo, pop. < 2,500 inhabitants — Orozco 1997), is presented in Barber and Stuckey (1999). The Tenjo reactors were designed to give a liquid upflow velocity of $3\ m{\cdot}h^{-1}$ based on laboratory findings. However, despite following strict guidelines for start-up, the reactors experienced several practical problems during early operation. Hydraulic shocks increased solids washout, and poor screening of solid material caused the plastic packing media to float with gas production. These problems were overcome by using a by-pass pipe during the rainy season and improved screening facilities. The reactors performed well with approximately 70% COD reduction and 80% removal of suspended solids over a two-month period. Varying the volumetric load between 0.4 and $2\ kg{\cdot}m^{-3}{\cdot}d^{-1}$ had no effect on removal efficiency. However, the author concluded that a polishing lagoon was necessary to achieve discharge quality effluent. Although a detailed economic study was not presented, construction costs for the baffled reactor were 20% less than those for UASB reactors in Colombia running at ambient temperature, and five times less than a conventional activated sludge plant for a small town.

In 2000, a $30\ m^3$ pilot plant containing five separate compartments was built in the UK at Ellesmere Port in the UK running on screened sewage. At a 6 h HRT through a whole year, the reactor achieved around 50% COD removal and 50% solids removal. The effluent was primarily VFAs, and hence with a longer HRT higher COD removals could have been achieved. While little of the work has been published in academic journals, a German NGO has taken the ABR out into the field and used it for DEWATS (Decentralized Wastewater Treatment Systems). At present there are a total of 707 plants taking an average flow of $20–30\ m^3{\cdot}d^{-1}$ in South Africa (see Fig. 8.2 of pilot-scale plant in KwaZulu Natal), India, Nepal, Indonesia, China, Laos, Cambodia, Vietnam and Thailand, while the largest plant in India taking hospital wastewaters is designed for $840\ m^3{\cdot}d^{-1}$. A number of large-scale industrial plants (3) have been built in Thailand recently running on tapioca starch wastewaters ($>32\ g\text{-}COD{\cdot}l^{-1}$, TSS $> 15\ g{\cdot}l^{-1}$ and a pH of 3.8–4.2) using a low cost CIGAR ABR with a plastic covering (Waste Solutions Ltd., Dunedin, NZ). With a $7{,}000\ m^3{\cdot}d^{-1}$ flowrate and an unknown HRT these reactors are removing 99% of the COD, BOD and TSS, with an effluent pH of 7.1. This has resulted in a huge reduction in GHG emissions, and energy self-sufficiency.

Fig. 8.2. Installation of pilot-scale ABR in Durban, South Africa (picture courtesy of C. Buckley).

9. Conclusions

Based on this review of recent literature on the ABR it can be concluded that:

- The design is unique amongst anaerobic reactor designs in that its compartments (up to 8) form a "staged" reactor which separates different trophic groups, most specifically *Methanosarcina* sp. in the front of the reactor from *Methanosaeta* sp. due to their different kinetics and half-rate constants. This staging makes the ABR very robust to shock loads.
- The hydrodynamics of the reactor reveal that it approximates a series of CSTRs with low to medium (8%–37%) hydraulic and biological dead space depending on HRT, OLR, gas production and biomass concentration without necessarily being strongly correlated to any specific parameter. The reactor is very stable to both organic and hydraulic shock loads, and in the latter case channels form through the bed minimising washout of biomass.
- Recycle usually results in poorer COD removal, however, with toxins in the feed it can enhance performance.
- Start-up of the ABR should be carried out initially at low space loads (<1.2 kg-COD·m^{-3}·d^{-1}) until a stable population has been established, and it is better to do this with a constant feed strength and reducing the HRT, rather than vice versa. The use of granules in the ABR enables higher initial loads to be used at start-up provided the granules used come from similar waste streams.
- The ABR is capable of treating both low strength (0.3 g-COD·l^{-1}) and high strength (115 g-COD·l^{-1}) wastewaters, and could achieve 95% COD removal at 10 hours HRT with low strength and even at 5 g-COD·l^{-1} could remove 94% COD at 6 hours HRT.
- Low temperature performance (10–15 °C) can still result in reasonable removal (83% at 15 °C at 10 h, with 60%–43% at 10 °C), and the unmetabolised COD is a mixture of VFAs and SMPs.
- High levels of sulfate can be tolerated well due to the staged nature of the reactor, and COD:SO_4^{2-} ratios as low as 1 can treated with very high activities of sulfate reducers in the front of the reactor.

- The addition of granules to an ABR (GRABBR) result in increased stability and potentially higher organic loads.
- Soluble microbial products (SMPs) are produced in the ABR and vary down the reactor. SMPs set the lower limit on COD removal, and their production is enhanced with decreasing temperature, HRT, reduced nutrient supply, and increasing OLR and toxins. Their MW is bimodal with some as high as 300 kDa, while many are below 1 kDa.
- Models of ABR performance have been developed based on microbial kinetics, mass transfer (through the floc surface), and hydrodynamics, and while the later ones do a good job predicting COD removal, we often need to fit the biomass surface area to get good predictions.
- Over 700 small (20–30 $m^3 \cdot d^{-1}$) ABRs are now being used for low cost sanitation (DEWATS) in SE Asia, and at least four large-scale (7,000 $m^3 \cdot d^{-1}$) plants are running in Thailand treating starch wastewaters and pig slurry.

References

Akunna, J. C. and Clark, M. (2000) Performance of a granular-bed anaerobic baffled reactor (GRABBR) treating whisky distillery wastewater. *Bioresour. Technol.* **74**(3), 257–261.

Alves, M. M., Pereira, M. A., Mota, M., Novais, J. M., and Colleran, E. (1997) Staged and non staged anaerobic filters: Microbial selection, hydrodynamic aspects and performance. In: *Proceedings of the 8th International Conference on Anaerobic Digestion*, Vol. 2, Sendai, Japan, pp. 56–63.

Angenent, L. T., Zheng, D., Sung, S., and Raskin, L. (2000) *Methanosaeta* fibers in anaerobic migrating blanket reactors. *Wat. Sci. Technol.* **41**(4–5), 35–39.

Angenent, L. T. and Sung, S. W. (2001) Development of anaerobic migrating blanket reactor (AMBR), a novel anaerobic treatment system. *Wat. Res.* **35**(7), 1739–1747.

Angenent, L. T., Banik, G. C., and Sung, S. W. (2001) Anaerobic migrating blanket reactor treatment of low-strength wastewater at low temperatures. *Wat. Environ. Res.* **73**, 5567–5574.

Aquino, S. F. and Stuckey, D. C. (2002) Characterization of soluble microbial products (SMP) in effluents from anaerobic reactors. *Wat. Sci. Technol.* **45**(10), 127–132.

Aquino, S. F. and Stuckey, D. C. (2003) Production of soluble microbial products (SMP) in anaerobic chemostats under nutrient deficiency. *J. Environ. Eng.* **129**(11), 1007–1014.

Aquino, S. F. and Stuckey, D. C. (2004) Soluble microbial products formation in anaerobic chemostats in the presence of toxic compounds. *Wat. Res.* **38**(2), 255–266.

Arnirfakhri, J., Vossoughi, M., and Soltanieh, M. (2006) Assessment of desulfurization of natural gas by chemoautotrophic bacteria in an anaerobic baffled reactor (ABR). *Chem. Eng. Process.* **45**(3), 232–237.

Bachmann, A., Beard, V. L., and McCarty, P. L. (1983) Comparison of fixed film reactors with a modified sludge blanket reactor. In: Wu, Y. C. and Smith, E. D. (eds.), *Fixed Film Biological Processes for Wastewater Treatment*. N.J: Noyes Data Corp.

Bachmann, A., Beard, V. L., and McCarty, P. L. (1985) Performance characteristics of the Anaerobic Baffled Reactor. *Wat. Res.* **19**(1), 99–106.

Baloch, M. I. and Akunna, J. C. (2003) Granular bed baffled reactor (GRABBR): Solution to a two-phase anaerobic digestion system. *J. Environ. Eng.* **129**(11), 1015–1021.

Baloch, M. I., Akunna, J. C., Kierans, M., and Collier, P. J. (2008) Structural analysis of anaerobic granules in a phase separated reactor by electron microscopy. *Bioresour. Technol.* **99**(5), 922–929.

Barber, W. P. and Stuckey, D. C. (1998) Influence of start-up strategies on the performance of an anaerobic baffled reactor. *Environ. Technol.* **19**, 489–501.

Barber, W. P. and Stuckey, D. C. (1999) The use of the anaerobic baffled reactor (ABR) for wastewater treatment: A review. *Wat. Res.* **33**(7), 1559–1578.

Barber, W. P. and Stuckey, D. C. (2000a) Nitrogen removal in a modified anaerobic baffled reactor (ABR): 1, denitrification. *Wat. Res.* **34**(9), 2413–2422.

Barber, W. P. and Stuckey, D. C. (2000b) Nitrogen removal in a modified anaerobic baffled reactor (ABR): 2, nitrification. *Wat. Res.* **34**(9), 2423–2432.

Barber, W. P. and Stuckey, D. C. (2000c) Effect of sulphate reduction on chemical oxygen demand removal in an anaerobic baffled reactor. *Wat. Environ. Res.* **72**(5), 593–601.

Barber, W. P. and Stuckey, D. C. (2000d) Metal bioavailability and trivalent chromium removal in the ABR. *J. Environ. Eng.* **126**(7), 649–656.

Barker, D. J. and Stuckey, D. C. (1999) A review of soluble microbial products (SMP) in wastewater treatment systems. *Wat. Res.* **33**(14), 3063–3082.

Barker, D. J., Mannucchi, G. A., Salvi, S. M. L., and Stuckey, D. C. (1999) Characterisation of soluble residual chemical oxygen demand (COD) in anaerobic wastewater treatment effluents. *Wat. Res.* **33**(11), 2499–2510.

Barker, D. J., Salvi S. M. L., Langenhoff, A. A. M., and Stuckey, D. C. (2000) Soluble microbial products in ABR treating low-strength wastewater. *J. Environ. Eng.* **126**(3), 239–249.

Bayrakdar, A., Sahinkaya, E., Gungor, M., Uyanik, S., and Atasoy, A. D. (2009) Performance of sulfidogenic anaerobic baffled reactor (ABR) treating acidic and zinc-containing wastewater. *Bioresour. Technol.* **100**(19), 4354–4360.

Bell, J., Plumb, J. J., Buckley, C. A., and Stuckey, D. C. (2000) Treatment and decolorization of dyes in an anaerobic baffled reactor *J. Environ. Eng.* **126**(11), 1026–1032.

Bell, J. and Buckley, C. A. (2003) Treatment of a textile dye in the anaerobic baffled reactor. *Water SA* **29**(2), 129–134.

Bodkhe, S. Y. (2009) A modified anaerobic baffled reactor for municipal wastewater treatment. *J. Environ. Manage.* **90**(8), 2488–2493.

Boopathy, R. and Tilche, A. (1991) Anaerobic-digestion of high-strength molasses waste-water using a hybrid anaerobic baffled reactor. *Wat. Res.* **25**(7), 785–790.

Boopathy, R. and Tilche, A. (1992) Pelletization of biomass in a hybrid anaerobic baffled reactor (HABR) treating acidified wastewater. *Bioresour. Technol.* **40**(2), 101–107.

Briones, A. and Raskin, L. (2003) Diversity and dynamics of microbial communities in engineered environments and their implications for process stability. *Curr. Opin. Biotechnol.* **14**(3), 270–276.

Cavicchioli, R. (2006) Cold-adapted archaea. *Nat. Revs.-Microbiol.* **4**, 331–343.

Chynoweth, D. P., Srivastra, V. J., and Conrad, J. R. (1980) *Research Study to Determine the Feasibility of Producing Methane Gas from Sea Kelp.* Annual Report for General Electric Company, IGT Project 30502, Institute of Gas Technology, IL, USA.

Cohen, A., Breure, A. M., van Andel, J. G., and van Deursen, A. (1980) Influence of phase separation on the anaerobic digestion of glucose — I. Maximum COD-turn-over-rate during continuous operation. *Wat. Res.* **14**, 1439–1448.

Cohen, A., Breure, A. M., van Andel, J. G., and van Deursen, A. (1982) Influence of phase separation on the anaerobic digestion of glucose-II. Stability and kinetic responses to shock loadings. *Wat. Res.* **16**, 449–455.

El-Mamouni, R., Rouleau, D., Mayer, R., Guiot, S. R., and Samson, R. (1992) Comparison of the novel multiplate anaerobic reactor with the upflow anaerobic sludge blanket reactor. In: *46th Purdue Industrial Waste Conference Proceedings*, Chelsea, MI, USA: Lewis Publishers, Inc.

Fannin, K. F., Srivastra, V. J., Conrad, J. R., and Chynoweth, D. P. (1981) *Marine Biomass Program: Anaerobic Digester System Development.* Annual Report for General Electric Company, IGT Project 65044, Institute of Gas Technology, Chicago, IL, USA.

Fannin, K. F., Srivastra, V. J., Mensinger, J., Conrad, J. R., and Chynoweth, D. P. (1982) *Marine Biomass Program: Anaerobic Digester System Development.* Annual Report for General Electric Company, IGT Project 65044 and 30547, Institute of Gas Technology, Chicago, IL, USA.

Feng, H. J., Hu, L. F., Mahmood Q., Fang, C. R., Qiu, C. D., and Shen, D. S. (2009) Effects of temperature and feed strength on a carrier anaerobic baffled reactor treating dilute wastewater. *Desalination* **239**(1–3), 111–121.

Feng, H. J., Hu, L. F., Shan, D., Fang, C. R., and Shen, D. S. (2008a) Effects of temperature and hydraulic residence time (HRT) on treatment of dilute wastewater in a carrier anaerobic baffled reactor. *Biomed. Environ. Sci.* **21**(6), 460–466.

Feng, H. J., Hu, L. F., Shan, D., Fang, C. R., He, Y. H., and Shen, D. S. (2008b) Effects of operational factors on soluble microbial products in a carrier anaerobic baffled reactor treating dilute wastewater. *J. Environ. Sci. China* **20**(6), 690–695.

Fox, P. and Venkatasubbiah, V. (1996) Coupled anaerobic/aerobic treatment of high-sulphate wastewater with sulphate reduction and biological sulphide oxidation. *Wat. Sci. Technol.* **34**(5–6), 359–366.

Freese, L. H. and Stuckey, D. C. (2000) Influence of seed inoculum on the start-up of an anaerobic baffled reactor. *Environ. Technol.* **21**(8), 909–918.

Freese, L. H. and Stuckey, D. C. (2004) Anaerobic treatment of sulphate-enriched wastewaters. *Proc. Institut. Civil Eng. Wat. Management* **157**(4), 187–195.

Ghaniyari-Benis, S., Borja, R., Monemian, S. A., and Goodarzi, V. (2009) Anaerobic treatment of synthetic medium-strength wastewater using a multistage biofilm reactor. *Bioresour. Technol.* **100**(5), 1740–1745.

Ghosh, S., Conrad, J. R., and Klass, D. L. (1975) Anaerobic acidogenesis of sewage sludge. *JWPCF* **47**, 30–45.

Grobicki, A. M. W. (1989) *Hydrodynamic Characteristics and Performance of the Anaerobic Baffled Reactor*. PhD Dissertation, Department of Chemical Engineering, Imperial College, London, UK.

Grobicki, A. M. W. and Stuckey, D. C. (1991) Performance of the anaerobic baffled reactor under steady state and shock loading conditions. *Biotechnol. Bioeng.* **37**, 344–355.

Grobicki, A. M. W. and Stuckey, D. C. (1992) Hydrodynamic characteristics of the anaerobic baffled reactor. *Wat. Res.* **26**, 371–378.

Grover, R., Marwaha, S. S., and Kennedy, J. F. (1999) Studies on the use of an anaerobic baffled reactor for the continuous anaerobic digestion of pulp and paper mill black liquors. *Process Biochem.* **34**(6–7), 653–657.

Guiot, S. R., Safi, B., Frigon, J. C., Mercier, P., Mulligan, C., Tremblay, R., and Samson, R. (1995) Performances of a full-scale novel multiplate anaerobic reactor treating cheese whey effluent. *Biotechnol. Bioeng.* **45**, 398–495.

Gujer, W., Zehnder, A. J. B. (1983) Conversion processes in anaerobic digestion. *Wat. Sci. Technol.* **15**(8–9), 127–167.

Hassouna, S. (1998) Personal Communication.

He, S. J., Wang, J. L., Luo, Y., and Sun, L. L. (2009) The performance of a modified anaerobic baffled reactor (ABR) treating high strength synthetic wastewater. *Intl. J. Environ. Pollut.* **37**(2–3), 309–318.

Henze, M. and Harremoës, P. (1983) Anaerobic treatment of wastewater in fixed film reactors — a literature review. *Wat. Sci. Technol.* **15**(8/9), 1–101.

Holt, C. J., Matthew, R. G. S., and Terzis, E. (1997) A comparative study using the anaerobic baffled reactor to treat a phenolic wastewater. In: *Proceedings of the 8th International Conference on Anaerobic Digestion*, Vol. 2, Sendai, Japan, pp. 40–47.

Intrachandra, N. (2001) *Anaerobic Digestion of a Complex Dilute Colloidal Wastewater Using the Anaerobic Baffled Reactor (ABR)*. PhD thesis, Department of Chemical Engineering, Imperial College London, UK.

Iza, J., Colleran, E., Paris, J. M., and Wu, W.-M. (1991) International workshop on anaerobic treatment technology for municipal and industrial wastewaters: Summary paper. *Wat. Sci. Technol.* **24**(8), 1–16.

Ji, G. D., Sun, T. H., Ni, J. R., and Tong, J. J. (2009) Anaerobic baffled reactor (ABR) for treating heavy oil produced water with high concentrations of salt and poor nutrient *Bioresour. Technol.* **100**(3), 1108–1114.

Kato, M. T., Field, J. A., and Lettinga, G. (1997) The anaerobic treatment of low strength wastewaters. In: *Proceedings of the 8th International Conference on Anaerobic Digestion*, Vol. 1, Sendai, Japan, pp. 356–363.

Kennedy, J. F., Panesar, P. S., Grover, R., and Marwaha, S. S. (2006) Continuous methanogenesis of black liquor of pulp and paper mills in an anaerobic baffled reactor using an immobilized cell system. *J. Chem. Technol. Biotechnol.* **81**(7), 1277–1281.

Kennedy, K. and Barriault, M. (2007) Treatment kinetics of aircraft deicing fluid in an anaerobic baffled reactor. *J. Environ. Eng. Sci.* **6**(1), 11–17.

Krishna, G. V. T. G., Kumar, P., and Kumar, P. (2008) Treatment of low strength complex wastewater using an anaerobic baffled reactor (ABR). *Bioresour. Technol.* **99**(17), 8193–8200.

Langenhoff, A. A. M., Intrachandra, N., and Stuckey, D. C. (2000) Treatment of dilute soluble and colloidal wastewater using an anaerobic baffled reactor: Influence of hydraulic retention time. *Wat. Res.* **34**(4), 1307–1317.

Langenhoff, A. A. M. and Stuckey, D. C. (2000) Treatment of dilute wastewater using an anaerobic baffled reactor: Effect of low temperature. *Wat. Res.* **34**(15), 3867–3875.

Lawrence, A. W. and McCarty, P. L. (1969) Kinetics of methane fermentation in anaerobic treatment. *J. WPCF*, Supplement R2.

Lettinga, G., Van Velsen, A. F. M., Hobma, S. W., de Zeeuw, W., and Klapwijk, A. (1980) Use of the upflow sludge blanket (USB) reactor concept for biological wastewater-treatment, especially for anaerobic treatment. *Biotechnol. Bioeng.* **22**(4), 699–734.

Levenspiel, O. (1972) *Non Ideal Flow. In: Chemical Reaction Engineering*. 2nd ed. New York: John Wiley and Sons, pp. 253–308.

Manariotis, I. D. and Grigoropoulos, S. G. (2002) Low-strength wastewater treatment using an anaerobic baffled reactor. *Wat. Environ. Res.* **74**(2), 170–176.

McCarty, P. L. (1981) One hundred years of anaerobic treatment. In: Hughes *et al.* (eds.), *Anaerobic Digestion*. Elsevier Biomedical Press B.V.

Nachaiyasit, S. (1995) *The Effect of Process Parameters on Reactor Performance in an Anaerobic Baffled Reactor*. PhD Dissertation, Department of Chemical Engineering, Imperial College, London, UK.

Nachaiyasit, S. and Stuckey, D. C. (1995) Microbial response to environmental changes in an anaerobic baffled reactor (ABR). *Antonie van Leeuwenhoek* **67**, 111–123.

Nachaiyasit, S. and Stuckey, D. C. (1997a) The effect of low temperature on the performance of an anaerobic baffled reactor (ABR). *J. Chem. Technol. Biotechnol.* **69**, 276–284.

Nachaiyasit, S. and Stuckey, D. C. (1997b). The effect of shock loads on the performance of an anaerobic baffled reactor (ABR): 2, step and transient hydraulic shocks at constant feed strength. *Wat. Res.* **31**, 2747–2755.

Noguera, D. R., Araki, N., and Rittmann, B. E. (1994) Soluble microbial products (SMP) in anaerobic chemostats. *Biotechnol. Bioeng.* (9), 1040–1047.

Orozco, A. (1988) Anaerobic wastewater treatment using an open plug flow baffled reactor at low temperature. In: *5th International Symposium on Anaerobic Digestion*, Bologna, Italy, pp. 759–762.

Orozco, A. (1997) Pilot and full-scale anaerobic treatment of low-strength wastewater at sub-optimal temperature (15 °C) with a hybrid plug flow reactor. In: *Proceedings of the 8th International Conference on Anaerobic Digestion*, Vol. 2, Sendai, Japan, pp. 183–191.

Owen, W. F., Stuckey, D. C., Healey, J. B., Young, L. Y., and McCarty, P. L. (1979) Bioassay for monitoring biochemical methane potential and anaerobic toxicity. *Wat. Res.* **13**(6), 485–492.

Patidar, S. K. and Tare, V. (2004) Effect of micro-nutrients in anaerobic degradation of sulphate laden organics. *Can. J. Civil Eng.* **31**(3), 420–431.

Patidar, S. K. and Tare, V. (2008) Soluble microbial products formation and their effect on trace metal availability during anaerobic degradation of sulphate laden organics. *Wat. Sci. Technol.* **58**(4), 749–755.

Pavlostathis, S. G. and Giraldo-Gomez, E. (1991) Kinetics of anaerobic treatment: A critical review. *Crit. Rev. Environ. Control* **21**(5,6), 411–490.

Pillay, S., Foxon, K. M., and Buckley, C. A. (2008) An anaerobic baffled reactor/membrane bioreactor (ABR/MBR) for on-site sanitation in low income areas. *Desalination* **231**, 91–98.

Plumb, J. J., Bell, J., and Stuckey, D. C. (2001) Microbial populations associated with treatment of an industrial dye effluent in an anaerobic baffled reactor. *Appl. Environ. Microbiol.* **67**(7), 3226–3235.

Pohland, F. G. and Ghosh (1971) Developments in anaerobic treatment processes. In: Canale, R. P. (ed.), *Biological Waste Treatment*. pp. 85–106, New York: Interscience.

Polprasert, C., Kemmadamrong, P., and Tran, F. T. (1992) Anaerobic baffled reactor (ABR) process for treating a slaughterhouse wastewater. *Environ. Technol.* **13**, 857–865.

Rittmann, B. E. and McCarty, P. L. (1978) Variable-order model of bacterial-film kinetics. *J. Environ. Eng.* **104**, 889–900.

Rostron, W. M., Stuckey, D. C., and Young, A. A. (2001) Nitrification of high strength ammonia wastewaters: Comparative study of immobilisation media. *Wat. Res.* **35**(5), 1169–1178.

Roy, C. S., Talbot, G., Topp, E., Beaulieu, C., Palin, M. F., and Masse, D. I. (2009) Bacterial community dynamics in an anaerobic plug-flow type bioreactor treating swine manure. *Wat. Res.* **43**(1), 21–32.

Saritpongteeraka, K. and Chaiprapat, S. (2008) Effects of pH adjustment by parawood ash and effluent recycle ratio on the performance of anaerobic baffled reactors treating high sulphate wastewater. *Bioresour. Technol.* **99**(18), 8987–8994.

Schiener, P., Nachaiyasit, S., and Stuckey, D. C. (1998) Production of soluble microbial products (SMP) in an anaerobic baffled reactor: Composition, biodegradability, and the effect of process parameters. *Environ. Technol.* **19**(4), 391–399.

Skiadas, I. V. and Lyberatos, G. (1998) The periodic anaerobic baffled reactor. *Wat. Sci. Technol.* **38**(8–9), 401–408.

Skiadas, I. V., Gavala, H. N., and Lyberatos, G. (2000) Modelling of the periodic anaerobic baffled reactor (PABR) based on the retaining factor concept. *Wat. Res.* **34**(15), 3725–3736.

Stamatelatou, K., Lokshina, L., Vavilin, V., and Lyberatos, G. (2003) Performance of a glucose fed periodic anaerobic baffled reactor under increasing organic loading conditions: 2. Model prediction. *Biotechnol. Bioeng.* **88**(2), 137–142.

Stuckey, D. C. (1983) Anaerobic digestion in developing countries. *Adv. Ferment.*, SCI meeting at Chelsea College, London.

Switzenbaum, M. S. and Jewell, W. J. (1980) Anaerobic attached-film expanded-bed reactor treatment. *JWPCF* **52**(7), 1953–1965.

Tait, S. J. and Freidman, A. A. (1980) Anaerobic rotating biological contactor for carbonaceous wastewaters. *JWPCF* **52**(8), 2257–2269.

Tilche, A. and Yang, X. (1987) Light and scanning electron microscope observations on the granular biomass of experimental SBAF and HABR reactors. In: *Proceedings of GASMAT Workshop*, Netherlands, pp. 170–178.

Tilche, A. and Vieira, S. M. M. (1991) Discussion on reactor design of anaerobic filters and sludge bed reactors. *Wat. Sci. Technol.* **24**(8), 193–206.

Uyanik, S., Sallis, P. J., and Anderson, G. K. (2002) The effect of polymer addition on granulation in an anaerobic baffled reactor (ABR). Part II: compartmentalization of bacterial populations. *Wat. Res.* **36**(4), 944–955.

van Lier, J. B., Boersma, F., Debets, M. M. W. H., and Lettinga, G. (1994) High rate thermophilic anaerobic wastewater treatment in compartmentalized upflow reactors. *Wat. Sci. Technol.* **30**(12), 251–261.

van Lier, J. B., Groeneveld, N., and Lettinga, G. (1996) Development of thermophilic methanogenic sludge in compartmentalized upflow reactors. *Biotechnol. Bioeng.* **50**, 115–124.

Vossoughi, M., Shakeri, M., and Alemzadeh, I. (2003) Performance of anaerobic baffled reactor treating synthetic wastewater influenced by decreasing COD/SO_4^{2-} ratios. *Chem. Eng. Process.* **42**(10), 811–816.

Wang, R. M., Wang, Y., Ma, G. P., He, Y. F., and Zhao, Y. Q. (2009) Efficiency of porous burnt-coke carrier on treatment of potato starch wastewater with an anaerobic-aerobic bioreactor. *Chem. Eng. J.* **148**(1), 35–40.

Weiland, P., and Rozzi, A. (1991) The start-up operation and monitoring of high-rate anaerobic treatment systems: Discusser's report. *Wat. Sci. Technol.* **24**(8), 257–277.

Williamson, K., and McCarty, P. L. (1976) A model of substrate utilization by bacterial films. *JWPCF* **48**(1), 9–24.

Xing, J., Boopathy, R., and Tilche, A. (1991) Model evaluation of hybrid anaerobic baffled reactor treating molasses wastewater. *Biomass Bioenergy* **1**(5), 267–274.

Yang, P. Y. and Moengangongo, T. H. (1987) Operational stability of a horizontally baffled anaerobic reactor for diluted seine wastewater in the tropics. *Trans. ASAE* **30**(4), 1105–1110.

Yang, X., Garuti, G., Farina, R., Parisi, V., and Tilche, A. (1988) Process differences between a sludge bed filter and an anaerobic baffled reactor treating soluble wastes. In: *5th International Symposium on Anaerobic Digestion*, Bologna, Italy, pp. 355–360.

Young, H. W. and Young, J. C. (1988) Hydraulic characteristics of upflow anaerobic filters. *J. Environ. Eng.* **114**(3), 621–638.

Chapter 9

Anaerobic Treatment of Phenolic Wastewaters

Dawei Liang[1] and Herbert H.P. Fang[2]

[1]*Beihang University (BUAA), School of Chemistry and Environment, Beijing, PR China*
liangdw@buaa.edu.cn

[2]*The University of Hong Kong, Department of Civil Engineering, Hong Kong SAR, China*
hrechef@hkucc.hku.hk

Phenolic pollutants are commonly found in wastewaters from synthetic chemicals, pesticides, coal conversion, pulp-paper, oil-refining, plastics, and pharmaceuticals industries. They are of significant environmental concerns because of their toxicity and potential carcinogenic property. The upflow anaerobic sludge blanket (UASB) process has been found effective for the degradation of many recalcitrant pollutants, including phenols. However, phenolic pollutants at high concentrations in wastewater may also inhibit the bioactivities of the granular sludge and consequently lower the treatment efficiency. This paper is to review literature on the anaerobic treatment of phenolic pollutants with emphasis on the following aspects: (1) phenolic inhibitory effect to the anaerobes, and the tolerance levels; (2) treatment of phenolic wastewater in anaerobic reactor of various designs; (3) effects of operating factors, including hydraulic retention time, effluent recirculation, temperature, loading shock and co-substrate on the biodegradation of phenols; and (4) the anaerobes and the degradation pathway involved in anaerobic degradation of phenols. Overall, phenolic pollutants in wastewater, when operated below the inhibitory concentrations and avoiding the loading shocks, can be effectively degraded under anaerobic processes with proper sludge acclimation, effluent recirculation and, if necessary, co-substrates.

1. Introduction

Phenol is a chemical that is commonly used as a raw material in the production of phenolic resins, aromatic chemicals, and pesticides. Phenol and its derivatives, generally termed "phenols", are also found in the industrial wastewaters from the manufacturing of phenolic resins, synthetic chemicals, pesticides, coal conversion, pulp-paper, oil-refining, and pharmaceuticals. Industrial wastewater may contain up to 17,000 $mg \cdot l^{-1}$ of phenols. They may contribute 30–80% of chemical oxygen demand (COD) in the wastewater of coal gasification, lignite gasification, coal liquefaction, coke plant, petroleum refinery, oil refineries, general pharmaceuticals, herbicide manufacturing, plastics factory, and phenolic resin productions (Veeresh *et al.* 2005). Phenols are dangerous organic pollutants because of their potential carcinogenic and mutagenic properties; most of the phenols are priority pollutants identified by United States Environmental Protection

Agency (USEPA). Thus, the release of phenols into the environment is of great concern (IARC 1999), and their discharge limit is below 0.5 mg·l^{-1} in some cases (Tay *et al.* 2001).

Biodegradation of phenols in wastewater, when operated properly, is generally more cost-effective than chemical degradation. Biodegradation can be conducted either aerobically or anaerobically. Between the two, anaerobic treatment is preferred since it does not require the energy for the costly aeration and produces substantially lower amount of sludge, plus the production of methane which is a readily useful fuel. As early as 1964, Chemielowski *et al.* (1964) found that phenol could be degraded under methanogenic condition, which was then confirmed by Healy and Young (1978). Later, Boyd *et al.* (1983) examined the biodegradation of twelve monosubstituted phenols, including phenol, cresols, nitrophenols, and chlorophenols, and found that most of phenols were biodegradable under methanogenic condition. Since the introduction of upflow anaerobic sludge blanket (UASB) reactor by Lettinga in early 1980s, anaerobic process has been developed into a cost-effective process for the treatment of a variety of high-strength industrial wastewaters. UASB is also successfully used to treat a series of recalcitrant phenolic compounds in laboratory studies. This paper is to review literature related to anaerobic treatment of phenolic pollutants in wastewater, including phenolic inhibitory effect on the anaerobes, degradation of phenols in reactors of various designs, effects of operating parameters, phenol-degrading anaerobes, and degradation pathways.

2. Phenols Inhibition Effect on the Anaerobes and Restoration of the Bioactivity

2.1. *Phenols inhibition effect*

As a disinfectant, phenol is inhibitory to bacteria at the concentration of 46 mg·l^{-1} (Heipieper *et al.* 1991), and is recalcitrant to biodegradation. The toxicity is caused by phenol's interference toward the electron transport involved in the energy production or toward the electrochemical proton gradient developed across cell membranes (Escher *et al.* 1996). Under anaerobic condition, phenols may disturb the trophic chain of anaerobes, affecting the production of fatty acids and H_2, and consequently affecting the complete mineralization and methane production. In a study of effects of seven phenols to the anaerobic biogranules from a UASB reactor treating starch wastewater, Fang and Chan (1997) found that the toxicity of phenols depended on the nature of the chemical functional group, i.e., the more hydrophobic the functional group, the higher the toxicity. Generally, the toxicity of phenols to the anaerobes is in the following descending order: chlorophenols > alkylphenols > phenol > hydroxyphenols (catechol, hydroquinone, and resorcinol). Slight differences in toxicity among steric isomers have also been observed.

2.2. *The tolerance of anaerobes to phenolic toxicity*

The reported IC_{50} values, i.e., the concentration at which the bioactivity is reduced by 50%, of phenol varied substantially for different anaerobes, from 1,099 to 4,596 mg-phenol·g-VSS^{-1} (Hernandez and Edyvean 2008). Similarly, the IC_{50} of phenol for anaerobic sludge also varied greatly, from 850 to 1,100 mg·l^{-1} for disintegrated sludge treating distillery, propionate, and benzoate (Fang and Chan 1997; Fang *et al.* 1997). However, granular sludge can tolerate higher concentrations of phenol; the respective IC_{50} of phenol were 1,750, 1,000, 1,370 and 1,700 mg·l^{-1} for acetate-, propionate-, starch-, and benzoate-degrading granular sludges. The varied resistance toward phenolic toxicity exhibited by biogranules could probably be attributed to their microstructures. The fermentative bacteria shield the toxicity-sensitive methanogens, which are mostly populated in the interior of granules from being exposed to the phenols (Fang *et al.* 1997).

Cresols (i.e., methyl phenols) are more toxic than phenol. For example, *m*-creosol has a threshold toxicity of 600–800 mg·l^{-1} in a continuous reactor, and a much lower IC_{50} of 330 mg·l^{-1} in a batch reactor (Zhou and Fang 1997). Chlorophenols are extremely toxic. Among all phenols, pentachlorophenol (PCP) is inhibitory to methanogenesis at 0.2–0.4 mg·l^{-1} for an un-acclimated digester sludge and 1.0–2.5 mg·l^{-1} for a granular sludge (Wu *et al.* 1993).

Whether the inhibitory effect of phenols to anaerobic sludge is reversible is concentration dependent. The inhibitory effect does not increase progressively with phenol concentration; instead, the effect is noticeable when phenol concentration surpasses certain threshold level (Fang and Chan 1997). Fang *et al.* (1997) found the phenol threshold toxicities for continuous reactors treating benzoate, propionate, and acetate were 1,050–1,600 mg·l^{-1}, 850 mg-phenol·l^{-1}, and 1,100–1,700 mg-phenol·l^{-1}, respectively; all of these threshold values were generally higher than the corresponding IC_{50} values found in batch tests. Below these threshold levels, phenol did not show inhibitory effect to the bioactivity of anaerobic biogranules; but above which, the inhibitory effect was 100% and irreversible. Collins *et al.* (2005a) also studied the IC_{50} of phenol to different anaerobes and concluded that syntrophic propionate-utilizing microorganisms were the most sensitive to phenol, while acetotrophic methanogens were more sensitive to phenol than the hydrogenotrophic methanogens.

On the other hand, anaerobes are able to increase their tolerance to phenol within certain concentrations through acclimation. Many studies showed that anaerobic sludge could tolerate phenol at concentrations of 630–1,260 mg·l^{-1} (Fang *et al.* 2006, 2004; Razo-Flores *et al.* 2003; Scully *et al.* 2006; Tay *et al.* 2001; Veeresh *et al.* 2006), and even up to 3,764 mg·l^{-1} according to a recent study (Bajaj *et al.* 2009).

2.3. *Recovery of bioactivity after inhibition*

Most studies showed that the bioactivity of an anaerobic sludge, after being inhibited by high concentrations of phenol, could be fully recovered when the phenol concentration is lowered under its threshold level. However, the sludge which had been in contact with high concentrations of phenol for a long duration would require a lengthy recovery. In addition to lowering the influent phenol concentration to the reactor, the inhibited bioactivity could be recovered by effluent recycling, lowering influent flow, or feeding with readily degradable co-substrates.

3. Reactors Treatment of Phenolic Wastewater

Most of the anaerobic reactors today for phenolic wastewater treatment are designed according to UASB (Droste 1998; Fang *et al.* 2006, 2004; Fang and Zhou 2000; Ke *et al.* 2004; Majumder and Gupta 2008; Razo-Flores *et al.* 2003; Subramanyam and Mishra 2008a,b; Tay *et al.* 2000, 2001; Veeresh *et al.* 2006; Ye *et al.* 2004) and its modifications, such as hybrid UASB (Fernandez *et al.* 2001; Ramakrishnan and Gupta 2006; 2008; Sreekanth *et al.* 2009) and the enhanced granular sludge blanket (EGSB) (Almendariz *et al.* 2005; Chou *et al.* 2008; Collins *et al.* 2005a; Scully *et al.* 2006). Some other reactors were also tested for the treatment of phenolic wastewater; these included using granular activated carbon (GAC) in anaerobic filter (Suidan *et al.* 1983a), fixed film reactor (Bajaj *et al.* 2009; Latkar *et al.* 2003; Lin and Lee 2006), expanded bed (Wang *et al.* 1986), and fluidized bed reactors (Garibay-Orijel *et al.* 2005; Lao 2002). Recently, some new reactor designs have also been utilized for phenolic wastewater treatment, including anaerobic immobilized biomass reactor (Bolaños *et al.* 2001) and microbial fuel cell (MFC) (Luo *et al.* 2009).

3.1. *UASB*

UASB has become the most common anaerobic industrial wastewater treatment process since its introduction in early 1980s, due to its simplicity in design and operation, low cost and high efficiency. In a UASB reactor, wastewater flows upwards through a dense blanket of anaerobic granular sludge while organic pollutants are being degraded. Although it has been widely used for treating high-strength wastewater from food/beverage/agricultural industries, UASB has also been successfully applied for the treatment of various types of phenolic wastewaters; these include wastewaters containing solely phenol (Fang *et al.* 1996, 2006, 2004; Fang and Zhou 2000; Lay and Cheng 1998; Tay *et al.* 2001), solely cresols (Fang and Zhou 1997a,b, 2000; Hwang and Cheng 1991b; Razo-Flores *et al.* 2003), and mixtures of phenol and various cresols, such as *m*-cresol (Fang and Zhou 1999; Zhou and Fang 1997), *p*-cresol (Fang and Zhou 2000), mixed cresol isomers (Guiot *et al.* 2000; Hajji *et al.* 2000; Razo-Flores *et al.* 2003). Meanwhile, UASB has also been used in the treatment of hydroxyphenols, including catechol (Subramanyam and Mishra 2008a,b), and resorcinol (Subramanyam and Mishra 2008a), as well as nitrophenols and chlorophenols (Droste 1998; Wu *et al.* 1993; Ye *et al.* 2004; Majumder and Gupta 2008). Table 9.1 summarizes the details of these studies.

3.2. *Hybrid UASB*

A hybrid of UASB and anaerobic filter has been used for wastewater treatment, by adding packing carriers (such as PVC rings) in the upper zone of the conventional UASB reactor in order to enhance sludge retention. Hybrid UASB has exhibited effective degradation capability to aromatic compounds, such as phthalic acid and phenolic wastewater. Ramakrishnan and Gupta (2006, 2008) studied the performance of a hybrid UASB reactor treating a synthetic wastewater with a mixture of phenolic pollutants, including phenol (490 $mg \cdot l^{-1}$), *m*-, *o*-, *p*-cresols (123, 58.6, 42 $mg \cdot l^{-1}$, respectively), and 2,4- 2,5-, 3,4-, and 3,5-dimethyl phenols (6.3, 6.3, 4.4, and 21.3 $mg \cdot l^{-1}$, respectively). They demonstrated that the reactor was able to remove 96% of phenols and 86% of COD with 24 h of HRT. Fernandez *et al.* (2001) showed that a pilot-scale hybrid UASB was effective in treating a fiberboard manufacturing wastewater, containing 10% of phenol and *p*-cresol, at an OLR of 8 $g\text{-}COD \cdot l^{-1} \cdot d^{-1}$. Recently, Sreekanth *et al.* (2009) reported that with glucose as co-substrate, a hybrid UASB was able to degrade chloro-nitrophenol and chloro-methylphenol, individually, at 30 $mg \cdot l^{-1}$ of concentration.

3.3. *EGSB*

EGSB is modified from UASB by using effluent recirculation combining with a taller reactor configuration, resulting in a high superficial velocity (>4 $m \cdot h^{-1}$) and thus an improved sludge-wastewater contact (Seghezzo *et al.* 1998). EGSB reactors have been successfully used to treat phenolic wastewater, especially at low temperature. Collins *et al.* (2005a) reported that EGSB

Table 9.1. The overall performance of UASB reactor treating phenolic wastewater.

Phenolic pollutant	Influent concentration ($mg \cdot l^{-1}$)		Co-substrate	Temp. (°C)	HRT (h)	OLR ($g\text{-}phenol \cdot l^{-1} \cdot d^{-1}$)	Removal efficiency (%)		CH_4-COD[a] (%)	Remarks	References
	Phenol	Total COD					Phenol	Total COD			
Phenol	500	1,200	—	—	9.6	1.25	100	100	100	Inhibition occurred when phenol was increased to 800 $mg \cdot l^{-1}$.	Chang *et al.* (1995)
Phenol	1,176	4,000	Molasses	30 ± 2	12	2.35	88–99	74–91.3	72	Bioactivity was deteriorated beyond 70% phenol COD.	Veeresh *et al.* (2006)
Phenol	1,260	3,000	—	37	12	2.52	97	97	94.7	The seed sludge took 7 weeks to develop phenol-degrading activity.	Fang *et al.* (1996)
Phenol	1,260	3,000	—	26 ± 1	12	2.52	98	98	—	Benzoate was the key intermediate in phenol anaerobic degradation.	Fang *et al.* (2004)
Phenol	1,226	3,000	—	26 ± 1	24	1.23	99	99	78	Phenol degradation efficiency decreased with HRT, from 95% at 16 h to 93% at 12 h.	Ke *et al.* (2004)
Phenol	630	1,500	—	55 ± 2	40	0.38	99	99	—	Degradation pathway of phenol under thermophilic condition was likely via caproate, instead of benzoate.	Fang *et al.* (2006)
Phenol	1,260	3,000	—	35	12	2.52	86	86	—	Lower removal efficiency could be attributed to inhibition by high influent phenol loading.	Tay *et al.* (2000)
Phenol	1,260	3,000	Glucose	35	12	2.52	98	92	81	Glucose supplement promoted phenol degradation.	Tay *et al.* (2000)

(Continued)

Table 9.1. *(Continued)*

Phenolic pollutant	Influent concentration ($mg \cdot l^{-1}$) Phenol	Total COD	Co-substrate	Temp. (°C)	HRT (h)	OLR (g-phenol$\cdot l^{-1} \cdot d^{-1}$)	Removal efficiency (%) Phenol	Total COD	CH_4-COD[a] (%)	Remarks	References
Phenol, *p*-Cresol	Phenol 800; *p*-Cresol 300;	2,650	—	37	8–12	2.20–3.30	Phenol (95); *p*-Cresol (65)	85	73.9	Phenol was preferentially removed over *p*-Cresol.	Fang and Zhou (2000)
Phenol *p*-Cresol	Phenol 1100; *p*-Cresol 550	4,000	—	30	14	2.82	Phenol (100); Cresol (93)	94	75.8[b]	Concentration of each phenol should not exceed 600 $mg \cdot l^{-1}$ in order to prevent inhibition.	Razo-Flores *et al.* (2003)
Phenol, *p*-, *o*-cresol	Phenol 550 *p*-Cresol 132 *o*-ocresol 132	2000	—	30	16	1.22	Phenol (100) *p*-Cresol (100) *o*-Cresol (80)	82	90[b]		Razo-Flores *et al.* (2003)
Phenol *m*-Cresol	Phenol 900; *m*-Cresol 320	—	—	37	24	1.22	Phenol (98); *m*-Cresol (20)	—	—	*m*-Cresol had a threshold toxicity concentration of 600–800 $mg \cdot l^{-1}$.	Zhou and Fang (1997)
p-Cresol	650	2,500	VFA	35	16	0.98	100	—	—	The addition of VFA allowed higher *p*-Cresol concentration removal.	Kennes *et al.* (1997)
Catechol	1,000	2,500	glucose	35 ± 2	8	3.00	95	95	—	Small quantity of glucose (500 $mg l^{-1}$) was enough to substantially improve the reduction of catechol.	Subramanyam and Mishra (2007)

(Continued)

Table 9.1. *(Continued)*

Phenolic pollutant	Influent concentration ($mg \cdot l^{-1}$) Phenol	Total COD	Co-substrate	Temp. (°C)	HRT (h)	OLR ($g\text{-}phenol \cdot l^{-1} \cdot d^{-1}$)	Removal efficiency (%) Phenol	Total COD	CH_4-COD[a] (%)	Remarks	References
Catechol/ resorcinol	Total 1,000, in different R/C ratio	2,000	—	35	8	3.00	92 (at $R/C = 1/4$)	92	40–76	An increase in the *R/C* ratio (>1/4) induced resorcinol toxicity and inhibited the bioactivity.	Subramanyam and Mishra (2008a)
Nitrophenol (NP)	2-NP (300) 4-NP (260) 2,4-DNP (30)	~4,000	VFAs	30 ± 2	8	1.77	2-NP (94), 4-NP (99) 2,4-DNP (68)	—	—	2-NP was largely mineralized, while 4-NP and 2,4-DNP were mostly biotransferred into aminophenol.	Donlon *et al.* (1996)
NP	2-NP (60) 4-NP (60) 2,4-DNP (60)	2,000	Methanol	27 ± 4	24	0.18	2-NP (99) 4-NP (97) 2,4-DNP (93)	93	—	2-Aminophenol (AP), 4-AP, and 2,4-AP were recovered in the range of 17–28%.	Karim and Gupta (2006)
Pentachloro-phenol (PCP)	40–60	10,000	VFAs	28	10.8–15	~0.10	99	—	52% of PCP	The maximum specific PCP removal rate was $18\ mg \cdot (g\text{-}VSS \cdot d)^{-1}$.	Wu *et al.* (1993)
PCP	150	7,500	Synthetic wastewater	28	24	0.15	99	98	—	—	Ye *et al.* (2004)
PCP	180	5,000	Sucrose	28	20–22	0.15	100	94	—	—	Shen *et al.* (2006)

[a] CH_4-COD/influent COD.
[b] Calculated from the original data.

reactor was effective for phenolic wastewater containing 1,200 mg·l^{-1} of phenol at 15–18 °C and a loading rate of 1.21 g-phenol·l^{-1}·d^{-1}, using volatile fatty acids (VFAs) as co-substrate after a lengthy startup of 135 days. Collins *et al.* (2005b) also demonstrated that EGSB with a fixed-film in the upper section of reactor was effective for the degradation of 2,4,6-trichlorophenol (TCP) as well as di-CPs and mono-CPs, both of which were degraded at higher rates than 2,4,6-TCP. Recently, Scully *et al.* (2006) reported that EGSB was able to remove 88–98% of phenol at high loading rates of 1.0–2.0 g-phenol·l^{-1}·d^{-1} even under 10–15 °C.

3.4. *Granular activated carbon (GAC)-fluidized bed reactor*

Fluidized bed reactor has the merits of improved mixing and preventing short-circuiting and dead-zone formation inside the reactor. Using GAC as carriers in anaerobic fluidized bed reactors provides the extra advantages of large surface area for biomass attachment, and the adsorption of toxic phenolic pollutants. Garibay-Orijel *et al.* (2005) showed in a laboratory study that a fluidized bed reactor with GAC was able to remove 99% of both 80 mg·l^{-1} of 2,4,6 TCP and 20 mg·l^{-1} of phenol from a wastewater, coupled with a post-aeration to completely remove the residues.

3.5. *Fixed-film reactor (anaerobic filter)*

Suidan *et al.* (1983a,b) studied the anaerobic biodegradation of phenol and *o*-, *m*-, and *p*-cresols in a synthetic coal gasification wastewater in fixed-film reactors containing GAC. The reactor was able to remove 100% phenol, 90% *o*-cresol, 57% *m*-cresol, and 89% *p*-cresol for a wastewater containing 180 mg·l^{-1} of phenol, 55 mg·l^{-1} of *o*-cresol, 40 mg·l^{-1} of *m*-cresol and 70 mg·l^{-1} of *p*-cresol, with corresponding volumetric loading rates of 22, 6.8, 2.4, and 8.6 mg·l^{-1}·d^{-1}, respectively. Using a 2-l anaerobic reactor packed with "Liapor" clay beads, Bajaj *et al.* (2009) showed a 94% phenol removal efficiency at 1.5 g-phenol·l^{-1}·d^{-1} of loading rate and 60 h of HRT. Lin and Lee (2006) reported a 98.5% phenol removal efficiency in a fixed-film reactor treating a wastewater containing 1,500 mg·l^{-1} phenol operated at 60 h or HRT after 40 days of acclimation.

Latkar *et al.* (2003) used upflow fixed-film bed reactors to treat wastewaters containing three dihydric phenols individually. For the wastewater containing resorcinol (750 mg·l^{-1}), the removal efficiency was 98% at the OLRs of 2.05 g-phenol·l^{-1}·d^{-1}. The corresponding values were 93% for treating catechol (711 mg·l^{-1}) at 2.01 g-phenol·l^{-1}·d^{-1} and 83% for treating hydroquinone (589 mg·l^{-1}) at 1.61 g-phenol·l^{-1}·d^{-1}.

3.6. *Other types of reactor*

Bolaños *et al.* (2001) studied the treatment of a phenolic wastewater containing 1,200 mg·l^{-1} of phenol using a horizontal-flow anaerobic immobilized biomass reactor filled with polyurethane foam matrices. The reactor was able to remove 99% of phenol at 12 h of HRT with a short startup period. Further, the immobilized biomass was able to tolerate phenol at concentrations as high as 1,500 mg·l^{-1}.

Using graphite-packed microbial fuel cells with a ferricyanide cathode treating a wastewater containing phenol (400 mg·l^{-1}) as the sole carbon source, Luo *et al.* (2009) reported that 90% of phenol was removed at the peak voltage output. When using a mixture of glucose and phenol (1,000 mg·l^{-1}) as feed, the cells were able to remove phenol completely in 60 h.

4. Affecting Factors for the Anaerobic Wastewater Treatment of Phenols

4.1. *Hydraulic retention time (HRT)*

In biological treatment, increase of HRT generally results in an increase of removal efficiency. Fang and Zhou (1997a) reported that with 12 h of HRT, only 11% of *m*-cresol and 8.3% of *o*-cresol were removed in a continuous reactor; however, the efficiencies were increased to 46.2% for *m*-cresol and 37.4% for *o*-cresol with 100 h of retention in batch tests.

Although sludge acclimation can improve the degradation of phenols, yet there is a certain minimum requirement for HRT. In a thermophilic UASB reactor, Fang *et al.* (2006) reported that over 96% of phenol could be effectively degraded by gradually reducing the HRT from 60 h to 48 h and further to 40 h from a wastewater containing 630 mg·l^{-1} of phenol. In two separate attempts to reduce the HRT from 40 to 28 h, the phenol removal efficiency was reduced to only 70% and it took 7 weeks for the sludge to fully recover its phenol-degrading efficiency. Fang and Zhou (2000) also reported that a UASB consistently removed 95% of phenol, 65% of *p*-cresol, and 85% of total COD within 8–12 h of HRT, whereas the COD removal efficiency decreased to 75% when the HRT was lowered to 6 h, and further down to 56% at 4 h of HRT.

Ramakrishnan and Gupta (2008) studied the anaerobic treatment of a complex phenolic wastewater in a hybrid UASB reactor under various HRTs. They found that, by lowering the HRT from 36 to 18 h, degradation of *o*-cresol (58.6 mg·l^{-1} in the wastewater) and 3,4-dimethylphenol (4.4 mg·l^{-1}) was seriously affected; the removal efficiencies, which were 99% both at 36 h of HRT, were dropped to 68% and 81%, respectively. However, removals of phenol, *m*-, *p*-cresol, 2,4-, 2,5-, and 3,5-dimethylphenol were only slightly affected by such a change of HRT. Sreekanth *et al.* (2009) reported that, in a hybrid UASB reactor, the removal efficiencies of 4-chloro-2-nitrophenol (30 mg·l^{-1} in wastewater) and 2-chloro-4-nitrophenol (30 mg·l^{-1}) dropped, respectively, from 95% to 90% and from 92% to 85% as HRT was lowered from 30 h to 8 h, whereas the removal efficiency of 2-chloro-5-methylphenol was lowered from 88% to 74% for the same change of HRT.

4.2. *Loading shock*

Loading shock of phenol could be caused by a sudden increase of phenol concentration in wastewater and/or a sudden reduction of HRT. The shock could inhibit the bioactivity of sludge, resulting in an acute increase of phenol concentration in the effluent, and the accumulation of VFAs, and sometimes benzoate, in the mixed liquor (Tay *et al.* 2001). Fang *et al.* (1996) found that when the loading was increased from 1.26 to 2.52 g-phenol·l^{-1}·d^{-1}, the COD removal efficiency dropped substantially from 83 to 53%. After the sludge fully recovered its bioactivity by reducing the loading to 1.68 g-phenol·l^{-1}·d^{-1}, a further increase of the loading to 2.1 g-phenol·l^{-1}·d^{-1} resulted in another sudden decrease of COD removal efficiency. Subsequently, they also found that after the reactor operated steadily at 1.68 g-phenol·$l^{-1}d^{-1}$ for a long period, the loading rate was raised for the third time to 2.23 g-phenol·l^{-1}·d^{-1} and then to 2.94 g-phenol·l^{-1}·d^{-1}, the UASB could be readily recovered to 88% of removal efficiency in about 2 weeks (Fang *et al.* 1996).

Subramanyam and Mishra (2007) reported that a well acclimated sludge in a UASB reactor could effectively degrade 1,000 mg·l^{-1} of catechol as the sole carbon source with a high OLR of 3.0 g-cresol·l^{-1}·d^{-1}. The removal efficiency was lowered to 67% when the catechol concentration was increased to 1,250–1,500 mg·l^{-1}.

Generally, the inhibitory effect caused by loading shock to the anaerobes is reversible, meaning the bioactivity could be recovered by lowering the influent phenol concentration, by increasing the HRT to reduce the organic load, and/or by feeding with readily degradable co-substrate. Fang *et al.* (2004) reported that a stable phenol-degrading UASB operated under ambient temperature was shocked by a high concentration of influent phenol (1,470 mg·l^{-1}) resulting in a decrease of COD removal efficiency to 78% in 20 days. However, by restoring the phenol influent concentration back to 1,260 mg·l^{-1} and by doubling the HRT to 24 h (and later back to 12 h), the UASB was able to remove again 98% of phenol within 52 days. Tay *et al.* (2001) found that, a loading shock resulting from the increase of phenol concentration in wastewater from 1,680 mg·l^{-1} to 2,100 mg·l^{-1}, lowered the phenol removal efficiency severely from 90% to 36%, and the efficiency was recovered slowly only back to 51%. By supplementing glucose (1 g-COD·l^{-1}) and restoring the phenol influent concentration to 1,260 mg·l^{-1}, the bioactivity of phenol-degrading sludge was effectively recovered in 22 days. However, without glucose supplement, the recovery took 61 days. Bajaj *et al.* (2009) recently conducted a loading shock study for a phenol-degrading fixed-film reactor by increasing phenol concentration in wastewater from 3,680 to 4,600 mg·l^{-1}, and found a sharp decline of phenol removal efficiency from 98.5% to 43% and an accumulation of fatty acids within 2 weeks. Recovery of the sludge bioactivity could not be achieved by increasing the HRT to lower the organic loading, and it was achieved only by stopping feeding phenol for 1 month before gradually increasing the phenol concentration in wastewater.

4.3. *Temperature*

Compared with readily degradable substrates, anaerobic degradation of phenols is highly sensitive to the temperature change. Fang *et al.* (1996) reported that a temperature drop from 37 to 20 °C for 48 h in a mesophilic UASB reactor resulted in a sharp decrease of the bioactivity of the phenol-degrading granules, as the biogas production reduced by 64%. It took 40 days for the sludge to fully recover its bioactivity when the temperature was back to 37 °C. In a parallel study, granular sludge in UASB reactor degrading mixed VFA, peptone and benzoate, fully recovered their bioactivities in only 2 days after the temperature returned to 37 °C.

The bioactivity of anaerobic sludge under different temperatures varies greatly, since temperature exerts a strong selective pressure on the microbial communities and can affect the degradation of organic compounds through direct effects on enzyme activity (Kohring *et al.* 1989). Most of the phenolic wastewater treatments were conducted under the mesophilic condition (30–37 °C) since this condition has demonstrated to be most effective. Only a limited number of treatments were conducted under thermophilic (55 °C) and psychrophilic conditions (<20 °C). Thermophilic anaerobic treatment of wastewater has several merits, including reducing the need for a cooling system, high metabolic rates and thus short retention time, and the removal of pathogenic microorganisms. Wiegant (1985) was the first to demonstrate that thermophilic UASB could be effective for wastewater treatment. However, anaerobic phenol degradation under thermophilic condition was inferior to those under mesophilic and ambient temperature conditions. Fang and his co-workers have conducted phenol anaerobic degradation under thermophilic (Fang *et al.* 2006), mesophilic (Fang *et al.* 1996), and ambient temperature (26 °C) conditions (Fang *et al.* 2004). They reported that the maximum specific methanogenic activity (SMA) of phenol-degrading sludge using phenol as substrate was highly temperature dependent, varying from 0.09 g-CH_4-COD·$gVSS^{-1}$·d^{-1} at 55 °C to 0.19 g-CH_4-COD·$gVSS^{-1}$·d^{-1} at 26 °C, and 0.24 g-CH_4-COD·$gVSS^{-1}$·d^{-1} at 37 °C.

Levén and Schnürer (2005) reported that phenol, *o*-, *m*-, *p*-cresol, resorcinol, PCP, 4-nitrophenol and 4-ethylphenol, could be degraded at 37 °C, but not at 55 °C. They suggested that channeling reaction of phenols to the central intermediate benzoyl-CoA was inactive in the thermophilic microbial community; however, this enzyme can be reactivated by lowering the temperature from 55 °C to 48 °C.

Although anaerobic treatment of wastewater could be achieved under psychrophilic conditions (<20 °C) (Lettinga 1995), the rate is slow and additional energy is needed (Lettinga *et al.* 2001). Several studies were conducted for the feasibility of treating phenolic wastewater under psychrophilic conditions (Collins *et al.* 2005a,b; Scully *et al.* 2006). Efficient phenol removal was observed at the loading rate of 0.4–1.2 g-phenol·l^{-1}·d^{-1} in an EGSB reactor at 15–18 °C. Collins *et al.* (2005a) reported that the SMA of phenol-degrading sludge using phenol as substrate at 15 °C was up to 0.25 g-CH_4·$gVSS^{-1}$·d^{-1}, which was still lower than the 0.34 g-CH_4-COD·$gVSS^{-1}$·d^{-1} at 37 °C for the same sludge. Interestingly, they reported that the anaerobic population under psychrophilic conditions were the same as those under mesophilic conditions.

4.4. *Effluent recirculation*

Recirculating reactor effluent to the reactor inlet may dilute the wastewater phenol concentration, and thus reduce phenol toxicity to the anaerobic sludge. Recirculation also increases the superficial velocity, resulting in increased substrate transfer for biodegradation and the retention of bigger sludge granules in the reactor (Lay and Cheng 1998). Fang *et al.* (1996) conducted two parallel studies on treating a wastewater containing 840 mg-phenol·l^{-1} at a loading rate of 4.g-phenol·l^{-1}·d^{-1}: one with effluent recirculation and the other without recirculation. They reported that the reactor without recirculation removed 81% of COD, whereas the reactor with 1:1 recycle rate removed 92.5% COD.

Lay and Cheng (1998) evaluated the effect of hydraulic flow in UASB reactor in the treatment of concentrated phenolic wastewater by using the effluent recirculation. They reported that increasing the flow by recirculation may reduce the adverse impact of concentrated phenols on the sludge bioactivity and improve the substrate transfusion, resulting in an increase of overall methanogenic activity. At the same phenolic loading rate, the relative bacterial activity decreased from 80% to 50% when the flow rate to the UASB reactor decreased from 2.5 to 1.6 m·h^{-1}.

For effluent recirculation, complete mixing of recycle effluent and wastewater could be crucial. Chang *et al.* (1995) evaluated the effect of effluent recirculation on the treatment of an industrial phenolic wastewater, which contained 400 mg·l^{-1} of phenol and other unknown phenolic compounds with a total COD of 9,300 mg·l^{-1}. By using a two fold dilution of wastewater and an effluent recycle ratio of 7.3, 90% of phenol and 80% of COD were removed. However, by increasing the effluent recycle ratio to 28.4 but without diluting the wastewater, only 43–50% of phenol was removed. This could be resulted from the inadequate mixing of recycled effluent and incoming wastewater.

4.5. *Co-substrate*

Due to the inhibitory effect of phenol, anaerobic treatment of phenolic wastewater has to be conducted at low phenol concentrations and requires a lengthy acclimation period, ranging from 2 months up to a year (Chang *et al.* 1995). This problem could be alleviated by adding more biodegradable co-substrates to the wastewater. Co-metabolism is one of the preferred means for the degradation of recalcitrant pollutants. A number of co-substrates have been studied in phenolic wastewater treatment, including glucose (Hwang and Cheng 1991b; Tay *et al.* 2001), volatile fatty acids (VFAs) (Kennes *et al.* 1997), and synthetic readily degradable wastewater (Veeresh *et al.* 2005, 2006). With a properly selected co-substrate, most of the startup periods of UASB reactors treating phenolic wastewater were reported in the range of 30–148 days (Fang *et al.* 1996, 2004; Fang and Zhou 1997a, 2000; Hwang and Cheng 1991a; Razo-Flores *et al.* 2003; Tay *et al.* 2001; Zhou and Fang 1997), except in one study of treating *p*-cresol which took 219 days (Hwang and Cheng 1991b). Tay *et al.* (2001) conducted two parallel studies on phenolic wastewater treatment: one was added with 1,000 $mg \cdot l^{-1}$ of glucose as co-substrate and the other without glucose. Results showed that the addition of glucose shortened the startup and granulation period from 210 days to 120 days.

In some cases, co-substrate was used only during the startup for sludge acclimation, and its dosage was gradually reduced afterward and eliminated eventually. Razo-Flores *et al.* (2003) studied the treatment of a wastewater containing phenol, *p*-cresol, and *o*-cresol. After 55 days of startup using 1 $g\text{-}COD \cdot l^{-1}$ of acetate as co-substrate, the reactor continued to perform satisfactorily after the removal of acetate. In another study, Fang *et al.* (2004) used mesophilic phenol-degrading granules as seed sludge to treat a phenolic wastewater at ambient temperature by gradually increasing phenol concentration in wastewater from 100 to 1,200 $mg \cdot l^{-1}$ jointly with decreasing glucose concentration from 1,000 $mg \cdot l^{-1}$ to nil. Fang *et al.* (2006) used similar strategy to successfully start up a thermophilic UASB reactor treating phenolic wastewater also using mesophilic phenol-degrading granules as seed sludge using sucrose as co-substrate. The COD ratio of phenol and sucrose concentrations was changed from 1/4 to 2/3, 3/2, 4/1, and finally without feeding sucrose, leaving phenol as sole substrate to complete the reactor startup.

The use of co-substrate can also increase the removal efficiency of phenols. Tay *et al.* (2001) reported that the UASB reactor operated at the loading rate of 2.52 $g\text{-}phenol \cdot l^{-1} \cdot d^{-1}$ using glucose as co-substrate removed 98% of phenol, as compared to 88% in another reactor without glucose addition. Similarly, Kennes *et al.* (1997) demonstrated that using VFA as co-substrate, 650 $mg \cdot l^{-1}$ of *p*-cresol was completely degraded at 16 h of HRT in a UASB reactor, but only 80% was removed without VFA under the otherwise same operation condition.

With co-substrate, reactor treating phenolic wastewater is more resilient to loading shock and is faster to recover from shocks. Veeresh *et al.* (2006) reported that when the UASB reactor performance deteriorated due to the increased concentration of phenol (3,200 $mg\text{-}phenol\text{-}COD \cdot l^{-1}$), only feeding readily degradable co-substrate was able to restore the bioactivity, while other attempts failed, including increasing HRT and decreasing the organic loading.

The effect of co-substrate is concentration dependent. Tay *et al.* (2001) showed that in batch tests, glucose at 500–2,000 $mg \cdot l^{-1}$ enhanced the phenol (420 $mg \cdot l^{-1}$) degradation, but not at 4,000 and 8,000 $mg \cdot l^{-1}$ due to the accumulation of intermediate VFAs resulting from the high organic loadings. Sreekanth *et al.* (2009) reported a similar phenomenon, i.e., glucose at 3000 $mg \cdot l^{-1}$ was optimum for the degradation of chloro-nitrophenols and chloro-methylphenol (both at 30 $mg \cdot l^{-1}$), and glucose at 5,000 $mg \cdot l^{-1}$ actually decreased the phenol removal efficiency. Hwang and Cheng (1991a,b) reported the slow degradation of *p*-cresol using 4,500 $mg \cdot l^{-1}$ of glucose as co-substrate; but the removal efficiency of *p*-cresol increased when glucose concentration was decreased stepwise.

In addition to glucose and sucrose, other co-substrates have also been used for phenolic wastewater treatment, including molasses (Veeresh *et al.* 2006), whey (Hajji *et al.* 2000), and proteose peptone (Charest *et al.* 1999; Kobayashi *et al.* 1989). Charest *et al.* (1999) reported that the proteose peptone is essential for anaerobic degradation of phenol and *o*-cresol from a petrochemical refinery wastewater using an upflow fixed-film anaerobic reactor inoculated with phenol-degrading consortium. The bioreactor removed almost completely 350 $mg{\cdot}l^{-1}$ of phenol and 89 $mg{\cdot}l^{-1}$ of *o*-cresol, respectively, from the wastewater with 7.8 h of HRT. Hajji *et al.* (2000) found whey could be used as a substitute of proteose peptone for the effective treatment of phenolic wastewater. In an anaerobic upflow fixed-film reactor inoculated with phenol- and *p*-cresol-degrading sludge, over 98% of 150 $mg{\cdot}l^{-1}$ of phenol and 35 $mg{\cdot}l^{-1}$ each of *o*- and *p*-cresol were removed with the addition of 0.025% whey at 6 h of HRT and volumetric loading rate of 880 $mg{\cdot}l^{-1}{\cdot}d^{-1}$.

For anaerobic degradation of chlorophenols, the initial step is reductive dechlorination, which requires co-substrate fermentation to provide H as electron donor for the chlorine substitute. Hendriksen *et al.* (1992) studied the PCP anaerobic degradation in 2-l UASB reactors and found that adding glucose (0.9 $g{\cdot}l^{-1}$) as co-substrate increased the removal of PCP (3.0–4.5 $mg{\cdot}l^{-1}$) from 32–77% to 99%.

4.6. *Degradation of mixed phenols*

Wastewaters from industries often contain more than one type of phenolic pollutant. Those with more complex structures are often more toxic and recalcitrant than the simple phenol. Fang and Zhou were the first to study the degradation of mixed phenols (Zhou and Fang 1997; Fang and Zhou 2000). They reported that phenol was more biodegradable than *m*- and *p*-cresol in UASB reactors. Treating a wastewater containing 900 $mg{\cdot}l^{-1}$ of phenol, 320 $mg{\cdot}l^{-1}$ of *m*-cresol, and 250 $mg{\cdot}l^{-1}$ of *p*-cresol with 24 h of HRT, the removal efficiencies were 98% for phenol, 20% for *m*-cresol and 80% for *p*-cresol. Farooqi *et al.* (2007) also reported the successful degradation of phenol and *m*-cresol under anaerobic condition by properly keeping the phenol/*m*-cresol concentration ratio within the range of 10/1 (500/50, in $mg{\cdot}l^{-1}$) to 1.2/1 (300/250, in $mg{\cdot}l^{-1}$).

Razo-Flores *et al.* (2003) studied the degradation of phenol and *p*-cresol in a UASB reactor, and demonstrated that keeping the phenol/*p*-cresol ratio at 2/1 (1,000/500, in $mg{\cdot}l^{-1}$) or 3/1 (1,200/400 $mg{\cdot}l^{-1}$), the reactor was able to effectively remove 94% and 91% of the COD under the OLR of 2.82 g-phenol$\cdot l^{-1}{\cdot}d^{-1}$. In addition, 80% of *o*-cresol could also be degraded at an average influent concentration of 110 $mg{\cdot}l^{-1}$.

Constitutes of mixed phenols in wastewater have different degradation rates. Ramakrishnan and Gupta (2008) studied the anaerobic degradation of mixed phenols in coal wastewater and reported that the relative degradation rates are in the following order: phenol > *p*-cresol > 3,5-dimethylphenol (DMP) > 2,4-DMP = 2,5-DMP > 3,4-DMP > *m*-cresol > *o*-cresol. They suggested that various anaerobes were involved in the degradation of the phenolic isomers, each following a different anaerobic degradation pathway.

Anaerobes degrading mixed phenols are substrate specific. Latkar *et al.* (2003) reported that when catechol (200 $mg{\cdot}l^{-1}$) was added to a reactor treating resorcinol-containing wastewater, the resorcinol removal efficiency by the reactor was reduced from 94% to 35% due to the toxic effect of catechol. However, resorcinol could be readily and completely degraded in a reactor treating catechol (Swaminathan *et al.* 1999). Similar phenomenon was also observed by Godbole and Chakrabarti (1991). They studied the cross-feeding of phenol, catechol, and resorcinol in multi-substrate matrices and found that phenol and catechol adversely interacted each other's metabolism in the resorcinol-acclimated reactor.

Phenol may also be served as a co-substrate to facilitate the anaerobic degradation of other phenols. Zhou and Fang (1997) reported that in treating a wastewater containing phenol and *m*-cresol in a UASB reactor, phenol degradation was independent of the *m*-cresol concentration. However, the UASB sludge could not degrade any *m*-cresol in the absence of phenol in the batch tests. In another study, catechol was poorly degraded by a phenol-degrading sludge, but catechol degradation could be improved using phenol or resorcinol as co-substrate (Godbole and Chakrabarti 1991).

5. Phenols-degrading Granular Sludge

The success of effective anaerobic treatment of wastewater relies to a large degree on the formation of granular sludge in the reactor, resulting in high biomass concentration and sludge retention due to improved settleability. Phenols-degrading granule in anaerobic reactors was first characterized by Chang *et al.* (1995), and later by Fang *et al.* (1996) and Tay *et al.* (2000). The granulation of phenol-degrading sludge generally takes months or up to a year due to the lengthy acclimation towards phenol as required by the sludge. Based on electron microscopic observations, Chang *et al.* (1995) demonstrated that phenol-degrading granule was 0.66–0.75 mm in size with numerous cavities. Filamentous methanogens (resembling *Methanosaeta*, previously known as *Methanothrix*) were predominantly located on the surface and in the interior of the granule. Fang *et al.* (1996) reported that a typical phenol-degrading granule in UASB reactors was 1–2 mm in diameter with good settleability. The phenol-degrading granule does not have a layered structure, unlike those granules degrading readily degradable substrates (MacLeod *et al.* 1990). This was attributed to the slow initial step of phenol degradation converting phenol to benzoate (Fang *et al.* 1996). As a consequence, a substantial fraction of phenol diffuse into, and was degraded in, the interior of the granule. Thus all anaerobes involved in phenol degradation, including acidogens, acetogens and methanogens, are rather uniformly distributed within the granules.

Formation of granular sludge in an anaerobic reactor signals a successful reactor startup. The granulation of phenol-degrading sludge can be facilitated with easily degradable co-substrate. Tay *et al.* (2000) reported that the granulation of phenol-degrading sludge was initially observed after 3 months of startup with low phenol loadings, and the granules took another 6 months to become fully mature under increased phenol loadings. They also demonstrated that the startup of a UASB reactor fed with the 1,000 $mg{\cdot}l^{-1}$ glucose as co-substrate was 3 months shorter than another reactor running parallel but without glucose supplement (Tay *et al.* 2001).

By sludge degradation more complex phenols were also able to form granules. Catechol-degrading granules were developed in a UASB reactor treating wastewater containing catechol as sole substrate at a loading of 3 $g{\cdot}l^{-1}{\cdot}d^{-1}$ (Subramanyam and Mishra 2007, 2008b). Using VFAs as co-substrate, granular sludge was formed in a hybrid UASB reactor treating wastewater containing phenol, cresols, and dimethylphenols with high efficiencies (Ramakrishnan and Gupta, 2006). PCP-degrading granules were also developed in UASB reactors removing 99% of PCP and 98% of COD from a wastewater containing 150 $mg{\cdot}l^{-1}$ of PCP and 7.5 $g{\cdot}l^{-1}$ of COD with readily degradable co-substrate (Ye *et al.* 2004).

6. Anaerobic Phenols-degrading Microorganisms

Anaerobic degradation of phenols under methanogenic conditions has long been studied using enriched sludge without knowing the individual anaerobes responsible (Gallert *et al.* 1991; Knoll and Winter 1989; Sharak Genthner *et al.* 1990; Young and Rivera 1985). Knowing the identities of anaerobes involved in phenol degradation should be valuable for the better operation of the process. Four phenol-degrading pure cultures have so far been isolated and characterized. Li *et al.* (1996) reported a *Clostridium* isolate from a phenol-degrading methanogenic consortium; however, it was later found not a pure culture, but contained a small amount of another bacterium. This new bacterium was further isolated and identified as a novel bacterium, *Cryptanaerobacter phenolicus*, capable of converting phenol into benzoate (Juteau *et al.* 2005). Recently, an obligatory anaerobe, *Syntrophorhabdus aromaticivorans*, was isolated as a pure culture capable of degrading phenol to acetate syntrophically with hydrogenotrophic methanogens (Qiu *et al.* 2008). This isolate could also metabolize *p*-cresol, 4-hydroxybenzoate, isophthalate, and benzoate. Based on 16S rRNA gene sequences analysis, this new family, *Syntrophorhabdaceae*, is composed of a large number of bacteria found in anaerobic communities degrading aromatic compounds (Qiu *et al.* 2008). In addition to these two pure cultures, two sulfate-reducing isolates, *Desulfobacterium phenolicum* Ph01 (Bak and Widdel 1986) and *Desulfobacterium anilini* AK1 (Ahn *et al.* 2009), were also capable of degrading phenol.

In order to fully understand the microbial population and diversity in phenols-degrading sludge, a series of 16S rRNA gene-based molecular techniques, including polymerase chain reaction (PCR), cloning and sequencing, terminal restriction fragment length polymorphism, and denaturing gradient gel electrophoresis (DGGE), have been used to separate and analyze the genomic DNA. Fang and co-workers (Fang *et al.* 2004, 2006; Zhang *et al.* 2005; Liang *et al.* 2008) studied the microbial community of anaerobic granules degrading phenol at 26 °C and 55 °C using PCR–DGGE, quantitative real-time PCR (qRT-PCR) and cloning–sequencing techniques. They reported that, in the 26 °C sludge, *Desulfotomaculum* and *Clostridium* were most likely responsible to convert phenol into benzoate, which was then further degraded by *Syntrophus* to acetate, and hydrogen. Acetate was then converted into methane by *Methanosaetaceae*, and hydrogen was converted to hydrogen by *Methanomicrobiales* and *Methanobacteriaceae* (Fang *et al.* 2004; Zhang *et al.* 2005).

Fang *et al.* (2004) analyzed the microstructure of phenol-degrading granule using fluorescence *in situ* hybridization (FISH) in couple with confocal laser scanning microscope. The FISH images demonstrated that phenol-degrading granules did not have a layered microstructure and comprised more *Archaea* (74.1 ± 8.5%) than *Bacteria* (27.3 ± 5.8%), both of which were randomly distributed in the phenol-degrading granules. In the thermophilic phenol-degrading granular sludge, *Thermotogae* (38.9% of the clones population) was found to be the most predominant population (Fang *et al.* 2006). Based on qRT-PCR analysis, the population of *Thermotogae* bacterium TPDB in the phenol-degrading granules increased from 3.5% to 9.9% as the phenol loading increased from 0.25 to 0.51 g-phenol$\cdot l^{-1}\cdot d^{-1}$, indicating that phenol degradation increased specially with the increase of TPDB population, instead of the overall biomass in the reactor (Liang *et al.* 2008).

Chen *et al.* (2008) also investigated the microbial population in mesophilic and thermophilic phenol-degrading granules using 16S rRNA gene-based molecular techniques. Results of FISH showed that group TA in the δ-*Proteobacteria* (88% of bacterial cells detected by FISH) and

Pelotomaculum spp. in *Desulfotomaculum* family were the predominant bacteria in phenol-degrading granules. The group TA represented more than 68% of FISH-detected bacterial cells in a full-scale mesophilic reactor treating phenolic wastewater. In addition, most members in group TA have been found in anaerobic sludge degrading various aromatics, including phthalates and its isomers (Qiu *et al.* 2004; Wu *et al.* 2001), trichlorobenzene (von Wintzingerode *et al.* 1999) and phenols (Fang *et al.* 2004, 2006; Chen *et al.* 2008).

By using DGGE and FISH, Lanthier *et al.* (2005) found a halorespiring bacterium *Desulfitobacterium hafniense* (accounting for 19% of the community population) scattering in the anaerobic biofilm treating PCP-containing wastewater. In another FISH study, scattered colonies of *Desulfibacterium frappieri* PCP-1 was found on the outer layer of anaerobic granules degrading PCP, protecting other anaerobes in the interior from the toxic PCP (Guiot *et al.* 2002).

7. Anaerobic Degradation Pathway of Phenols

Anaerobic degradation of aromatic compounds consists of two phases. In the first phase, aromatic compounds are converted to intermediates for the subsequent reductive de-aromatization. In the second phase, these intermediates are converted to non-aromatic compounds followed by further degradation to acetyl-CoA and CO_2 (Heider and Fuchs 1997). Knoll and Winter (1989) and Kobayashi *et al.* (1989) reported that phenol was degraded by carboxylation to benzoate, via 4-hydroxybenzoate, 4-hydroxybenzoyal-CoA and benzoyl-CoA, by a methanogenic bacterial consortium under mesophilic conditions. Benzoate was then de-aromatized to form cyclohexane carboxylic acid, which was further cleaved to form heptanoate. Heptanoate was then degraded either through β-oxidation to form valerate, propionate, and acetate (Keith *et al.* 1978), or directly to form propionate and butyrate, both of which could be further degraded to acetate (Fina *et al.* 1978). This pathway was supported by the presence of enzymes performing carboxylation, decarboxylation, and dehydroxylation in the anaerobic degradation of phenol (Gallert and Winter 1992). An obligate syntrophy between the phenol-carboxylating, benzoate-degrading, and methane-producing bacteria was required for the complete phenol degradation to methane. However, Fang *et al.* (1996) found that the mesophilic phenol-degrading sludge had little ability to degrade propionate and butyrate, suggesting propionate and butyrate were not the degradation intermediates.

Phenol can also be reduced by denitrification to form cyclohexanone and then to *n*-caproate or to adipate, followed by β-oxidation to form lower VFAs (Bakker 1977). However, Fang *et al.* (2006) reported that thermophilic phenol-degrading sludge was able to convert caproate into methane, but could not degrade benzoate, cyclohexanone, and adipate. This suggests that the pathway of thermophilic phenol degradation does not go through benzoate, cyclohexanone, and adipate.

Schink *et al.* (2000) reported that anaerobic degradation of cresols has diverse pathway for each isomer. *p*-cresol is degraded through hydroxylation at the methyl group to form 4-hydroxybenzoate. *o*-cresol is carboxylated into 3-methyl-4-hydroxybenzoate for further degradation, or through methyl group hydroxylation to form salicylic acid. The pathway of anaerobic *m*-cresol degradation is, on the other hand, as follows: the methyl group of *m*-cresol adds to fumarate to form 3-hydroxybenzyl succinate, which is then through β-oxidation to release succinyl-CoA forming 3-hydroxybenzoyl-CoA.

For the anaerobic degradation of chlorophenols and nitrophenols, the first step is, respectively, reductive dechlorination and reduction of nitro group to amino, followed by phenol degradation (Kulkarni and Chaudhari 2007; Razo-Flores *et al.* 1996).

References

Ahn, Y. B., Chae, J. C., Zylstra, G. J., and Haggblom, M. M. (2009) Degradation of phenol via phenylphosphate and carboxylation to 4-hydroxybenzoate by a newly isolated strain of the sulfate-reducing bacterium *Desulfobacterium anilini*. *Appl. Environ. Microbiol.* **75**(13), 4248–4253.

Almendariz, F. J., Meraz, M., Olmos, A. D., and Monroy, O. (2005) Phenolic refinery wastewater biodegradation by an expanded granular sludge bed reactor. *Wat. Sci. Technol.* **52**(1–2), 391–396.

Bajaj, M., Gallert, C., and Winter, J. (2009) Treatment of phenolic wastewater in an anaerobic fixed bed reactor (AFBR) — recovery after shock loading. *J. Hazard. Mater.* **162**, 1330–1339.

Bak, F. and Widdel, F. (1986) Anaerobic degradation of phenol and phenol derivatives by *Desulfobacterium phenolicum* sp. nov. *Arch. Microbiol.* **146**(177–180).

Bakker, G. (1977) Anaerobic degradation of aromatic-compounds in presence of nitrate. *FEMS Microbiol. Lett.* **1**, 103–107.

Bolaños, R. M. L., Varesche, M. B. A., Zaiat, M., and Foresti, E. (2001) Phenol degradation in horizontal-flow anaerobic immobilized biomass (HAIB) reactor under mesophilic conditions. *Wat. Sci. Technol.* **44**(4), 167–174.

Boyd, S. A., Shelton, D. R., Berry, D., and Tiejie, J. M. (1983) Anaerobic biodegradation of phenolic compounds in digested sludge. *Appl. Environ. Microbiol.* **46**(1), 50–54.

Chang, Y. J., Nishio, N., and Nagai, S. (1995) Characteristics of granular methanogenic sludge grown of phenol synthetic waste water in UASB reactor. *J. Ferment. Bioeng.* **79**(4), 348–353.

Charest, A., Bisaillon, J. G., Lepine, F., and Beaudet, R. (1999) Removal of phneolic compounds from a petrochemical effluent with a methanogenic consortium. *Can. J. Microbiol.* **35**, 235–241.

Chemielowski, J., Grossman, A., and Wegrzynowska, T. (1964) The anaerobic decomposition of phenol during methane fermentation. *Zesz. Nauk. Politech. Slask. Inz. Sanit.* **8**, 97–122

Chen, C. L., Wu, J. H., and Liu, W. T. (2008) Identification of important microbial populations in the mesophilic and thermophilic phenol-degrading methanogenic consortia. *Wat. Res.* **42**(8–9), 1963–1976.

Chou, H. H., Huang, J. S., Jheng, J. H., and Ohara, R. (2008) Influencing effect of intra-granule mass transfer in expanded granular sludge-bed reactors treating an inhibitory substrate. *Biores. Technol.* **99**(9), 3403–3410.

Collins, G., Foy, C., McHugh, S., Mahony, T., and O'Flaherty, V. (2005a) Anaerobic biological treatment of phenolic wastewater at 15–18 °C. *Wat. Res.* **39**(8), 1614–1620.

Collins, G., Foy, C., McHugh, S., and O'Flaherty, V. (2005b) Anaerobic treatment of 2,4,6-trichlorophenol in an expanded granular sludge bed-anaerobic filter (EGSB-AF) bioreactor at 15 degree C. *FEMS Microbiol. Ecol.* **53**, 167–178.

Donlon, B. A., Razo-Flores, E., Lettinga, G., and Field, J. A. (1996) Continuous detoxification, transformation and degradation of nitrophenols in upflow anaerobic sludge blanket (UASB) reactor. *Biotechnol. Bioeng.* **51**, 439–449.

Droste, R. L. (1998) Removal of chlorinated phenols in upflow anaerobic sludge blanket reactors. *Wat. Sci. Technol.* **38**, 359–367.

Escher, B. I., Snozzi, M., and Schwarzenbach, R. P. (1996) Uptake, speciation, and uncoupling activity of substituted phenols in energy transducing membranes. *Environ. Sci. Technol.* **30**, 3071–3079.

Fang, H. H. P. and Chan, O. C. (1997) Toxicity of phenol towards anaerobic biogranules. *Wat. Res.* **31**(9), 2229–2242.

Fang, H. H. P., Chen, T., Li, Y. Y., and Chui, H. K. (1996) Degradation of phenol in wastewater in an upflow anaerobic sludge blanket reactor. *Wat. Res.* **30**(6), 1353–1360.

Fang, H. H. P., Lau, I. W. C., and Chung, D. W. C. (1997) Inhibition of methanogenic activity of starch-degrading granules by aromatic pollutants. *Wat. Sci. Technol.* **35**(8), 247–253.

Fang, H. H. P., Liang, D. W., Zhang, T., and Liu, Y. (2006) Anaerobic treatment of phenol in wastewater under thermophilic condition. *Wat. Res.* **40**, 427–434.

Fang, H. H. P., Liu, Y., Ke, S. Z., and Zhang, T. (2004) Anaerobic degradation of phenol in wastewater at ambient temperature. *Wat. Sci. Technol.* **49**(1), 95–102.

Fang, H. H. P. and Zhou, G. M. (1997a) Anaerobic degradation of benzoate and cresol isomers in sulfate-rich wastewater. *Wat. Sci. Technol.* 36(6–7), 7–14.

Fang, H. H. P. and Zhou, G. M. (1997b) Denitrification of phenolic wastewater by immobilized sludge. *Environ. Technol.* **18**, 827–834.

Fang, H. H. P. and Zhou, G. M. (1999) Interactions of methanogens and denitrifiers in degradation of phenols. *J. Environ. Eng. ASCE* **125**(1), 57–63.

Fang, H. H. P. and Zhou, G. M. (2000) Degradation of phenol and *p*-cresol in reactors. *Wat. Sci. Technol.* **42**(5–6), 237–244.

Farooqi, I. H., Basheer, F., and Isa, M. H. (2007) Co-degradation of phenol and *m*-cresols by upflow anaerobic sludge blanket reactor. *Wat. Sci. Technol.* **56**(7), 73–79.

Fernandez, J. M., Omil, F., Mendez, R., and Lema, J. M. (2001) Anaerobic treatment of fibreboard manufacturing wastewaters in a pilot scale hybrid USBF reactor. *Wat. Res.* **35**(17), 4150–4158.

Fina, L. R., Bridges, R. L., Coblentz, T. H., and Roberts, F. F. (1978) Anaerobic decomposition of benzoic-acid during methane fermentation 3. Fate of carbon four and identification of propanoic acid. *Archive Microbiol.* **118**, 169–172.

Gallert, C., Knoll, G., and Winter, J. (1991) Anaerobic carboxylation of phenol to benzoate: Use of deuterated phenols revealed carboxylation exclusively in the C4-position. *Appl. Microbiol. Biotechnol.* **36**, 124–129.

Gallert, C. and Winter, J. (1992) Comparison of 4-hydroxybenzoate decarboxylase and phenol carboxylase activities in cell-free-extracts of a defined, 4-hydroxybenzoate and phenol-degrading anaerobic consortium. *Appl. Microbiol. Biotechnol.* **37**, 119–124.

Garibay-Orijel, C., Rios-Leal, E., Garcia-Mena, J., and Poggi-Varaldo, H. M. (2005) 2,4,6-Trichlorophenol and phenol removal in methanogenic and partially-aerated methanogenic conditions in a fluidized bed bioreactor. *J. Chem. Technol. Biotechnol.* **80**(10), 1180–1187.

Godbole, A. and Chakrabarti, T. (1991) Biodegradation in upflow anoxic fixed film-fixed bed reactors of resorcinol, catechol and phenol in mono and binary substrate matrices. *Wat. Res.* **25**(9), 1113–1120.

Guiot, S. R., Tawfiki-Hajji, K., and Lepine, F. (2000) Immobilization strategies for bioaugmentation of anaerobic reactors treating phenolic compounds. *Wat. Sci. Technol.* 42(5–6), 245–250.

Guiot, S. R., Tartakovsky, B., Lanthier, M., Levesque M. J., Manuel, M. F., Beaudet, R., Greer, C. W., and Villemur, R. (2002) Strategies for augmenting the pentachlorophenol degradation potential of UASB anaerobic granules. *Wat. Sci. Technol.* **45**(10), 35–41.

Hajji, K. T., Lepine, F., Bisaillon, J. G., Beaudet, R., Hawari, J., and Guiot, S. R. (2000) Effects of bioaugmentation strategies in UASB reactors with a methanogenic consortium for removal of phenolic compounds. *Biotechnol. Bioeng.* **67**(4), 417–423.

Healy, J. B. and Young, L. Y. (1978) Catechol and phenol degradation by a methanogenic population of bacteria. *Appl. Environ. Microbiol.* **35**, 216–218.

Heider, J. and Fuchs, G. (1997) Microbial anaerobic aromatic metabolism. *Anaerobe* **3**, 1–22.

Heipieper, H. J., Keweloh, H., and Rehm, H. (1991) Influence of phenols on growth and membrane permeability of free and immobilized *Escherichia coli. Appl. Environ. Microbiol.* **57**, 1213–1217.

Hendriksen, H. V., Larsen, S., and Ahring, B. K. (1992) Influence of a supplemental carbon source on anaerobic dechlorination of pentachlorophenol in granular sludge. *Appl. Environ. Microbiol.* **58**, 365–370.

Hernandez, J. E. and Edyvean, R. G. J. (2008) Inhibition of biogas production and biodegradability by substituted phenolic compounds in anaerobic sludge. *J. Hazard. Mater.* **160**, 20–28.

Hwang, P. C. and Cheng, S. S. (1991a) The influence of glucose supplement on the degradation of catechol. *Wat. Sci. Technol.* **23**, 1201–1209.

Hwang, P. C. and Cheng, S. S. (1991b) Treatment of *p*-cresol with a recirculating UASB reactor using the concept of kinetic control. *Wat. Sci. Technol.* **24**(5), 133–140.

IARC, International Agency for Research on Cancer (1999) Summaries and evaluations. *Phenol* **71**, 749.

Juteau, P., Cote, V., Duckett, M. F., Beaudet, R., Lepine, F., Villemur, R., and Bisaillon, J. G. (2005) *Cryptanaerobacter phenolicus* gen. nov., sp nov., an anaerobe that transforms phenol into benzoate via 4-hydroxybenzoate. *Int. J. Syst. Evol. Microbiol.* **55**, 245–250.

Karim, K. and Gupta, S. K. (2006) Effect of shock and mixed nitrophenolic loadings on the performance of UASB reactors. *Wat. Res.* **40**(5), 935–942.

Ke, S. Z., Shi, Z., Zhang, T., and Fang, H. H. P. (2004) Degradation of phenol in an upflow anaerobic sludge blanket (UASB) reactor at ambient temperature. *J. Environ. Sci. China* **16**(3), 525–528.

Keith, C. L., Bridges, R. L., Fina, L. R., Iverson, K. L., and Cloran, J. A. (1978) Anaerobic decomposition of benzoic-acid during methane fermentation 4. Dearomatization of ring and volatile fatty-acids formed on ring rupture. *Archive Microbiol.* **118**, 173–176.

Kennes, C., Mendez, R., and Lema, J. M. (1997) Methanogenic degradation of *p*-cresol in batch and in continuous UASB reactors. *Wat. Res.* **31**(7), 1549–1554.

Knoll, G. and Winter, J. (1989) Degradation of phenol via carboxylation to benzoate by a defined, obligate syntrophic consortium of anaerobic bacteria. *Appl. Environ. Microbiol.* **30**, 318–324.

Kobayashi, T., Hashinaga, T., Mikami, E., and Suzuki, T. (1989) Methanogenic degradation of phenol and benzoate in acclimated sludge. *Wat. Sci. Technol.* **21**(1), 55–65.

Kohring, G.-W., Rogers, J. E., and Wiegel, J. (1989) Anaerobic biodegradation of 2,4-dichlorophenol in freshwater lake sediments at different temperatures. *Appl. Environ. Microbiol.* **55**, 348–353.

Kulkarni, M. and Chaudhari, A. (2007) Microbial remediation of nitro-aromatic compounds: An overview. *J. Environ. Manage.* **85**(2), 496–512.

Lanthier, M., Juteau, P., Lepine, F., Beaudet, R., and Villemur, R. (2005) *Desulfitobacterium hafniense* is present in a high proportion within the biofilms of a high-performance pentachlorophenol-degrading, methanogenic fixed-film reactor. *Appl. Environ. Microbiol.* **71**(2), 1058–1065.

Lao, S. G. (2002) Mechanisms of granular activated carbon anaerobic fluidized-bed process for treating phenols wastewater. *J. Environ. Sci. China* **14**(1), 132–135.

Latkar, M., Swaminathan, K., and Chakrabarti, T. (2003) Kinetics of anaerobic biodegradation of resorcinol catechol and hydroquinone in upflow fixed film-fixed bed reactors. *Biores. Technol.* **88**(1), 69–74.

Lay, J. J. and Cheng, S. S. (1998) Influence of hydraulic loading rate on UASB reactor treating phenolic wastewater. *J. Environ. Eng. ASCE* **124**(9), 829–837.

Lettinga, G. (1995) Anaerobic digestion and wastewater treatment systems. *Antonie Van Leeuwenhoek* **6**, 3–28.

Lettinga, G., Rebac, S., and Zeeman, G. (2001) Challenge of psychrophilic anaerobic wastewater treatment. *Trends Biotechnol.* **19**, 363–370.

Levén, L. and Schnürer, A. (2005) Effects of temperature on biological degradation of phenols, benzoates and phthalates under methanogenic condtions. *Int. Biodeter. Biodegrad.* **55**(1), 153–160.

Li, T., Biaillon, J.-G., Villemur, R., Létourneau, L., Bernard, K., Lépine, F., and Beaudet, R. (1996) Isolation and characterization of a new bacterium carboxylating phenol to benzoic acid under anaerobic conditions. *Appl. Environ. Microbiol.* **178**(9), 2551–2558.

Liang, D. W., Zhang, T., and Fang, H. H. P. (2008) Real-time quantifications of dominant anaerobes in an upflow reactor by polymerase chain reaction using a TaqMan probe. *Wat. Sci. Technol.* **57**(11), 1851–1855.

Lin, Y. H. and Lee, K. K. (2006) Kinetics of phenol degradation in an anaerobic fixed-biofilm process. *Wat. Environ. Res.* **78**(6), 598–606.

Luo, H. P., Liu, G. L., Zhang, R. D., and Jin, S. (2009) Phenol degradation in microbial fuel cells. *Chem. Eng. J.* **147**(2–3), 259–264.

MacLeod, F. A., Guiot, S. R., and Costerton, J. W. (1990) Layered structure of bacterial aggregates produced in an upflow anaerobic sludge bed and filter reactor. *Appl. Environ. Microbiol.* **56**, 1598–1607.

Majumder, P. S. and Gupta, S. K. (2008) Degradation of 4-chlorophenol in UASB reactor under methanogenic conditions. *Biores. Technol.* **99**(10), 4169–4177.

Qiu, Y. L., Hanada, S., Ohashi, A., Harada, H., Kamagata, Y., and Seikiguchi, Y. (2008) *Syntrophorhabdus aromaticivorans* gen. nov., sp. nov., the first cultured anaerobe capable of degrading phenol to acetate in obligate syntrophic associations with a hydrogenotrophic methanogen. *Appl. Environ. Microbiol.* **74**(7), 2051–2058.

Qiu, Y. L., Sekiguchi, Y., Imachi, H., Kamagata, Y., Tseng, I. C., Cheng, S. S., Ohashi, A., and Harada, H. (2004) Identification and isolation of anaerobic, syntrophic phthalate isomer-degrading microbes from methanogenic sludges treating wastewater from terephthalate manufacturing. *Appl. Environ. Microbiol.* **70**(3), 1617–1626.

Ramakrishnan, A. and Gupta, S. K. (2006) Anaerobic biogranulation in a hybrid reactor treating phenolic waste. *J. Hazard. Mater.* **137**(3), 1488–1495.

Ramakrishnan, A. and Gupta, S. K. (2008) Effect of hydraulic retention time on the biodegradation of complex phenolic mixture from simulated coal wastewater in hybrid UASB reactors. *J. Hazard. Mater.* **153**(1–2), 843–851.

Razo-Flores, E., Iniestra-Gonzalez, M., Field, J. A., Olguin-Lora, P., and Puig-Grajales, L. (2003) Biodegradation of mixtures of phenolic compounds in an upward-flow anaerobic sludge blanket reactor. *J. Environ. Eng. ASCE* **129**(11), 999–1006.

Razo-Flores, E., Donlon, B., Lettinga, G., and Field, J. A. (1996) Biotransformation and biodegradation of N-substituted aromatics in methanogenic granular sludge. *FEMS Microbiol. Rev.* **20**(3–4), 523–538.

Schink, B., Philipp, B., and Muller, J. (2000) Anaerobic degradation of phenolic compounds. *Naturwissenschaften* **87**, 12–23.

Scully, C., Collins, G., and O'Flaherty, V. (2006) Anaerobic biological treatment of phenol at 9.5–15 degrees C in an expanded granular sludge bed (EGSB)-based bioreactor. *Wat. Res.* **40**(20), 3737–3744.

Seghezzo, L., Zeeman, G., van Lier, J. B., Hamelers, H. V. M., and Lettinga, G. (1998) A review: The anaerobic treatment of sewage in UASB and EGSB reactors. *Biores. Technol.* **65**, 175–190.

Sharak Genthner, B. R., Townsend, G. T., and Chapman, P. J. (1990) Effect of fluorinated analogues of phenol and hydroxybenzoates on the anaerobic transformation of phenol to benzoate. *Biodegradation* **1**, 65–74.

Shen, D. S., He, R., Liu, X. W., and Long, Y. (2006) Effect of pentachlorophenol and chemical oxygen demand mass concentrations in influent on operational behaviors of upflow anaerobic sludge blanket (UASB) reactor. *J. Hazard. Mater.* **136**, 645–653.

Sreekanth, D., Sivaramakrishna, D., Himabindu, V., and Anjaneyulu, Y. (2009) Thermophilic degradation of phenolic compounds in lab scale hybrid up flow anaerobic sludge blanket reactors. *J. Hazard. Mater.* **164**(2–3), 1532–1539.

Subramanyam, R. and Mishra, I. M. (2007) Biodegradation of catechol (2-hydroxy phenol) bearing wastewater in an UASB reactor. *Chemosphere* **69**(5), 816–824.

Subramanyam, R. and Mishra, I. M. (2008a) Co-degradation of resorcinol and catechol in an UASB reactor. *Biores. Technol.* **99**(10), 4147–4157.

Subramanyam, R. and Mishra, I. M. (2008b) Treatment of catechol bearing wastewater in an upflow anaerobic sludge blanket (UASB) reactor: Sludge characteristics. *Biores. Technol.* **99**(18), 8917–8925.

Suidan, M. T., Siekerka, G. L., Kao, S. W., and Pfeffer, J. T. (1983a) Anaerobic filters for the treatment of coal gasification wastewater. *Biotechnol. Bioeng.* **26**, 1581–1596.

Suidan, M. T., Strubler, C. E., Kao, S. W., and Pfeffer, J. T. (1983b) Treatment of coal gasification wastewater with anaerobic filter technology. *J. Wat. Pollut. Control Fed.* **55**, 1263–1270.

Swaminathan, K., Chakrabarti, T., and Subramanyam, P. V. R. (1999) Substrate–substrate interaction of resorcinol and catechol in upflow anaerobic fixed film-fixed bed reactors in mono and multisubstrate matrices. *Bioprocess. Eng.* **20**, 349–353.

Tay, J. H., He, Y. X., and Yan, Y. G. (2000) Anaerobic biogranulation using phenol as the sole carbon source. *Wat. Environ. Res.* **72**(2), 189–194.

Tay, J. H., He, Y. X., and Yan, Y. G. (2001) Improved anaerboic degradation of phenol with supplemental glucose. *J. Environ. Eng. ASCE* **127**(1), 38–45.

Veeresh, G. S., Kumar, P., and Mehrotra, I. (2006) Biodegradation of phenol with wastewater as cosubstrate in upflow anaerobic sludge blanket. *J. Environ. Eng. ASCE* **132**(11), 1539–1542.

Veeresh, S. G., Kumar, P., and Mehrotra, I. (2005) Treatment of phenol and cresols in upflow anaerobic sludge blanket (UASB) process: A review. *Wat. Res.* **39**, 154–170.

von Wintzingerode, F., Selent, B., Hegemann, W., and Gobel, U. B. (1999) Phylogenetic analysis of an anaerobic, trichlorobenzene-transforming microbial consortium. *Appl. Environ. Microbiol.* **65**(1), 283–286.

Wang, Y. T., Suidan, M. T., and Rittman, B. E. (1986) Anaerobic treatment of phenol by an expanded-bed reactor. *J. Wat. Pollut. Control Fed.* **58**(3), 227–233.

Wiegant, W. M. (1985) Thermophilic anaerobic digestion for wastewater treatment, Agricultural University, Wageningen, The Netherlands.

Wu, J. H., Liu, W. T., Tseng, I. C., and Cheng, S. S. (2001) Characterization of microbial consortia in a terephthalate-degrading anaerobic granular sludge system. *Microbiol.* **147**(2), 373–382.

Wu, W. M., Bhatnagar, L., and Zeikus, G. (1993) Performance of anaerobic granules for degradation of pentachlorophenol. *Appl. Environ. Microbiol.* **59**, 389–397.

Ye, F. X., Shen, D. S., and Feng, X. S. (2004) Anaerobic granule development for removal of pentachlorophenol in an upflow anaerobic sludge blanket (UASB) reactor. *Process Biochem.* **39**(10), 1249–1256.

Young, L. Y. and Rivera, M. D. (1985) Methanogenic degradation of four phenolic compounds. *Wat. Res.* **19**, 1325–1332.

Zhang, T., Ke, S. Z., Liu, Y., and Fang, H. H. P. (2005) Microbial characteristics of a methanogenic phenol-degrading sludge. *Wat. Sci. Technol.* **52**(1–2), 73–78.

Zhou, G. M. and Fang, H. H. P. (1997) Co-degradation of *m*-cresol and phenol in a UASB reactor. *Biores. Technol.* **61**, 47–52.

Chapter 10

Application of Molecular Methods for Anaerobic Technology

Tong Zhang

Environmental Biotechnology Laboratory,
Department of Civil Engineering, The University of Hong Kong,
Pokfulam Road, Hong Kong
zhangt@hkucc.hku.hk

Anaerobic technology has been applied to sludge digestion for over 50 years and to wastewater treatment for nearly 30 years. This chapter reviews the nucleic acids-based molecular methods for the analysis of microbial communities, and their applications for the microbial communities in anaerobic reactors. At first, nucleic acid extraction and selection of nucleic acid biomarkers are discussed with emphasis on special aspects related to anaerobic sludge samples. Second, the common nucleic acids-based methods are introduced, including cloning and sequencing for characterization, quantitative real-time polymerase chain (qRT-PCR) for quantification, fluorescence *in situ* hybridization (FISH) for visualization, denaturing gradient gel electrophoresis (DGGE) and terminal restriction fragment length polymorphism (T-RFLP) for screening/typing/monitoring, and stable isotope probing (SIP) and microautoradiography–FISH to link the microbial identity with the specific functions. Finally, applications of the aforementioned molecular methods in anaerobic technology are summarized, focusing on identification of new species in the anaerobic process, characterization of microbial compositions in the anaerobic reactors, visualization of layered structure of anaerobic granular sludge, etc.

1. Introduction

Application of anaerobic technology for wastewater treatment was first introduced by MaCarty (1964) nearly half a century ago, and it has been gradually accepted by industries due to the technical development led by Lettinga *et al.* (1980) three decades ago. Compared to aerobic treatment of wastewater, anaerobic process not only saves the energy needed for aeration and generates much less sludge, but also produces an energy-rich biogas (about 60% methane). In recent years, anaerobic technology has attracted increasing attention from both academia and industries, as global warming has become a serious issue.

Anaerobic degradation of organic matter in wastewater is conducted syntrophically by microbial communities consisted of both *Bacteria* and *Archaea* species. The degradation may be divided into three steps. In the first step, hydrolysis bacteria degrade polymeric organic matter into monomers, such as sugar and amino acid, which are further degraded in the second step by acetogenic bacteria into volatile fatty acids (VFAs), such as acetic acid. In the last step, methanogens, including hydrogenotrophs, methylotrophs, and acetotrophs, produce carbon dioxide and methane from formate and hydrogen, and acetate, respectively. In the presence of sulfate, sulfate-reducing bacteria (SRB), a diverse group of prokaryotes which use sulfate as the terminal electron acceptor for the metabolism of organic substrates, may act as a competitor of methanogens in terms of VFA utilization.

Identification and quantification of microbial species involved in anaerobic degradation are essential to the study of anaerobic technology and to the optimization of operational conditions of the reactors. Conventionally, microorganisms are identified by isolating pure cultures, and then characterizing their physiological/biochemical/morphological properties (Rao *et al.* 2000). However, these traditional methods have serious biases. First, many microorganisms may not be properly isolated, because the artificial growth media may not exactly simulate the environment in the anaerobic reactors. Second, many microorganisms require syntrophic interactions with others, especially for the degradation of refractory pollutants, and thus cannot be cultured individually (Wagner *et al.* 1993). Third, some microorganisms share similar physiological, biochemical and/or morphological characteristics, and thus cannot be distinguished from one another accordingly. As a result, many estimated that no more than 1% of the existing bacteria had been isolated and characterized (Amann 1995).

Thanks for the development of Internet, bioinformatics, and florescence chemistry, nucleic acid-based molecular methods has advanced greatly in the recent years (Fig. 10.1). Currently, microorganisms can be unambiguously distinguished by the DNA sequences of their genes, e.g., ribosomal RNA (rRNA) genes, which are used as biomarkers. With application of nucleic acid-based molecular techniques, it has been found that microbial communities in the anaerobic process are highly diverse, and the interactions among different species and regulation of metabolic functions mostly remain unknown. Further studies on the effects of operational factors (including temperature, pH, loading rate, etc.) on microbial communities using molecular techniques will help optimization of the anaerobic process, increase of the reactor stability, and maximization of methane yield.

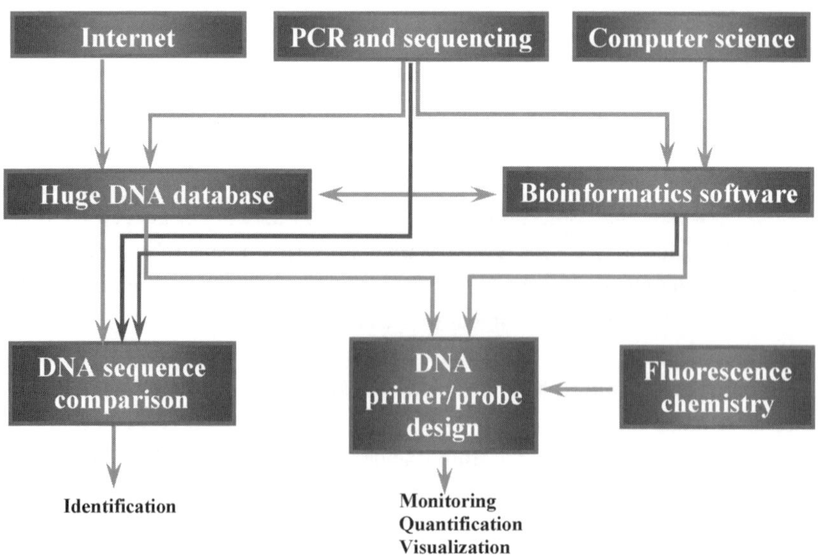

Fig. 10.1. Diagram of development of the nucleic acid-based molecular technology.

This chapter is to review the state-of-the-art knowledge of nucleic acid-based molecular techniques and their application in the analysis of microbial communities in anaerobic reactors. At first, two basic subjects, i.e., nucleic acid extraction and selection of nucleic acid biomarkers, are discussed with emphasis on special pre-cautions in handling anaerobic sludge samples. Second, the common methods for different applications are introduced sequentially, i.e., identification, screening/typing/monitoring, visualization, quantification, etc. To characterize the community in an anaerobic reactor, cloning–sequencing after polymerase chain reaction (PCR) is the most commonly used method (Sekiguchi *et al.* 1998; Zhang and Fang 2001), while microarray (Franke-Whittle *et al.* 2009) and metagenome (Schlüter *et al.* 2008) have also been applied very recently. Combined with other techniques, such as stable isotope probing (SIP) (Hori *et al.* 2007), the selectivity and effectiveness of the above characterization methods could be further improved. For screening/typing/monitoring, denaturing gradient gel electrophoresis (DGGE) (Muyzer *et al.* 1993; Zhang and Fang 2000) and terminal restriction fragment length polymorphism (T-RFLP) (Liu *et al.* 1997) are the two mainstream techniques followed by single-strand conformation polymorphism (SSCP) (Lee *et al.* 1996), denaturing high-performance liquid chromatography (dHPLC) (Wagner *et al.* 2009), and ribosomal intergenic spacer analysis (RISA) (Borneman and Triplett 1997). To visualize the different cells in the anaerobic sludge, fluorescence *in situ* hybridization (FISH) is the basic method and may be enhanced to generate many variants by combining with other techniques, including CLSM (confocal laser scanning microscopy)–FISH (Zhang *et al.* 2005), MAR–FISH (microautoradiography–FISH) (Lee *et al.* 1999), CARD–FISH (catalyzed reporter deposition–FISH) (Pernthaler *et al.* 2002), etc. Although not used as commonly as FISH, *in situ* PCR is also an alternative to visualize cells (Hodson *et al.* 1995). For quantification of microorganism and elucidation gene expression associated with specific activities, quantitative PCR (qPCR) is almost always the first choice due to its high reliability and broad dynamic ranges (up to 10^7–10^9) (Zhang and Fang 2006). FISH (Amann 1995), membrane hybridization (Raskin *et al.* 1995), RNAse H method (Ueno *et al.* 2004), etc. are also proved to be useful to quantify microorganisms. To achieve a direct link of metabolism function and phylogenetic classification, isotope technology is essential by using either radioactive isotope in MAR–FISH (Lee *et al.* 1999) or stable isotope in stable isotope probing (Hori *et al.* 2007) and Raman–FISH (Huang *et al.* 2007). Finally, the applications of the above methods in anaerobic technology will be discussed in detail, focusing on identification of new species in the process, characterization of microbial compositions in the reactors, visualization of layered structure of granular sludge, etc.

2. Extraction of Nucleic Acids

The reliability of the molecular techniques depends on quality and representation of RNA/DNA extracted from sludge samples in the reactors. The starting amount samples for nucleic acid extraction should be quantified in terms of cell number, dry mass of sludge (or sediment/soil), or volume of slurry. Such information is important for the absolute quantification of specific microorganisms. If there is no previous knowledge of variation within an anaerobic reactor, replicate must be taken to enable comparison between reactors and to evaluate the reproducibility in the same reactor. Nucleic acids in environmental samples are easily degraded due to the activity of endogenous nucleases. For RNA analyses, nuclease activity should be inhibited by freezing the sample immediately using liquid nitrogen and then stored at -80 °C, whereas for DNA analyses, sample may be stored at either -20 °C or -80 °C. After freezing, thawing the sample before extraction should be avoided.

Extracting DNA or RNA quantitatively from sludge in an anaerobic reactor without bias is not easy. The DNA extraction process is composed of cell lysis, contamination removal, solvent extraction, precipitation, and purification (Miller *et al.* 1999). Bacteria cells can be lysed by several methods, including bead beating homogenization (Bruce *et al.* 1992), ultrasonication (Picard *et al.* 1992), freeze–thawing (Erb and Wagner-Döbler 1993), grinding under liquid nitrogen (Volossiouk *et al.* 1995), chemical lysis using detergents and enzymes (Bruce *et al.* 1992), and the combination of the above methods. Mechanical disruption of microbial cells with phenol is more convenient than lysis by enzymes as the nuclease activities are inhibited immediately (Raskin *et al.* 1995).

A few protocols for the extraction of DNA or RNA from various environmental samples have been published (Griffiths *et al.* 2000). The efficiency and reliability of these methods depend on many factors, such as cell wall structure, cell size, molecular weight of the target DNA, etc. The extraction methods have to be optimized according to the type of samples. Generally, spores, Gram-positive cells, and small cells are less readily lysed than non-spores, Gram-negative cells, and bigger cells. To effectively disrupt the wall of cells that are hard to break, such as methanogens and Gram-positive bacteria, the method which may be used is bead-beating in citrate buffered, low-pH phenol and chloroform extractions (Libonati and Sorrentino 2001). Mild extraction methods, such as freeze–thawing and chemical lysis, tend to underestimate the microbial diversity of the samples, while harsh extraction methods, such as extensive bead beating and ultrasonication, may cause contamination and shear genomic DNA. In general, mechanical methods have less bias than lysis by enzymes. The combination of repeated freeze–thawing, lysis with detergents, and short time bead beating, was able to break approximately 96% of cells and endospores without significant shearing of DNA (Miller *et al.* 1999). The biases of different DNA and/or RNA extraction methods have been reported and summarized by Martin-Laurent *et al.* (2001). The extraction efficiency can be checked with the spiked internal standard (Bachoon *et al.* 2001). But this is hardly applied in works involving a large number of samples.

Cells of some methanogens, the major population in anaerobic reactors, lyse easily, but others may not. Jarrell *et al.* (1992) reported a disruption method for methanogens by crushing the collected pellet of culture in liquid N_2 to break cells. This method could be applied to all methanogens yielding DNA of high quality. For more information about other extraction methods, readers may refer to the works of Frischer *et al.* (2000).

The materials interfering with extraction of nucleic acid or downstream manipulation should be removed. For example, humic substances in the extract of nucleic acids may interfere with PCR reaction and membrane hybridization. It was reported that humic acids could be removed by repeatedly washing samples with low-pH citrate solution before cell disruption (Hofman-Bang *et al.* 2003). A number of more complicated methods have also been developed to remove the humic substances, including gel purification, polyvinylpolypyrrolidone (PVPP) adsorption, and dilution (Frischer *et al.* 2000). Humic substances could also be easily conducted by binding to resin which is often provided in commercially available kits, then removing contaminants by eluting, and finally releasing to collect the DNA/RNA.

Furthermore, DNA and RNA may both be present in the extract from environmental samples. To avoid interference, RNase and DNase should be used, if necessary, to remove RNA and DNA, respectively, although partial degradation of RNA and DNA due to impurities in RNAse and

DNase should be concerned. Purified RNA should be stored at -80 °C and DNA at -20 °C or -80 °C. Additionally, unlike DNA which is relatively stable, RNA should always be kept on ice in the downstream operations to minimize RNA degradation.

The amount of nucleic acid (as expressed by A_{260}) and purity (as expressed by the ratios of A_{260}/A_{280} for protein contamination and A_{260}/A_{230} for salt contamination) is measured spectrophotometrically using the conventional UV–visible spectrometer or the more advanced spectrometers, such as Nanodrop. The integrity of rRNA (containing 5S, 16S, and 23S) and genomic DNA could be visually checked by conducting electrophoresis using agarose gel.

3. Selection of Biomarker

Although applications of rRNA internal transcribed spacer (ITS), functional genes, and σ factor *rpo*B have also been reported for microbial analysis, the 5S rRNA, 16S rRNA, and 23S rRNA genes are the most common phylogenetic biomarkers because they are: (1) key cellular elements and functionally as well as evolutionarily homologous for all organisms, (2) conserved in overall structure, (3) abundant in most cells (10^3–10^5 copies), (4) non-transferable among organisms, and (5) having various regions at different conservation levels (Hofman-Bang *et al.* 2003).

The earliest efforts to analyze the diversity of microbial communities relied upon 5S rRNA gene (5S rDNA), which, however, contains only about 120 nucleotides and thus has limited resolution in phylogenetic analysis and thus can only be used for the characterization of communities of relatively low diversity (Bruce *et al.* 1992). Compared to 5S rDNA, 16S rRNA gene (16S rDNA) gives more genetic information for species identification and probe/primer design, due to its longer sequence (up to 1540 nucleotides) comprising highly conserved parts interspersed with more variable regions. Based on the analysis of 16S rDNA, all cellular life is divided into three domains, i.e., *Archaea*, *Bacteria*, and *Eukarya* (Woese *et al.* 1990). Although 23S rRNA gene (23S rDNA) has also been used in molecular studies as it may give even longer sequence (2,900 nucleotides) (Jeng *et al.* 2001), the 16S rDNA-based techniques are still the most widely used methods due to the huge database available (over 1,740,000 of 16S rDNA sequences in GenBank as of January, 2010).

16S rDNA contains conserved and variable regions according to their genetic stability. Based on variations of DNA sequences in these regions, organisms can be classified into different taxonomy levels. Organisms within a domain often share the same DNA sequence in the most conserved region, whereas species of the same genus may be discriminated from DNA sequence in the variable regions (Woese *et al.* 1990). Hybridization probes and PCR primers may be designed based on those sequences available in the genetic databases, such as GenBank, RDP (Ribosomal Database Project), and ARB. The *Bacteria*- and *Archaea*-specific primers, such as the set of EUB8F and UNIV1492R for *Bacteria* (Lane 1991), and the set of A109F (Grosskopf *et al.* 1998) and A934b (Stahl and Amann 1991) for *Archaea*, were commonly used for the diversity analysis of anaerobic microbial community.

Although the phylogenetic properties of 16S rDNA and the huge amount of sequences available offer a considerable advantage, the molecular techniques based on 16S rDNA gene still have some disadvantages. For instance, the heterogeneity of 16S rDNA among multiple copies in one species (Klappenbach *et al.* 2001) hampers pattern analysis, and may confuse the interpretation of microbial diversity obtained from clone libraries or sequences of the cut DGGE bands. Additionally, 16S rDNA sequence may lack resolution at the species level for some microbial groups.

23S rDNA (about 49,000 sequences in GenBank as of January, 2010, i.e., 3% of those of 16S rDNA) may become more popular as it is much longer than 16S rDNA, and may resolve very closely related species. Another limitation of the use of 16S rDNA for the microbe quantification is that little information is available on the copy number of rRNA operons on microbial genomes. The rrn operon copy number (Y_{rrn}) is species specific; e.g., the number is 1 for *Rhodothermus marinus*, but 15 for *Clostridium paradoxum* (Rainey *et al.* 1996). According to the Ribosomal RNA Operon Copy Number Database (*rrndb*) (Klappenbach *et al.* 2001), the Y_{rrn} values so far are only available for ~400 species among ~5,000 for known prokaryotic species. A few methods to measure Y_{rrn} have been proposed recently (Zhang *et al.* 2009).

4. Polymerase Chain Reaction (PCR)

For many molecular methods introduced in this chapter, the first step after nucleic acid extraction is PCR or RT (reverse transcription)-PCR to amplify a target fragment of DNA or RNA, respectively. The PCR cycle takes place in three steps: denaturing, annealing, and extension. In the denaturing step, the DNA template is denatured by heating to break apart the double-stranded molecule into the single strands. In the annealing step, primers, which are short DNA fragments complementary to specific sequences, bind to the single-stranded DNA during cooling. In the extension step, the temperature is raised and DNA polymerase is used to add complementary bases to the 3′ end of the primer along the rest of the single-stranded DNA templates from 5′ end to 3′ end, resulting in a new copy of the DNA fragment. The above three-step process is repeated for a number of cycles (usually 30–50) to generate a huge quantity of the desired DNA fragment.

The crucial part of a PCR is the selected primer set targeting genes of interest which cover special taxonomic or functional groups. These primers may be individually designed based on the alignments of relevant DNA sequences, or simply adapted from the literature. Web-based software such as ProbeMatch at RDP website (Cole *et al.* 2005) and those which may be installed to a personal computer, such as ARB (Ludwig *et al.* 2004), may facilitate the primer/probe design.

The selected primer sets may target the sequences at various taxonomic levels, from domain to division, subdivision, class, family, genus, species, and even strain. The PCR products contain a mixture of multiple copies of the same segment amplified at the selected taxonomic level. The PCR products can be cloned and then sequenced to identify species. They can also be analyzed by various techniques, such as DGGE, T-RFLP, SSCP, or temperature gradient gel electrophoresis (TGGE), which may separate PCR products originating from different DNA sequences of various species in the sample.

Bias may be introduced in PCR amplification steps, no matter what kind of gene is targeted. All downstream techniques, including cloning, fingerprint analysis, hybridization, and sequencing, will be affected by the PCR biases. Since the amount of DNA generated by PCR increases exponentially in the amplification, any error happening early in the PCR process will be amplified (Amann 1995). The first bias is preferential amplification as PCR may not amplify all DNA to the same extent (Amann 1995). It has been found that the original template in the DNA extract is amplified in the beginning 5–6 cycles of the PCR reaction (Kurata 2001). The amplification after that only happens on the PCR products generated earlier. Thus the sequences with high abundance and matching well with a primer will be selected in the initial PCR cycles and be amplified preferably in the consequent cycles. This may result in the overestimation of the predominant population and the overlook of the minor ones.

The specificity of primers is one of the major factors for a successful PCR, especially when a mixture of homologous target sequences is to be quantified. "Universal" primers designed for various microbial groups may not be able to amplify all target genes since the primer sites are not totally conserved. It has been proven that designing general primers is a difficult job, and even a single mismatch base in the middle section of the primer may cause a preferential selection. For example, a previously commonly used universal primer EUB968F has many mismatches, indicating that it may not be proper to be used as the universal primer for the PCR amplification of 16S rDNA of diverse bacterial communities.

PCR at an inappropriate annealing condition may also have bias and amplify genes that are not targeted. Low annealing temperature tolerates more mismatches and increases diversity of PCR products; however, it might also generate unwanted PCR products, so-called non-specific amplification.

Chemicals and enzymes may be contaminated by sequences of some microorganisms (Liu *et al.* 2002). A few specific 16S rDNA have been found to be highly correlated with PCR contamination in the clone libraries of environmental samples (Liu *et al.* 2002). Among them, sequences repetitively reported were closely affiliated with *Xanthomonas* spp. and *Stenotrophomonas maltophilia*.

Chimeric sequences are often produced due to partially reverse transcribed DNA when performing RT-PCR, during which premature PCR products acting as primers in a subsequent PCR cycle or the presence of partial fragments of rDNA in DNA extracts (Amann 1995). Additionally, primer dimmer may be formed between primer pairs, leading to preferential amplification of certain targets and poor sensitivity (Markoulatos *et al.* 2002).

Multiplex PCR (i.e., using multiple primer pairs in one reaction) could be conducted to simultaneously amplify a few different sequences to save time and effort. But, its application is limited as the combined primer pairs cannot always function properly, usually simply because of the different annealing temperatures required by individual primers. Although multiplex PCR has such drawbacks, it is still a simple, rapid and convenient screening tool (Kong *et al.* 2002).

5. Characterization of Microbial Community

5.1. *Cloning*

Cloning of PCR products is to separate the PCR fragments of the same length but of different sequences. Cloning is composed of three steps: ligation, transformation, and host cell reproduction. Several commercial kits are available for DNA cloning, such as the pGEM-T cloning kit and the TA cloning kit. Taking the TA cloning kit as an example, PCR products are first inserted into the plasmids under the action of the ligase. After ligation, the plasmids with PCR product inserts are transformed into *Escherichia coli* competent cells. Each cell carrying the plasmid with the PCR product insert forms a single white colony (called the clone) on the solid discriminative medium. Whereas, each cell carrying the plasmid without the PCR product insert forms a blue one, and each cell carrying no plasmid does not form any colony. A DNA library consists of the selected white colonies. The insert in plasmids of the white colonies may be recovered using PCR with a primer set targeting the sequence on the plasmids which locate at the two sides of the insert or the original primer set which is used to generate PCR products initially. The whole plasmid could be extracted after further cultivation of the colonies in liquid medium. The PCR product obtained using the first method, or the plasmids obtained using the second method, will be sequenced for further identification of the sequences.

Sequencing of the full 16S rDNA (about 1,540 bp) is preferred for microbial identification, especially for oligonucleotide probe design and classification of the pure culture. Double check by repeated sequencing increases the reliability of obtained sequences.

Once a sequence database of a clone library is established, the microbial diversity can be determined with reference to the published sequences of the pure cultures and environmental samples. The similarity analysis of DNA sequences is greatly facilitated by a number of rDNA sequence databases, such as GenBank (http://www.ncbi.nlm.nih.gov/blast/) and the ribosomal database project (RDP, http://rdp.cme.msu.edu/), and powerful software packages, including MEGA 2.1 (Kumar *et al.* 1993) and ARB (Strunk and Ludwig 1997). By using the Blast program in GenBank, species closely related to the obtained sequences are listed in the similarity order (Altschul *et al.* 1990). A primary taxonomy position of the species represented by obtained sequences may also be given. More accurate taxonomy analysis of DNA sequences could be conducted by construction of phylogenetic trees. To do this, the first step is always the careful alignment of the sequences (Thompson *et al.* 1994). It is a relatively simple job for highly conserved sequences, but considerably problematic for variable regions. If regions of some sequences cannot be unambiguously aligned, they are usually excluded in the subsequent phylogenetic analyses (Mora and Amann 2001). RDP10 released on July 28, 2009 by the Center for Microbial Ecology (Michigan State University) contains more than one million aligned and annotated 16S rRNA gene sequences of *Bacteria* and *Archaea*. The hierarchy model used by the RDP Classifier is in agreement with those in the most updated (Release 6.0) *Bergey's Manual of Systematic Bacteriology*. Additionally, the newly trained Classifier may be used to classify both archaeal and bacterial 16S rDNA sequences, making it valuable for the analysis of microbial community of an anaerobic reactor which is populated with both *Bacteria* and *Archaea* species.

After alignment, phylogenetic trees can be constructed by three approaches: distance, maximum likelihood, and parsimony analyses. For the distance method, which is the simplest and the most common among the three, a distance matrix is constructed by pairwise comparison of a series of aligned DNA sequences. The distances in the matrix are calculated based on a model of base substitution at a single base site, such as the Jukes and Cantor model. Then, the distance matrix can be used to construct a bifurcating tree by grouping the most closely related pairs of sequences (Saitou and Nei 1987).

The obtained sequence information in a clone library can be used to evaluate species diversity or richness of microbial communities in anaerobic reactors, and to design specific probes/primers for characterization/quantification of a certain microorganism. The cloning–sequencing method was first developed to estimate the microbial diversity of bacterial plankton in the Sargasso Sea using 16S rDNA (Giovannoni *et al.* 1990), and is still the most widely used molecular method in the field of anaerobic technology treating various wastewaters, including those containing phenol (Zhang *et al.* 2005), dimethyl phthalate (DMP) (Liang *et al.* 2007, 2009), *cis*-1,2-dichloroethylene (Hata *et al.* 2004), terephthalate (Chen *et al.* 2004), etc. Zhang *et al.* (2005) demonstrated the cloning approach in phenol-degrading anaerobic sludge combined with FISH analysis, to give a detailed structure of the microbial population, and to identify the species responsible for phenol degradation.

Clone libraries can be used to characterize a substantial part of the microbial community in an environmental sample and have the following advantages: (1) precise taxonomic results and high resolution phylogenetic trees, (2) useful information for primers/probes design, (3) recovery of most microorganisms if library size is big enough, including minority groups, and (4) identification of uncultured species. But the cloning method is time-demanding and less applicable for analysis of a larger set of samples, such as monitoring the changes of a microbial community over time (Sanz and Köchling 2007).

5.2. *Diversity estimation*

The stability of an anaerobic reactor tends to increase with the microbial diversity; increased diversity improves the reactor's resistance toward process disturbance and shortens the time for recovery. Thus, monitoring diversity in anaerobic reactors may give the engineers basic information to predict the reactor performance and take action to prevent possible failure.

However, there is no simple solution on how to estimate the accurate microbial diversity in a sample. It is widely accepted that the diversity is measured according to the number of species as well as their abundances (McCune and Grace 2002). The sequences obtained using the cloning–sequencing method can be compared with those in international databases to find the closest relatives. Usually, two 16S rDNA sequences with $>97\%$ similarity, corresponding to DNA:DNA hybridization values of $>70\%$, are considered likely to be the same species (Rossello-Mora and Amman 2001); and those with <93–95% similarity are considered definitely different genus (Hofman-Bang *et al.* 2003). The frequently used term, operational taxonomy unit (OTU), which is usually equivalent to the species level, can be defined using 1% (Sekiguchi *et al.* 1999), 2.5% (Bull 2004) or even 5% (Zhang *et al.* 2009) cut-off of 16S rDNA sequence. For fingerprint methods, like DGGE and T-RFLP, a band or a peak could also be defined as an OTU. For cloning and sequencing method, a rarefaction curve or other coverage estimators of species richness are usually used to decide whether more sequences should be obtained to evaluate diversity correctly (Zhang *et al.* 2009). For an anaerobic reactor with high diversity, a large number of sequences (usually >100 sequences) are required for a correct estimation of the diversity.

5.3. *Metagenome*

Another method which may be used to obtain sequence information and study potential functions of uncultivated bacteria in environmental samples is metagenomics, or so-called environmental genomics. Generally, metagenomics considers genomes of all species in a microbial community as a whole. Using the shotgun sequencing, a huge amount of sequences may be acquired from environmental samples, and need a great effort to systematically analyze the information using bioinformatics concept and expertise with the help of powerful software (Pachter 2007). However, sometime it is not easy to link the genes obtained using the shotgun sequencing approach to any taxonomy or function, due to the lack of existing referencing genomes (Falkowski and de Vargas 2004). Although it may have a huge potential, the application of metagenomics at present is still at its very early stage. For example, the "heavy" (containing C^{13}) genomic DNA from SIP experiments may be analyzed by metagenomic approaches and substrate uptake may thus be associated to phylotypes (Dumont and Murrell 2005). So far, there is only very limited application of metagenomics in studies of anaerobic treatment (Meyerdierks *et al.* 2009).

6. Fingerprint Methods

6.1. *DGGE*

PCR amplification produces DNA segments with the same size but different sequences. These segments may be separated in an acrylamide gel, but not in an agarose gel, having a linear ascending gradient of denaturants, usually urea and formamide (Muzyer *et al.* 1993). This so-called DGGE method is based on the differences of the electrophoresis mobility of the partially denatured double-stranded DNA fragments in the polyacrylamide gel. Double-stranded DNA molecule may melt or separate at a specific denaturing condition (temperature, salt concentration, denaturant concentration, etc.), depending on the hydrogen bonds between the two strands. GC-rich, double-stranded fragment melts at more strict denaturing condition (high temperatures, salt, or denaturant concentrations). Double-stranded DNA fragments initially migrate with the same mobility through the denaturing gradient gel. As the electrophoresis proceeds, some fragments partially denature to form a composite structure containing both double- and single-stranded regions. Partial denaturation of the double-stranded DNA fragments occurs at a unique denaturant concentration corresponding to the special sequence of the DNA resulting in a sharp decrease in its electrophoresis mobility. Sequence variants of fragments with the same length will therefore stop migrating at different denaturing gradients and thus can be separated effectively. To prevent complete separation of two strands, which is undesirable for DGGE application, a GC clamp (a high-melting DNA fragment) is added to one of the two primers. Small differences in the sequences, ranging from 1 to 10 nucleotides, are usually sufficient to separate PCR products of the same length. The universal primer for *Bacteria* domain, 341F–518R (Liang *et al.* 2009), and the universal primer set 662F–934R for *Archaea* domain (Liu *et al.* 2002), are usually used in anaerobic studies. After staining, DNA fragments will appear as separated bands on the gel. Each DGGE band is derived from one specific species in the original samples. Thus the band number of a DGGE profile provides a quick estimate of species richness. The intensity of a band may be used as a rough indication of the relative abundance of a species (Zhang and Fang 2001).

The separated DNA fragments may be recovered from the DGGE bands in the gel and sequenced to identify the corresponding microbial species (Nielsen *et al.* 1999). Combined with sequencing and similarity-based phylogenetic analysis, DGGE can give an overview of the composition of a microbial community. Sequencing of the excised DGGE bands is usually more convenient and effective (Yu and Morrison 2004) than other fingerprinting methods which require cloning–sequencing to identify peaks or bands of interest, such as T-RFLP. However, it is not easy to slice the bands and purify the carried DNA fragments in some cases. For a complex community, to get DNA sequence of a cut band often demands another cloning–sequencing step due to co-migration or poor band separation. The separation could be optimized by changing the electrophoresis conditions, such as the denaturing gradient range, running time, voltage, etc.

Comparing patterns across the gel, especially for those including a large number of bands, is a difficult job. Different combinations of samples should be carefully designed if numerous samples are being investigated and multiple gels are required. DNA fragments of different sequences may co-migrate, and minor populations in the samples may be overlooked due to the limited detection sensitivity (Vallaeys *et al.* 1997). Additionally, the short sequences of the excised bands (usually 200–400 bp) are less informative for probe and primer designs, and make the resulting phylogenetic analysis less reliable than those using the full 16S rDNA sequences from cloning, especially for those novel sequences having <85% similarity to known sequences (Hugenholtz *et al.* 1998). Single-stranded DNA fragments generated in PCR may appear as bands and result in an overestimation of microbial diversity in a sample (Gilbride *et al.* 2006). Moreover, DGGE band intensity does not always quantitatively correlate with the abundance of a specific species, as DNA copy number in PCR product depends on both the abundance of species and the ease of amplification. It is also difficult to analyze the DGGE profile with too many bands, and only the predominant species could be obtained. Lastly, the reproducibility of DGGE is relative low, compared to other fingerprint methods, such as T-RFLP.

However, DGGE method has been extensively applied to study microbial diversity (Muyzer *et al.* 1993; Zhang and Fang 2001), population dynamics (Santegoeds *et al.* 1998), culture enrichment (Overmann and Tuschak 1997), DNA extraction protocols evaluation (Liesack *et al.* 1997), and clone library screening (Liang *et al.* 2007), due to its following advantages: (1) rapid and simple monitoring of microbial communities variation based on band patterns, (2) quick overview of the predominant populations in a sample, and (3) cost-effective for a large number of samples. DGGE has also been used in anaerobic wastewater treatment to evaluate microbial diversity of the granular sludge treating phenol (Zhang *et al.* 2005) and dimethyl phthalate (DMP) (Liang *et al.* 2007), etc.

6.2. *T-RFLP*

Terminal-restriction fragment length polymorphism (T-RFLP) analysis is a fingerprinting technique based on restriction digestion of PCR products obtained using a fluorescently labeled primer (Liu *et al.* 1997). To apply this method, one primer fluorescently labeled at 5′ end enables a single species to generate a specific fluorescent terminal restriction fragments (T-RFs) of a given size after enzymatic restrictive digestion. The segments are then separated by non-denaturing polyacrylamide gel electrophoresis or capillary electrophoresis, and distinguished by laser-induced fluorescence detection. The detailed principles of T-RFLP have been extensively described by Liu *et al.* (1997).

The choices of the primers and restriction enzymes are very important in the application of T-RFLP (Lueders and Friedrich 2003). Traditionally, T-RFLP relies on one fluorescently labeled primer and the digestion of PCR fragments using a restriction enzyme to generate specific T-RFs. Recently two primers labeled with different fluorescence dyes and digestion using several restriction enzymes have also been applied to improve the identification resolution (Padmasiri *et al.* 2007).

With high reproducibility, T-RFLP could be applied to conduct both quantitative and qualitative analyses of a gene (such as 16S rDNA) in a microbial community. Its advantage is the ability to detect even rare population of a sample. In addition, phylogenetic information can be inferred from the T-RF sizes of the sequences of known bacteria in the databases, including TRFMA (Nakano *et al.* 2006), T-Align (Smith *et al.* 2005), PAT (Kent *et al.* 2003), and TAP (Marsh *et al.* 2000). More important, T-RFLP can be standardized and be used to compare results between different researchers.

However, compared to DGGE, T-RFLP is more expensive and time-consuming, as PCR fragments need to be purified prior to enzymatic restriction digestion. Another disadvantage of T-RFLP is that the experimental fragment size may not be exactly same as the theoretical length, a 1–4 bps difference is commonly observed. As a result, T-RFLP seems to be useful in characterizing microbial communities with low-to-intermediate diversity, but may not for samples with high diversity (Engebretson and Moyer 2003). Additionally, the T-RFLP reproducibility may be affected by incomplete restriction digestion. Formation of pseudo T-RFs was also reported (Egert and Friedrich 2003), resulting in overestimation of biodiversity. More information about the limitation of T-RFLP could be found in the paper of Nocker *et al.* (2007).

Liu *et al.* (1997) first demonstrated that T-RFLP was a useful way to analyze and compare the bacterial community structure in an anaerobic–aerobic batch reactor. Since then, T-RFLP has been applied to study stressful conditions of anaerobic operation (Scully *et al.* 2005), and qualitative/quantitative analysis of anaerobic mixtures (Lueders and Friedrich 2003). The T-RFLP analysis of temporal methanogen shifts in a low-temperature anaerobic reactor showed that performance deterioration was preceded by the population decrease of aceticlastic *Methanosaeta* and the population increase of hydrogenotrophic *Methanomicrobiales* (McHugh *et al.* 2006).

7. FISH

7.1. *Introduction*

FISH is a visualization technique based on microscopic examination of a given species or groups of bacteria after staining cells using specific fluorogenic oligonucleotide probes which bind RNA molecules in the cells (Amann *et al.* 1990). As a method without DNA extraction and PCR, FISH is an excellent means to overcome problems associated with PCR-based molecular methods, such as DGGE, T-RFLP, cloning, and sequencing.

FISH probes are short DNA sequences (about 20 nucleotides) labeled with one or two fluorescent dyes. The probe may specifically hybridize *in situ* with 16S rRNA, 23S rRNA, or mRNA in cells according to DNA–RNA complementary matching. The cells are typically examined by epifluorescence microscope or laser scanning microscopes after staining. The specificity of the probe enables detection/identification on any desired taxonomic level, from domain down to species, depending on the probe applied. Figure 10.2 is a close-up image of *Eubacteria* and methanogen cells in a typical anaerobic granular phenol-degrading sludge.

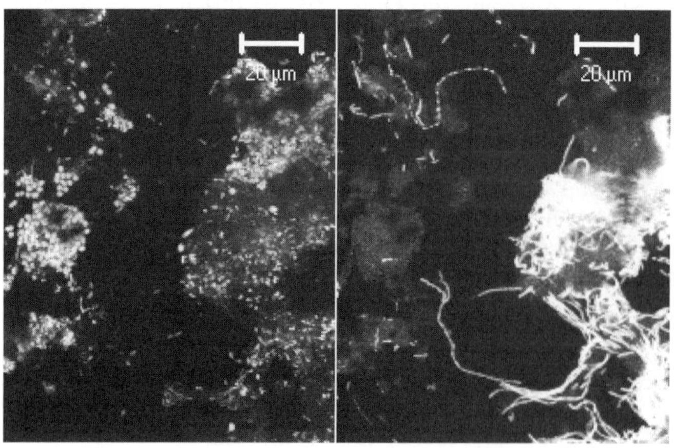

(a) (b)

Fig. 10.2. FISH images illustrating the individual cells of (a) *Eubacteria* and (b) methanogens in UASB granule degrading phenol (Zhang *et al.* 2005).

A specific probe targeting a particular population is usually designed according to the specific sequence at species, genus, family, subdivision, division, or domain levels. Although the probes developed by Raskin *et al.* (1994) are still being used to target most of the defined phylogenetic groups of methanogens in *Archaea*, a re-evaluation of these probes was reported by Crocetti *et al.* (2006) using the new SSU rRNA gene sequence data. In their report, the FISH conditions were evaluated for probes targeting methanogenic *Archaea* at the order, family and subgroup levels, including seven previous probes and seven new probes, using environmental samples from mesophilic anaerobic digesters and pure cultures of methanogens. The information about most of those existing probes commonly used in the anaerobic studies is available at ProbeBase (http://www.microbial-ecology.net/probebase/) which totally contains 1,516 probes. To design new probes, ARB/SILVA is a good tool to perform online analysis and design (http://www.arb-silva.de/fish-probes/).

7.2. *Quantification using FISH*

In addition to visualization, FISH is also widely used for microbial quantification in the anaerobic reactor, basing on either the fluorescence-emitting area or the number of individual cells. However, the cell count method is only applicable to evenly distributed and homogeneous microbial samples (Sanz and Köchling 2007). Usually, the bacterial count in a microscopic view field is required to be in the range of 30–150. Cell counts have to be performed at more than ten regions in order to have statistical significance.

"Spike FISH" was proposed as an extended FISH protocol to help quantification of a specific population in a mixed culture. In brief, a known amount of reference cells (for example, *E. coli*) could be added to a sample as an internal standard. After FISH and microscopic examination, automatic image analysis could be used to quantify the cells in the sample. The hybridized cells are then calculated in terms of relative area (or volume) and converted to cell number using the conversion factor between area (or volume) and cell number of the spiked reference bacteria, assuming the conversion factor is the same.

In a UASB granule sludge treating phenol wastewater, based on FISH results, the majority of microbes in the sludge were methanogens (74%) and *Bacteria* were only 27% (Zhang *et al.* 2005). For comparison, Wu *et al.* (2001b) found that methanogens accounted for 47% of the microbial population in a methanogenic terephthalate-degrading sludge. Using probe EUB338 (specific for *Bacteria*) and probes MX825 (specific to *Methanosaetaceae*), MG1200 (*Methanomicrobiales*) and MB1174 (*Methanobacteriaceae*), respectively, the FISH results revealed that 73% of the archaeal population in the sludge were the acetotrophic *Methanosaetaceae*, whereas the hydrogenotrophic *Methanomicrobiales* accounted for only 19% and *Methanobacteriacea* for 4% (Zhang *et al.* 2005).

7.3. *Variation of FISH*

As a widely applied technique, FISH has been upgraded with some new ideas and combined with other techniques to overcome its pitfalls. Some of such improvements are especially helpful in the analysis of microbial community in the anaerobic digestion, including:

- FISH flow cytometry which is used to conduct fast analysis and sensitive determination of rare populations by combining FISH with flow cytometry (Simon *et al.* 1995).

- TSA-FISH (tyramide signal amplification of FISH) which may boost the fluorescent intensity from hybridized cells by 20–40-folds (Biegala *et al.* 2003).
- CARD-FISH which may overcome the problems associated with low signal and/or the background signal swamping emission from the hybridized cells by labeling the probe with horseradish peroxidase. Horseradish peroxidase may catalyze the formation of fluorochrome-labeled tyramide radicals which strongly bind to tyrosine-rich regions of the ribosome and make detection of a single ribosome in a cell possible (Pernthaler *et al.* 2002). CARD-FISH was applied to determinate *Thiobacillus denitrificans* responsible for autotrophic denitrification in UASB reactors (Sanz and Köchling 2007).
- MAR-FISH combining FISH with microautoradiography using radioactively labeled substrates to determine metabolic activities and simultaneously identify the microorganisms involved (Lee *et al.* 1999).

7.4. *Advantages and disadvantages of FISH*

Compared to PCR-based molecular techniques, FISH has the following advantages: (1) fast and simple to have results in a couple of hours, (2) direct observation of uncultured microbes, (3) semi-quantitative, (4) preferential/differential examination of populations, (5) structural analysis of aggregates (flocs, granular sludge, biofilms) when combined with a CLSM, (6) easy for routine application, requiring only basic knowledge of microscope, and (7) possible to measure metabolic activity of cells by analyzing the intensity of fluorescence of positive cells. However, for the mixed culture in the anaerobic reactor, strong background autofluorescence often affects the interpretation of the FISH images, subject to individual judgment and experience about the samples; the problem may partially resolved by using a standardized and automatic procedure. Additionally, *in situ* detection of methanogens is sometimes known to be unsuccessful due to the low permeability of oligonucleotide probes in the cell wall. This might be partially resolved by using pseudomurein endoisopeptidase, an enzyme specifically acting on the cell wall and making it more permeable, prior to FISH. More disadvantages of FISH include: (1) RNA gene sequence must be known for a target microorganism if the special probe has not yet been reported yet; (2) not always possible to design an unambiguously restrictive probe for a microbial group, especially for those groups classified based on metabolic criteria (e.g., acidogenic bacteria); (3) difficult to optimize hybridization conditions of a newly designed probe; and (4) hard to quantify accurately due to subjective nature involved.

As an alternative method in visualization of the target population, *in situ* polymerase chain reaction (IS-PCR) with a fluorescent primer could also be used (Hodson *et al.* 1995). It does not require RNA molecules with a high-copy number, and theoretically may amplify and detect single-copy genes. However, non-specific amplification might be a potential problem associated with this method.

Overall, FISH is a useful method in combination with either fingerprint methods (T-RFLP or DGGE) or cloning–sequencing methods to quantify and visualize the OTUs of interest (Collins *et al.* 2006). This is so-called "full-cycle rRNA approach" which may enhance the understanding about anaerobic degradation. Details of the FISH technique have been reviewed by Wagner *et al.* (2003).

8. Microbial Quantification Using qPCR

Hybridization-based techniques, such as FISH (Okabe *et al.* 1999) and membrane hybridization (Raskin *et al.* 1995), and PCR-based techniques, such as DGGE (Muyzer *et al.* 1993) and cloning–sequencing (Zhang and Fang 2001), have long been used to quantify microorganisms in the anaerobic reactors. Hybridization methods, which have detection limits in the order of 10^5 DNA/RNA copies or greater, are in general less sensitive. They can thus only be used for environmental samples of relatively high microbial concentrations. PCR-based methods, on the other hand, are capable of detecting DNA/RNA at low concentrations. However, the precision of PCR-based methods may be compromised due to a number of factors, including reagent depletion, competition of amplicons with primers, and the loss of polymerase activity as the number of amplification cycle increases (Schneegurt and Kulpa 1998).

Due to the advanced development of fluorogenic chemistry, qPCR has become an emerging technique for the detection and quantification of microorganisms in the environment. Compared to the conventional hybridization- and PCR-based techniques, qPCR not only has better sensitivity and reproducibility, but it is also quicker to perform and has a minimum risk of amplicon carryover contamination.

During PCR, the target DNA sequence is amplified over a number of denaturation–annealing–extension cycles. In a conventional PCR, only the final concentration of the amplicon may be monitored using a DNA-binding fluorescent dye. However, in the qPCR, the concentration of the amplicon is monitored throughout the amplification cycles using a group of new fluorescent reagents. These reagents bind with the amplicon without causing damage at the end of each cycle so that amplification may continue to proceed. The fluorescence intensity emitted during this process reflects the amplicon concentration in real time. Fluorescent reagents used for qPCR can be classified into three categories: (1) dyes that bind with double-stranded DNA (dsDNA) (Hernández *et al.* 2004), (2) DNA-sequence-specific probes, including TaqMan probe (Haugland *et al.* 1999), and molecular beacon (Briones and Raskin 2003), and (3) DNA-sequence-specific primers, such as Amplifluor primer (Hernández *et al.* 2004). Among these fluorescent reagents, SYBR Green I, TaqMan probe, and molecular beacon are the most widely used in qPCR applications of environmental samples. During the extension phase in the PCR cycle, dsDNA are synthesized from the denatured single-stranded DNA. The fluorescence intensity of a dsDNA-binding dye increases, as a result, when the dye binds with the increased amount of amplified dsDNA. Such an increase can be monitored in real time.

Figure 10.3a shows the plots of fluorescence intensity throughout the amplification cycles in whole cell PCR, using a series of five standard solutions. These solutions contained *Microcystis aeruginosa* PCC 7820 at concentrations varied from 4.6×10^3 to 4.6×10^7 cells/ml. The threshold cycle, C_t, is defined as the cycle at which the fluorescence intensity crosses over a level where the amplification enters a logarithmic growth phase. C_t is inversely proportional to the log value of the initial concentration of the target, as illustrated in Fig. 10.3b for *Microcystis aeruginosa* PCC 7820. Based on the standard curve of Fig. 10.3b, concentration of the target DNA in an environmental sample can be estimated from its C_t measurement in qPCR. Figure 10.3a also shows that as the number of PCR cycle increases, the amplification efficiency reduces gradually. This is due to a number of factors, including the depletion of reagents, the inhibition caused by the increased product concentration, and the gradual loss of polymerase activity. Thus amplicon concentration and the corresponding fluorescence intensity

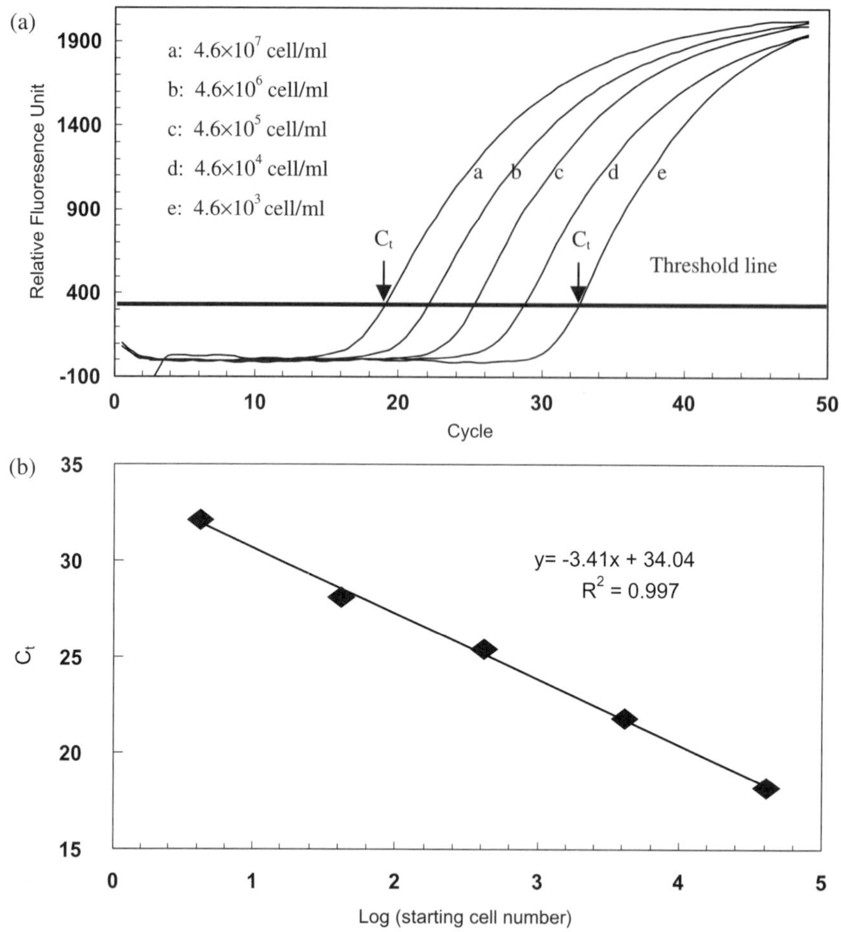

Fig. 10.3. Quantification of *Microcystis aeruginosa* PCC 7820 using qPCR: (a) relative fluorescence intensity of five standard solutions of PCC 7820 throughout amplification cycles, where C_t represents the threshold cycle number; and (b) the standard curve of PCC 7820 (Zhang and Fang 2006).

beyond certain number of amplification cycles cannot be used to estimate the initial concentration of the target DNA (Schneegurt and Kulpa 1998). Figure 10.3a shows that the fluorescent intensities of the five standard solutions after 45 cycles were nearly the same despite of their initial concentrations.

For a qPCR analysis, a standard curve like Fig. 10.3b is essentially needed for each target gene. Such a curve may be constructed from serial dilutions of a given source, which can be either cells, extracted genomic DNA of a pure culture, plasmids with the target DNA insert, or PCR products. Cells in the serial standard solutions can be quantified by plate counting or microscopic counting. Genomic DNA, plasmids and PCR products may be quantified by spectrophotometry at 260 nm, or by kits using dsDNA-binding dyes, such as PicoGreen®. The precision of microbial quantification using qPCR relies on the assumption that the environmental sample and the standard solutions share the same PCR efficiency. It is thus crucial to check the PCR efficiency in the analysis of both standard solutions and environmental samples. For a PCR process,

$$N_n = N_o \times (1 + \eta/100\%)^n \quad (10.1)$$

where N_n is number of amplified target at the end of nth cycle of amplification, N_o the initial number of target, and η (%) the PCR efficiency (Abell and Bowman 2005). At 100% efficiency, two DNA segments are produced from a target segment in each PCR cycle. The value of threshold cycle, C_t, can then be expressed as:

$$C_t = (\log N_o - \log N_t)/\log(1 + \eta/100\%) \quad (10.2)$$

where N_t is the number of amplified target after the threshold cycle, C_t. The slope (S) of a standard curve of C_t vs $\log N_o$, as illustrated in Fig. 10.3b, equals to $-1/\log(1+\eta/100\%)$. The slope of an ideal PCR amplification thus equals to -3.32 (i.e., $-1/\log 2$). In practice, a reliable standard curve should have an R^2 value of over 0.95 and a slope between -3.0 and -3.9, corresponding to PCR efficiencies of 80–115%. The standard cure in Fig. 10.3b shows a slope of -3.41, corresponding to an efficiency of 96%, and an R^2 of 0.997.

The dynamic range refers to the range in which the target quantity is linear with C_t in the standard curve. Probes, primers, and PCR conditions should be optimized not just for a low detection limit but also for a broad dynamic range. Under the most optimum conditions, SYBR Green I and TaqMan probe may have a detection limit as low as 2 cells per PCR sample and a dynamic range of 10^7 (Chen *et al.* 1997). The precision of qPCR analysis may be affected by PCR inhibitors present in the environmental sample. Ideally, each sample should be serially diluted and tested to determine the PCR efficiency to see if there are any PCR inhibitors present (Harms *et al.* 2003). The effect of the PCR inhibitor may be reduced by conducting the test at higher degrees of dilution. Careful handling and storage of sample, primers, probes, and enzymes are crucial to the accuracy of microbial quantification. Additionally, the DNA extraction efficiency, which may vary from batch to batch, has a significant effect on the reliability of qPCR.

Sawayama *et al.* (2006) showed that *Methanosarcina* spp. and *Methanobacterium* spp. immobilized in the polyurethane foam in a reactor were 7.6×10^9 and 2.6×10^8 DNA-segment·ml^{-1}, respectively, about 1,000 times higher than those in the original anaerobically digested sludge. The qPCR analysis for Gram-positive SRB *Desulfotomaculum* lineage 1 had the detection limit of 100 DNA target segment per PCR sample, equivalent to 10^6 DNA-segments·g^{-1} of soil (Stubner 2002), while the detection limits for Gram-negative SRB were between 2×10^5 and 4×10^3 DNA-segments·g^{-1} of soil (Stubner and Meuser 2000). Additionally, qPCR primer set for dissimilatory sulfite reductase gene has also been developed (Tang *et al.* 2004).

A TaqMan probe was designed targeting a novel group of thermophilic bacteria (named as TAB) in the anaerobic phenol-degrading sludge (Liang *et al.* 2008). By applying that probe in qPCR, it was showed that the abundance of TAB in biomass increased from 3.5% to 4.6%, and lastly to 9.9%, as the phenol loading increased from 0.25, 0.38, and 0.51 g-phenol·l^{-1}·d^{-1}. This implies that the rate of phenol degradation was specifically due to the increase of TAB population, instead of the overall biomass in the reactor. Results also showed that the TAB population ranged 3.5–9.9% in the 55 °C phenol-degrading anaerobic sludge, but only 0.0044% in the 37 °C sludge and 0.000086% in the 26 °C sludge.

9. Applications of Molecular Techniques in Anaerobic Technology Studies

The application of the DNA/RNA-based techniques helps to address some fundamental questions, i.e., who is present, how many are they, which substrate do they utilize, where are they located, what are their function in the reactor, and when will they start to become active? The answers to these questions may give new insight into the function, structure, and composition of anaerobic sludge and build bridges for the gaps between microbiology and anaerobic wastewater treatment. Some significant achievements of anaerobic technology with the application of nucleic acid-based techniques are summarized as follows.

9.1. *Identification of new species*

Based on 16S rDNA cloning analyses, the majority of microbial communities have not yet been cultivated and well characterized (Zhang *et al.* 2005). Using a range of techniques, including membrane hybridization, the clone library analysis, SSCP, DGGE, FISH, etc., it has been found that anaerobic reactors contain methanogens, syntrophic bacteria, sulfate-reducing bacteria (SRB), green non-sulfur bacteria, Low G+C gram-positive bacteria, and a number of unknown populations. The common species obtained from anaerobic digesters using 16S rDNA as fingerprint are mainly from the archaeal classes of *Methanomicrobia*, *Methanobacteria*, *Methanosarcinales*, and *Thermoplasmata*, plus bacterial populations, including *Proteobacteria* (particularly delta-*Proteobacteria*), *Firmicutes*, *Chloroflexi*, *Spirochaetes*, and *Bacteroides*. Additionally, anaerobic sludge may also contain a few uncultured candidate phyla, including BA024, EM3, OD1, OP8, OP10, OSK, TM6, WSA2 (novel methanogens), and the *Crenarchaeota* clone cluster C2 (also known as rice Cluster IV) (Sekiguchi *et al.* 2001). An enriched anaerobic sludge effectively degrading over 99% of dimethyl phthalate in a UASB reactor was characterized and quantified by 16S rDNA-based molecular methods. A total of 78 *Bacteria* clones were classified into 22 OTUs in 9 divisions, including *Firmicutes*, *Proteobacteria, Chloroflexi*, *Thermotogae*, *Bacteroidetes/Chlorobi*, *Spirochaetes*, *Acidobacteria* and 2 candidate divisions (Liang *et al.* 2009). Ueno and Tatara (2008) investigated microbial community in thermophilic methane production from organic acids in a packed-bed reactor using 16S rDNA clone analysis showed that microbial populations changed with hydraulic retention time (HRT), and *Thermoanaerobacteriaceae* was mainly present in the biofilm with hydrogenotrophic methanogens whereas clones that are affiliated to *Clostridiaceae* were mainly found in the fermentation broth. It was also found that a certain amount of unknown or unidentified organisms were present in the microflora. Using increasing improved sequencing ability, e.g., pyrosequencing, researchers have found more and more unknown species in the anaerobic digesters (Krober *et al.* 2009).

9.2. *Characterization of microbial compositions*

9.2.1. *Methanogens*

According to Boone *et al.* (1993), there are five orders of methanogenic *Archaea*, including *Methanobacteriales*, *Methanococcales*, *Methanomicrobiales*, *Methanosarcinales*, and *Methanopyrales* (extreme thermophiles), based on taxonomic information obtained using sequencing of 16S rDNA, DNA/DNA hybridization, and phenotypic characterization. On the basis of carbon source utilization, methanogens may be classified into three major groups: (1) aceticlastic methanogens (or acetotrophs) utilizing solely acetate as substrate, (2) methylotrophs utilizing methanol, methylamines, or dimethylsulfide, and (3) hydrogenotrophs oxidizing H_2 and reducing CO_2 to form methane (Garcia *et al.* 2000).

Most researches showed that the predominant methanogen genera in anaerobic reactors are limited to *Methanosaeta* (formerly known as *Methanothrix*), *Methanosarcina*, *Methanothermobacter* (formerly *Methanobacterium*), *Methanobrevibacter*, and *Methanobacterium* (Sekiguchi *et al.* 2001). Compared with bacteria populations, methanogens are well-characterized microorganisms in the anaerobic digesters. Among these methanogens, *Methanosaeta* spp. can only utilize acetate while *Methanosarcina* spp. are the most diversified in terms of substrate utilization.

Using membrane hybridization, Zheng and Raskin (2000) found that the archaeal 16S rRNA accounted for around 40% of total RNA in UASB granules using domain-specific probes and aceticlastic *Methanosarcinales* were the dominant methanogens using group-specific probes, and *Methanosaeta concilii* was found to be the predominant species using the species-specific probes. This finding concurred with the observations of dominant *Methanosaeta*-like filaments in anaerobic granular sludge (Hofman-Bang *et al.* 2003). Sekiguchi *et al.* (1998) characterized the microbial communities of UASB granular sludges treating synthetic wastewater containing sucrose, propionate, and acetate under mesophilic (35 °C) and thermophilic (55 °C) conditions, respectively. 16S rDNA clone library analysis showed that 19% of the clones from the mesophilic granules and 22% from the thermophilic granules were affiliated with methanogens, while the remaining were *Bacteria* populations. The results of Hansen *et al.* (1998) on an anaerobic digester treating cow and swine manure showed that *Methanomicrobiales* were the most abundant hydrogenotrophic methanogens, constituting 10% of the prokaryotes, the only other methanogens present being *Methanosarcina* spp. Raskin *et al.* (1995) quantified methanogens in a few single-phase mesophilic full-scale sludge anaerobic digesters using 16S rDNA probe and found that methanogens accounted for 8–12% of microbial population, of which *Methanosarcinales* and *Methanomicrobiales* constituted the majority; *Methanobacteriales* and *Methanococcales* only played a minor role.

16S rDNA-based DGGE and sequence analysis revealed that species related to the genera *Methanolobus* and *Methanosarcina* dominated the archaeal community in the UASB reactor treating methanethiol wastewater (van Leerdam *et al.* 2008). The most dominant methanethiol-degrading archaeon was enriched from the reactor and was found to be most closely related to *Methanolobus taylorii*.

Tang *et al.* (2008) investigated methanogens in a hyperthermophilic (65–80 °C) chemostat treating synthetic wastewater containing glucose conditions based on clone library analysis and DGGE of the 16S rRNA gene. They found that aceticlastic methanogens related to *Methanosarcina thermophila* at 65 °C and hydrogenotrophilic methanogens related to *Methanothermobacter thermautotrophicus* at 70–77.5 °C were dominant. Their results also suggested that, at temperatures of 70 °C and above, methane production via the aceticlastic pathway was negligible and indirect methanogenesis from acetate was dominant.

9.2.2. *Syntrophs*

Syntrophs may oxidize the intermediates products, including butyrate, propionate, acetate, and benzoate, during anaerobic degradation through their syntrophic association with hydrogenotrophic methanogens. Although syntrophic bacteria were found to be the major populations in many UASB granular sludges (Zhang *et al.* 2005), little has been known about their taxonomy, diversity, and physiology, due to the isolation difficulties. Nucleic acid-based methods have been greatly used to characterize and isolate syntrophic oxidizers of butyrate, propionate and acetate from anaerobic reactors.

Up to now, four genera of syntrophic bacteria have been described, i.e., *Syntrophobacter* and *Syntrophus* in delta-*Proteobacteria*, and *Syntrophospora* and *Syntrophomonas* in *Syntrophomonadaceae* of *Clostridiales*, LGC. Among them, *Syntrophus* has been found in many types of UASB sludges treating sucrose, propionate, acetate (Sekiguchi *et al.* 1998), terephthalate (Wu *et al.* 2001b), and phenol (Zhang *et al.* 2005), while *Syntrophobacter* and *Syntrophomonas* were only detected so far in the pentachlorophenol-degrading UASB granule (Tartakovsky *et al.* 2001). The type of the *Syntrophus*-like species found in many UASB granules could also be diverse depending on the substrates. The *Syntrophus*-like species in UASB granule sludge degrading sucrose, propionate, and acetate (Sekiguchi *et al.* 1999) are different from those degrading aromatic compounds (Wu *et al.* 2001b; Zhang *et al.* 2005). Jackson *et al.* (1999) isolated from an anaerobic reactor *Syntrophus aciditrophicus* which is a versatile syntroph with the ability to oxidize fatty acids and benzoate. A thermophilic propionate-oxidizing bacterium was isolated by Imachi *et al.* (2000) in coculture with a hydrogenotrophic methanogens. Using FISH with a 16S rRNA-targeted DNA probe, Imachi *et al.* (2000) found that this syntroph was one of the dominant populations in the anaerobic reactor, forming syntrophic microcolonies with hydrogenotrophic methanogens.

9.2.3. *SRB*

Due to their diversity on substrate utilization, SRB may compete with methanogens, syntroph and other SRB species in anaerobic reactors. According to the study of Harmsen *et al.* (1996) on anaerobic granule treating potato-processing industry wastewater, the population of *Desulfobulbus* spp. in the granules treating propionate and sulfate increased from 2% in the inoculum up to 35%, while a syntroph SYN7 increased from <1% to 10% in the granules treating only propionate. Hristova *et al.* (2000) applied genus- and subgenus-specific probes of SRB to probe a thermophilic anaerobic reactor and found *Desulfotomaculum* population decreased from 2% at the start-up to 0.3% after 31 days, whereas *Desulfovibrio* accounted for 1.7% as it was more competitive than *Desulfotomaculum* under the low sulfate concentrations in that reactor.

9.2.4. *Others*

In addition to methanogen, syntrophs and SRB, other populations may also exist in the anaerobic reactors. The phylogenetic analysis of a DMP-degrading anaerobic UASB sludge showed the presence of species of *Firmicutes*, delta-*Proteobacteria*, and *Chloroflexi*, similar to microbial sludges degrading aromatics, such as phenol (Fang and Zhang 2004), 4-methylbenzoate (Wu *et al.* 2001a), terephthalate (Wu *et al.* 2001b), etc. This suggests that *Firmicutes*, delta-*Proteobacteria*, and *Chloroflexi* are prevalent in the aromatic organics anaerobic degradation environment. Among them, *Chloroflexi* is a typical microbial group in anaerobic environments (Yamada *et al.* 2005), and were found in various anaerobic sludges degrading hydrocarbons (Sekiguchi *et al.* 1998), aromatics (Wu *et al.* 2001a,b), and halogenated compounds (von Wintzingerode *et al.* 1997). Satoh *et al.* (2007) reported that in an anaerobic granule with layered structure, filamentous *Chloroflexi*-like bacteria were found distributed in the outer layer, presumably degrading complex organic compounds.

9.3. *Visualization of structure of granular sludge*

FISH combined with fluorescence microscopy (either conventional or CLSM) has been applied to investigate the spatial distribution of microorganisms in anaerobic granules after they are sliced into thin sections.

Similar to the FISH results of the terephthalate-degrading UASB granular sludge (Wu *et al.* 2001b), the so-called non-layered architecture was also observed in the study of Zhang *et al.* (2005) about phenol-degrading granule. However, the UASB granule utilizing the mixture of sucrose, acetate, propionate, and peptone or yeast extract (Sekiguchi *et al.* 1999) and brewery (Liu *et al.* 2002), formed the layered structure. For some refractory compounds, such as DMP, layered structure was also be confirmed by electron microscopic and FISH observation (Liang *et al.* 2009). Since DMP is degraded anaerobically through de-esterification and de-aromatization to fatty acids, which are further degraded via β-oxidation, such a layered microstructure suggests that the rate of DMP de-esterification is faster than its inward diffusion rate, resulting in the dense concentration of de-esterifying bacteria in the outer layer with little population of other species. The de-esterified metabolites diffused inward and were then further degraded into phthalate and fatty acids (Liang *et al.* 2009).

9.4. *Quantification of various microbial groups*

As introduced before, FISH, cloning, qPCR, and DGGE can be applied in the quantification of microbial populations. Functional *Bacteria* and *Archaea* community structures of a full-scale anaerobic sludge digester were investigated by using a full-cycle 16S rRNA approach followed by MAR-FISH (Ariesyady *et al.* 2007). FISH analysis with a comprehensive set of 16S and 23S rRNA-targeted oligonucleotide probes based on 16S rRNA clone libraries revealed that the Gram-positive bacteria, represented by probe-HGC69A-hybridized *Actinobacteria* ($8.5 \pm 1.4\%$ of total 4′,6-diamidino-2-phenylindole (DAPI)-stained cells) and probe-LGC354-hybridized *Firmicutes* ($3.8 \pm 0.8\%$), were the major phylogenetic bacterial phyla, followed by *Bacteroidetes* ($4.0 \pm 1.2\%$) and *Chloroflexi* ($3.7 \pm 0.8\%$). The probe-MX825-hybridized *Methanosaeta* ($7.6 \pm 0.8\%$) was the most abundant archaeal group, followed by *Methanomicrobiales* ($2.8 \pm 0.6\%$) and *Methanobacteriaceae* ($2.7 \pm 0.4\%$). The functional community structures (diversity and relative abundance) of major trophic groups were quantitatively analyzed by MAR-FISH. The results revealed that glucose-degrading microbial community had higher abundance ($10.6 \pm 4.9\%$ of total DAPI-stained cells) and diversity (at least seven phylogenetic groups) as compared with fatty acid-utilizing microbial communities, which were more specialized to a few phylogenetic groups. Despite the dominance of *β-Proteobacteria*, members of *Chloroflexi*, *Smithella*, *Syntrophomonas*, and *Methanosaeta* groups dominated the glucose-, propionate-, butyrate- and acetate-utilizing microorganism community, and accounted for $27.7 \pm 4.3\%$, $29.6 \pm 7.0\%$, $34.5 \pm 7.6\%$ and $18.2 \pm 9.5\%$, respectively.

Franke-Whittle *et al.* (2009) designed a microarray, named ANAEROCHIP, targeting methanogens in anaerobic digesters, and to apply this chip together with a cloning approach to quantify the methanogens in an anaerobic digester. Oligonucleotide probes were designed based on sequence differences in the 16S rRNA genes in order to target microorganisms *in situ*. For microarray hybridizations, DNA was subjected to PCR amplification of the 16S rRNA gene and Cy5-labeled. The microarray was tested with pure cultures, and of the 1,854 individual probe-target hybridization reactions performed, there were only 28 false-positive (1.5%) and 16 false-negative signals (0.86%). The sensitivity of the array was also tested, and it was found that when 0.4 pg of DNA from a pure culture was subjected to PCR amplification, signals above the detection limit were obtained. Also, the application of 25 ng of PCR product from a pure culture to an array resulted in detectable signals. The ANAEROCHIP was hybridized with DNA from an anaerobic sludge. Strong hybridization signals were obtained for *Methanoculleus*, and weaker signals were obtained for, in descending order, *Methanosarcina*, *Methanobacterium*, *Methanobrevibacter*, and *Methanosphaera*. In order to check the results obtained with the microarray, the archaeal community structure of the same digester was analyzed by 16S rRNA gene cloning and sequencing. Community structure was determined by restriction digestion of almost 200 clones and by sequencing of the 15 different resulting patterns. *Methanoculleus* was the dominant (84.1%) microorganism in the anaerobic sludge, and *Methanobrevibacter* (5.8%), *Methanobacterium* (3.7%), *Methanosarcina* (2.1%), *Methanosphaera* (1.6%), an uncultured archaeon (1.6%) and *Methanothermobacter* (1%) were also detected. These results showed the microarray to be a suitable tool for quantification of methanogens in anaerobic sludge.

Quantitative changes in methanogenic community associated with reactor performance were investigated in three anaerobic batch digesters treating synthetic glucose medium, whey permeate, and liquefied sewage sludge (Lee *et al.* 2009). All digesters were initially seeded with anaerobic sludge obtained from a local municipal wastewater treatment plant. The dynamics of methanogenic populations was monitored, at order and family levels, using qPCR based on the 16S rDNA. The results showed that in each digester the quantitative structure of methanogenic community varied continuously over time, and the variation corresponded well to the changes in chemical profiles. Biphasic production of methane, associated with successive increases in aceticlastic (mainly *Methanosarcinaceae*) and hydrogenotrophic (mainly *Methanomicrobiales*) methanogens, was observed in each digester. The non-metric multidimensional scaling analysis of the quantification results demonstrated that the community shift patterns in three digesters were totally different from one another. Considering that the operating conditions in all trials were identical except substrates, the differences in quantitative shift profiles were suggested to be due to the substrate compositions, implying that the composition of wastewater could affect the evolution of quantitative methanogenic community structure in an anaerobic process.

In order to identify microorganisms within an anaerobic CSTR fed with fodder beet silage, a 16S rDNA clone library was constructed by PCR amplification applying a prokaryote-specific primer set (Klocke *et al.* 2007). A total of 147 clones were obtained, and subsequently characterized by amplified rDNA restriction analysis (ARDRA), producing 60 unique ARDRA patterns (also called as OTU). Among them, four (7%) were assigned to the domain *Archaea* and 56 (93%) to the domain *Bacteria*. Within the domain *Archaea*, all clones showed a close relationship to methanogens. Major bacterial groups represented in the clone library were the class *Clostridia* of the phylum *Firmicutes* (22% of all 16S rDNA clones), the class *Deltaproteobacteria* of the phylum *Proteobacteria* (24%), the class *Bacilli* of the phylum *Firmicutes* (22%) and members of the phylum *Bacteroidetes* (21%). Within these major groups, the highest biodiversity was found within the class *Clostridia* (35% of OTUs).

Song *et al.* (2010) quantified the domain *Archaea* in a pilot-scale UASB treating swine wastewater using qPCR based on 16S rDNA concentrations to evaluate the process performance and methanogenic population dynamics. A high abundance of hydrogenotrophic methanogens was observed with *Methanobacteriales* as the major group, suggesting that hydrogenotrophic methanogenesis, with the syntrophic oxidation of VFAs, was a major route of methane formation mainly due to the high ammonium concentration in swine wastewater which has a severe inhibitory effect mainly on the aceticlastic methanogens. Although there was no significant growth of *Methanosaetaceae*, its abundance might have contributed to the formation and maintenance of granule.

9.5. *Evaluation of degradation of refractory compounds*

Molecular techniques have been applied to study the microbial communities of anaerobic sludge treating refractory compounds, such as tetraphthalate (Wu *et al.* 2001b), phenol (Zhang *et al.* 2005), and DMP (Liang *et al.* 2009). The results of Zhang *et al.* (2005) indicated that the degradation of phenol to benzoate was likely carried out by joint action of the second and third most abundant OTUs, i.e., PD-UASB-10 and PD-UASB-29; this concurred with the observations by Letowski *et al.* (2001) that the degradation of phenol to benzoate was not carried out by individual strains, but by the co-culture of the anaerobic bacterium 'strain 7' (with 97% similarity to PD-UASB-29) and a *Clostridium*-like 'strain 6' (with 98% similarity to PD-UASB-10). Benzoate was then further degraded by *Syntrophus* species, such as PD-UASB-5, to acetate and H_2/CO_2. Methanogens lastly converted acetate and H_2/CO_2 into methane (Zhang *et al.* 2005).

At ambient temperature and mesophilic condition, phenol was presumably degraded through the benzoate pathway (Fang and Zhang 2004) as the sludge has high specific methanogenic activity (SMA) using benzoate as substrate, as the benzoate-degrading *Syntrophus*-like bacteria accounted for about 49% of the population (Zhang *et al.* 2005). However, judging from the SMA data of another study (Fang *et al.* 2006), the thermophilic phenol-degrading sludge could hardly utilize benzoate to produce methane and it had higher SMA in degrading caproate than degrading benzoate. Additionally, only one *Syntrophus*-like OTU (TPD-86, 1.9% of the population) was detected in this study using the 16S rDNA method. Thus, the thermophilic degradation pathway of phenol was likely via caproate, instead of benzoate, judging from the sludge incapability of degrading benzoate and the absence of benzoate-degrading *Syntrophus* bacteria (Fang *et al.* 2006).

Palatsi *et al.* (2010) investigated anaerobic sludge in a thermophilic digester fed with manure and exposed to successive inhibitory pulses of long-chain fatty acids using 16S rDNA DGGE, and found that there was no significant shift of *Bacteria* and *Archaea* populations upon exposure to long-chain fatty acid. Sequencing of the predominant DGGE bands revealed close phylogenetic relation to specific β-oxidation bacterial genera of *Syntrophomonas* and *Clostridium*, while a single predominant syntrophic *Archaea* species was related with the *Methanosarcina* genus.

Anaerobic benzene-degrading cultures performing methanogenesis were enriched from a non-contaminated soil. Stable isotope probing with ^{13}C-labeled benzene was used to detect the bacteria that were involved in benzene degradation. DGGE of fractionated samples exhibited an obvious shift of some DGGE bands to a heavier DNA fraction. An almost full-length 16S rRNA gene sequence corresponding to the DGGE band, namely Hasda-A, was obtained by constructing a clone library of the heavier fraction. The Hasda-A sequence showed only 85.1% identity with the closest identified bacterium, *Syntrophus gentianae*. Hasda-A may be an important bacterium involved in the initial steps of benzene degradation under methanogenic conditions, as it was the most prominent bacterium that assimilated labeled benzene early in the process of benzene degradation. A primer set was designed to quantify the gene copies of Hasda-A by qPCR. Hasda-A was present at a concentration of $(3.5 \pm 0.8) \times 10^6$ copies·ml^{-1}, representing 8.4% of gene copies among bacteria in the enrichment culture. The enrichment culture consisted of three dominant bacterial groups: Hasda-A, plus both aceticlastic and hydrogenotrophic methanogens. Methane is believed to be produced from benzene via a sequential degradation by fermenting bacteria, hydrogen-producing acetogens, and ultimately methanogens.

9.6. *Investigation of microbial communities in anaerobic digestion of special wastewater*

Anaerobic sludge granules were obtained from laboratory-scale anaerobic bioreactors used to treat pharmaceutical-like (methanol-, acetone-, and propanol-contaminated) wastewater under low-temperature conditions (15 °C) (Enright *et al.* 2007). The microbial diversity and diversity changes of the sludge samples were ascertained by applying 16S rDNA cloning and TRFLP analyses, respectively. Data from genetic fingerprinting correlated well with those from physiological activity assays of the reactor biomass. TRFLP profiles indicated the dominance of hydrogenotrophic methanogens within the archaeal community, supporting the findings of specific methanogenic activity measurements. TRFLP data supported the hypothesis that the deviation between the replicated reactors, in terms of treatment efficiency, was associated with succession within the microbial communities present, indicating that community development was linked to both operating temperature and wastewater composition. FISH was also applied, to quantitatively assess the abundance of selected microbial groups, and revealed the underestimation of the abundance *Methanosarcina* by gene cloning analysis and demonstrated the spatial arrangement of these organisms within the architecture of the low-temperature solvent-degrading anaerobic biofilms.

Tang *et al.* (2007) investigated the microbial structure and spatial distribution of microorganisms on the nonwoven fabric material in thermophilic upflow anaerobic filter reactors treating distillery wastewater from awamori making using molecular methods. The reactor steadily achieved a high TOC loading rate of 18 g·l^{-1}·d^{-1} with ~80% TOC removal efficiency. Methanogens closely related to *Methanosarcina thermophila* and *Methanoculleus bourgensis* and bacteria in the phyla *Firmicutes* and *Bacteroidetes* were predominant in the reactor, as indicated by 16S rRNA gene clone analysis. FISH results showed that a large quantity of bacterial cells adhered throughout the fabric, and *Methanosarcina*-like microbes were present mostly close to the surface while *Methanoculleus* cells were mostly at the interior of the fabric.

Roy *et al.* (2009) investigated shifts in a bacterial community structure and metabolic activities in a plug-flow type anaerobic reactor degrading swine manure at 25 °C based on the *Bacteria* 16S rDNA- and rRNA-based fingerprints of amplicon length heterogeneity PCR (LH-PCR). Results of multivariate statistical analysis of the fingerprints showed that the rDNA-based fingerprints described a more temporal than compartmentalized distribution of similar bacterial communities. By contrast, the rRNA-based multivariate analyses described a distribution that was linked more to reactor performance parameters, especially during short time periods. Diversity indices calculated from fingerprint data were used to assess overall diversity shifts.

The decrease in rRNA-based diversity observed through the reactor compartments was greater than the decrease in rDNA-based diversity. This finding indicates that the analysis of metabolically active bacteria diversity was more discriminative than the analysis based on the mere presence of bacteria. All these findings suggest that rRNA-based analyses are more appropriate for monitoring reactor performance.

The bacterial communities in three continuously stirred tank reactors co-digesting cow manure with grass silage, oat straw, and sugar beet tops, respectively, were investigated by 16S rRNA gene-based fingerprints and clone libraries (Wang *et al.* 2009). The analyses revealed both clearly distinct but similar phylotypes in the bacterial communities among the reactors. The major groups were *Clostridia*, unclassified *Bacteria*, and *Bacteroidetes*. Phylotypes affiliated with *Bacilli* or *Deltaproteobacteria* were unique to the sugar beet and straw reactor, respectively. Unclassified *Bacteria* dominated in sugar beet reactor, while in the straw and grass reactors *Clostridia* was dominant. An increase in organic loading rate from 2 to 3 $kg\text{-}VS{\cdot}m^{-3}{\cdot}d^{-1}$ resulted in a larger change in the bacterial community in the straw than in the grass reactor. The study shed more light on the evolution of bacterial community during anaerobic co-digestion of different crops and manure to methane.

Microbial communities were analyzed in six full-scale anaerobic digesters and a lab-scale digester using the random cloning method (Cheon *et al.* 2008). They compared the difference in the dominant microbes in different operational conditions of temperature (mesophilic and thermophilic) and substrate (garbage, sewage sludge, and livestock waste). Each sample was analyzed in triplicate. In one of the samples, 373 clones were analyzed to study the composition of microbial community in the digester. Time course analysis was conducted from the start-up period for approximately one year in one of the digesters. Similar microbial diversity was obtained corresponding to the type of substrate change (sewage sludge to municipal solid waste). Microbial diversity in digesters is strongly affected by the operational conditions, and similar microbial diversity can be obtained in triplicate analysis and under similar operational conditions.

Yang *et al.* (2008) studied anaerobic biodegradation of glycerol-containing synthetic wastes in a semi-continuous fixed-bed bioreactor packed with polyurethane under mesophilic and thermophilic conditions. 16S rDNA sequences indicated that the archaeal clones in the thermophilic reactor were affiliated with two main groups, i.e., the hydrogenotrophic *Methanobacterium* sp. (17 out of 20 clones) and the aceticlastic *Methanosarcina* sp. (3 out of 20 clones). The bacterial clones were mostly affiliated with *Bacillus* sp., *Clostridium* sp., *Desulfotomaculum* sp. and *Ruminococcus* sp.

The microbial ecology in full-scale anaerobic digesters was investigated when the methane production rate was significantly increased by focused-pulsed pre-treatment (Zhang *et al.* 2009). Using clone libraries and qPCR, methanogenic population was shifted from the H_2-oxidizing *Methanoculleus* to the aceticlastic *Methanosaeta*. In addition, the acetate concentration in the effluent was very low, probably due to the dominance of *Methanosaeta*, which is capable of scavenging acetate at low concentrations. By analyzing 36,797 pyrosequencing tags from the V6 region of the bacterial 16S rDNA, along with archaeal and bacterial clone libraries and qPCR, the microbial community composition was compared before and after the focused-pulsed treatment. The bacterial community became more diverse after focused-pulsed pre-treatment and was populated more by phylotypes associated with cellulose fermentation (*Ruminococcus*), scavenging of biomass-derived organic carbon (*Chloroflexi*), and homo-acetogenesis (*Treponema*). It was interpreted that, as the overall activity of the community was stimulated by addition of more bioavailable organic matter, the bacterial community became more phylogenetically diverse to take advantage of the added input of biodegradable material and in response to the more efficient utilization of acetate by *Methanosaeta*.

The phylogenetic structure of the microbial community residing in a fermentation sample from a full-scale biogas plant fed with maize silage, green rye, and liquid manure was analyzed by an integrated approach using clone library sequences and metagenome sequence data obtained by 454-pyrosequencing (Krober *et al.* 2009). Sequencing of 109 clones from a bacterial and an archaeal 16S-rDNA amplicon library revealed that the obtained nucleotide sequences are similar but not identical to 16S-rDNA database sequences derived from different anaerobic environments, including digesters and bioreactors. Most of the bacterial 16S-rDNA sequences could be assigned to the phylum *Firmicutes*, the most abundant of which was the class *Clostridia*, and to the class *Bacteroidetes*, whereas most archaeal 16S-rDNA sequences cluster close to the *Methanoculleus bourgensis*. Further sequences of the archaeal library most probably represent the not yet characterized species within the genus *Methanoculleus*. A similar result was derived from phylogenetic analysis of *mcr*A clone sequences. The *mcr*A gene product encodes the α-subunit of methyl-coenzyme-M reductase involved in the final step of methanogenesis. BLASTn analysis applying stringent settings resulted in assignment of 16S-rDNA metagenome sequence reads to 62 16S-rDNA amplicon sequences, thus enabling frequency of abundance estimations for 16S-rDNA clone library sequences. RDP Classifier processing of metagenome 16S-rDNA reads revealed the abundance of the phyla *Firmicutes*, *Bacteroidetes*, and *Euryarchaeota* and the orders *Clostridiales*, *Bacteroidales*, and *Methanomicrobiales*. Moreover, a large fraction of 16S-rDNA metagenome reads could not be assigned to lower taxonomic ranks, demonstrating that numerous microorganisms in the biogas plant are still unclassified or unknown.

10. Perspectives

The studies of microbial communities in anaerobic treatment have advanced significantly with the application of various molecular techniques. However, such information of microbial communities still has limited practical value in improving treatment effectiveness and efficiency. The phylogenetic positions of the microorganisms are of little meaning for engineers unless they can be associated with the performance or the functions of the reactor. Even the new developments of molecular techniques, such as microarray, Phylochip, SIP, Raman-FISH, and Pyrosequencing, may generate more information to link the function of microorganism and their phylogenetic positions, the combined efforts of environmental microbiologists, who work on fundamental researches, and environmental engineers, who have expertise on the practical work are still very essential to ensure that the further advancement of environmental molecular microbiology may be incorporated into the design and operation of anaerobic reactors. This is a great opportunity, but also a big challenge, for the further development of anaerobic technology.

Acknowledgments

The author wishes to thank the Hong Kong General Research Fund (HKU 7195/06E) for the financial support of this study.

References

Abell, G. C. J. and Bowman, J. P. (2005) Colonization and community dynamics of class *Flavobacteria* on diatom detritus in experimental mesocosms based on Southern Ocean seawater. *FEMS Microbiol. Ecol.* **53**, 379–391.

Altschul, S. F., Gish, W., Miller, W., Myers, E., and Lipman, D. J. (1990) Basic local alignment search tool. *J. Mol. Biol.* **215**, 403–410.

Amann, R. I. (1995) Phylogenetic identification and *in situ* detection of individual microbial cells without cultivation. *Microbiol. Rev.* **59**, 143–169.

Amann, R. I., Krumholz, L., and Stahl, D. A. (1990) Fluorescent-oligonucleotide probing of whole cells for determinative, phylogenetic and environmental studies in microbiology. *J. Bacteriol.* **172**, 762–770.

Ariesyady, H. D., Ito, T., and Okabe, S. (2007) Functional bacterial and archaeal community structures of major trophic groups in a full-scale anaerobic sludge digester. *Wat. Res.* **41**, 1554–1568.

Bachoon, D. S., Otero, E., and Hodson, R. E. (2001) Effects of humic substances on fluorometric DNA quantification and DNA hybridization. *J. Microbiol. Methods* **47**, 73–82.

Biegala, I. C., Not, F., Vaulot, D., and Simon, N. (2003) Quantitative assessment of picoeukaryotes in the natural environment using taxon specific oligonucleotide probes in association with tyramide signal amplification-fluorescence *in situ* hybridization and flow cytometry. *Appl. Environ. Microbiol.* **69**, 5519–5529.

Boone, D. R., Whitman, W. B., and Rouvière, P. (1993) Diversity and taxonomy of methanogens. In: *Methanogenesis*. New York: Chapman and Hall Co., pp. 35–80.

Borneman, J. and Triplett, E. W. (1997) Molecular microbial diversity in soils from eastern Amazonia-evidence for unusual microorganisms and microbial population shifts associated with deforestation. *Appl. Environ. Microbiol.* **63**, 2647–2653.

Briones, A. and Raskin, L. (2003) Diversity and dynamics of microbial communities in engineered environments and their implications for process stability. *Curr. Opin. Biotechnol.* **14**, 270–276.

Bruce, K. D., Hiorns, W. D., Hobman, J. L., Osborn, A. M., Strike, P., and Ritchie, D. A. (1992) Amplification of DNA from native populations of soil bacteria by using the polymerase chain reaction. *Appl. Environ. Microbiol.* **58**, 3413–3416.

Bull, A. T. (2004) *Microbial Diversity and Bioprospecting*. Washington, DC: ASM Press.

Chen, C. L., Macarie, H., Ramirez, I., Olmos, A., Ong, S. L., Monroy, O., and Liu, W. T. (2004) Microbial community structure in a thermophilic anaerobic hybrid reactor degrading terephthalate. *Microbiology* **150**, 3429–3440.

Chen, S., Yee, A., Griffiths, M., Larkin, C., Yamashiro, C. T., Behari, R., Paszko-Kolva, C., and Grandis, S. A. D. (1997) The evaluation of a fluorogenic polymerase chain reaction assay for the detection of *Salmonella* species in food commodities. *Int. J. Food. Microbiol.* **35**, 239–250.

Cheon, J., Hidaka, T., Mori, S., Koshikawa, H., and Tsuno, H. (2008) Applicability of random cloning method to analyze microbial community in full-scale anaerobic digesters. *J. Biosci. Bioeng.* **106**, 34–140.

Cole, J. R., Chai, B., Farris, R. J., Wang, Q., Kulam, S. A., Mcgarrell, D. M., Garrity, G. M., and Tiedje, J. M. (2005) The Ribosomal Database Project (RDP-II): Sequences and tools for high-throughput rRNA analysis. *Nucleic Acids Res.* **33**, 294–296.

Collins, G., Kavanagh, S., McHugh, S., Connaughton, S., Kearny, A., Rice, O., Carrigg, C., Scully, C., Bhreathnach, N., Mahony, T., Madden, P., Enright, A. M., and O'Flaherty, V. (2006) Accessing the black box of microbial diversity and ecophysiology: Recent advances through polyphasic experiments. *J. Environ. Sci. Health Part A* **41**, 897–922.

Crocetti, G., Murto, M., and Bjornsson, L. (2006) An update and optimisation of oligonucleotide probes targeting methanogenic Archaea for use in fluorescence in situ hybridisation (FISH). *J. Microbiol. Methods* **65**, 194–201.

Dumont, C. G. and Murrell, J. C. (2005) Stable isotope probing — linking microbial identity to function. *Nat. Rev. Microbiol.* **3**, 499–504.

Egert, M. and Friedrich, M. W. (2003) Formation of pseudo-terminal restriction fragments, a PCR-related bias affecting terminal restriction fragment length polymorphism analysis of microbial community structure. *Appl. Environ. Microbiol.* **69**, 2555–2562.

Engebretson, J. J. and Moyer, C. L. (2003) Fidelity of select restriction endonucleases in determining microbial diversity by terminal-restriction fragment length polymorphism. *Appl. Environ. Microbiol.* **69**, 4823–4829.

Enright, A. M., Collins, G., and O'Flaherty, V. (2007) Temporal microbial diversity changes in solvent-degrading anaerobic granular sludge from low-temperature (15 °C) wastewater treatment bioreactors. *Syst. Appl. Microbiol.* **30**, 471–482.

Erb, R. W. and Wagner-Döbler, I. (1993) Detection of polychlorinated biphenyl genes in polluted sediments by direct DNA extraction and polymerase chain reaction. *Appl. Environ. Microbiol.* **59**, 4065–4073.

Falkowski, P. G. and de Vargas, C. (2004) Shotgun sequencing in the sea: A blast from the past? *Science* **304**, 58–60.

Fang, H. H. P., Liang, D. W., Zhang, T., and Liu Y. (2006) Anaerobic treatment of phenol in wastewater under thermophilic condition. *Wat. Res.* **40**, 427–434.

Fang, H. H. P. and Zhang, T. (2004) Quantification of microbial diversity in biofilms using denaturing gradient gel electrophoresis profile. *Environ. Biotechnol.* Taylor & Francis Group plc, London, UK. 563–565.

Franke-Whittle, I. H., Goberna, M., Pfister, V., and Insam, H., (2009) Design and development of the ANAEROCHIP microarray for investigation of methanogenic communities. *J. Microbiol. Methods* **79**, 279–288.

Frischer, M. E., Danforth, J. M., Newton Healy, M. A., and Saunders, F. M. (2000) Whole cell versus total RNA extraction for the analysis of microbial community structure using 16S rRNA targeted oligonucleotide probes in saltmarsh sediments. *Appl. Environ. Microbiol.* **66**, 3037–3043.

Garcia, J. L., Patel, B. K. C., and Ollivier, B. (2000) Taxonomic phylogenetic and ecological diversity of methanogenic *Archaea*. *Anaerobe*. **6**, 205–226.

Gilbride, K. A., Lee, D. Y., and Beaudette, L. A. (2006) Molecular techniques in wastewater: Understanding microbial communities, detecting pathogens, and real-time process control. *J. Microbiol. Methods* **66**, 1–20.

Giovannoni, S. J., Britschgi, T. B., Moyer, C. L., and Field, K. G. (1990) Genetic diversity in Sargasso Sea bacterioplankton. *Nature* **345**, 60–63.

Griffiths, R. I., Whiteley, A. S., O'Donnell, A. G., and Bailey, M. J. (2000) Rapid method for coextraction of DNA and RNA from natural environments for analysis of ribosomal DNA- and rRNA-based microbial community composition. *Appl. Environ. Microbiol.* **66**, 5488–5491.

Grosskopf, R., Janssen, P. A., and Liesack, W. (1998) Diversity and structure of the methanogenic community in anoxic rice paddy soil microcosms as examined by cultivation and direct 16S rRNA gene sequence retrieval. *Appl. Environ. Microbiol.* **64**, 960–969.

Hansen, K. H., Angelidaki, I., and Ahring, B. K. (1998) Anaerobic digestion of swine manure: Inhibition by ammonia. *Wat. Res.* **32**, 5–12.

Harms, G., Layton, A. C., Dionisi, H. M., Gregory, I. M., Garrett, V. M., Hawkins, S. A., Robinson, K. G., and Sayler, G. S. (2003) Real-time PCR quantification of nitrifying bacteria in a municipal wastewater treatment plant. *Environ. Sci. Technol.* **37**, 343–351.

Harmsen, H. J. M., Kengen, H. M. P., Akkermans, A. D. L., Stams, A. J. M., and De Vos, W. M. (1996) Detection and localization of syntrophic propionate-oxidizing bacteria in granular sludge by *in situ* hybridization using 16S rRNA-based oligonucleotide probes. *Appl. Environ. Microbiol.* **62**, 1656–1663.

Hata, J., Miyata, N., Kim, E. S., Takamizawa, K., and Iwahori, K. (2004) Anaerobic degradation of *cis*-**1**,2-dichloroethylene and vinyl chloride by *Clostridium* sp. strain DC1 isolated from landfill leachate sediment. *J. Biosci. Bioeng.* **97**, 196–201.

Haugland, R. A., Vesper, S. J., and Wymer, L. J. (1999) Quantitative measurement of *Stachybotrys chartarum* conidia using real time detection of PCR products with TaqMan fluorogenic probe system. *Mol. Cell Probes.* **13**, 329–340.

Hernández, M., Estev, Te., Prat, S., and Pla, M. (2004) Development of real-time PCR systems based on SYBR® Green I, Amplifluor™ and TaqMan® technologies for specific quantitative detection of the transgenic maize event GA21. *J. Cereal. Sci.* **39**, 99–107.

Hodson, R. E., Dustman, W. A., Garg, R. P., and Moran, M. A. (1995) *In situ* PCR for visualization of microscale distribution of specific genes and gene products in prokaryotic communities. *Appl. Environ. Microbiol.* **61**, 4074–4082.

Hofman-Bang, J., Zheng, D., Westermann, P., Ahring, B. K., and Raskin, L. (2003) Molecular ecology of anaerobic reactor systems. *Adv. Biochem. Eng. Biotechnol.* **81**, 151–203.

Hori, T., Noll, M., Igarashi, Y., Friedrich, M. W., and Conrad, R. (2007) Identification of acetate-assimilating microorganisms under methanogenic conditions in anoxic rice field soil by comparative stable isotope probing of RNA. *Appl. Environ. Microbiol.* **73**, 101–109.

Hristova, K. R., Mau, M., Zheng, D., Aminov, R. I., Mackie, R. I., Gaskins, H. R., and Raskin, L. (2000) *Desulfotomaculum* genus- and group-specific small-subunit rRNA hybridization probes for environmental studies. *Environ. Microbiol.* **2**, 143–159.

Huang, W. E., Stoecker, K., Griffiths, R. I., Newbold, L., Daims, H., Whiteley, A. S., and Wagner, M. (2007) Raman-FISH: Combining stable-isotope Raman spectroscopy and fluorescence in situ hybridization for the single cell analysis of identity and function. *Environ. Microbiol.* **9**, 1878–1889.

Hugenholtz, P., Goebel, B. M., and Pace, N. R. (1998) Impact of culture-independent studies on the emerging phylogenetic view of bacterial diversity. *J. Bacteriol.* **180**, 4765–4774.

Imachi, H., Sekiguchi, Y., Kamagata, Y., Ohashi, A., Harada, H. (2000) Cultivation and in situ detection of a thermophilic bacterium capable of oxidizing propionate in syntrophic association with hydrogenotrophic methanogens in a thermophilic methanogenic granular sludge. *Appl. Environ. Microbiol.* **66**, 3608–3615.

Jackson, B. E., Bhupathiraju, V. K., Tanner, R. S., Woese, C. R., and McInerney, M. J. (1999) *Syntrophus aciditrophicus* sp. nov., a new anaerobic bacterium that degrades fatty acids and benzoate in syntrophic association with hydrogen-using microorganisms. *Arch. Microbiol.* **171**, 107–114.

Jarrell, K. F., Faguy, D., Hebert, A. M., and Kalmokoff, M. L. (1992) A general method of isolating high molecular weight DNA from methanogenic archaea (archaebacteria). *Can. J. Microbiol.* **38**, 65–68.

Jeng, R. S., Svircev, A. M., Myers, A. L., Beliaeva, L., Hunter, D. M., and Hubbes, M. (2001) The use of 16S and 16S-23S rDNA to easily detect and differentiate common Gram-negative orchard epiphytes. *J. Microbiol. Met.* **44**, 69–77.

Kent, A. D., Smith, D. J., Benson, B. J., and Triplett, E. W. (2003) Web-based phylogenetic assignment tool for analysis of terminal restriction fragment length polymorphism profiles of microbial communities. *Appl. Environ. Microbiol.* **69**, 6768–6776.

Klappenbach, J. A., Saxman, P. R., Cole, J. R., and Schmidt, T. M. (2001) rrndb: The ribosomal RNA operon copy number database. *Nucleic Acids Res.* **29**, 181–184.

Klocke, M., Mahnert, P., Mundt, K., Souidi, K., and Linke, B. (2007) Microbial community analysis of a biogas-producing completely stirred tank reactor fed continuously with fodder beet silage as monosubstrate. *Syst. Appl. Microbiol.* **30**, 139–151.

Kong, R. Y. C., Lee, S. K. Y., Law, T. W. F., Law, S. H. W., and Wu, R. S. S. (2002) Rapid detection of six types of bacterial pathogens in marine waters by multiplex PCR. *Wat. Res.* **36**, 2802–2812.

Krober, M., Bekel, T., Diaz, N. N., Goesmann, A., Jaenicke, S., Krause, L., Miller, D., Runte, K. J., Viehover, P., Puhler, A., and Schluter, A. (2009) Phylogenetic characterization of a biogas plant microbial community integrating clone library 16S-rDNA sequences and metagenome sequence data obtained by 454-pyrosequencing. *J. Biotechnol.* **142**, 38–49.

Kumar, S., Tomura, K., and Nei, M. (1993) *MEGA: Molecular Evolution Genetics Analysis*, Pennsylvania State University, Philadelphia.

Kurata, S. (2001) Fluorescent quenching-based quantitative detection of specific DNA/RNA using a BODIPY® FL-labeled probe or primer. *Nucl. Acids Res.* **29**, (34)

Lane, D. J. (1991) *16S/23S rRNA sequencing. Nucleic Acid Techniques in Bacterial Systematics*. Chichester: Wiley, pp. 115–175.

Lee, C., Kim, J., Hwang, K., O'Flaherty, V., and Hwang, S. (2009) Quantitative analysis of methanogenic community dynamics in three anaerobic batch digesters treating different wastewaters. *Wat. Res.* **43**, 157–165.

Lee, D. H., Zo, Y. G., and Kim, S. J. (1996) Nonradioactive method to study genetic profiles of natural bacterial communities by PCR-single-stranded-conformation polymorphism. *Appl. Environ. Microbiol.* **62**, 3113–3120.

Lee, N., Nielsen, P. H., Andersen, K. H., Juretschko, S., Nielsen, J. L., Schleifer, K. H., and Wagner, M. (1999) Combination of fluorescent *in situ* hybridization and microautoradiography — A new tool for structure-function analyses in microbial ecology. *Appl. Environ. Microbiol.* **65**, 1289–1297.

Letowski, J., Juteau, P., Villemur, R., Beaudet, R., Lépine, F., and Bisaillon, J. G. (2001) Separation of a phenol carboxylating organism from a two-member, strict anaerobic coculture. *Can. J. Microbiol.* **47**, 373–381.

Lettinga, G., van Velsen, A. F. M., Hobma, S. W., Zeeuw, W. J., and de Klapwijk, A. (1980) Use of the upflow sludge blanket (USB) reactor concept for biological wastewater treatment, especially for anaerobic treatment. *Biotechnol. Bioeng.* **22**, 699–734.

Liang, D. W., Fang, H. H. P., and Zhang, T. (2009) Microbial characterization and quantification of an anaerobic sludge degrading dimethyl phthalate. *J. Appl. Microbiol.* **106**, 296–305.

Liang, D. W., Zhang, T., and Fang, H. H. P. (2007) Anaerobic degradation of dimethyl phthalate in wastewater in a UASB reactor. *Wat. Res.* **41**, 2879–2884.

Liang, D. W., Zhang, T., and Fang, H. H. P. (2008) Real-time quantifications of dominant anaerobes in an upflow reactor by polymerase chain reaction using a TaqMan probe. *Wat. Sci. Technol.* **57**, 1851–1855.

Libonati, M. and Sorrentino, S. (2001) Degradation of double-stranded RNA by mammalian pancreatic-type ribonucleases. *Methods Enzymol.* **341**, 234–248.

Liesack, W., Janssen, P. H., Rainey, F. A., Ward-Rainey, N. L., Stackebrandt, E. (1997) Microbial diversity in soil: The need for a combined approach using molecular and cultivation techniques. *Modern Soil Microbiology* New York, N.Y.: Marcel Dekker, Inc., pp. 375–439.

Liu, W. T., Chan, O. C., and Fang, H. H. P. (2002) Characterization of microbial community in granular sludge treating brewery wastewater. *Wat. Res.* **36**, 1767–1775.

Liu, W. T., Marsh, T. L., Cheng, H., and Forney, L. J. (1997) Characterization of microbial diversity by determining terminal restriction fragment length polymorphisms of genes encoding 16S rRNA. *Appl. Environ. Microbiol.* **63**, 4516–4522.

Ludwig, W., Strunk, O., Westram, R., Richter, L., Meier, H., Yadhukumar Buchner, A., Lai, T., Steppi, S., *et al.* (2004) ARB: A software environment for sequence data. *Nucleic Acids Res.* **32**, 1363–1371.

Lueders, T. and Friedrich, M. W. (2003) Evaluation of PCR amplification bias by T-RFLP analysis of SSU rRNA and *mcrA* genes using defined template mixtures of methanogenic pure cultures and soil DNA extracts. *Appl. Environ. Microbiol.* **69**, 320–326.

MaCarty, P. L. (1964) Anaerobic waste treatment fundamentals — Part two, environment requirements and control. *Public Works* **95**, 123–126.

Markoulatos, P., Siafakas, N., and Moncany, M. (2002) Multiplex polymerase chain reaction: A practical approach. *J. Clin. Lab. Anal.* **16**, 47–51.

Marsh, T. L., Saxman, P., Cole, J., and Tiedje, J. (2000) Terminal restriction fragment length polymorphism analysis program, a web-based research tool for microbial community analysis. *Appl. Environ. Microbiol.* **66**, 3616–3620.

Martin-Laurent, F., Philippot, L., Hallet, S., Chaussod, R., Germon, J. C., Soulas, G., and Catroux, G. (2001) DNA extraction from soils: Old bias for new microbial diversity analysis methods. *Appl. Environ. Microbiol.* **67**, 2354–2359.

McCune, B. and Grace, J. B. (2002) *Analysis of Ecological Communities*. Gleneden Beach, Oregon: MjM Software, 304 pp.

McHugh, S., Collins, G., and O'Flaherty, V. (2006) Long-term, high-rate anaerobic biological treatment of whey wastewaters at psychrophilic temperatures. *Bioresour. Technol.* **97**, 1669–1678.

Meyerdierks, A., Kube, M., Kostadinov, I., Teeling, H., Glöckner, F. O., Reinhardt, R., and Amann, R. (2009) Metagenome and mRNA expression analyses of anaerobic methanotrophic archaea of the ANME-1 group. *Environ. Microbiol.* **12**, 422–439.

Miller, D. N., Bryant, J. E., Madsen, E. L., and Ghiorse, W. C. (1999) Evaluation and optimization of DNA extraction and purification procedures for soil and sediment samples. *Appl. Environ. Microbiol.* **65**, 4715–4724.

Mora, R. R. and Amann, R. (2001) The species concept for prokaryotes. *FEMS Microbiol. Rev.* **25**, 39–67.

Muyzer, G., de Waal, E. C., and Uitterlinden, A. G. (1993) Profiling of complex microbial population by DGGE analysis of polymerase chain reaction amplified genes encoding for 16S rRNA. *Appl. Environ. Microbiol.* **62**, 2676–2680.

Nakano, Y., Takeshita, T., and Yamashita, Y. (2006) TRFMA: A web-based tool for terminal restriction fragment length polymorphism analysis based on molecular weight. *Bioinformatics* **22**, 1788–1789.

Nielsen, A. T., Liu, W. T., Filipe, C., Grady, L., Molin, S., and Stahl, D. A. (1999) Identification of a novel group of bacteria in sludge from a deteriorated biological phosphorus removal reactor. *Appl. Environ. Microbiol.* **65**, 1251–1258.

Nocker, A., Burr, M., and Camper, A. K. (2007) Genotypic microbial community profiling: A critical technical review. *Microb. Ecol.* **54**, 276–289.

Okabe, S., Satoh, H., and Watanabe, Y. (1999) *In situ* analysis of nitrifying biofilms as determined by the *in situ* hybridisation and the use of microelectrodes. *Appl. Environ. Microbiol.* **65**, 3182–3191.

Overmann, J. and Tuschak, C. (1997) Phylogeny and molecular fingerprinting of green sulfur bacteria. *Arch. Microbiol.* **167**, 302–309.

Pachter, L. (2007) Interpreting the unculturable majority. *Nat. Methods* **4**, 479–480.

Padmasiri, S. I., Zhang, J. Z., Fitch, M., Norddahl, B., Morgenroth, E., and Raskin, L. (2007) Methanogenic population dynamics and performance of an anaerobic membrane bioreactor (AnMBR) treating swine manure under high shear conditions. *Wat. Res.* **41**, 134–144.

Palatsi, J., Illa, J., Prenafeta-Boldu, F. X., Laureni, M., Fernandez, B., Angelidaki, I., and Flotats, X. (2010) Long-chain fatty acids inhibition and adaptation process in anaerobic thermophilic digestion: Batch tests, microbial community structure and mathematical modeling. *Bioresour. Technol.* **101**, 2243–2251.

Pernthaler, A., Pernthaler, J., and Amann, R. (2002) Fluorescence *in situ* hybridization and catalyzed reporter deposition for the identification of marine bacteria. *Appl. Environ. Microbiol.* **68**, 3094–3101.

Picard, C., Ponsonnet, C., Paget, E., Nesme, X., and Simonet, P. (1992) Detection and enumeration of bacteria in soil by direct DNA extraction and polymerase chain reaction. *Appl. Environ. Microbiol.* **58**, 2717–2722.

Rainey, F. A., Ward-Rainey, N. L., Janssen, P. H., Hippe, H., and Stackebrandt, E. (1996) *Clostridium paradoxum* DSM 7308T contains multiple 16S rRNA genes with heterogeneous intervening sequences. *Microbiology* **142**, 2087–2095.

Rao, T. S., Sairam, T. N., Viswanathan, B., and Nair, K. V. K. (2000) Carbon steel corrosion by iron oxidizing and sulphate reducing bacteria in a freshwater cooling system. *Corros. Sci.* **42**, 1417–1431.

Raskin, L., Amann, R. I., Poulsen, L. K., Rittmann, B. E., and Stahl, D. A. (1995) Use of ribosomal RNA-based molecular probes for characterization of complex microbial communities in anaerobic biofilms. *Wat. Sci. Technol.* **31**, 261–272.

Raskin, L., Poulsen, L. K., Noguera, D. R., Rittmann, B. E., and Stahl, D. A. (1994) Quantification of methanogenic groups in anaerobic biological reactors by oligonucleotide probe hybridization, *Appl. Environ. Microbiol.* **60**, 1241–1248.

Rossello-Mora, R. and Amman, R. (2001) The species concept for prokaryotes. *FEMS Microbiol. Rev.* **25**, 39–67.

Roy, C. S., Talbot, G., Topp, E., Beaulieu, C., Palin, M. F., and Masse, D. I. (2009) Bacterial community dynamics in an anaerobic plug-flow type bioreactor treating swine manure. *Wat. Res.* **43**, 21–32.

Saitou, N. and Nei, M. (1987) The neighbor-joining method: A new method for constructing phylogenetic trees. *Mol. Biol. Evol.* **4**, 406–425.

Santegoeds, C. M., Ferdelman T.G., Muyzer G., and de Beer D. (1998) Structural and functional dynamics of sulfate-reducing populations in bacterial biofilms. *Appl. Environ. Microbiol.* **64**, 3731–3739.

Sanz, J. L. and Köchling, T. (2007) Molecular biology techniques used in wastewater treatment: An overview. *Process. Biochem.* **42**, 119–133.

Satoh, H., Miura, Y., Tsushima, I., and Okabe, S. (2007) Layered structure of bacterial and Archaeal communities and their *in situ* activities in anaerobic granules. *Appl. Environ. Microbiol.* **73**, 7300–7307.

Sawayama, S., Tsukahara, K., and Yagishita, T. (2006) Phylogenetic description of immobilized methanogenic community using real-time PCR in a fixed-bed anaerobic digester. *Bioresour. Technol.* **97**, 69–76.

Schlüter, A., Bekel, T., Diaz, N. N., Dondrup, M., Eichenlaub, R., Gartemann, K. H., Krahn, I., Krause, L., Krömeke, H., Kruse, O., Mussgnug, J. H., Neuweger, H., Niehaus, K., Pühler, A., Runte, K. J., Szczepanowski, R., Tauch, A., Tilker, A., Viehöver, P., and Goesmann, A. (2008) The metagenome of a biogas-producing microbial community of a production-scale biogas plant fermenter analysed by the 454-pyrosequencing technology. *J. Biotechnol.* **136**, 77–90.

Schneegurt, M. A. and Kulpa, C. F. (1998) The application of molecular techniques in environmental biotechnology for monitoring microbial systems. *Biotechnol. Appl. Biochem.* **27**, 73–79.

Scully, C., Collins, G., and O'Flaherty, V. (2005) Assessment of anaerobic wastewater treatment failure using terminal restriction fragment length polymorphism analysis. *J. Appl. Microbiol.* **99**, 1463–1471.

Sekiguchi, Y., Kamagata, Y., and Harada, H. (2001) Recent advances in methane fermentation technology, *Curr. Opin. Biotechnol.* **12**, 277–282.

Sekiguchi, Y., Kamagata, Y., Nakamura, K., Ohashi, A., and Harada, H. (1999) Fluorescence *in situ* hybridization using 16S rRNA-targeted oligonucleotides reveals localization of methanogens and selected uncultured bacteria in mesophilic and thermopilic sludge granules. *Appl. Environ. Microbiol.* **65**, 1280–1288.

Sekiguchi, Y., Kamagata, Y., Syutsubo, K., Ohashi, A., Harada, H., and Nakamura, K. (1998) Phylogenetic diversity of mesophilic and thermophilic granular sludge determined by 16S rRNA gene analysis. *Microbiology* **144**, 2655–2665.

Simon, N., Lebot, N., Marie, D., Partensky, F., and Vaulot, D. (1995) Fluorescent *in situ* hybridization with rRNA-targeted oligonucleotide probes to identify small phytoplankton by flow cytometry. *Appl. Environ. Microbiol.* **61**, 2506–2513.

Smith, C. J., Danilowicz, B. S., Clear, A. K., Costello, F. J., Wilson, B., and Meijer, W. G. (2005) T-Align, a web-based tool for comparison of multiple terminal restriction fragment length polymorphism profiles. *FEMS Microbiol. Ecol.* **54**, 375–380.

Song, M., Shin, S. G., and Hwang, S. (2010) Methanogenic population dynamics assessed by real-time quantitative PCR in sludge granule in upflow anaerobic sludge blanket treating swine wastewater. *Bioresour. Technol.* **101**, s23–s28.

Stahl, D. A. and Amann, R. I. (1991) *Development and Application of Nucleic Acid Probes in Bacterial Systematics. Nucleic Acid Techniques in Bacterial Systematics*. Chichester, England: John Wiley & Sons, pp. 205–248.

Strunk, O. and Ludwig, W. (1997) ARB software program package. http://www.biolchemietu-muenchende/pub/ARB.

Stubner, S. and Meuser, K. (2000) Detection of *Desulfotomaculum* in an Italian rice paddy soil by 16S ribosomal nucleic acid analyses. *FEMS Microbiol. Ecol.* **34**, 73–80.

Stubner, S. (2002) Enumeration of 16S rDNA of *Desulfotomaculum* lineage 1 in rice field soil by real-time PCR with SybrGreen™ detection. *J. Microbiol. Methods* **50**, 155–164.

Tang, Y., Shigematsu, T., Morimura, I. S., and Kida, K. (2004) The effects of micro-aeration on the phylogenetic diversity of microorganisms in a thermophilic anaerobic municipal solid-waste digester. *Wat. Res.* **38**, 2537–2550.

Tang, Y. Q., Fujimura, Y., Shigematsu, T., Morimura, S., and Kida, K. (2007) Anaerobic treatment performance and microbial population of thermophilic upflow anaerobic filter reactor treating awamori distillery wastewater. *J. Biosci. Bioeng.* **104**, 281–287.

Tang, Y. Q., Matsui, T., Morimura, S., Wu, X. L., and Kida, K. (2008) Effect of temperature on microbial community of a glucose-degrading methanogenic consortium under hyperthermophilic chemostat cultivation. *J. Biosci. Bioeng.* **106**, 180–187.

Tartakovsky, B., Manuel, M. F., Beaumier, D., Greer, C. W., and Guiot, S. R. (2001) Enhanced selection of an anaerobic pentachlorophenol-degrading consortium. *Biotechnol. Bioeng.* **73**, 476–483.

Thompson, J. D., Higgins, D. G., and Gibson, T. J. (1994) CLUSTAL W: Improving the sensitivity of progressive multiple sequence alignment through sequence weighting, position-specific gap penalties and weight matrix choice. *Nucleic Acids Res.* **22**, 4673–4680.

Ueno, Y., Sekiguchi, Y., Sunaga, A., Yoshida, H., and Kamagata, Y. (2004) Sequence-specific cleavage of small-subunit (SSU) rRNA with oligonucleotides and ribonuclease H: A rapid and simple approach to the SSU rRNA-based quantitative detection of microorganisms. *Appl. Environ. Microbiol.* **70**, 3650–3663.

Ueno, Y. and Tatara, M. (2008) Microbial population in a thermophilic packed-bed reactor for methanogenesis from volatile fatty acids. *Enzyme Microb. Technol.* **43**, 302–308.

Vallaeys, T., Topp, E., and Muyzer, G. (1997) Evaluation of DGGE in the detection of 16S rDNA sequence variation in rhizobia and methanotrophs. *FEMS Microbiol. Ecol.* **61**, 1444–1450.

van Leerdam, R. C., de Bok, F. A. M., Bonilla-Salinas, M., van Doesburg, W., Lomans, B. P., Lens, P. N. L., Stams, A. J. M., and Janssen, A. J. H. (2008) Methanethiol degradation in anaerobic bioreactors at elevated pH (>8): Reactor performance and microbial community analysis. *Bioresour. Technol.* **99**, 8967–8973.

Volossiouk, T., Robb, E. J., and Nazar, R. N. (1995) Direct DNA extraction for PCR-mediated assays of soil organisms. *Appl. Environ. Microbiol.* **61**, 3972–3976.

von Wintzingerode, F., Göbel, U. B., and Stackebrandt, E. (1997)Determination of microbial diversity in environmental samples: Pitfalls of PCR-based rRNA analysis. *FEMS Microbiol. Rev.* **21**, 213–229.

Wagner, A. O., Malin, C., and Illmer, P. (2009) Application of denaturing high performance liquid chromatography (dHPLC) in microbial ecology: Community profiling in fermenter sludge, compost, and soil. *Appl. Environ. Microbiol.* **75**, 956–964.

Wagner, M., Amann, R., Lemmer, H., and Schleifer, K. H. (1993) Probing activated sludge with oligonucleotides specific for proteobacteria: Inadequacy of culture-dependent methods for describing microbial community structure. *Appl. Environ. Microbiol.* **59**, 1520–1525.

Wagner, M., Horn, M., and Daims, H. (2003) Fluorescence *in situ* hybridisation for the identification and characterisation of prokaryotes, *Curr. Opin. Microbiol.* **6**, 302–309.

Wang, H., Lehtomaki, A., Tolvanen, K., Puhakka, J., and Rintala, J. (2009) Impact of crop species on bacterial community structure during anaerobic co-digestion of crops and cow manure. *Bioresour. Technol.* **100**, 2311–2315.

Woese, C. R., Kandler, O., and Wheelis, M. L. (1990) Towards a natural system of organisms: Proposal for the domains *Archaea, Bacteria*, and *Eucarya. Proc. Natl. Acad. Sci. U.S.A.* **87**, 4576–4579.

Wu, J. H., Liu, W. T., Tseng, I. C., and Cheng, S. S. (2001a) Characterization of a 4-methylbenzoate-degrading methanogenic consortium as determined by small-subunit rDNA sequence analysis. *J. Biosci. Bioeng.* **91**, 449–455.

Wu, J. H., Liu, W. T., Tseng, I. C., and Cheng, S. S. (2001b) Characterization of microbial consortia in a terephthalate-degrading anaerobic granular sludge system. *Microbiology* **147**, 373–382.

Yamada, T., Sekiguchi, Y., Imachi, H., Kamagata, Y., Ohashi, A., and Harada, H. (2005) Diversity, localization, and physiological properties of filamentous microbes belonging to *Chloroflexi* subphylum I in mesophilic and thermophilic methanogenic sludge granules. *Appl. Environ. Microbiol.* **71**, 7493–7503.

Yang, Y., Tsukahara, K., and Sawayama, S. (2008) Biodegradation and methane production from glycerol-containing synthetic wastes with fixed-bed bioreactor under mesophilic and thermophilic anaerobic conditions. *Process Biochem.* **43**, 362–367.

Yu, Z. and Morrison, M. (2004) Improved extraction of PCR-quality community DNA from digesta and fecal samples. *BioTechniques* **36**, 808–812.

Zhang, H., Banaszak, J. E., Parameswaran, P., Alder, J., Krajmalnik-Brown, R., and Rittmann, B. E. (2009) Focused-Pulsed sludge pre-treatment increases the bacterial diversity and relative abundance of acetoclastic methanogens in a full-scale anaerobic digester. *Wat. Res.* **43**, 4517–4526.

Zhang, T. and Fang, H. H. P. (2000) Digitization of DGGE (denaturing gradient gel electrophoresis) profile and cluster analysis of microbial communities. *Biotechnol. Lett.* **22**, 399–405.

Zhang, T. and Fang, H. H. P. (2001) Phylogenetic diversity of a SRB-rich marine biofilm. *Appl. Microbiol. Biotechnol.* **57**, 437–440.

Zhang, T. and Fang, H. H. P. (2006) Applications of real-time polymerase chain reaction for quantification of microorganisms in environmental samples. *Appl. Microbiol. Biotechnol.* **70**, 281–289.

Zhang, T., Jin, T., Yan, Q. M., Shao, M. F., Wells, G., Criddle, C., and Fang, H. H. P. (2009) Occurrence of ammonia-oxidizing archaea (AOA) in activated sludges of a laboratory scale reactor and two wastewater treatment plants. *J. Appl. Microbiol.* **107**, 970–977.

Zhang, T., Ke, S. Z., Liu, Y., and Fang, H. H. P. (2005) Microbial characteristics of a methanogenic phenol-degrading sludge. *Wat. Sci. Technol.* **52**, 73–78.

Zhang, T., Shao, M. F., and Fang, H. H. P. (2009) A qRT-PCR based method for the measurement of *rrn* operon copy number. *Lett. Appl. Microbiol.* **49**, 26–30.

Zheng, D. and Raskin, L. (2000) Quantification of *Methanosaeta* Species in anaerobic bioreactors using genus- and species-specific hybridization probes. *Microb. Ecol.* **39**, 246–262.

Chapter 11

Application of Mathematical Models to Anaerobic Digestion Process

Hidenari Yasui[1] and Rajeev Goel[2]

[1]*The University of Kitakyushu, 1-1, Hibikino, Wakamatsu, Kitakyushu 808-0135, Japan*
hidenari-yasui@env.kitakyu-u.ac.jp

[2]*Hydromantis Inc., 1 James Street South, Suite 1601, Hamilton, Ontario L8P 4R5, Canada*

The application of mathematical models has become a standard practice in wastewater treatment plant design, optimization, and operational control. There are many state-of-the-art models currently available for different unit processes applied for wastewater treatment plants. These models have been implemented in various commercial process simulators and are readily available for uses by researchers, engineers, and consultants. Although, the commercial process simulators are easy to use, the magnitude and complexity of biochemical, and physico-chemical reactions involved in wastewater treatment make it challenging for the first time practitioner to adequately grasp the intricate details of the model. As the benefits of modeling in wastewater treatment are well accepted, the intention of this chapter is first to familiarize the first-time user to important basic fundamentals and terminologies used in present day wastewater treatment models. This is followed by detailed discussions on a few important issues which shall be helpful in practical application of Anaerobic Digestion Model No. 1 (ADM1).

1. Classification of Mathematical Models

Figure 11.1 shows a conceptual illustration of the relationship between input and output in a given system. The input 'm' is transformed by a function $f(m)$ to produce an output 'p'. For a given system, a mathematical model is a set of one or more equations incorporating the input/output transformation functions. Depending on the formulation of the problem, the model may contain algebraic, differential, or partial differential equations. Although the input/output models are useful and sufficient in various circumstances, they do not consider the mechanisms by which the inputs are related to outputs. In order to overcome this deficiency, a state-space approach is normally used for formulating models with known internal mechanism (Carstensen *et al.* 1997).

The most important feature of a state-space model is the introduction of state variables which act as mediators between the inputs and outputs. These state variables are additional model constituents and are helpful in defining the internal mechanistic behavior of the system. By definition, state variables may or may not be measurable. The required number of state variables is equal to or more than the number of unique input–output variables. The conceptual bases of a state-space model are as shown in Fig. 11.2. The state-space models require definition

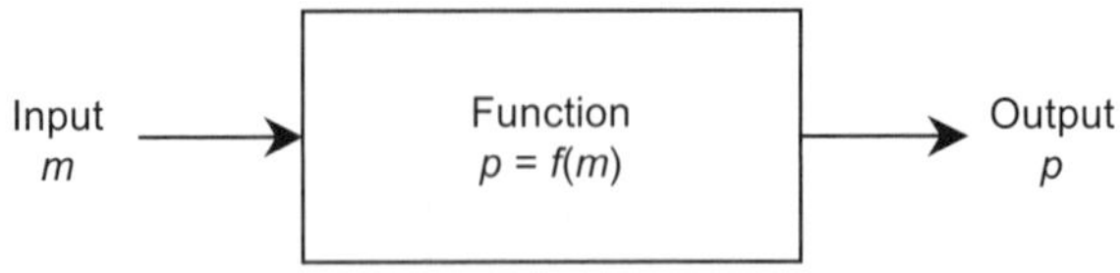

Fig. 11.1. Relationship between input and output in a given system.

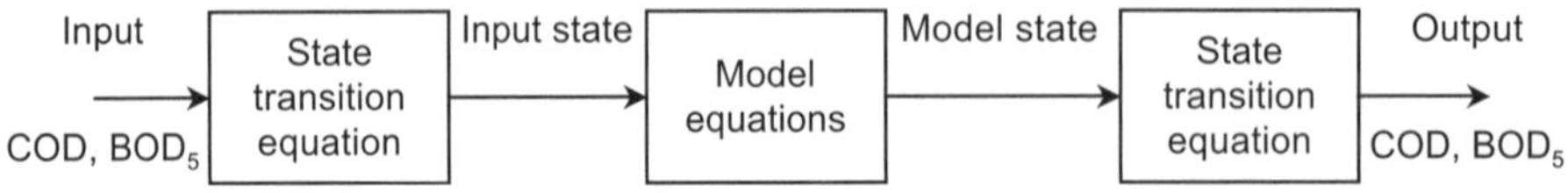

Fig. 11.2. Conceptual diagram for state-space model as modified from Carstensen *et al.* (1997).

Table 11.1. Classification of mathematical model.

Type	Black-box model	Gray-box model	White-box model
Relationship between m and p	Direct	Semi-mechanistic	Mechanistic
Function of f	Empirical	Empirical/theoretical	Theoretical
Complexity of f	Generally simple	Moderate to complex	Simple to complex
Applicability	Specific	Limited general	Limited general

of functions to:

(1) estimate input states from given inputs;
(2) describe the internal model processes; and
(3) convert system states to outputs.

The present day Activated Sludge Models (ASMs) and Anaerobic Digestion Model No. 1 (ADM1) are based on the state-space formulation (Henze *et al.* 2000; Batstone *et al.* 2002). Depending on the formulation of a model, the mathematical model may be classified as a black-box, gray-box, or a white-box model. As shown in Table 11.1, a black-box model is formulated based on experimentally determined relationship between inputs and outputs. Thus, a black-box model also known as empirical model mainly focuses on the apparent relationship between inputs and outputs. For an anaerobic digestion process, a model consisting of simple relationship between the sludge flow rate (Q_S) and biogas production rate (Q_G) will be considered as a black-box model (Buswell 1939). The generality of a black-box model depends on the amount and type of experimental data used in its construction. As most of the black-box models are based on limited site-specific data, the model applications are limited to small range of operational parameters (temperature, HRT, *etc.*) and plants inputs (sludge composition, sludge concentration, *etc.*). The simplicity of the black-box models makes them very useful tool in formulating rule-of-thumb for design and operation. However, as the black-box models do not consider the underlying mechanism of biological, chemical, and physical reactions in the system, they are not suitable for in-depth analysis of the system.

In contrast to a black-box model, a white-box (mechanistic) model is intended to generalize system response by identifying the underlying reactions affecting the system response. The white-box model is formulated by considering the kinetics and stoichiometry of each identified reaction. A state-space model formulation approach is normally used in white-box model to define the internal structure of the complex biochemical, physico-chemical processes. Unfortunately, due to the complexity of biochemical reactions, it is not easy to develop a perfect white-box model and sometimes several assumptions are required to be made during model formulation.

The gray-box model seems to lie somewhere in between the black-box and white-box model. These models are used when theoretical understanding is not enough to establish a white-box model but a practical generalization of an overall reaction scheme based on theoretical and experimental knowledge can be formulated.

Based on the above discussion, the Anaerobic Digestion Model No. 1 (ADM1) presented by IWA task group in 2002 can be classified as a white-box, state-space model (Batstone *et al.* 2002). The theoretical concepts of material conservation and biochemical reactions are used in formulating the model structure. As pointed out earlier, several assumptions are made with respect to the constituent microbial population and processes to keep the model as simple as possible.

2. Essentials of Mechanistic Models

2.1. *State Variables*

In state-space models, the state variables are the principal components of the model based on which conversion processes are defined. As indicated earlier, some of the state variables may be same as the measurable inputs, e.g., NH_3-N, while some other may need to be estimated from composite measurements, e.g., estimation of degradable and non-degradable substrates from COD and $BOD_{ultimate}$ measurements. The state variables in the model are considered to have a fixed elemental composition. For example, an elemental composition of $C_5H_7O_2N$ or $CH_{1.8}O_{0.5}N_{0.2}$ is usually considered for the biomass species in the model (Hoover and Porgens 1952; Roels 1983). The fixed elemental composition of the state variables makes it feasible to define process reactions by observing the conservation laws.

2.2. *Formulation of conservation equations*

In mechanistic models, it is crucial to define the conversion reactions between state variables using the principles of mass, charge, and energy conservation. Application of these principles is illustrated by using Eq. (11.1) (adapted from Thauer *et al.* 1977), which is written for conversion of glucose to ethanol in anaerobic fermentation.

$$\begin{aligned} &a\cdot C_6H_{12}O_6 + b\cdot NH_4^+ + c\cdot H_2O \\ &\quad \rightarrow d\cdot H^+ + e\cdot C_2H_6O + f\cdot HCO_3^- + 1\cdot CH_{1.8}O_{0.5}N_{0.2} + \Delta H^o + \Delta G^o \end{aligned} \tag{11.1}$$

For the above reactions, the elemental, charge, enthalpy and Gibbs energy balance equations can be formulated as below.

C conservation	$6a\text{-}2e\text{-}f\text{-}1 = 0$
H conservation	$12a + 4b + 2c\text{-}d\text{-}6e\text{-}f\text{-}1.8 = 0$
O conservation	$6a + c\text{-}e\text{-}3f\text{-}0.5 = 0$

N conservation	$b\text{-}0.2 = 0$
Charge conservation	$b(+1)\text{-}d(+1)\text{-}f(\text{-}1) = 0$
Enthalpy conservation	$\sum(\alpha_m \cdot \Delta H^o_{f,m})\text{-}\sum(\alpha_p \cdot \Delta H^o_{f,p}) = \Delta H^o$
Gibbs energy conservation	$\sum(\alpha_m \cdot \Delta G^o_{f,m})\text{-}\sum(\alpha_p \cdot \Delta G^o_{f,p}) = \Delta G^o$

where α represents one of the stoichiometric coefficients (= one of a, b, c, d, e, or f), ΔH^o_f is the standard enthalpy of formation and ΔG^o_f is the standard Gibbs energy of formation. The subscripts m and p symbolize the material (reactant) and product, respectively.

In the above set of seven equations, there are eight unknowns ($a, b, c, d, e, f, \Delta H^o$, and ΔG^o). Therefore, to solve this set of equations, one of the unknown is required to be experimentally or theoretically determined. In biological growth systems, biomass yield defined as the ratio of substrate consumed per unit of biomass produced (namely $Y = a^{-1}$) is normally determined experimentally and is used in solving the above set of equations. As discussed later, correlations to estimate the ΔG^o for working energy of biomass synthesis on a given substrate are also available and may be used to solve the above set of equations.

2.3. *Selection of model units*

The mechanistic models follow the law of elemental conservation as shown above, therefore molar concentration of each element (mol-C·m^{-3}, mol-N·m^{-3}, *etc.*) could be an easy choice for model development. However, in wastewater engineering, several measures of concentration like 5-day biochemical oxygen demand (BOD_5), chemical oxygen demand (COD), total organic carbon (TOC), and volatile suspended solids (VSS) for organic material have been traditionally used. Also, these concentrations have been expressed as g·m^{-3} and use of mol·m^{-3} is not common. To keep the traditional way of expressing the strength of organic matter, while still be able to describe the reactions using the mass conservation approach, the unit of COD expressed as g·m^{-3} was adopted by task groups involved in the developments of ASMs and ADM1. The COD unit is considered to be superior as it is conserved and provides a link between the electron equivalents in the organic substrate, the biomass, and the oxygen utilized. According to the definition of COD, theoretical COD of any compound can be estimated by finding the oxygen consumption in which the organic compound is oxidized to the reference compounds of CO_2, H_2O, and NH_3. For example, an oxidation of an organic compound $C_nH_XO_YN_Z$ with oxygen can be expressed as below in Eq. (11.2):

$$C_nH_XO_YN_Z + \left(n + \frac{X}{4} - \frac{Y}{2} - \frac{3Z}{4}\right)O_2 \rightarrow nCO_2 + \left(\frac{X}{2} + \frac{3Z}{2}\right)H_2O + ZNH_3 \tag{11.2}$$

Based on above reaction, the theoretical chemical oxygen demand for 1 mol of $C_nH_XO_YN_Z$ can be estimated as $(n + X/4\text{-}Y/2\text{-}3Z/4) \times 32$ g-COD. As indicated earlier, the COD is a conserved quantity by definition assuming that all the reference products have zero COD, the COD of the oxygen consumed in the reaction has to be equal to $\text{-}(n + X/4\text{-}Y/2\text{-}3Z/4) \times 32$. Therefore, it is important to remember that the COD of 1 mol of O_2 is -32 g in all the reactions to avoid misunderstanding regarding COD conservation in biological reactions.

As mentioned earlier, the concept of COD conservation is quite useful to link substrate (electron donor), oxygen (electron acceptor), and product (biomass) in aerobic biological growth system. Similar COD link between electron donor, electron acceptor, and biomass exists in

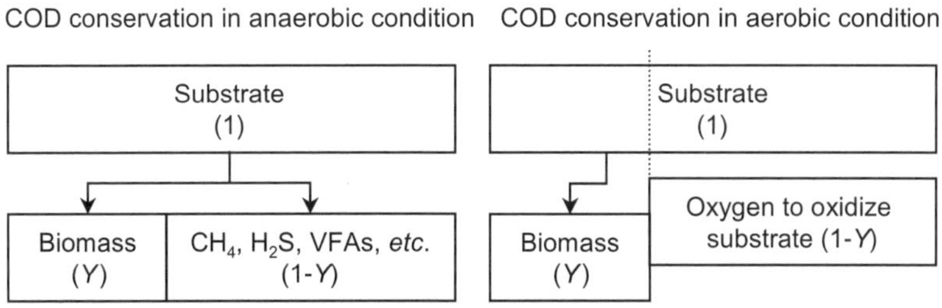

Fig. 11.3. COD balance in biological reaction.

anaerobic biological system. The principle of COD balance for anaerobic and aerobic system is illustrated in Fig. 11.3. It can be easily checked that the net COD consumption or production is zero in both the anaerobic and aerobic systems.

Using the COD conservation concept illustrated above, the expected amount of products (e.g., biomass, CH_4) can be easily obtained if the value of stoichiometry coefficient Y is known. Conversely, if the amount of biomass or CH_4 production at a plant is measured, the accuracy of the plant data can be evaluated by checking COD conservation. To express the mass balance of model components other than the organics, the models normally use elemental units for each component. For example, NH_3 is expressed as g NH_3-N·m^{-3} and PO_4^{-3} is expressed as g PO_4-P·m^{-3}. These unit selections again make it easy to write and check mass balances in the model.

2.4. *Development of model structure*

One of the important steps in formulating a model structure is to make a statement of the problem. For example, to formulate an anaerobic digestion process model, we may state the problem as 'estimate CH_4 production from the decomposition of organic solids'. In an input/output model, we may consider organic solids as the input and CH_4 production as the output variable and work out a transformation function. However, in mechanistic modeling, it is important to think about the underlying mechanisms and the state variables which might be involved in the above process. The details of the mechanisms may depend on the prevalent knowledge about the process. In the present case, for formulating a state-space model, we may start with the two obvious state variables of soluble methane (S_M) and organic solid (X_{OS}) and evaluate whether they can adequately describe the underlying mechanisms in the process. The mechanism in the present case is a biological conversion in which microorganisms grow over the biodegradable solids.

According to this mechanism, the model should incorporate the bacterial species participating in the conversion process. In the simplest form of the model, this point may be incorporated by considering a state variable representing the active anaerobic biomass concentration (X_B) in the model. Also the mechanism states that only the biodegradable solids are converted to methane. Therefore, it is required that the organic solids are represented by at least two fractions of biodegradable organic solid (X_S) and unbiodegradable organic solids (X_I).

Based on these considerations, the initially identified state variable of organic solid (X_{OS}) needs to be replaced by the three new state variables which are more suitable for defining the model mechanistically. Therefore, the simplest form of the model should contain four state

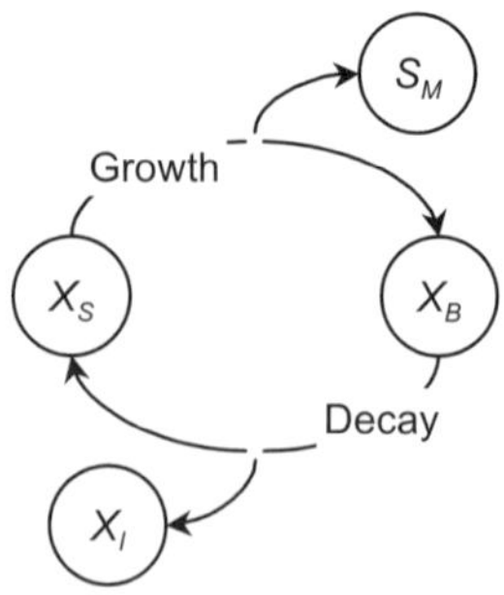

Fig. 11.4. Mechanisms for state conversion in proposed model.

variables of soluble methane (S_M), active anaerobic biomass concentration (X_B), biodegradable organic solid (X_S), and unbiodegradable organic solids (X_I). This approach in which constituent state variables are used instead of the apparent composite variable is also called 'structured modeling'.

After the relevant state variables for the model are identified, process reactions are formulated to define the kinetic and stoichiometric of inter-conversion of the state variables. In the present case, two process reactions of biological growth and decay may be formulated. In the first process, the anaerobic microorganisms (X_B) grow by decomposing the biodegradable solids (X_S) and producing CH_4. In the second process, the anaerobic microorganisms (X_B) are lost by decay, lysis, *etc.* producing biodegradable and non-biodegradable solids in the system. The non-biodegradable organic solids (X_I) do not react with the anaerobic microorganisms (X_B) and build up in the system unless removed. The proposed mechanism of the model is as shown in Fig. 11.4. It is appropriate to note here, that the state variables abbreviations in ASMs and ADM1 use X for solid components and an S for soluble components.

The above structured model with four state variables is certainly an improvement over the input/output model approach. However, it is evident that considering the complexity of microbial reactions in an anaerobic digester, it is an oversimplification. As long as these simplifications are understood and discounted in interpretation of the model results, there is little risk in using this simplified model. If the objectives of the modeling or the statement of problem require that the intermediate products in the anaerobic digestion be quantified, then extensions to this simple model can be made by including new state variables corresponding to the required intermediate products and microorganisms producing these products.

2.5. *Model definition using Petersen matrix*

In a bioreactor, the state variables are converted into other states with a specified reaction rate. As discussed above, in every conversion the conservation law is maintained. In general there are a large number of conversion processes proceeding simultaneously and/or sequentially in a bioreactor, therefore, it is important to have a logical way of expressing all of these reactions. The Petersen-matrix approach has been successfully used in developing ASMs and ADM1 models. This approach of model definition is illustrated by defining the model reactions worked out in previous section.

The Petersen matrix for the model with four state variables and associated conversions is as shown in Table 11.2. The defined model includes two unit processes of r_1: biomass growth process and r_2: biomass decay process. In the first unit process, 1 unit of X_S is consumed producing Y and (1-Y) unit of X_B and S_M respectively. In the second unit process, the 1 unit of biomass is lost producing f_U and (1-f_U) units of X_I and X_S, respectively. The coefficients of Y (g-COD·g-COD^{-1}) and f_U (g-COD·g-COD^{-1}) inter-relating the states in each unit process are known as stoichiometric coefficients.

The rate expressions for the reactions are expressed on the right-hand side. Depending on the type of unit process, the rate expression may be a function of more than one state variable and may contain several kinetics coefficients. In the present example, the rate expression for unit process, r_1, is considered to follow Monod-kinetics having two kinetic coefficients of k_m and K_S, while the unit process, r_2, is considered to follow first-order kinetics with only one kinetic coefficient of b.

The principle of conservation can be easily checked in a Petersen matrix by adding the cell entries in a row provided they have the same units. In this example, as all the states have the units of g COD·m^{-3}, entries in all the cells for a row must be added. For instance, in the first row for r_1, 1 unit of X_S is lost from the system while $+Y$ unit of X_B and (1-Y) unit of S_M are produced, result of adding all entries in all the cells -1 + Y + (1-Y) of this row is zero. Thus no COD is lost and no COD is produced and principle of conservation is maintained. In Petersen matrix, it is also easy to estimate the rate of change of each variable by adding the individual reaction rate in each cell of the state variable column. For example, the expression for rate of change of X_S in the system (r_{XS}) can be obtained by adding the reaction rates in vertical column. The corresponding expression for r_{XS} is as shown in Eq. (11.3):

$$r_{XS} = -1 \times r_1 + (1\text{-}f_u) \times r_2 = -1 \times \left\{k_m \left(\frac{X_S}{K_S + X_S}\right) X_B\right\} + (1\text{-}f_u) \times (bX_B) \qquad (11.3)$$

assuming that the system is operated under chemostat mode, the complete reaction for rate of change of X_S can be expressed by the differential equation as shown in Eq. (11.4):

$$\frac{dX_S}{dt} = \left(\frac{Q}{V}\right) X_{S0} + r_{XS}\text{-}\left(\frac{Q}{V}\right) X_S \qquad (11.4)$$

where Q: influent flow (m^3·d^{-1}), V: reactor volume (m^3), and X_{S0}: X_S concentration in influent (g-COD·m^{-3}).

As is evident from above example, there are numerous advantages of using Petersen matrix approach for model definition. The approach is specifically helpful in (1) checking the mass balance and (2) understanding the inter-relationship among various state variables. The ADM1 incorporates 24 state variables (carbohydrates, amino acids, acetate, acetate degraders, *etc.*) and defines 38 unit processes (disintegration, hydrolysis of lipids, uptake of hydrogen, *etc.*) with carefully selected rate expressions (Monod-type, first-order type, and some inhibition factors). Although the Petersen matrix of the model looks complicated due to many state variables and reactions, its concepts are fundamentally similar to that described in the above simple example.

The full implementation of ASMs (Henze *et al.* 2000) and ADM1 models is available in most of the commercialized process simulators. These simulators also provide a user friendly interface to simplify the modeling process and flexibility to extend or modify the existing models. Some of the wastewater process simulators are GPS-X (Hydromantis Inc., Canada), BIOWIN

Table 11.2. Example of mathematical expression using Petersen matrix.

Unit process	X_S	X_I	X_B	S_M	Reaction	Rate expression
r_1	-1		$+Y$	$+(1-Y)$	Growth of X_B and production of S_M from decomposition of X_S	Monod-type: $k_m\left(\frac{X_S}{K_S+X_S}\right)X_B$
r_2	$+(1-f_U)$	$+f_U$	-1		Decay of X_B and production of inert compounds	First order-type: bX_B
Description	Substrate (g-COD·m^{-3})	Inert (g-COD·m^{-3})	Anaerobic microorganism (g-COD·m^{-3})	Methane (g-COD·m^{-3})	Stoichiometric coefficients • Y: Biomass yield (-) • f_U: yield of inert compounds (-) Kinetic coefficients • k_m: maximum specific reaction rate (d^{-1}) • K_S: half-saturation constant (g-COD·m^{-3}) • b: specific decay rate (d^{-1})	

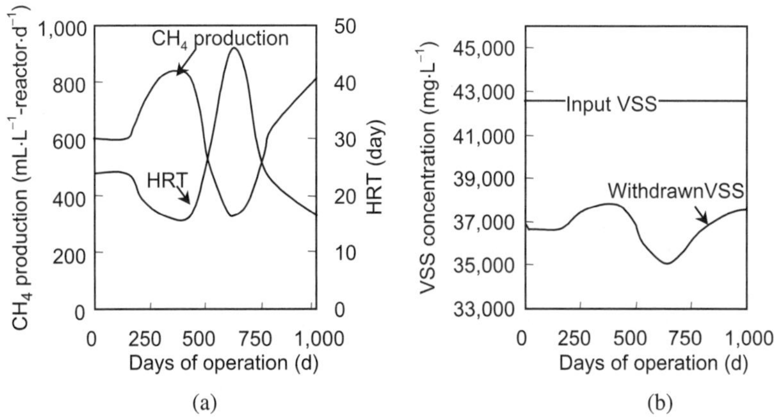

Fig. 11.5. Example of dynamic simulation using ADM1: (a) daily production of CH_4 per reactor volume, and hydraulic retention time of influent; and (b) input VSS: Organic fraction of sludge introduced into the bioreactor, Withdrawn VSS Organic fraction of sludge withdrawn from the bioreactor.

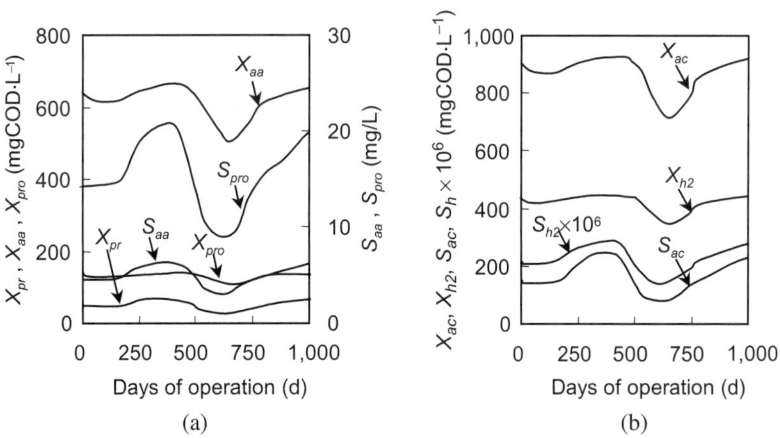

Fig. 11.6. Example of dynamic simulation using ADM1 (cond.): (a) X_{aa} (amno acid degraders) X_{pro} (propionate degraders), X_{pr} (proteins), S_{aa} (amino acids), and (b) S_{pro} (propionate); and (IV) X_{ac} (acetate degraders), X_{h2} (hydrogen degraders), S_{ac} (acetate), and S_{h2} (hydrogen).

(EnviroSim Associates Ltd., Canada), WEST (MOSTforWATER N.V., Belgium), and SIMBA (ifak system, GmbH). In all of these simulation platforms, it is possible to conduct both steady state and dynamic analyses. As an example the dynamic response of some of the state variables of ADM1 with respect to change in influent flow rate using GPS-X modeling platform is as shown in Figs. 11.5 and 11.6.

3. Considerations in Model Application

In above sections, a few important concepts used in formulating mechanistic models were discussed. It is clear that these models normally incorporate state-of-the-art process knowledge built into them and are open for extension depending on the objectives of modeling. If used properly, the models can help in better understanding and finding solution to a given problem, otherwise the model outputs may be misleading and sometimes wrong, if applied without paying attention to underlying assumptions. There are extensive modeling protocols which have been developed with the support of WERF, USA (Melcer *et al.* 2003), HSG, Austria, Switzerland, and Germany (Langergraber *et al.* 2004), STOWA, Netherlands (Roeleveld and van Loosdrecht 2002), BIOMATH, Belgium (Vanrolleghem *et al.* 2003) and JSWA, Japan (Itokawa *et al.* 2003). All of these protocols address different approaches for wastewater characterization, parameter estimation, and model calibration in activated sludge process modeling. Although, it is not the intention in this chapter to discuss the details of these protocols, a few insights are presented for some of the issues central to all of these protocols and applicable to anaerobic digestion modeling. In general, the good modeling practices require that an appropriate wastewater characterization and judicious selection of stoichiometry and kinetic coefficients are made. While, wastewater characterization normally requires experimental studies, the values of model parameters may be determined based on theoretical and/or experimental considerations. In the following sections, some of the theoretical and experimental bases are presented for identification of required model parameters. Finally, the issue of integration of ASMs and ADM which is normally encountered during whole plant modeling is briefly discussed.

3.1. *Stoichiometric and kinetics parameters*

3.1.1. *Theoretical considerations*

In a microbial growth system, correlations have been proposed to estimate the stoichiometry and kinetic parameters of biomass yield (Y) and maximum specific growth rate (μ_{max}), respectively. These correlations are based on the principles of thermodynamics and microbial growth data from a wide variety of microbial growth system. The detailed theory and basis of these correlations are presented in Heijnen *et al.* (1992) and Tijhuis *et al.* (1993). In essence, in a microbial growth system, a certain amount of Gibbs energy is dissipated during biomass formation. The minimum amount of Gibbs energy required to be dissipated per unit of biomass formation depends on the composition of the substrate. Based on a comprehensive dataset of biomass yield constants under aerobic, anoxic, and anaerobic condition with various substrates, the minimum Gibbs energy dissipation needed to synthesize 1 unit of biomass is expressed by the following correlation of Eq. (11.5).

$$\frac{1}{Y_G} = 200 + 18(6\text{-}C)^{1.8} + \exp[\{(3.8\text{-}\gamma)^2\}^{1.6}(3.6 + 0.4C)] \tag{11.5}$$

where Y_G^{-1}: Minimal Gibbs energy to synthesize 1-C mol biomass (kJ·C-mol$^{-1}_{\text{-Biomass}}$), C: number of C atoms of the substrate (-), γ: degree of reduction per C atom of the substrate (mol-e^-·mol^{-1}) (e.g., $\gamma_{Glucose} = 4$, $\gamma_{Methane} = 8$).

As discussed earlier in the section highlighting the principles of conservation, if the ΔG^o for a given reaction is known, all the other stoichiometric coefficients of the reaction can be calculated. For example for the biological reaction of Eq. (11.1), the minimal Gibbs energy ($\Delta G^o = Y_G^{-1}$) to synthesize 1-C mol biomass from Glucose is estimated to be 236 kJ·C-mol$^{-1}_{\text{-Biomass}}$ by using Eq. (11.5). Next, using a set of standard Gibbs energy of formation for Glucose = -917.22, NH_4^+ = -79.37, H_2O = -237.18, H^+ = -39.87, ethanol = -181.75, HCO^{3-} = -586.85, and biomass ($CH_{1.8}O_{0.5}N_{0.2}$) = -67 (Thauer *et al.* 1977) in Gibbs energy conservation equation

Table 11.3. Stoichiometry for production of 1 unit of biomass from ethanol fermentation (adopted from Heijnen and van Dijken 1992).

Item Coefficient	Glucose *a*	NH_4^+ *b*	H_2O *c*	H^+ *d*	Ethanol *e*	HCO_3^- *f*	ΔG^o	ΔH^o
Measured	-1.19	-0.2	-1.63	+2.28	+2.03	+2.08	+255	+118
Predicted	-1.11	-0.2	-1.47	+2.12	+1.87	+1.92	+236	+108

$a{\cdot}C_6H_{12}O_6 + b{\cdot}NH_4^+ + c{\cdot}H_2O \rightarrow d{\cdot}H^+ + e{\cdot}C_2H_6O + f{\cdot}HCO_3^- + 1.CH_{1.8}O_{0.5}N_{0.2} + \Delta H^o + \Delta G^o$

for reaction in Eq. (11.1), unknown stoichiometric coefficients of a-f in Eq. (11.6) can be solved by solving the C, H, N, O, charge and enthalpy (ΔH^o) balance equations simultaneously. In Table 11.3 the calculated coefficients are compared with those measured under anaerobic cultivation of *S. cervisiae*.

$$+a(\text{-}917.22) + b(\text{-}79.37) + c(\text{-}237.18) + d(\text{-}39.87) + e(\text{-}181.75)$$
$$+ f(\text{-}586.85) + 1(\text{-}67) + 236 = 0 \quad (11.6)$$

The importance of the above relationship is that for a given substrate, it is possible to estimate expected theoretical biomass yield constant of Y. In the typical anaerobic digestion process, Y (g-biomass-COD·g-substrate-COD^{-1}) of acidifiers from surplus activated sludge as substrate was calculated to be 0.107 which was in agreement with the observed data (Ekama *et al.* 2007; Yasui *et al.* 2008). A comparison of calculated and measured yield constants for biological reactions with different substrate is shown in Fig. 11.7 (Heijnen 1999).

The principles of thermodynamics may also be used to estimate the maximum specific growth rate (μ_{max}). The μ_{max} in a biological reaction is governed by electron flux in catabolic pathway in the cell (Heijnen and van Dijken 1992; Tijhuis *et al.* 1993; Heijnen 1999). The equation for estimation of μ_{max} is as shown in Eq. (11.7).

$$\mu_{\max} = \frac{[3(\text{-}\Delta G_{CAT})/\gamma\text{-}4.5]}{1/Y_G} \exp\left[\frac{\text{-}69}{R}\left(\frac{1}{T}\text{-}\frac{1}{298}\right)\right] \quad (11.7)$$

where μ_{max}: maximum specific growth rate (h^{-1}), ΔG_{CAT}: Gibbs energy produced from the biochemical reaction (kJ·mol$^{-1}_{\text{-Substrate}}$), R: Gas constant (0.008342 kJ·mol^{-1}·K^{-1}), T: temperature

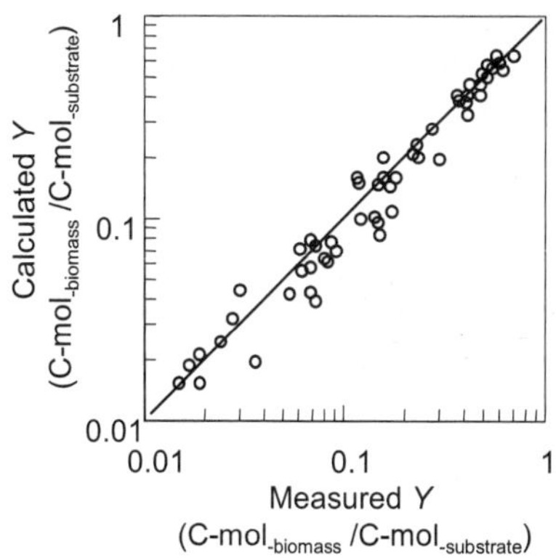

Fig. 11.7. The correlation between the calculated yields and the measured ones.

Table 11.4. Microbial growth properties.

Biochemical reaction	$-\Delta G_{CAT}/\gamma$ (kJ·mol-e^-)	Y_G^{-1} (kJ·C-mol^{-1})	μ_{max}(h^{-1}) Calculated	Measured	Reference
Oxic respiration for acetate	105.5	432	0.7	0.62	*Acinetobacter* sp. (Weon *et al.* 2002)
Methanogenesis from acetate	3.87	432	0.015	0.08	*Methanosarcina* sp. (Yang and Okos 2004)
Ethanol fermentation from glucose	9.39	236	0.10	0.5	*S.cervisiae* (McCarty 1971; Andersen and Meyenburg 1980)
Nitrification from ammonium	45.8	3,500	0.04	0.027	*Nitrosomonas* sp. (Vadivelua *et al.* 2006)

(K), 3: rate constant of electron flux in the cell (mol-e^-·C-mol$^{-1}_{\text{-Biomass}}$), γ: degree of reduction for C of the substrate (mol-e^-·mol$^{-1}_{\text{-Substrate}}$), 4.5: required electron to maintain 1-C mol biomass (kJ·C-mol$^{-1}_{\text{-Biomass}}$), and 69: Arrhenius constant (kJ·mol^{-1}).

The theoretical and measured values of μ_{max} for various biochemical reactions are compared and listed in Table 11.4. The calculated specific growth rates seem to compare reasonably well with the measured values. The discrepancy observed in the rates for fermentation and methanogenesis is suggested to be due to the difference in biochemical pathways used in these reactions (Heijnen 1999).

3.1.2. *Experimental methods*

In combination with theoretical methods, experiments may also be conducted to get information regarding model parameters. In the application of activated sludge models, the use of oxygen uptake rate (OUR), nitrate utilization rate (NUR), phosphorous release rate (PRR), and other batch experiments is very common to estimate model parameters. The OUR tests are also very useful tool for wastewater characterization. More information about these tests can be found elsewhere (Kappeler and Gujer 1992; Young *et al.* 1991).

In the application of anaerobic digestion model, the batch tests measuring methane production rate (MPR) under controlled conditions have been quite useful for wastewater characterization

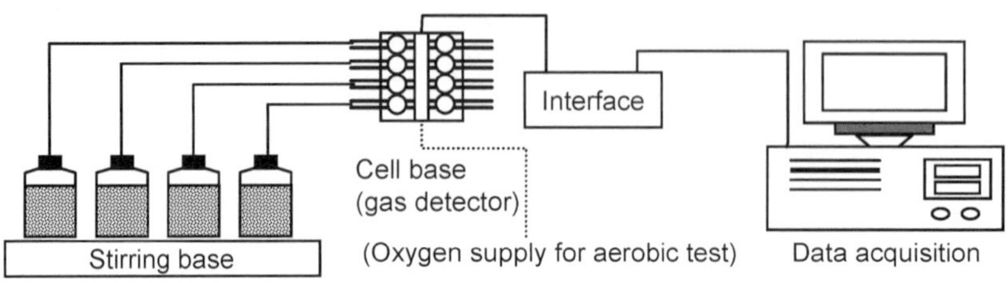

Fig. 11.8. Illustration of schematic diagram of the respirometry test apparatus and the major components.

and model parameter estimation. Figure 11.8 illustrates a typical diagram of a commercialized test apparatus (Respirometer, Challenge Technology Inc., USA) which can be used for conducting both OUR and MPR experiments. In this apparatus, several sets of batch test can be conducted simultaneously. The test apparatus consists of four components (Young *et al.* 1991; Yasui *et al.* 2006). The components are (1) a gas-tight incubation vessel on the magnetic stirring base for mixing sample to provide contact between microorganism and substrate, (2) a cell device measuring methane gas production (or oxygen gas consumption in case of aerobic test), (3) an interface module to convert gas production data (or oxygen uptake) to digital form, and (4) a computer for data acquisition. Temperature of the incubation vessel and the cell base is maintained ($\pm 0.2\,^{\circ}$C) in a temperature-controlled incubator. A small scrubber consisting of a caustic material is set between the incubation vessel and the cell base to absorb CO_2 from the headspace gas. The data regarding methane production is logged at regular interval in the computer. The experimental curve for methane production rate produced in these batch tests can be analyzed for wastewater characterization and estimating the expected values of kinetic and stoichiometric parameters.

To apply the simple model presented in Section 2.5, we are required to know the concentration of each state variable in the influent and the value of the model stoichiometry and kinetic parameters. Depending on the complexity of the model, several batch experiments may be necessary to obtain this information. It is also necessary to point out at this stage, that it is not always necessary to determine all of the model parameters. For example, the values of stoichiometry coefficients like Y and f_u are fairly well determined for biological reactions and, unless there is strong evidence, the default values of these parameters could be used without significantly compromising the model results. With this understanding, an approach to determine the influent state variable concentration using a methane production test is demonstrated.

In this analysis, two data sets from two batch experiments are used. The first experiment serving as control is conducted with only seed sludge (control), while the second test is conducted with seed sludge and surplus activated sludge (test). Net methane production can be obtained by subtracting the methane production of control from the test experiment. The typical output showing the net cumulative methane production test is as shown in Fig. 11.9. As the test progresses over time, the methane production reduces and cumulative methane production almost saturates. Assuming that the net cumulative methane production (P_M, mg-COD·L^{-1}) approximates the

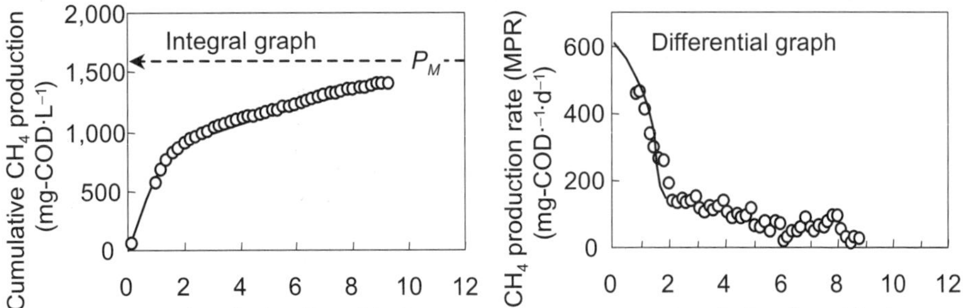

Fig. 11.9. Two kinds of graphs from batch test for anaerobic digestion of activated sludge (left: integral curve; right: differential curve), subtracted methane production from the seed sludge.

ultimate methane production, following relations can be formulated.

$$P_M = (1\text{-}Y)\cdot X_S + (1\text{-}f_U)\cdot Y\cdot X_S \tag{11.8}$$

$$X_S = \frac{P_M}{(1\text{-}Y\cdot f_U)} \tag{11.9}$$

$$X_I = X_{OS}\text{-}X_S \tag{11.10}$$

The first term $(1\text{-}Y)\cdot X_S$ in Eq. (11.8) accounts for the methane generation due to degradation of biodegradable solid (X_S). The second term $(1\text{-}f_U)\cdot Y\cdot X_S$ signifies the methane generation that would result if all the biomass generated due to degradation of X_S is decayed in the batch test. Equation (11.9) shows the final expression for estimating the biodegradable solid (X_S) in the influent. Once concentration of X_S is determined, X_I can be estimated by Eq. (11.10) in which X_{OS} is the total measured COD concentration of organic solids. In the above scheme of classification, it is assumed that the state variables of X_B and S_M are almost negligible in the influent. It is clear that the method involves several assumptions and accurate estimation of the states is not possible. However, depending on the accuracy requirement, more experiments may be conducted for improved accuracy. Nevertheless, for most of the practical application, the characterization method is able to provide reasonable information for model application.

To evaluate the process kinetic parameters, methane production rate data is useful. A typical methane production rate (MPR) curve from a batch test in which activated sludge is used as substrate is as shown in Fig. 11.10 (the right graph in Fig. 11.9). The MPR curves are useful as it is easy to visualize the change in process kinetics over time. Moreover, the analytical methods which are developed for analysis of oxygen uptake rate (OUR) curves for aerobic system could be used to analysis MPR curves (Kappeler and Gujer 1992; Yasui *et al.* 2006).

Using the analogy from OUR curve analysis, the MPR curve may be used to extract information about the concentration of individual state variables. In an MPR curve, an inflection point is normally associated with the depletion of a specific substrate. In Fig. 11.10, the inflection point, marked as "•", points to the depletion of biodegradable substrate (X_S) added to the batch test. The methane production rate after the inflection point may be attributed to the decay of biomass (X_B) in the activated sludge sample. Based on this identification, the approximate COD concentrations of the individual state variables could then be calculated from the area of the corresponding region (*A* and *B*) and a knowledge of yield constant (*Y*). It should be noted that the dimension of the area is mg-COD·L^{-1} (=mg-COD·L^{-1}·d^{-1} × d) and mathematically it is equivalent to 'substrate concentration ×(1-*Y*). The estimation of state variable concentration

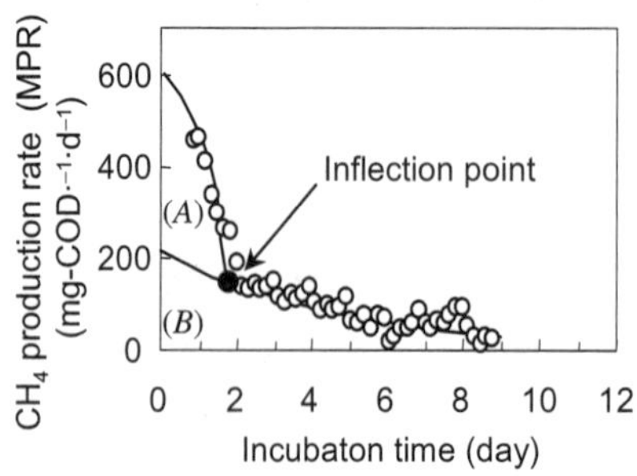

Step (i) Identification of state variables (*A and B*) (by visual inspection)
Step (ii) Concentarion of the individual state variable (from area of the corresponding region)
Step (iii) Rate expression (from shape of the curve)

Fig. 11.10. Procedure to identify state variables and estimation of the rate expressions from MPR graph.

from this approach is similar to that described using the cumulative methane production (the left graph in Fig. 11.9) (Angelidaki *et al.* 2009).

The MPR curve is also helpful in determining the suitability of a selected kinetic expression and estimate the kinetic parameters of the model processes. It is important to point out here that due to the dependence of kinetic expression on the non-measurable state variables, it is sometimes not possible to uniquely identify the value of a kinetic parameter. For example, for a proposed reaction kinetics of $k_m(X_S/K_S + X_S)X_B$, even if a test is conducted at such an X_S concentration that the $(X_S/K_S + X_S) \cong 1$, we can only determine the composite value of $k_m{\cdot}X_B$. Unless the value of X_B in the reactor is determined, the value of kinetic parameter k_m cannot be uniquely determined. Based on this understanding, it is sometimes required to conduct several batch tests under such conditions that maximum information regarding the model parameters is extracted. The designing of such experiments is discussed elsewhere (Dochain *et al.* 1995; Vanrolleghem and Daele 1995). For typical applications, it is not always necessary to estimate kinetic parameters as the default values of parameter provided in ASMs and ADM1 are reasonable estimates originating from the condensation of large amount of research data. However, for definition of new processes, it is still necessary to design an appropriate experimental approach to determine kinetic expression and its parameters. After the preliminary investigation about the wastewater characterization and process kinetics, the model can be tested on a full-scale plant and minor adjustments in the parameter values may be performed during model calibration.

3.2. *Integration of ASM and ADM1 models*

In model applications to municipal wastewater treatments, it is normally required to model the whole plant consisting of activated sludge and anaerobic digestion processes. One of the important issues which arises during these applications is the mapping of the state variable of one model to the other. The proposed mapping of ASMs and ADM1 state variables is as shown in Fig. 11.11 (Batstone *et al.* 2002).

More elaborate state variable mapping schemes have been recently proposed (Nopens *et al.* 2009). The advantages and disadvantages of this approach of state variable mapping from one biological model to another have been discussed (Volke *et al.* 2006, Grau *et al.* 2009). The approaches of state variable mapping are sometimes confusing and in general result in loss of information. Moreover, the approach does not provide a systematic platform for a generalized

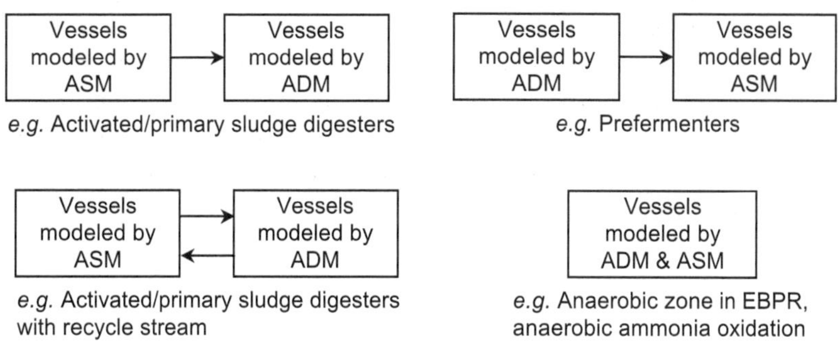

Fig. 11.11. Scenarios for integration of the ASM and ADM showing outputs from one model being used as inputs to the other model (EBPR: enhanced biological phosphorus removal) (Batstone *et al.* 2002).

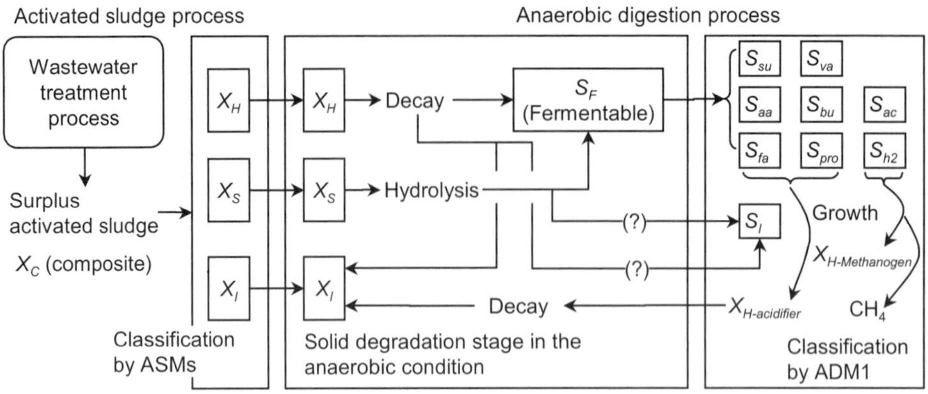

Fig. 11.12. A direct mapping between ASMs and ADM1.

model development. Based on these considerations an alternative approach in which a uniform set of state variables is used to define all of the biological reactions in a wastewater treatment plant is being proposed (Yasui *et al.* 2006; Grau *et al.* 2009). This approach does not require mapping interfaces between the individual biological models, however it results in a 'super model' with a large set of states (Jones and Takács 2004). To develop such a model, it is important to redefine the processes in ADM1 model using the state variables of ASMs and vice versa.

To elucidate this approach, Yasui *et al.* (2006) explained the experimental results of anaerobic digestion process by defining the processes using the state variables of ASMs. The proposed modified anaerobic digestion model structure for organic matter degradation in this study is as shown in Fig. 11.12. In this model structure, the ASM state variables of X_I, X_H (heterotrophs), and X_S are directly mapped to the anaerobic digester. The state variables of X_H and X_S are assumed to be decomposed under anaerobic conditions with the rate expressions similar to that used in the ASMs albeit with different kinetic parameters. For active biomass X_H, it is assumed that the degradation is like a self-disintegration of X_H under anaerobic conditions and may be termed as 'anaerobic decay'. Compatible to the decay mechanism in ASMs, part of X_H produce inert solids X_I, while rest is recycled back to the substrate pool. For biodegradable substrate, X_S, the degradation is governed by hydrolysis by the active biomass in the anaerobic digestion process, while small amount of soluble inert fraction (S_I) may be produced here.

As shown in Fig. 11.12, the biodegradable solids (X_S) are biologically converted into fermentable soluble compounds (S_F). This S_F is further metabolized to lower molecular weight compounds such as sugars, amino acids, and lipids. These low molecular weight compounds are fermented to acetate and hydrogen, which in turn are used to produce the end product of methane. This approach of model building is effective, as the underlying biological reactions can be expressed using a uniform set of state variables and maximum information is available at every point of the biological reactor.

References

Angelidaki, I., Alves, M., Bolzonella, D., Borzacconi, L., Campos, J. L., Guwy, A. J., Kalyuzhnyi, S., Jenicek, P., and van Lier, J. B. (2009) Defining the biomethane potential (BMP) of solid organic wastes and energy crops: A proposed protocol for batch assays. *Wat. Sci. Technol.* **59**(5), 927–934.

Andersen, K. B. and Meyenburg, von K. (1980) Are growth rates *of Escherichia coli* in batch cultures limited by respiration? *J. Bacteriol.* **144**(1), 114–123.

Batstone, D. J., Keller, J., Angelidaki, I., Kalyuzhnyi, S. V., Pavlostathis, S. G., Rozzi, A., Sanders, W. T. M., Siegrist, H., and Vavilin, V. A. (2002) *Anaerobic Digestion Model No.1.* (*ADM1*). IWA Scientific and Technical report No.13, *IWA*, ISBN: 1 900222 78 7.

Buswell, A. M. (1939) *Anaerobic Fermentations. State of Illinois Bull.* 32, 1–193, Dept. of Registration and Education, Div. of the State Water Survey, Urbana, IL, USA.

Carstensen, J., Vanrolleghem, P., Wolfgang, R., and Reichert, P. (1997) Terminology and methodology in modeling for water quality management — A discussion starter. *Wat. Sci. Technol.* **36**(5), 157–168.

Dochain, D., Vanrolleghem, P. A., and Daele, M. Van (1995) Structural identifiability of biokinetic models of activated sludge respiration. *Wat. Res.* **29**(11), 2571–2578.

Ekama, G. A., Sötemann, S. W., and Wentzel, M. C. (2007) Biodegradability of activated sludge organics under anaerobic conditions. *Wat. Res.* **41**, 244–252.

Grau, P., Copp, J., Vanrolleghem, P. A., Takács, I., and Ayes, E. (2009) A comparative analysis of different approaches for integrated WWTP modeling. *Wat. Sci. Technol.* **59**(1), 141–147.

Heijnen, J. J. and van Dijken, J. P. (1992) In search on a thermodynamic description of biomass yields for the chemotrophic growth of microorganisms. *Biotechnol. Bioeng.* **39**, 833–858.

Heijnen, J. J. (1999) *Encyclopedia of Bioprocess Technology: Fermentation, Biocatalysis and Bioseparation*, Flickinger, M. C. and Drew, S. W. (eds.), New York: Wiley, ISBN: 0471138223.

Henze, M., Gujer, W., Mino, T., and Loosdrecht, M. C. M. (2000) *Activated Sludge Models ASM1, ASM2, ASM2D, ASM3.* IWA Scientific and Technical report No. 9, *IWA*, ISBN: 1900222248.

Hoover, S. R., and Porgens, N. (1952) Assimilation of dairy wastes by activated sludge. *Sew. Indus. Wastes J.* **24**, 306–312.

Itokawa, H., Hashimoto, T., and Murakami, T. (2003) Characterization of municipal wastewater organic components for activated sludge process modeling. *Proceedings of Environmental Engineering Research* **40**, 41–52. (*in Japanese*).

Jones, R. M. and Takács, I. (2004) Importance of anaerobic digestion modeling on predicting the overall performance of waste-water treatment plants. *Proceedings of Anaerobic Digestion 2004*, 10th World Congress, Montreal, Canada, 29 Aug–2 Sep, 2004, pp. 1371–1375.

Kappeler, J. and Gujer, W. (1992) Estimation of kinetic parameters of heterotorophic biomass under aerobic conditions and characterization of wastewater for activated sludge modeling. *Wat. Sci. Technol.* **25**(6), 125–139.

Langergraber, G., Rieger, L., Winkler, S., Alex, J., Wiese, J., Owerdieck, C., Ahnert, M., Simon, J., and Maurer, M. (2004) A guideline for simulation studies of wastewater treatment plants. *Wat. Sci. Technol.* **50**(7), 131–138.

McCarty, P. L. (1971). *Organic Compounds in Aquatic Environments.* Faust, S. D. and Funter, J. V. (eds.), New York: Marcel Dekker, pp. 495–531.

Melcer, H., Dold, P., Jones, R. M., Bye, C. M., Takacs, I., Stensel, H. D., Wilson, A. W., Sun, P., and Bury, S. (2003) *Methods for Wastewater Characterization in Activated Sludge Modeling*, Report No. 99-WWF-3. *WERF*.

Nopens, I., Batstone, D. J., Copp, J. B., Jeppsson, U., Volcke, E., Alex, J., and Vanrolleghem, P. A. (2009) An ASM/ADM model interface for dynamic plant-wide simulation. *Wat. Res.* **43**(7), 1913–1923.

Roels, J. A. (1983) *Energetics and Kinetics in Biotechnology*, Elsevier, New York, ISBN: 0444804420.

Roeleveld, P. J. and van Loosdrecht, M. C. M. (2002) Experience with guidelines for wastewater characterization on the Netherlands. *Wat. Sci. Technol.* **45**(6), 77–87.

Thauer, R. K., Jungermann, K., and Decker, K. (1977) Energy conservation in chemotrophic anaerobic bacteria. *Bacteriol. Rev.* **41**, 100–180.

Tijhuis, L., van Loosdrecht, M. C. M., and Heijnen, J. J. (1993) A thermodynamically based correlation for maintenance Gibbs energy requirements in aerobic and anaerobic chemotrophic growth. *Biotechnol. Bioeng.* **42**, 509–519.

Vadivelua, V. M., Kellera, J., and Zhiguo, Yuan, Z. (2006) Stoichiometric and kinetic characterisation of *Nitrosomonas* sp. in mixed culture by decoupling the growth and energy generation processes. *J. Biotechnol.* **26**(3), 342–356.

Vanrolleghem, P. A., Insel, G., Gujer, B., Sin, G., De Pauw, D., Nopens, I., Doverman, H., Weijers, S., and Gernaey, K. (2003) A comprehensive model calibration procedure for activated sludge models, *Proceedings of the 76th Annual Water Environment Federation Technical Exposition and Conference*, Los Angeles, California, 11/Oct–15/Oct, WEF.

Vanrolleghem, P. A. and Daele, M. Van (1995) Practical identifiability of a biokinetic model of activated sludge respiration. *Wat. Res.* **29**(11), 2561–2570.

Volke, E. I. P., van Loosdrecht, M. C. M., and Vanrolleghem, P. A. (2006) Continuity-based model interfacing for plant-wide simulation: A general approach. *Wat. Res.* **40**, 2817–2828.

Weon, S.-Y., Lee, C.-W., Lee, S.-I., and Koopman, B. (2002) Nitrite inhibition of aerobic growth of *Acinetobacter* sp, *Wat. Res.* **36**(18), 4471–4476.

Yang, S. T. and Okos, M. R. (2004) Kinetic study and mathematical modeling of methanogenesis of acetate using pure cultures of methanogens. *Biotechnol. Bioeng*. **30**(5), 661–667.

Yasui, H., Sugimoto, M., Komatsu, K., Goel, R., Li, Y. Y., and Noike, T. (2006) An approach for substrate mapping between ASM and ADM1 for sludge digestion. *Wat. Sci. Technol.* **54**, 83–92.

Yasui, H., Komatsu, K., Goel, R., Li, Y. Y., and Noike, T. (2008) Evaluation of state variable interface between ASM and ADM1. *Wat. Sci. Technol.* **57**(6), 901–907.

Young, J. C., Kuss, M. L., and Nelson, M. A. (1991) Use of anaerobic resipirometers for measuring gas production in toxicity and treatability tests. In: *Proceedings of the 84th Annual Meeting of Air and Waste Management Association*, AWMA.

Chapter 12

Anaerobic Digestion of Lignocellulosic Wastes by Rumen Microorganisms: Chemical and Kinetic Analyses

Zhen-Hu Hu[1,2], Zheng-Bo Yue[1], Shao-Yang Liu[1], Guo-Ping Sheng[1] and Han-Qing Yu[1,*]

[1]*Laboratory of Environmental Engineering, Department of Chemistry, University of Science & Technology of China, Hefei 230026, China*

[2]*School of Civil Engineering, Hefei University of Technology, Hefei 230009, China*

**hqyu@ustc.edu.cn*

A huge volume of lignocellulosic wastes is currently produced worldwide. Anaerobic digestion is an effective disposal method for these wastes because of its low cost and environmental benefits. The utilization of the produced methane-rich biogas can reduce the dependence of energy on fossil fuels and reduce the emission of greenhouse gas. However, the recalcitrant structure of lignocelluloses in these wastes is the barrier for the hydrolysis of lignocelluloses by microorganisms. Many microbial resources have been used for improving the hydrolysis, among which, rumen microorganisms have shown some advantages because of their high cellulolytic activities. The characteristics and structure changes of lignocelluloses in biological conversion have been investigated with chemical and imaging analysis to determine which characteristics are responsible for limiting the microbial hydrolysis and which changes benefit for the bioconversion. This chapter describes the recent results of chemical analysis using atomic force microscopy (AFM), Fourier transform infrared (FTIR), X-ray photoelectron spectroscopy (XPS), GC/MS, and X-ray diffraction (XRD) on the biological conversion of lignocellulosic wastes, and especially on the anaerobic digestion of lignocellulosic wastes by rumen microorganisms, and kinetic analysis of such a bioconversion.

1. Introduction

Lignocellulosic wastes refer to the wastes containing high percentage of plant biomass, including agricultural wastes, forestry wastes, solid municipal wastes, and food industry residues, such as straw, stover, peelings, cobs, and wood residues (Adney *et al.* 1989; Gea *et al.* 2005; Kumar *et al.* 2009; Mtui 2009; Rani *et al.* 2008; Songulashvili *et al.* 2007; Zhang and He 2006). Each year a huge volume of lignocellulosic wastes is produced worldwide. In China, the annual production of agricultural wastes alone is over 6.0×10^8 tons.

Due to the concern to climate change caused by the greenhouse gas emission from the burning of fossil fuel, the use of renewable resources is greatly encouraged today (Mtui 2009; Rani *et al.* 2008). The re-utilization of lignocellulosic wastes to produce energy has thus attracted great interest, because of its large quantity and low cost (Ragauskas *et al.* 2006; Service 2007), including producing cellulosic ethanol by fermentation, gasification by pyrolysis (Babu 2008; Demirbas and Balat 2007; Mohan *et al.* 2006), and producing methane-rich biogas by anaerobic digestion (Barnes and Keller 2004; Hu and Yu 2006; Rivard *et al.* 1990). Currently, the production of cellulosic ethanol and gasification are still too expensive to realize a large-scale application. On the other hand, biogas production by anaerobic digestion has proved to be a good option because of its low cost and many benefits. For example, the recovered biogas can be used for heating and generating electricity (Espinosa-Solares *et al.* 2009), which will reduce the dependence of energy on fossil fuels and reduce the greenhouse gas emission. Therefore, anaerobic digestion is a sustainable and environmentally friendly technology.

Generally, the anaerobic digestion of lignocellulosic wastes includes the following main steps: (1) physical disruption of the recalcitrant structure of lignocellulosic wastes, exposing cellulose and hemicelluloses to improve the attachment of bacteria; (2) hydrolysis of cellulose and hemicelluloses, which is the rate-limiting step for the biological conversion of lignocellulosic wastes. During the hydrolysis, cellulose and hemicellulose are hydrolyzed into polysaccharide and monosaccharide by cellulolytic microorganisms; (3) acidification of the hydrolyzed saccharides. The sugars produced are very easy to be fermented into volatile fatty acids, such as acetic, propionic, and butyric acids, in the anaerobic digestion system; and (4) methane production from the volatile fatty acids. The produced volatile fatty acids will be consumed by methanogens to produce biogas, which can be recovered for heating and generating electricity (Hu *et al.* 2004; Russell and Rychlik 2001).

The recalcitrant structure of lignocelluloses is the barrier for hydrolysis (Himmel *et al.* 2007). Many technologies have been developed to improve the process. Among them, the most efficient way is to break down the recalcitrant structure by physical or physiochemical means (Lissens *et al.* 2004; Yue *et al.* 2008), but the cost of such pretreatment remains high for biogas production. Biogas production from anaerobic digestion of lignocellulosic wastes may also be improved by optimizing the operation parameters such as increasing retention time (Alvarez and Liden 2008; Hu *et al.* 2007) and acclimating microorganisms with high cellulolytic activity (Barnes and Keller 2003, 2004; Hu and Yu 2005, 2006). Various microbial resources, such as anaerobic sludge, rumen, and other microorganisms, have been applied to degrade lignocellulosic wastes (Buffiere *et al.* 2005; Nair *et al.* 2005; Qu *et al.* 2009). Results showed that rumen microorganisms have advantages over other microbial resources (Barnes and Keller 2003; Hu and Yu 2006). The mixed rumen cultures, predominately composed of bacteria plus fungi and protozoa, have abundant enzymatic components and high cellulolytic activities for the degradation of lignocellulosic wastes (Russell and Rychlik 2001).

Hydrolysis of lignocellulosic wastes by microorganisms mainly occurred on the surface of biomass (Zhang and Lynd 2004). Microorganisms firstly attach to the surface of fiber, and then the extracellular cellulases disrupt the cellulosic structure (Rogers *et al.* 2007). Many chemical and kinetic analyses have been conducted to elucidate the degradation mechanisms. This chapter reviews the composition and chemical analysis of lignocelluloses, and introduces the chemical and kinetic analyses of anaerobic digestion of lignocellulosic wastes by rumen microorganisms.

2. Composition of Lignocelluloses

Lignocellulose forms the cell wall of plants and consists of three major components: cellulose, hemicelluloses, and lignin. Typically, lignocelluloses contain 35–60% of cellulose, 20–35% of hemicelluloses, and 5–30% of lignin (Lynd *et al.* 2002; Zhang and Lynd 2004). Cellulose forms bundles of fibrils and is the skeleton of the plants, which provide the strength for the plant, whereas

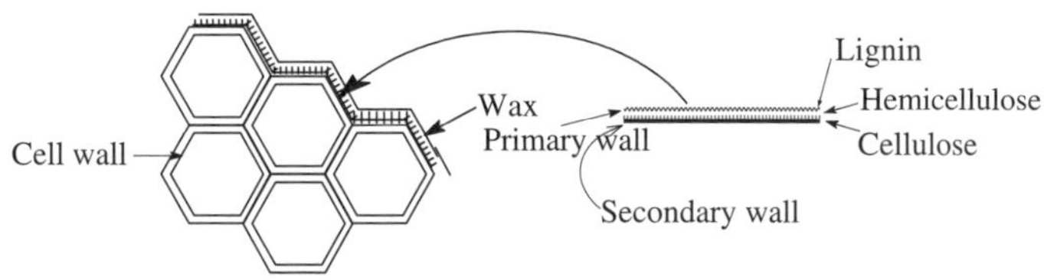

Fig. 12.1. Schematic diagram of cell structure of plant.

hemicelluloses and lignin are embedded in and bound the matrix of cellulose. Figure 12.1 shows the schematic diagram of plant cell wall.

The cell wall of plant consists of primary and secondary walls, of which the primary wall consists of randomly ranged microfibrils and the gaps between microfibrils are packed with, and linked by, pectin, hemicelluloses, and lignin. The secondary wall composes of regularly ranged microfibrils. Each cellulosic microfibril is bundled of 60–80 cellulosic molecules rigidly held together through intra- and intermolecular hydrogen bonds to form crystalline and amorphous segments. There is no obvious boundary between the crystalline and amorphous segments.

Cellulose is a linear polymer consisting of D-anhydroglucopyranose, which is the repeating unit of cellulose, joined together by β-(1–4) glycosidic bonds (Zhang and Lynd 2004). The chemical formulation of cellulose is $(C_6H_{10}O5)_n$, which contains 44.44% of carbon, 6.17% of hydrogen, and 49.39% of oxygen. The content of cellulose varies with the growth stage of plant. For a young plant, cell wall contains only 1–10% of cellulose, whereas for a mature or lignified plant, it can reach 50%.

Hemicellulose forms linkages between the cellulosic bundles and between cellulose and lignin. It consists of shorter, branched polymer chains of various C6 and C5 sugars. The C6 sugars consist mainly of glucose together with mannose and galactose. The C5 sugars consist of xylose and arabinose (Lange 2007). The steric hindrance caused by the side chain and the axial hydroxyl groups of the various sugars makes hemicelluloses to form amorphous domains.

Lignin is a three-dimensional polymer (3D) of propyl-phenol that is imbedded in and bound to cellulose. Lignin is very inert and provides rigidity to the plant structure. The phenol groups of lignin provide antibacterial activity that protects lignocelluloses against microorganisms (Lange 2007). The lignified cell wall is very difficult to be degraded by microorganisms. Hence, the disruption of lignin is the key step for the anaerobic digestion of lignocellulosic wastes.

3. Chemical Analyses

3.1. *AFM analysis*

Atomic Force Microscopy (AFM) is a powerful tool to provide topographical information of surface structures with dimensions of 10–100 nm (Koljonen *et al.* 2001). The changes in morphological characteristics of botanical surface during biodegradation could also be observed using scanning electron microscopy (SEM) and transmission electron microscopy (TEM). Compared to SEM and TEM, AFM provides a possibility to directly observe the micro- and nanostructure of fibrils without complex sample preparation, which makes it possible to obtain high-resolution images of surface microstructure, and of interactions on the surface between the microorganisms and the substrate, or between enzymes and the substrate.

Studies of microstructure of plant cell wall provided useful information for the further utilization of cellulosic materials. Yan *et al.* (2004) studied the nanostructure of wheat straw by AFM and observed that the outside surface of the straw was covered by compact layers of wax and a network structure of cellulose and hemicelluloses. The lignin grains, localized on the surface of the network, are about 30–100 nm in diameters and located at the boundary of the primary and secondary wallsz. Because of its advantage of non-destructive spectroscopic imaging, AFM is also a powerful tool to reveal the action mechanisms of cellulosic hydrolysis by cellulase. Lee *et al.* (2000) used AFM to observe the interaction between cotton fiber and two different cellulases, i.e., exoglucanase, CBH I and an endoglucanase, EG II, at high magnifications. The tracks of the action revealed that the action mechanisms of the two cellulases on cotton fibers were different and the hydrolysis needs the synergy of the two enzymes. With increasing time, the roughness and holes on the surface of cotton fiber was increased (Mu *et al.* 2006b).

The adhesion of bacteria on the substrate surface is a very key step for the degradation of solid substrate and the formation of biofilm. AFM can measure the interaction between the substrate and the bacteria surface, and provide a possibility to resolve the formation mechanism of biofilm. Bacterial surface contains proteins, polysaccharides, and other components, which will affect the adhesion process. Xu and Logan (2005, 2006) measured interaction forces between colloids and protein-coated surfaces using AFM. The properties of the surface and the environmental factors such as ionic strength significantly affected the adhesion forces (Zhu *et al.* 2005). Abu-Lail and Camesano (2003) found lipopolysaccharides on the surface of *Escherichia coli* JM109 that increased the interaction between the bacterial walls and the AFM tips. The forces between the AFM tip and bacterial polymers were correlated with bacterial attachment and retention.

3.2. *FTIR analysis*

Cellulose, hemicelluloses, and lignin are also composed of functional groups of alkanes, esters, aromatics, ketones, and alcohols (Moran *et al.* 2008). Each functional group has a characteristic absorbance band at Fourier Transform Infrared (FTIR) spectrum. The absorbance characteristics of various functional groups have been extensively used for identifying the structural changes of lignocelluloses in the pulping, pretreatment, or preparation, of lignocellulosic materials (Alvarez and Vazquez 2006; Moran *et al.* 2008; Oh *et al.* 2005).

The characteristic FTIR bands of cellulose, hemicellulose, and lignin have been well documented (Moran *et al.* 2008; Sun *et al.* 2004). For example, the absorption at 3,423 cm^{-1} is attributed to the stretching of O–H groups and that of 2,890 cm^{-1} is attributed to the C–H stretching. The absorption at 1,640 cm^{-1} is due to the bending mode of the absorbed water. The spectra at 1,440 cm^{-1} relates to the H–C–H bending and at 1,387 cm^{-1} relates to O–H bending. The absorbance at 1,328 cm^{-1} is originated from the absorption of the bands of C–C and C–O. Table 12.1 summarizes the characteristic FTIR bands of cellulose, hemicelluloses, and lignin (Moran *et al.* 2008).

FTIR has been applied to analyze the change in components and structure of lignocelluloses. Schultz *et al.* (2002) developed a method to predict the composition of lignocelluloses by FTIR, and got a very good fit between the absorbance and the content of lignin, glucose, and xylose in lignocellulose. Cuetos *et al.* (2009) co-digested poultry blood and municipal solid wastes, and the FTIR analysis proved that a reduction in volatile content and aliphatic structures of biowastes along with an increase in aromaticity during the digestion process. Marcato *et al.* (2009) used

Table 12.1. Absorption bands for functional groups of cellulose, hemicelluloses, and lignin (Moran *et al.* 2008).

Fiber component	Wavenumber (cm^{-1})	Functional group	Compound
Cellulose	4,000–2,995	OH	Acid, methanol
	2,890	H–C–H	Alkyl, aliphatic
	1,640	Fiber-OH	Adsorbed water
	1,270–1,280	C–O–C	Aryl-alkyl ether
	1,170–1,082	C–O–C	Pyranose ring skeletal
	1,048	OH	C–OH
Hemicellulose	4,000–2,995	OH	Acid, methanol
	2,890	H–C–H	Alkyl, aliphatic
	1,765–1,715	C=O	Ketone and carbonyl
	1,048	OH	C–OH
Lignin	4,000–2,995	OH	Acid, methanol
	2,890	H–C–H	Alkyl, aliphatic
	1,730–1,700, 1,632	C=C	Aromatic, Benzene stretching ring
	1,613, 1,450	C=C	Aromatic skeletal vibrations
	1,430	O–CH_3	Methoxyl-O–CH_3
	1,270–1,232	C–O–C	Aryl-alkyl ether
	1,158	C–O	Phenol
	1,048	OH	C–OH
	700–900	C–H	Aromatic hydrogen

FTIR to analyze the structural changes of organic matter in pig wastes during anaerobic digestion and found some changes of chemical structures of organic matter with a reduction in the aliphatic chain, lipid and polysaccharide levels. The FTIR analysis on the anaerobic digestion of *Spartina alterniflora* showed that co-digestion with potato improved the degradation of hemicellulose in *S. alterniflora*, but had little effect on degradation of lignin (Yang *et al.* 2009). FTIR was also used for measuring the structural changes of *Achras zapota* lignocellulose after solid fermentation by *Phanerochaete chrysosporium*; the results showed that lignin was very difficult to be converted (Ganesh Kumar *et al.* 2006).

3.3. *XPS analysis*

X-ray photoelectron spectroscopy (XPS) yields information on the elemental and chemical compositions of solid materials surface and is mainly applied for characterizing the surface property (Koljonen *et al.* 2001). Dorris and Gray (1978a,b) first applied XPS to analyze pulp fibers and found that the elemental surface composition and the bulk composition of both lignin and cellulose showed good agreement and the deconvolution of C1s in cellulose surfaces provided reasonable accordance with the oxygen-to-carbon ratios. Since then XPS has been widely used to evaluate the lignin of surface and extractives contents of fiber (Haensel *et al.* 2009; Koljonen *et al.* 2001) and to assess structural changes in lignocellulosic surfaces after special treatments (Inari *et al.* 2006; Ostenson *et al.* 2006).

The relative amounts of differently bound carbons can be determined from the high-resolution C1s spectra, using symmetric Gaussians (Koljonen *et al.* 2001). The overlapping peaks in the C1s region were resolved into their individual components by using a curve-fitting program (XPSPEAK v4.0), a Tougaard background and symmetric Gaussian curves (Hu *et al.* 2008). The main peak was assigned at 284.8 eV to aromatic and aliphatic carbons C–(C, H), referred to as C1.

The chemical shifts relative to C–C (C1) or C–O (C2), O–C–O or C=O (C3) and O=C–O (C4) are 1.7 ± 0.1, 3.1 ± 0.1 and 4.4 ± 0.2 eV, respectively. The constant full-width-half-maximum (FWHM) values were used in the curve fitting of each spectrum (Gustafsson *et al.* 2003; Lee *et al.* 1999).

3.4. *GC/MS analysis*

Lignin is the main barrier to impede the hydrolysis of lignocelluloses. The breakdown of lignin into low molecular weight polymer will improve the hydrolysis of lignocelluloses, but it was very controversial for whether lignin can be degraded by bacteria. Kajikawa *et al.* (2000) certified that the lignin model compound, 4-methylumbelliferone veratryl ether dimer, was completely degraded within 8 h in the presence of fungicidal antibiotics by ruminal bacteria.

The products of anaerobic digestion of lignocellulosic wastes, such as volatile fatty acids and biogas, can be quantitatively analyzed by gas chromatography. However, for determining whether lignin is degraded by rumen microorganisms or other microbes, the detection of lignin cleavage is necessary. GC/MS can analyze the low molecular products originated from the cleavage of lignin (Fu *et al.* 2002).

3.5. *XRD analysis*

Crystalline structure of cellulose is an important factor affecting the hydrolysis rate of lignocelluloses. The amorphous structure dramatically increased the initial hydrolysis rate of cellulose by cellulases (Yang *et al.* 2009). The crystallinity index (I_c) can be measured by X-ray diffraction (XRD) and estimated using the following equation (Moran *et al.* 2008):

$$I_c = \frac{I_{(002)} - I_{(am)}}{I_{(002)}} \times 100 \qquad (12.1)$$

where $I_{(002)}$ is the counter reading at peak intensity at a 2θ angle close to 26° representing crystalline material and $I_{(am)}$ is the counter reading at peak intensity at a 2θ angle close to 18° representing amorphous material in cellulosic fibers.

Any reduction in crystallinity of cellulose would contribute to an increased fragility of the cell wall as well as susceptibility to attach by cellulolytic microorganisms (Hoshino *et al.* 1993). Yang *et al.* (2009) investigated the change of crystallinity index of *S. alterniflora* during the anaerobic digestion and found that the crystallinity index decreased from 0.510 of undigested sample to 0.479 of digested sample, indicating that the anaerobic digestion destroyed the crystalline structure.

4. Chemical Analysis on Anaerobic Digestion of Wheat Straw by Rumen Microorganisms

4.1. *Materials and methods*

Wheat straw was collected from a farm in the suburb of Hefei city, China. After cleaned and dried, the straw was cut into 1 cm stalks. The fragments of the stem were dewaxed with toluene/ethanol (2:1, v/v) in a Soxhlet apparatus for 10 h. The dewaxed fraction was dried in a vacuum desiccator for use.

The ruminal digesta, obtained from a fistulated goat, was strained through four-layer gauze under CO_2 protection and rumen fluid was collected in a CO_2-filled flask. The residual digesta was agitated under CO_2 for 60 s in an electronic blender after the addition of equal volume of buffer solution. Fluid from the agitated digesta was filtered through gauze again and combined with rumen fluid in equal volumes as inocula. This procedure ensured that the obtained microbial suspensions contained free-floating and fiber-associated ruminal microbes.

Fifteen milliliters of inocula was automatically poured into Hungate tubes under CO_2 protection, each of which contained three pieces of stem fragments. For the control, the diluted rumen fluid autoclaved at 121 °C for 20 min was added into the Hungate tubes. Another group of control only contained 15 ml buffer solution. Each group had four tubes, in which the stem fragments came from same straw. Tubes were placed in a water-heating incubator at 39 ± 1 °C and agitated at 150 rpm. Initial pH was set at 6.9. Samples were removed at pre-determined intervals. The samples after incubation were washed five times with distilled water and then re-suspended in distilled water for 5 min. They were then washed with distilled water again. After cleaning, the samples were dried in a vacuum desiccator for analysis.

AFM investigation was performed using a NanoScope IIIa multimode scanning probe microscope (Digital Instruments, Inc., Santa Barbara). The images were scanned in tapping mode in air using an NP-S silicon nitride cantilever probe (DI, spring constant ca. $0.06\,N{\cdot}m^{-1}$) with a resonance frequency of 250–360 kHz. The scanning rate ranged from 0.5 to 3.0 Hz, and the scanning size varied from 1 to 10 μm. Both height and phase images were captured simultaneously and all images were measured at 512×512 pixels. For each sample, many locations were measured, but only the representative AFM images are presented here.

XPS measurements were conducted with an ESCALAB MK II electron spectrometer using a monochromated Al Kα X-ray source at a base pressure below 5×10^{-8} Torr. Scanning was carried out over a wide binding energy range (0–1,100 eV) at electron take-off angle 90 ° from sample areas less than 1 mm in diameter. The relative amounts of differently bound carbons were determined from high-resolution C1s spectra. FTIR analysis was carried out using a Magna-IR 750 (Nicolet Instrument Co., USA). The dried straw sample was grinded into powder, grounded with KBr and pressed at 40 MPa to form a uniform disk.

Three parallel experiments were performed to confirm the degradation of lignin in the straw by rumen microorganisms. After 5 days of incubation, the solution of each sample was filtrated with 0.45-μm filters and extracted with redistilled methyl *tert*-butyl ether. The extracts were merged and dried with anhydrous magnesium sulfate and concentrated under reduced pressure. GC/MS analysis was performed using a GC (Model 6890, Angilent Inc., USA) coupled with a GCT-MS (Micromass Ltd., UK) through an electron impact interface.

4.2. *Results of chemical analysis*

4.2.1. *AFM results*

Figure 12.2 shows typical AFM 3D and phase images for the stem outside surface of wheat straw after dewaxing. Part of the surface was covered by thin flakelets, which might be the coverage of plant wax, and aggregates with size ranging from 20 to 130 nm, which were likely to be lignin (Lee *et al.* 2000). No fibers were observed on the straw surface after dewaxing.

Fig. 12.2. Surface topography of the outside surface of straw after dewaxing.

The AFM images for the surface of stem after 5-day microbial treatment are similar to the dewaxing (Fig. 12.3a). However, holes and cracks were observed on the outside of the stem and at their bottom, and exposed microfibrils could also be observed. These exposed fibers located in the outer layers with a diameter of 15–30 nm (arrow in Fig. 12.3a). Since the amorphous materials covering the microfibrils were found to be hydrophobic in the AFM measurement, these structures probably consisted of wax or lignin, rather than hydrophilic hemicelluloses.

After 9-day anaerobic degradation, the AFM images showed that the majority of the surface consisted of non-fibrillar amorphous materials as described previously. Typical structures of the exposed surface are shown in Fig. 12.3b. Holes in the microfibril with a diameter of approximately 600 nm were observed, which might be penetrated by rumen microorganisms. On the stem surface, amorphous materials covered the microfibril. The granules covering the surface were found to be significantly reduced (Fig. 12.3b). The blending of amorphous fibril and grains suggests that the lignin granules on the outer surface had been gradually degraded or disrupted and that the amorphous materials might be the residues of hemicelluloses or lignin–hemicellulose complex (Koljonen *et al.* 2001).

After 13-d anaerobic treatment, the outer surface of the stem was still partially covered by granules. Simultaneously, microfibrils with a diameter of 20–60 nm occupied a larger area of the outer surface (Fig. 12.3c). This was substantially different from the fibrils in the samples after 9-d fermentation. These microfibrils were oriented in the same direction, and ranged in layers. This image implies that the degradation of different types of microfibrils might also occur. Such degradation occurred from outer surface to interior and layer by layer.

Figure 12.4 illustrates the surface of the control sample that soaked in the mixture of buffer or sterile rumen fluid for 13 days. The stem surface was fully covered by granules and flakelets. Comparison between Figs. 12.2 and 12.4 suggests that these granules and flakelets were very likely to be lignin and residual wax, respectively. This shows that, without the presence of rumens, the lignin on the stem surface could not become dissolved in the buffer under the same conditions.

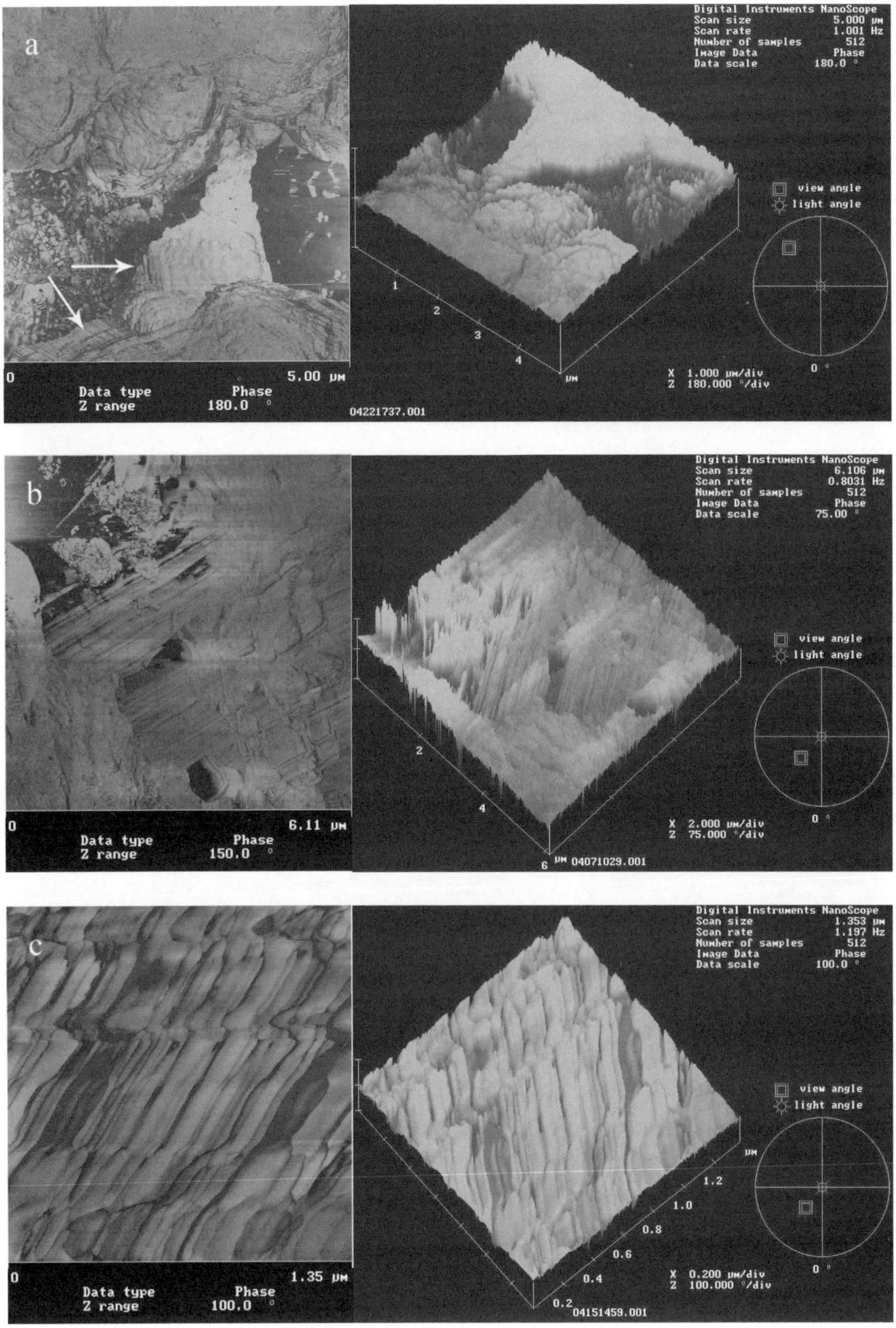

Fig. 12.3. Surface topography of straw after anaerobic treatment of: (a) 5 d; (b) 9 d; and (c) 13 d.

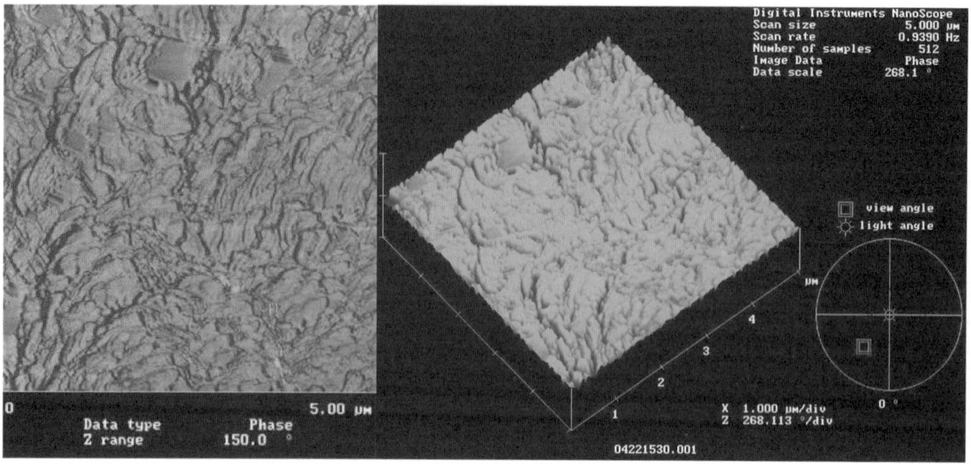

Fig. 12.4. Surface topography of straw soaked in sterile rumen fluid.

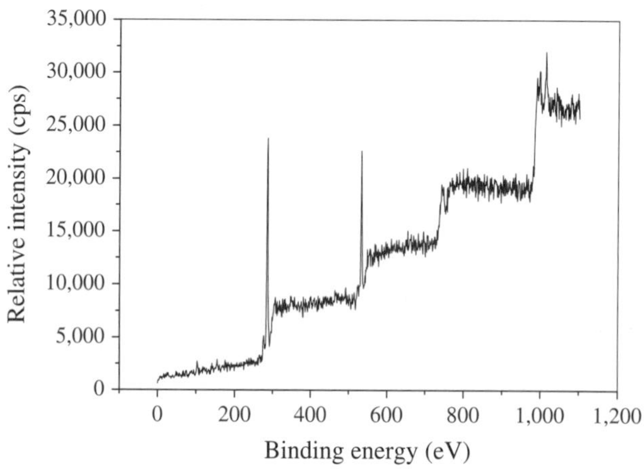

Fig. 12.5. Typical XPS wide scan (0–1,100 eV) spectrum of straw after dewaxing.

4.2.2. *XPS results*

Figure 12.5 shows a typical XPS scanning (0–1,100 eV) spectrum for straw after dewaxing. Except a small amount of nitrogen (N), the surface of the straw composed of oxygen (O) and carbon (C). The main components of wheat straw surface had different theoretical O/C atomic ratios and relative C1–C3 peak areas, as shown in Table 12.2. Therefore, the evolution of the O/C ratios and relative abundance of C1–C3 in the biodegradation could be used to characterize the changes in components on the straw surface, which is similar to the use of XPS on the structural change of pulping. Figure 12.6 illustrates the XPS spectra of C1s of the straws after dewaxing and after 13-d anaerobic degradation. The peak-fitting results and O/C atomic ratios are summarized in Table 12.2.

After the anaerobic treatment, the O/C ratios and relative abundance of C2 and C3 increased remarkably, while that of C1 decreased significantly. Comparison with the known theoretic

Table 12.2. O/C atomic ratios and relative C1–C3 peak areas of straw and its main components.

	Material	O/C	FWHM	Relative area (%)		
				C1	C2	C3
Theoretical[a]	Lignin	0.33	/	49	49	2
	Cellulose	0.83	/	0	83	17
	Hemicelluloses	0.80	/	0	83	17
	Wax	0.04 ~ 0.11	/	94 ~ 100	0	0 ~ 6
Measured	Wheat straw[b]	0.164	1.64	81.9	13.5	4.6
	Wheat straw[c]	0.352	2.08	65.7	25.5	8.8

Notes: [a]Cited from Gustafsson *et al.* 2003.
[b]Untreated wheat straw.
[c]Wheat straw after 13-d anaerobic treatment.

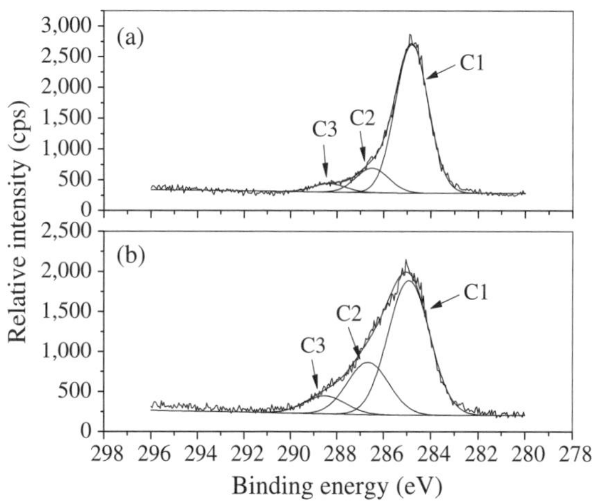

Fig. 12.6. XPS spectra of carbon 1s of straws: (a) after dewaxing; and (b) after 13-d anaerobic treatment.

values presented in Table 12.2 indicates that the lignin and wax on the stem surface were partially degraded by ruminal microbes and that the cellulose was exposed on the surface. This was in good agreement with the AFM observation. The O/C ratio in the dewaxed sample was lower and the relative amount of C1 was higher than the theoretical values of lignin, suggesting that the surface of the stem after dewaxing still contained a large amount of wax, which was consistent with the result from the AFM images.

4.2.3. *FTIR results*

The changes in relative absorbance of the characteristic bands were also applied to determine the degradation of the components of wheat straw during anaerobic digestion by rumen microorganisms. The FTIR spectra of the straw after dewaxing and after 13-d anaerobic treatment by rumen microorganisms are illustrated in Fig. 12.7. The corresponding assignments are listed in Table 12.3. The relative absorbances of the characteristic bands of lignin at wavenumbers of 1,459 cm^{-1}, 1,432 cm^{-1}, 1,323 cm^{-1}, 1,252 cm^{-1}, and 1,158 cm^{-1} decreased, whereas the relative absorbances of the characteristic bands of cellulose at wavenumbers of 1,378 cm^{-1} and 1,048 cm^{-1}

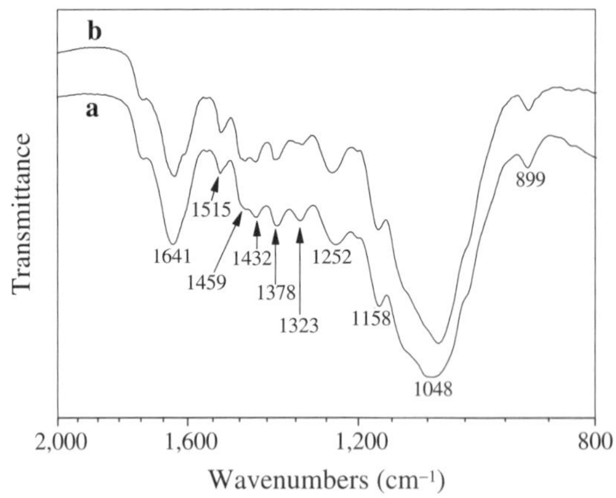

Fig. 12.7. FTIR spectra of wheat straw: (a) after dewaxing; and (b) after 13-d anaerobic treatment.

Table 12.3. Characteristic FTIR bands of wheat straw.

Material	Wavenumber reported (cm^{-1})	Wavenumber picked (cm^{-1})	Assignment
Cellulose	1,666	1,641	Absorbed water bending
	1,387	1,378	O–H bending
	1,049	1,048	C–O–C pyranose ring skeletal vibrations
Lignin	1,505	1,515	Aromatic skeletal vibrations
	1,460	1,459	Aromatic methyl group vibrations
	1,440	1,432	Aromatic skeletal vibrations
	1,325	1,323	Syringyl ring breathing with CO stretching
	1,280	1,252	Guaiacyl ring breathing with CO stretching
	1,155	1,158	Aromatic CH in-plane deformation, guaiacyl type

slightly increased, indicating a decrease in the relative fraction of lignin. These results confirmed that lignin was degraded by rumen microorganisms during the anaerobic digestion.

4.2.4. *GC/MS results*

By analyzing the effluent of anaerobic digestion of wheat straw by rumen microorganisms with GC/MS, it was found that some chemicals with aromatic bond were detected in the effluent, such as 3,4-dimethoxybenzoic acid (Fig. 12.8) and 3,4-dimethoxyphenol (Fig. 12.9), indicating that lignin could be disrupted and degraded by rumen microorganisms, which might be one of the reasons why rumen microorganisms improved the anaerobic digestion of lignocellulosic wastes (Hu *et al.* 2008).

4.3. *Discussion*

Anaerobic digestion of lignocellulosic wastes is a very complex bioprocess. In this work, combination of morphological observation by AFM with chemical analysis by XPS, FTIR, and GC/MS, plus component measurement, was successfully used to study the bioconversion process. Such

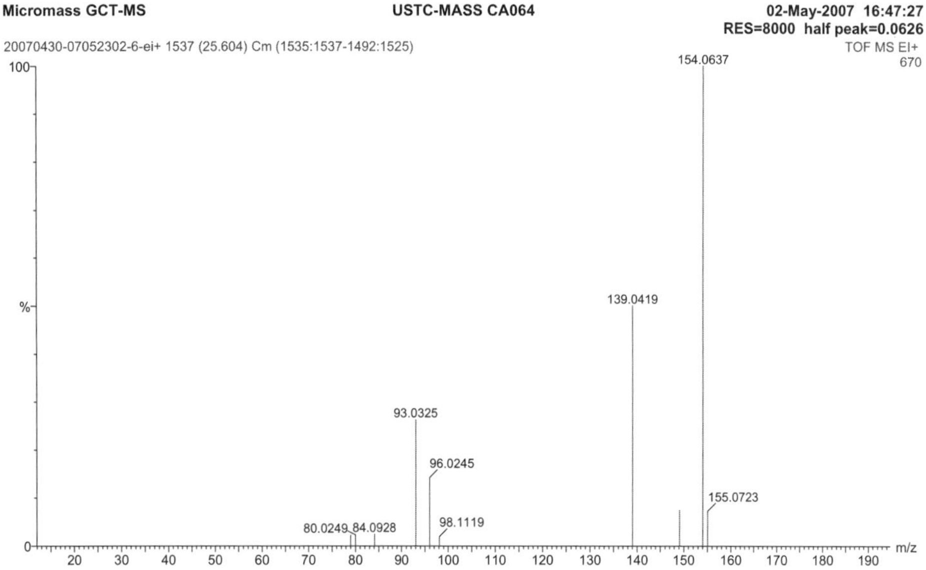

Fig. 12.8. Mass spectra of 3,4-dimethoxybenzoic acid.

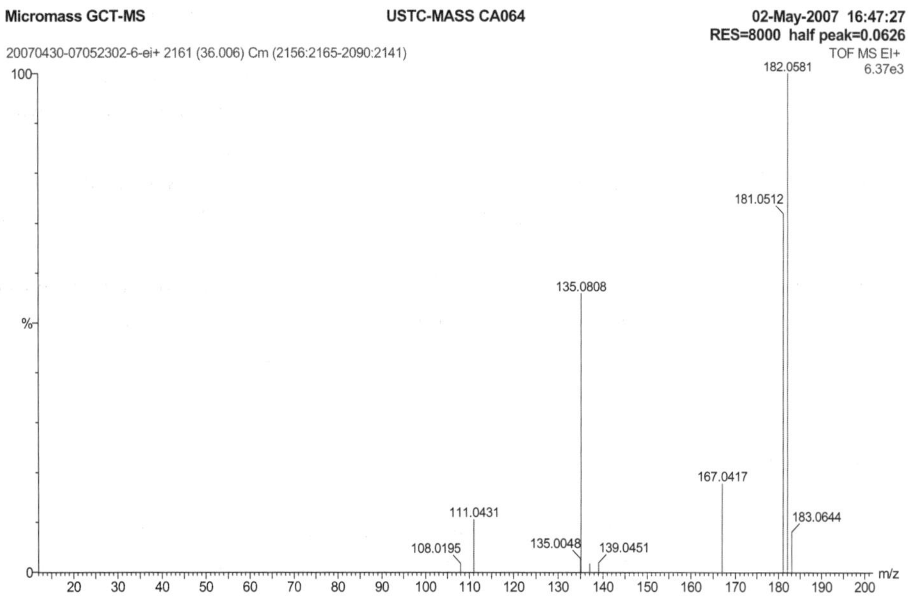

Fig. 12.9. Mass spectra of 3,4-dimethoxyphenol.

a combination provides unique and useful information on the straw biodegradation by rumen microorganisms.

In the present work, 25.5% of lignin in straw was removed by the rumen microorganisms within 13-day treatment. This further demonstrates the potential utilization of rumen microorganisms for the bioconversion of lignocellulosic wastes. However, the loss of lignin does not necessarily mean its biodegradation by the rumen microorganisms, and it could be attributed to the solubilization. For instance, on the basis of ^{14}C-lignocellulose test results, Akin and Benner (1988) showed that most losses of lignin were attributed to solubilization from fiber. In our present work, the AFM images directly show the gradual removal of lignin granules from the external straw surface (Fig. 12.3), while the GC/MS results indicate that the rumen microorganisms did decompose the lignin in straw, based on the comparison with ruminal treatment and the control. However, a further investigation using molecular biological techniques is warranted to examine the dominate microbes in the rumen microorganisms which are mainly responsible for the biodegradation of lignin in straw.

The 3D images for the straw surface at different fermentation times in Fig. 12.3 show that the surface traces and height of surface features changed significantly in the rumen-mediated digestion. Straw surface exposed to ruminal fluid displayed non-uniform surface traces, indicating the gradual falling off of the granules and flakelets. The changes in surface topography shown in the AFM 3D images are attributed to the deterioration, degradation, or falling off of the surface materials because of the microbial attack. Phase images of samples are less commonly reported in the literature than height (topographic) images, because the different components that give rise to phase shifts of the cantilever are more complex. Despite of limited ability to interpret the complicated phase images in a quantitative manner, phase images are useful because they provide a measure of sample heterogeneity (Camesano *et al.* 2000). The phase images in Fig. 12.3 clearly illustrate the change of the straw surface heterogeneity in anaerobic digestion.

Usually XPS is used for the analysis of the elemental and chemical compositions of solid surface. In the most common approach, XPS combined with solvent extraction has been used to evaluate the surface of lignin and extractives contents of pulp (Koljonen *et al.* 2001). In our work, combination of AFM, XPS, and FTIR was applied to characterize the straw surface in anaerobic digestion. AFM reveals surface morphology, while XPS and FTIR give chemical characteristics of surface regions.

The combination of morphological and chemical information yields better insight into the characteristics of straw surface. Such information led to a better understanding of the performance and mechanisms of the biodegradation of lignocelluloses by microorganisms.

5. Kinetics Analysis

Kinetic analysis is an effective way to determine the key step in the anaerobic digestion of lignocellulosic wastes. It also provides information for reactor design and operation, and automated digester control, which may result in more efficient process performance and reduce reliance on skilled operators (Page *et al.* 2008). Kinetic models are usually divided into two classes: structural models and un-structural models. Structural models take metabolic pathways into consideration and are generally complicated. The un-structural models are much simpler (Mu *et al.* 2006a, 2007). Here we will mainly review the application of un-structural models, such as Gompertz equation and Monod equation, on the anaerobic digestion of lignocellulosic wastes by rumen microorganisms.

5.1. *Gompertz model*

Gompertz model has been used to describe the anaerobic digestion of lignocellulosic wastes by rumen microorganisms (Hu and Yu 2006). Zwietering *et al.* (1990) applied Gompertz model to describe the growth of pure strains, e.g., *Lactobacillus plantarum*. The modified Gompertz model can be used to describe microbial growth, substrate degradation, and production formation (Zwietering *et al.* 1990).

For describing microbial growth, the Gompertz model can be modified as follows (Zwietering *et al.* 1990):

$$X = X_0 + \int_0^t r_X d_t = X_0 + \Delta X_{\max} \times \exp\left\{-\exp\left[\frac{R_{\max,X} \times e}{\Delta X_{\max}}(\lambda\text{-}t) + 1\right]\right\} \quad (12.2)$$

The rate of microbial growth can be estimated by differentiating Eq. (12.2):

$$r_X = r_{\max,X} \times \exp\left\{-\exp\left[\frac{R_{\max,X} \times e}{\Delta X_{\max}}(\lambda\text{-}t) + 1\right] + \left[\left[\frac{R_{\max,X} \times e}{\Delta X_{\max}}(\lambda\text{-}t) + 1\right] + 1\right]\right\} \quad (12.3)$$

where X is the microbial concentration, X_0 is the initial microbial concentration, $\Delta X_{\max}$ is the maximum concentration of microorganism growth, r_X is microbial growth rate, and $R_{\max,X}$ is maximum rate of microbial growth.

The substrate degradation can be calculated using the following modified Gompertz model (Mu *et al.* 2007):

$$S_0\text{-}S = \Delta S_{\max} \times \exp\left\{-\exp\left[\frac{R_{\max,S} \times e}{S_{\max}}(\lambda\text{-}t) + 1\right]\right\} \quad (12.4)$$

By differentiating Eq. (12.3), the degradation rate of substrate can be obtained as

$$r_s = R_{\max,s} \times \exp\left\{-\exp\left[\frac{R_{\max,s} \times e}{S_{\max}}(\lambda\text{-}t) + 1\right] + \left[\left[\frac{R_{\max,S} \times e}{S_{\max}}(\lambda\text{-}t) + 1\right] + 1\right]\right\} \quad (12.5)$$

where S is the substrate concentration, S_0 is the initial substrate concentration, $\Delta S_{\max}$ is the maximum concentration of substrate degradation, $R_{\max,s}$ is the maximum rate of substrate degradation, r_s is the substrate degradation rate.

The production of volatile fatty acids can be expressed as follows:

$$Pi = \Delta P_{\max,i} \times \exp\left\{-\exp\left[\frac{R_{\max,pi} \times e}{\Delta P_{\max,i}}(\lambda\text{-}t) + 1\right]\right\} \quad (12.6)$$

By differentiating Eq. (12.6), the formation rate of volatile fatty acids can be obtained as:

$$r_{pi} = R_{\max,Pi} \times \exp\left\{-\exp\left[\frac{R_{\max,Pi} \times e}{\Delta P_{\max,i}}(\lambda\text{-}t) + 1\right] + \left[\left[\frac{R_{\max,Pi} \times e}{\Delta P_{\max,i}}(\lambda\text{-}t) + 1\right] + 1\right]\right\} \quad (12.7)$$

where P_i is the concentration of volatile fatty acid, $\Delta P_{\max,i}$ is the maximum concentration of volatile fatty acids, $R_{\max,pi}$ is the maximum rate of volatile fatty acid formation, and r_{pi} is the formation rate of volatile fatty acids.

The kinetics of methane production in anaerobic digestion can also been described using the modified Gompertz model (Lay *et al.* 1997). Hu and Yu (2006) and Yue and Yu (2009) described the kinetics of volatile fatty acids formation in the anaerobic digestion of cattail by rumen microorganisms. The coefficients (R^2) between the experimental and predicted values were more than 0.98, indicating that the modified Gompertz equation could well describe the kinetics of products formation. Gompertz equation can also fit well to gas production from crystalline cellulose by rumen microorganisms (Schofield *et al.* 1994).

5.2. *Monod model*

When microorganisms grow on a grow-limiting substrate, the microorganism's growth rate is usually expressed by the Monod equation:

$$\mu = \mu_M S/(Ks + S) \tag{12.8}$$

where μ is the specific growth rate, which equals to $(1/X)(dX/dt)$, μ_M is the maximum growth rate, S is the substrate concentration, K_s is the half-saturation constant, and X is the microbial concentration.

The substrate consumption can be expressed as:

$$dS/dt = \text{-}(1/Y)\mu X = \mu_M SX/Y(Ks + S) \tag{12.9}$$

where Y is the microbial growth yield.

In a microbial system, the microbial apparent growth rate can be described using first-order equation as:

$$dX/dt = \mu X \tag{12.10}$$

In a microbial system, part of the substrate is consumed to maintain the endogenous metabolism and maintenance requirements. Therefore, the net microbial growth rate is the difference between the apparent growth rate and the maintenance rate, which is expressed using the following equation:

$$dX/dt = (\mu\text{-}\alpha)X = [\mu_M SX/(Ks + S)]\text{-}\alpha X \tag{12.11}$$

where α is the specific maintenance rate.

The product formation rate is usually expressed as:

$$dP/dt = (\mu\text{-}\alpha)X = [\mu_M SX/(Ks + S)]\text{-}\beta X\left\{\left[\mu MS/(Ks + S)\right]\text{-}\alpha\right\} \tag{12.12}$$

where P is the product concentration, β is a factor used to convert microbial biomass to equivalent substrate unit.

Pavlostathis *et al.* (1988) investigated the kinetics of the fermentation of insoluble cellulose by *Ruminococcus albus* in a continuous bioreactor system. They found that the utilization rate of cellulose depended on the steady-state concentration of cellulose and the first-order rate constant was 1.18 day^{-1}. The true microbial growth yield (Y) was 0.11 g·g^{-1}, and the maintenance coefficient was 0.10 g·g^{-1}·h^{-1}. We used Monod equation in our previous study to describe the kinetics of anaerobic digestion of cattail by mixed rumen microorganisms in a UASB reactor, and found that the first-order rate constant was 1.50 day^{-1} (Hu *et al.* 2007).

Table 12.4 presents the values of the specific growth rate and first-order rate constant for microbial cellulose and cellulosic material utilization in literature. Comparing with those growing

Table 12.4. Specific growth rate and first-order rate constant for microbial cellulose utilization.

Organism	Substrate, cultivation mode	Specific growth rate (h^{-1})	First-order rate constant (h^{-1})	References
Clostridium thermocellum	α-Cellulose batch, 60 °C	0.183	n.a.	Islam *et al.* (2009)
Trichoderma reesei RUT C30	Wastepaper, continuous, 25 °C	0.26	n.a.	Ju and Afolabi (1999)
Mixed rumen microorganisms	Cattail, continuous, 39.5 °C	0.12	0.063	Hu *et al.* (2007)
Ruminococcus albus	Avicel, continuous, 37 °C	0.095	0.05	Pavlostathis *et al.* (1988)
Clostridium thermocellum	Avicel, PH105, continuous, 60 °C	0.17	0.16	Lynd *et al.* (1989)

on crystalline cellulose and wastepaper, microorganisms growing on the cattail had a lower specific growth rate and first-order rate constant. This was because of the more recalcitrant structure of cattail.

6. Conclusions

In conclusion, the anaerobic digestion of lignocelluloses by microorganisms was limited by their recalcitrant structure. Acclimating microorganisms with high cellulolytic activities will improve the hydrolysis of lignocelluloses. The chemical and imaging analyses by AFM, XPS, FTIR, and GC/MS at micro-scale and molecular level revealed that lignin is one of the main barriers for the hydrolysis of lignocelluloses. Rumen microorganisms improved the cleavage of lignin during the anaerobic digestion and enhanced the degradation of lignocellulosic wastes. The kinetics of anaerobic digestion of lignocellulosic wastes could be analyzed by the Gompertz model and Monod model. Results from the chemical and kinetics analyses demonstrate that rumen microorganisms are the potential culture resources for improving anaerobic digestion of lignocellulosic wastes.

Acknowledgments

The authors wish to thank the NSFC (50625825 and 50738006), the CAS (KSCX2-YW-G-001), and National 863 Program (2006AA06Z340) for the partial support of this study.

References

Abu-Lail, N. I. and Camesano, T. A. (2003) Role of lipopolysaccharides in the adhesion, retention, and transport of *Escherichia coli* JM109. *Environ. Sci. Technol.* **37**, 2173–2183.

Adney, W. S., Rivard, C. J., Grohmann, K., and Himmel, M. E. (1989) Detection of extracellular hydrolytic enzymes in the anaerobic-digestion of municipal solid-waste. *Biotechnol. Appl. Biochem.* **11**, 387–400.

Akin, D. E. and Benner, R. (1988) Degradation of polysaccharides and lignin by ruminal bacteria and fungi. *Appl. Environ. Microbiol.* **54**, 1117–1125.

Alvarez, R. and Liden, G. (2008) Anaerobic co-digestion of aquatic flora and quinoa with manures from Bolivian Altiplano. *Waste Manage.* **28**, 1933–1940.

Alvarez, V. A. and Vazquez, A. (2006) Influence of fiber chemical modification procedure on the mechanical properties and water absorption of MaterBi-Y/sisal fiber composites. *Composites Part A* **37**, 1672–1680.

Babu, B. V. (2008) Biomass pyrolysis: A state-of-the-art review. *Biofuels Bioprod. Biorefin.* **2**, 393–414.

Barnes, S. P. and Keller, J. (2003) Cellulosic waste degradation by rumen-enhanced anaerobic digestion. *Water Sci. Technol.* **48**, 155–162.

Barnes, S. P. and Keller, J. (2004) Anaerobic rumen SBR for degradation of cellulosic material. *Water Sci. Technol.* **50**, 305–311.

Buffiere, P., Loisel, D., Bernet, N., and Delgenes, J. P. (2005) Towards new indicators for the prediction of solid waste anaerobic digestion properties. 4th International Symposium on Anaerobic Digestion Aug 31–Sep 02; Copenhagen, DENMARK. 233–241.

Camesano, T. A., Natan, M. J., and Logan, B. E. (2000) Observation of changes in bacterial cell morphology using tapping mode atomic force microscopy. *Langmuir* **16**, 4563–4572.

Cuetos, M. J., Moran, A., Otero, M., and Gomez, X. (2009) Anaerobic co-digestion of poultry blood with OFMSW: FTIR and TG-DTG study of process stabilization. *Environ. Technol.* **30**, 571–582.

Demirbas, M. F. and Balat, M. (2007) Biomass pyrolysis for liquid fuels and chemicals: A review. *J. Sci. Ind. Res.* **66**, 797–804.

Dorris, G. M. and Gray, D. G. (1978a) The surface analysis of paper and wood fibers by ESCA (II). Surface composition of mechanical pulps. *Cellul. Chem. Technol.* **12**, 721–734.

Dorris, G. M. and Gray, D. G. (1978b) The surface analysis of paper and wood fibres by ESCA (I). Application to cellulose and lignin. *Cellul. Chem. Technol.* **12**, 9–23.

Espinosa-Solares, T., Valle-Guadarrama, S., Bombardiere, J., Domaschko, M., and Easter, M. (2009) Effect of heating strategy on power consumption and performance of a pilot plant anaerobic digester. *Appl. Biochem. Biotechnol.* **156**, 465–474.

Fu, S. Y., Lucia, L. A., and Ragauskas, A. J. (2002) Oxygen delignification of low Kappa softwood kraft pulp. In: Liu, H. B., Zhan, H. Y., and Xie, Y. I. (eds.), Proceedings of the 2nd International Symposium on Emerging Technologies of Pulping & Papermaking Oct 9–11; Guangzhou, People R China. pp. 394–401.

Ganesh Kumar, A., Sekaran, G., and Krishnamoorthy, S. (2006) Solid state fermentation of *Achras zapota* lignocellulose by *Phanerochaete chrysosporium*. *Bioresour. Technol.* **97**, 1521–1528.

Gea, T., Artola, A., Sort, X., and Sanchez, A. (2005) Composting of residuals produced in the Catalan wine industry. *Compost Sci. Util.* **13**, 168–174.

Gustafsson, J., Ciovica, L., and Peltonen, J. (2003) The ultrastructure of spruce kraft pulps studied by atomic force microscopy (AFM) and X-ray photoelectron spectroscopy (XPS). *Polymer* **44**, 661–670.

Haensel, T., Comouth, A., Lorenz, P., Ahmed, S. I. U., Krischok, S., Zydziak, N., Kauffmann, A., and Schaefer, J. A. (2009) Pyrolysis of cellulose and lignin. *Appl. Surf. Sci.* **255**, 8183–8189.

Himmel, M. E., Ding, S. Y., Johnson, D. K., Adney, W. S., Nimlos, M. R., Brady, J. W., and Foust, T. D. (2007) Biomass recalcitrance: Engineering plants and enzymes for biofuels production. *Science* **315**, 804–807.

Hoshino, E., Sasaki, Y., Okazaki, M., Nisizawa, K., and Kanda, T. (1993) Mode of action of exo-type and endo-type cellulases from irpex-lacteus in the hydrolysis of cellulose with different crystallinities. *J. Biochem.* **114**, 230–235.

Hu, Z. H., Liu, S. Y., Yue, Z. B., Yan, L. F., Yang, M. T., and Yu, H. Q. (2008) Microscale analysis of *in vitro* anaerobic degradation of lignocellulosic wastes by rumen microorganisms. *Environ. Sci. Technol.* **42**, 276–281.

Hu, Z. H., Wang, G., and Yu, H. Q. (2004) Anaerobic degradation of cellulose by rumen microorganisms at various pH values. *Biochem. Eng. J.* **21**, 59–62.

Hu, Z. H. and Yu, H. Q. (2005) Application of rumen microorganisms for enhanced anaerobic fermentation of corn stover. *Process Biochem.* **40**, 2371–2377.

Hu, Z. H. and Yu, H. Q. (2006) Anaerobic digestion of cattail by rumen cultures. *Waste Manage.* **26**, 1222–1228.

Hu, Z. H., Yu, H. Q., Yue, Z. B., Harada, H., and Li, Y. Y. (2007) Kinetic analysis of anaerobic digestion of cattail by rumen microbes in a modified UASB reactor. *Biochem. Eng. J.* **37**, 219–225.

Inari, G. N., Petrissans, M., Lambert, J., Ehrhardt, J. J., and Gerardin, P. (2006) XPS characterization of wood chemical composition after heat-treatment. *Surf. Interface Anal.* **38**, 1336–1342.

Islam, R., Cicek, N., Sparling, R., and Levin, D. (2009) Influence of initial cellulose concentration on the carbon flow distribution during batch fermentation by *Clostridium thermocellum* ATCC 27405. *Appl. Microbiol. Biotechnol.* **82**, 141–148.

Ju, L. K. and Afolabi, O. A. (1999) Wastepaper hydrolysate as soluble inducing substrate for cellulase production in continuous culture of *Trichoderma reesei*. *Biotechnol. Prog.* **15**, 91–97.

Kajikawa, H., Kudo, H., Kondo, T., Jodai, K., Honda, Y., Kuwahara, M., and Watanabe, T. (2000) Degradation of benzyl ether bonds of lignin by ruminal microbes. *FEMS Microbiol. Lett.* **187**, 15–20.

Koljonen, K., Osterberg, M., Johansson, L. S., and Stenius, P. (2001) Surface chemistry and morphology of different mechanical pulps determined by ESCA and AFM. Jun 04–08; Helsinki, Finland. pp. 143–158.

Kumar, R., Mago, G., Balan, V., and Wyman, C. E. (2009) Physical and chemical characterizations of corn stover and poplar solids resulting from leading pretreatment technologies. *Bioresour. Technol.* **100**, 3948–3962.

Lange, J. P. (2007) Lignocellulose conversion: An introduction to chemistry, process and economics. *Biofuels Bioprod. Biorefin.* **1**, 39–48.

Lay, J. J., Li, Y. Y., Noike, T., Endo, J., and Ishimoto, S. (1997) Analysis of environmental factors affecting methane production from high-solids organic waste. *Water Sci. Technol.* **36**, 493–500.

Lee, I., Evans, B. R., and Woodward, J. (2000) The mechanism of cellulase action on cotton fibers: Evidence from atomic force microscopy. *Ultramicroscopy* **82**, 213–21.

Lissens, G., Thomsen, A. B., De Baere, L., Verstraete, W., and Ahring, B. K. (2004) Thermal wet oxidation improves anaerobic biodegradability of raw and digested biowaste. *Environ. Sci. Technol.* **38**, 3418–3424.

Lynd, L. R., Grethlein, H. E., and Wolkin, R. H. (1989) Fermentation of cellulosic substrates in batch and continuous culture by clostridium-thermocellum. *Appl. Environ. Microbiol.* **55**, 3131–3139.

Lynd, L. R., Weimer, P. J., van Zyl, W. H., and Pretorius, I. S. (2002) Microbial cellulose utilization: Fundamentals and biotechnology. *Microbiol. Mol. Biol. Rev.* **66**, 506–577.

Marcato, C. E., Mohtar, R., Revel, J. C., Pouech, P., Hafidi, M., and Guiresse, M. (2009) Impact of anaerobic digestion on organic matter quality in pig slurry. *Int. Biodeterior. Biodegrad.* **63**, 260–266.

Mohan, D., Pittman, C. U., and Steele, P. H. (2006) Pyrolysis of wood/biomass for bio-oil: A critical review. *Energy Fuels* **20**, 848–889.

Moran, J. I., Alvarez, V. A., Cyras, V. P., and Vazquez, A. (2008) Extraction of cellulose and preparation of nanocellulose from sisal fibers. *Cellulose* **15**, 149–159.

Mtui, G. Y. S. (2009) Recent advances in pretreatment of lignocellulosic wastes and production of value added products. *Afr. J. Biotechnol.* **8**, 1398–1415.

Mu, Y., Wang, G., and Yu, H. Q. (2006a) Kinetic modeling of batch hydrogen production process by mixed anaerobic cultures. *Bioresource Technol.* **97**, 1302–1307.

Mu, Y., Wang, G., and Yu, H. Q. (2006b) Response surface methodological analysis on biohydrogen production by enriched anaerobic cultures. *Enzyme Microbiol. Technol.* **38**, 905–913.

Mu, Y., Yu, H. Q., and Wang, G. (2007) A kinetic approach to anaerobic hydrogen-producing process. *Water Res.* **41**, 1152–1160.

Nair, S., Kuang, Y., and Pullammanappallil, P. (2005) Enhanced degradation of waste grass clippings in one and two stage anaerobic systems. *Environ. Technol.* **26**, 1003–1011.

Oh, S. Y., Yoo, D. I., Shin, Y., and Seo, G. (2005) FTIR analysis of cellulose treated with sodium hydroxide and carbon dioxide. *Carbohydr. Res.* **340**, 417–428.

Ostenson, M., Jarund, H., Toriz, G., and Gatenholm, P. (2006) Determination of surface functional groups in lignocellulosic materials by chemical derivatization and ESCA analysis. *Cellulose* **13**, 157–170.

Page, D. I., Hickey, K. L., Narula, R., Main, A. L., and Grimberg, S. J. (2008) Modeling anaerobic digestion of dairy manure using the IWA Anaerobic Digestion Model no. 1 (ADM1). *Water Sci. Technol.* **58**, 689–695.

Pavlostathis, S. G., Miller, T. L., and Wolin, M. J. (1988) Kinetics of insoluble cellulose fermentation by continuous cultures of *Ruminococcus-Albus*. *Appl. Environ. Microbiol.* **54**, 2660–2663.

Qu, X., Vavilin, V. A., Mazeas, L., Lemunier, M., Duquennoi, C., He, P. J., and Bouchez, T. (2009) Anaerobic biodegradation of cellulosic material: Batch experiments and modeling based on isotopic data and focusing on aceticlastic and non-aceticlastic methanogenesis. *Waste Manage.* **29**, 1828–1837.

Ragauskas, A. J., Williams, C. K., Davison, B. H., Britovsek, G., Cairney, J., Eckert, C. A., Frederick, W. J., Hallett, J. P., Leak, D. J., and Liotta, C. L. (2006) The path forward for biofuels and biomaterials. *Science* **311**, 484–489.

Rani, P., Kalyani, N., and Prathiba, K. (2008) Evaluation of lignocellulosic wastes for production of edible mushrooms. *Appl. Biochem. Biotechnol.* **151**, 151–159.

Rivard, C. J., Himmel, M. E., Vinzant, T. B., Adney, W. S., Wyman, C. E., and Grohmann, K. (1990) Anaerobic-digestion of processed municipal solid-waste using a novel high solids reactor — Maximum solids levels and mixing requirements. *Biotechnol. Lett.* **12**, 235–240.

Rogers, P. L., Jeon, Y. J., Lee, K. J., and Lawford, H. G. (2007) *Zymomonas mobilis* for fuel ethanol and higher value products. *Biofuels* **108**, 263–288.

Russell, J. B. and Rychlik, J. L. (2001) Factors that alter rumen microbial ecology. *Science* **292**, 1119–1122.

Schofield, P., Pitt, R. E., and Pell, A. N. (1994) Kinetics of fiber digestion from *in-vitro* gas-production. *J. Anim. Sci.* **72**, 2980–2991.

Schultz, T. P., Templeton, M. C., and McGinnis, G. D. (2002) Rapid determination of lignocellulose by diffuse reflectance Fourier transform infrared spectrometry. *Anal. Chem.* **57**, 2867–2869.

Service, R. F. (2007) Cellulosic ethanol — Biofuel researchers prepare to reap a new harvest. *Science* **315**, 1488.

Songulashvili, G., Elisashvili, V., Wasser, S. P., Nevo, E., and Hadar, Y. (2007) Basidiomycetes laccase and manganese peroxidase activity in submerged fermentation of food industry wastes. *Enzyme Microbiol. Technol.* **41**, 57–61.

Sun, J. X., Sun, R. C., Sun, X. F., and Su, Y. Q. (2004) Fractional and physico-chemical characterisation of hemicelluloses from ultrasonic irradiated sugarcane bagasse. *Carbohydr. Res.* **339**, 291–300.

Xu, L. C. and Logan, B. E. (2005) Interaction forces between colloids and protein-coated surfaces measured using an atomic force microscope. *Environ. Sci. Technol.* **39**, 3592–3600.

Xu, L. C. and Logan, B. E. (2006) Interaction forces measured using AFM between colloids and surfaces coated with both dextran and protein. *Langmuir* **22**, 4720–4727.

Yan, L. F., Li, W., Yang, J. L., and Zhu, Q. S. (2004) Direct visualization of straw cell walls by AFM. *Macromol. Biosci.* **4**, 112–118.

Yang, S., Li, J., Zheng, Z., and Meng, Z. (2009) Lignocellulosic structural changes of *Spartina alterniflora* after anaerobic mono- and co-digestion. *Int. Biodeterior. Biodegrad.* **63**, 569–575.

Yue, Z. B., Liu, R. H., Yu, H. Q., Chen, H. Z., Yu, B., Harada, H., and Li, Y. Y. (2008) Enhanced anaerobic ruminal degradation of bulrush through steam explosion pretreatment. *Ind. Eng. Chem. Res.* **47**, 5899–5905.

Yue, Z. B. and Yu, H. Q. (2009) Anaerobic batch degradation of cattail by rumen cultures. *Int. J. Environ. Pollution* **38**, 299–308.

Zhang, Y. and He, Y. (2006) Co-composting solid swine manure with pine sawdust as organic substrate. *Bioresour. Technol.* **97**, 2024–2031.

Zhang, Y. H. P. and Lynd, L. R. (2004) Toward an aggregated understanding of enzymatic hydrolysis of cellulose: Noncomplexed cellulase systems. *Biotechnol. Bioeng.* **88**, 797–824.

Zhu, S. D., Wu, Y. X., Yu, Z. N., Zhang, X. A., Wang, C. W., Yu, F. Q., Jin, S. W., Zhao, Y. F., Tu, S. Y., and Xue, Y. P. (2005) Simultaneous saccharification and fermentation of microwave/alkali pre-treated rice straw to ethanol. *Biosystems Eng.* **92**, 229–235.

Zwietering, M. H., Jongenburger, I., Rombouts, F. M., and Vantriet, K. (1990) Modeling of the bacterial-growth curve. *Appl. Environ. Microbiol.* **56**, 1875–1881.

Chapter 13

Enzymatic Treatment of Lignocellulosic Wastes for Anaerobic Digestion and Bioenergy Production

Ganesh D. Saratale[1], Liang-Jung Chien[2] and Jo Shu Chang[1,3,4,*]

[1]*Department of Chemical Engineering, National Cheng Kung University, Tainan, Taiwan*

[2]*Graduate Institute of Biochemical Engineering, Ming Chi University of Technology, Taipei, Taiwan*

[3]*Sustainable Environment Research Center, National Cheng Kung University, Tainan, Taiwan*

[4]*Center for Bioscience and Biotechnology, National Cheng Kung University, Tainan, Taiwan*

**changjs@mail.ncku.edu.tw*

Active research has been conducted on the production of biofuels from the abundant and inexpensive lignocellulosic wastes of forestry, agriculture, and municipal solid wastes. However, degradation of lignocellulose is hindered by the recalcitrant nature of lignocellulose. Hence, hydrolysis/saccharification of lignocellulose becomes the rate-limiting step for the fermentative production of cellulosic biofuels (such as H_2 and ethanol). Hydrolysis of cellulosic materials by biological means is environmentally benign and could be achieved either by using cellulolytic microorganisms or cellulolytic enzymes collected from those microorganisms. Biofuels production could be achieved by direct fermentation of raw lignocellulosic wastes or by a two-stage process, in which the hydrolysis step and the anaerobic fermentation step are operated separately. In this chapter, we review the state-of-the-art of the following aspects related to the enzymatic hydrolysis of lignocellulosic wastes for anaerobic fermentation and bioenergy production: (1) structure of plant cell walls and their cellulose, hemicellulose, and lignin components; (2) lignocellulose-degrading microorganisms and their characterisitics; (3) production of enzymes degrading lignocellulose; (4) treatment of wastes using lignocellulose-degrading enzymes; and (5) anaerobic fermentation process for bioenergy production from enzymatically pretreated lignocellusic wastes.

1. Introduction

Fossil fuels are the main global energy resources for the industrialization and economic growth in the past century. However, reserves of fossil fuels are limited. Coal may last about 100 years or more, but demand for oil is expected to exceed production in 10–20 years depending on the production and consumption rates (Rifkin 2002; Goldemberg and Johansson 2004; Goldemberg 2007). In addition, the unfettered use of fossil fuels emits greenhouse gases (CO_2, CH_4, and CO), resulting in global warming and pollution (Koh and Ghazoul 2008; Saratale *et al.* 2008). For these reasons, large efforts are being made worldwide to develop technologies that generate clean, sustainable energy from renewable carbon resources (in particular, lignocellulosic biomass) which could substitute for fossil fuels (Ragauskas *et al.* 2006; Levin *et al.* 2006).

Biofuels are eco-friendly and sustainable. This makes them an important and promising alternative energy source for fossil fuels (Puppan 2002; Schubert 2006). Although the worldwide annual production of biofuels increased from 4.4 to 50.1 billion liters, the political and public support for biofuels has been countermined. Some recent reports blamed that use of food crops or croplands for biofuels production resulted in food shortages and the increased prices of staple food crops such as maize and rice (James *et al.* 2008; Keeney and Hertel 2008). Nevertheless, many scientists remain optimistic that biological processes, such as anaerobic production of hydrogen, methane, and ethanol, using renewable carbon sources, such as lignocellulosic waste biomass, may minimize the negative environmental and social impacts. Moreover, these resources do not compete directly with the production of food (Slade *et al.* 2009).

Lignocellulosic biomass in the form of wood and agricultural residues is virtually inexhaustible, since their production is based on the photosynthetic process which is about 60% of total biomass produced (Kuhad *et al.* 1997). It was estimated that terrestrial plants produce annually about 1.3×10^{10} metric tons, equivalent to about two-thirds of the world's energy requirement (Kim and Yun 2006). Moreover, over 1.8×10^{8} tons of agricultural residues or by-products are annually available as source of renewable energy (Kapdan and Kargi 2006). The most abundant lignocellulose agricultural residues are corncobs, corn stover, wheat, rice, barley straw, sorghum stalks, coconut husks, sugar cane bagasse, switchgrass, pineapple, and banana leaves (Demain *et al.* 2005). In addition, other biomass from cereal crops, pulse crops, and palm oil are being produced in a large amount worldwide (Rajaram and Verma 1990). The biomass production rate and chemical composition of some agricultural residues are indicated in Table 13.1. Worldwide Canada is the largest supplier of woody lignocellulosic biomass (about 2.00×10^{8} tons/yr) (Das and Singh 2004), whereas China and India produce about 1.0×10^{9} and 0.2×10^{9} tons, respectively, of agricultural and forest residues (Qu *et al.* 2005). Wood and paper industries also produce huge amount of lignocellulosic biomass. In addition lignocellulosic wastes are also derived from municipal solid wastes (paper, cloth, garden debris) and industrial wastes (paper, packing materials, textiles, bagasse, demolition wood), all of which cause the waste disposal problem. Recycled paper waste has become a resource for biofuels production (Duff and Murray 1996).

Table 13.1. Biomass production rate and chemical composition of some agricultural residues.

Agricultural waste	Amount (ton × 10^6/yr)	Cellulose (%)	Hemicellulose (%)	Lignin (%)	References
Corncobs	159–191	45	35	15	Howard *et al.* (2003); Prasad *et al.* (2007); McKendry (2002)
Wheat straw	154–185	33–38	26–32	17–19	Prasad *et al.* (2007); McKendry (2002)
Bagasse	317–380	32–44	27–32	19–24	Rowell (1992)
Rice bran		35	25	17	Howard *et al.* (2003)
Rice straw	157–188	32–47	19–27	5–24	Prasad *et al.* (2007)
Cotton fiber	17–20	70–85	5–20	0	Howard *et al.* (2003)
Banana waste	13–15	13.2	14.8	14	John *et al.* (2006)
Barley bran		23	32	21.4	Saratale *et al.* (2008)
Barley straw	35–4	31–45	27–38	14–19	Saratale *et al.* (2008)

Lignocellulosic biomass is composed of cellulose (insoluble fibers of β-1,4-glucan), hemicellulose (noncellulosic polysaccharides, including xylans, mannans, and glucans) and lignin (a complex polyphenolic structure). Lignocellulosic biomass is being considered as the largest renewable energy resource and an economically feasible carbohydrate source for producing the new generation of biofuels (Kim *et al.* 2000). However, cellulosic materials are usually not readily hydrolyzable for subsequent microbial fermentation. This is due to the low porosity (accessible surface area) and crystallinity of the cellulosic fiber (Zhang *et al.* 2006). In addition, the different components of the plant fiber cell wall (i.e., cellulose, hemicellulose, and lignin) also limit the accessibility of microorganisms and their enzymes to wood and its fiber components. For that purpose, pretreatment is required to remove lignin and hemicellulose, to reduce the crystallinity of cellulose and to increase the surface area of materials which can improve the formation of fermentable sugars (Zhang *et al.* 2009; Kumar *et al.* 2008). Formation of soluble sugars from cellulose in agricultural residues relies on the sequential/coordinated action of individual components of cellulase enzyme system derived from cellulolytic microorganisms (Adsul *et al.* 2007). To improve the hydrolysis of cellulosic materials, better understanding of their molecular architecture and enzymatic degradation mechanisms are essential.

In this chapter, we first describe the structure of plant cell wall, which is highly correlated to the poor biodegradability of lignocellulosic materials. Next, the microorganisms possessing cellulolytic activities and the related cellulolytic enzymes are introduced. Issues on production and utilization of the cellulolytic enzymes for pretreatment/hydrolysis of lignocellulosic materials are then discussed. Finally, the reported technologies on the conversion of enzymatically pretreated lignocellulosic wastes into biofuels are summarized.

2. Structure and Composition of Plant Cell Wall

The plant cell wall is a complicated natural composite with three main components, i.e., cellulose, hemicelluloses, and lignin, as illustrated in Fig. 13.1 (Terashima and Fukushima 1989; Crawford 1981). The structure and composition of cell wall vary greatly, depending on the plan species, organs, growth conditions, etc. (Zhang and Lynd 2003). Usually plants have a primary and a secondary cell wall. The primary cell wall gives mechanical strength mainly due to the microfibril scaffold. The microfibril scaffold consists of crystalline and amorphous paracrystalline cellulose

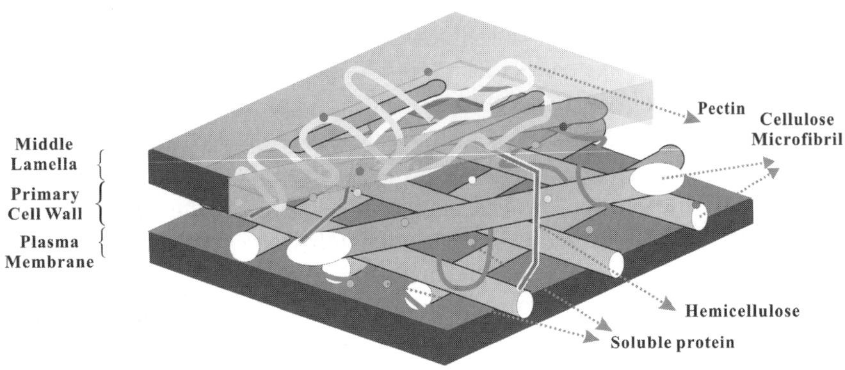

Fig. 13.1. The composition of plant cell wall.

core surrounded by hemicellulose, which is a branched polymer of pentose and hexose sugars. As growth progresses, lignin is successively incorporated into the secondary cell wall through primary cell wall which develops compact structure. The concentration of cellulose is highest in the primary and secondary cell walls, and decreases toward the middle lamella, whereas hemicelluloses and lignin are abundant in middle lamellae (Kuhad *et al.* 1997). The type of carbohydrate changes during the formation and development of cell wall layers (Saka and Goring 1985). The secondary cell wall constitutes the larger proportion of the total cell wall, containing 60–80% of the cell wall lignin. Lignocellulosic biomass has an extremely complex and well-designed nanoscale composite, making it resistant to microbial and enzymatic attacks. A detailed understanding of these chemical constituents of the cell wall components is helpful to develop the mechanistic model and strategy for lignocellulose conversion (Kotchoni *et al.* 2003).

2.1. *The cellulose component*

Cellulose is the most common organic compound on the Earth. About 33% of all plant matter is composed of cellulose (Crawford 1981). Cellulose with a molecular formula of $(C_6H_{10}O_5)_n$ is a linear condensation polymer consisting of 7,000–15,000 glucose monomers linked by β(1–4) glycosidic bonds (Kumar *et al.* 2008). It is water insoluble but recalcitrant to hydrolysis because of it's densely packed, highly crystalline structure with straight, stable supra-molecular fibers of great tensile strength and its low accessibility in polymer form (Demain *et al.* 2005). The multiple hydroxyl groups on the glucose residues from one chain form hydrogen bonds with the oxygen molecules on the same or on a neighbor chain, thereby holding the chains firmly together and forming microfibrils that make recalcitrant compact structure. The microfibrils are a group of about 30 individual cellulose chains, and about 100 microfibrils are packed to form fibrils, and these fibrils are further packed to form the cellulose fiber (Brown and Saxena 2000). It was observed that the degree of crystallinity of cellulose depends on its origin. For example, cotton cellulose has about 70% degree of crystallinity, while other celluloses are 30–70%. Compared to amorphous cellulose, crystalline cellulose is more resistant to microbial attack and enzymatic hydrolysis (Zhang *et al.* 2006; Kumar *et al.* 2008).

2.2. *The hemicellulose component*

Hemicelluloses account for 22% of softwood, 26% of hardwood, and 30% of various agricultural residues (Zhang *et al.* 2007). Hemicellulose consists of branched heteropolysaccharides containing 500–3,000 sugar monomers (Saha 2000; Kumar *et al.* 2008), and contributes to 25–35% of lignocellulosic biomass. Hemicellulose contains many kinds of sugar monomers such as pentoses (D-xylose, L-arabinose), hexoses (D-mannose, D-glucose, D-galactose), and sugar acids. The major hemicellulose components are xylan in hardwood, but glucomannans and galactomannans in softwood. The chemical composition and hemicellulose content depend on the plant materials, growth stage, and growth conditions (Niehaus *et al.* 1999). Xylan is the major ingredient of hemicelluloses, comprising about 15–30% of plants, 20–25% of hardwoods, and 7–12% of softwoods. It consists of 4-*O*-methyl-D-glucuronic acid, L-arabinose, or acetyl groups as substituents on the D-xylose backbone. Xylan is the major interface between lignin and other carbohydrate components as they are covalently linked to phenolic residues of lignin, thereby playing a major role in cell wall cohesion and protecting cellulose from enzymatic attack.

2.3. *The lignin component*

Lignin fills the spaces in the cell wall between cellulose, hemicellulose, and pectin components. It covalently links to hemicellulose, and thereby crosslinks different polysaccharides and gives mechanical strength to the cell wall. Lignin is mostly concentrated in the middle lamella, but may be most abundant in the secondary walls of some vascular plants (Kapdan and Kargi 2006). Lignin is a collection of various phenylpropanoid components having similar chemical properties with molecular weight over 100 kDa. The chemical structure of lignin depends on ecological and environmental factors, such as location, climate, sunlight, age of the wood, and plant sustenance (Crawford 1981). Lignin contains mainly three cinnamyl alcohol precursors, i.e., *p*-coumaryl alcohol, coniferyl alcohol, and sinapyl alcohol. Lignins are highly branched polymers containing phenylpropane-based monomeric units linked by alkyl–aryl, alkyl–alkyl, and aryl–aryl ether bonds (Kumar *et al.* 2008). During the polymerization, covalent bonding between xylan and lignin by ether linkage involving the L-arabinose side chains or xylose units was observed. In addition chemical substituents of the backbone of the hemicelluloses, such as arabinose, galactose, and 4-*O*-methylglucuronic acid, are covalently linked with lignin (Fromm *et al.* 2003), making the plant wall more resistant for microbial attack and degradation. Due to its insolubility in water, lignin is difficult to be penetrated and degraded by microorganisms.

2.4. *Other cell wall components*

Some structural proteins called extensins are also an important structural component of the plant cell wall. They are classified as hydroxyproline-rich glycoproteins, glycine-rich proteins and proline-rich proteins (Lagaert *et al.* 2009). Moreover, pectin, a structural polysaccharide, is abundant in sugar beet and in cell wall of some fruits (Brummell 2006; Bhat 2000). The pectin backbone consists of homo-galacturonic acid regions with neutral sugar side chains made from L-rhamnose, arabinose, galactose, and xylose (Kumar *et al.* 2008). Terpenes, resins, phenols, alkaloids, gums, and various other cytoplasmic constituents are also found in the plant cell wall.

3. Lignocellulose-Degrading Microorganisms

Cellulolytic microorganisms, including fungi and bacteria, produce cellulases to hydrolyze lignocellulosic materials. They can be either aerobic or anaerobic, mesophilic or thermophilic. Bacteria degrade cellulose by cell-bound or extracellular enzymes, while fungi secrete most of these enzymes into the surrounding growth medium (Zhang *et al.* 2007; Rabinovich *et al.* 2002b). The conversion of cellulosic biomass to fermentable sugars relies on the sequential/coordinated action of these multicomponent cellulase enzymes derived from these cellulolytic microorganisms (Bhat and Bhat 1997; Lynd *et al.* 2002).

3.1. *Cellulose-degrading fungi*

Fungi are the most studied organisms with respect to degradation of cellulose and production of cellulolytic enzymes. Among them, brown-, white-, and soft-rot fungi are more effective in breaking down lignin and the most recalcitrant component of the plant cell wall, so that they are generally used in the biological treatment processes. The well-studied soft-rot fungi include species of *Trichoderma*, *Fusarium*, *Aspergillus*, *Penicillium*, *Dactylomyces Papulaspora*,

Allescheria, *Hypoxylon*, *Xylaria*, and *Graphium* (Sánchez 2009; Shary *et al.* 2007). Soft-rot fungi typically attack high-moisture and low-lignin content materials. They attack on the biconical and cylindrical cavities within the secondary cell wall with their hyphae and can degrade lignin after they erode the secondary cell wall (Shary *et al.* 2007). They are effective in the degradation of wood carbohydrates, and can modify lignins in limited extent. Brown-rot fungi, including species of *Tyromyces, Gloeophyllum, Coniophora, Laetiporus*, and *Fomitopsis*, cause rapid and extensive degradation of cellulose. They are most effective in depolymerization of cellulose and hemicelluloses, but can only modify lignin (Yoon *et al.* 2007).

White-rot fungi produce different cellulolytic enzymes and have different modes of action from brown-rot fungi. White-rot fungi produce oxidative enzymes during the degradation of lignocellulose in the early stages of decay (Kimura *et al.* 1991). The basidomycetes white-rot fungi are wood rotting fungi which can attack all the components of the plant cell walls. The most studied fungi of this group are the species of *Phanerochaete*, *Trametes*, *Phlebia*, *Heterobasidium*, *Pleurotus*, *Ceriporiopsis*, and *Polyporus* (Leonowicsz *et al.* 1999). White-rot fungi attack first on the primary cell wall components, followed by the secondary cell wall and middle lamella. Their abilities to degrade lignin of secondary cell wall and middle lamellae make them suitable for industrial applications where lignin or other phenolic compounds need to be modified or removed. *Paneroachaete chrysosporium* and *Trametes versicolor*, can degrade cellulose, hemicellulose, and lignin, while some species of *Phlebia* can degrade the lignin component more selectively (Martinez *et al.* 2004). *Ceriporiopsis subvermispora* and *Cyathus stercoreus* were effective in the delignification of bermuda grass (Akin *et al.* 1995). The different modes of degradation patterns for plant fiber cell walls by several strains of *P. chrysosporium* and *Ceriporiopsis subvermispora*. *Pleurotus ostreatus*, *Phanerochaete sordida* 37 and *Pycnoporus cinnabarinus* 115 were studied in the pretreatment of wheat straw for effective reducing of sugar production (Hatakka 1983). Moreover, white-rot fungi were able to degrade not only lignin but also a variety of persistent environmental pollutants, such as chlorinated aromatic compounds, heterocyclic aromatic hydrocarbons, various dyes, and synthetic high polymers (Saratale *et al.* 2007). This high degradative ability of white-rot fungi is due to the strong oxidative activity and low substrate specificity of their ligninolytic enzymes. Degradation of highly recalcitrant materials, such as mestome sheath of leaf blades, palm press fibers, and wood, by some anaerobic fungi (e.g., *Neocallimastix frontalis*, and *Caecomyces (Sphaeromonas) communis*) has been reported (Bennett *et al.* 2002).

The genome sequences of many fungi, such as *Phanerochaete chrysosporium* strain RP8, *Coprinopsis cinerea*, *Postia placenta*, *P. ostreatus*, *Agaricus bisporus*, *Schizophyllum commune*, and *Serpula lacrymans* have been determined. This genomic information may greatly facilitate the understanding of the lignocellulose biodegradation process and could be useful for commercial cellulase enzymes production (Martinez *et al.* 2004; Sánchez 2009).

3.2. *Cellulose-degrading bacteria*

Bacteria degrade cellulosic biomass slower than fungi, because they lack the penetrating ability. Cellulolytic bacteria include *Clostridium*, *Cellulomonas*, *Bacillus*, *Thermomonospora*, *Ruminococcus*, *Bacteroides*, *Erwinia*, *Acetivibrio*, *Microbispora*, and *Streptomyces* (Bisaria, 1991; Saratale *et al.* 2010). Many cellulolytic bacteria, particularly the anaerobes such as *Clostridium thermocellum* and *Bacteroides cellulosolvens*, produce cellulases of high specific activity, but at slow rates. Both mesophilic and thermophilic aerobic bacteria (*Cellulomonas* sp., *Cellvibrio* sp., *Microbispora bispora*, and *Thermomonospora* sp.) and anaerobic bacteria (*Acetivibrio cellulolyticus*, *Bacteroides cellulosolvens*, *Bacteroides succinogenes*, *Ruminococcus albus*, *Ruminococcus flavefaciens*, and *C. thermocellum*) can produce cellulases effectively. Various species of *Streptomyces* are effective against grass lignocellulosics partial solubilization. *Streptomyces viridosporus* is effective in the oxidative depolymerization of lignin, and thus can degrade cellulose and hemicellulose components of plant residues (de Lima *et al.* 2005). Rumen bacteria, such as *Fibrobacter succinogenes*, *Ruminococcus albus*, and *R. flavefaciens*, can also hydrolyze cellulose by producing effective cellulase (Ohmiya *et al.* 1987). Wood pretreated with cellulase-less mutants of *P. chrysosporium* and *Phlebia pigantea* could be effectively degraded.

Thermophilic microorganisms are of particular interest, because of their ability to produce thermostable cellulases (up to 90 °C) which are generally stable over a wide range of pH (Turner *et al.* 2007). Most studied thermophilic cellulolytic microorganisms are *C. thermocellum*, *Thermomonospora fusca*, *Thermoascus aurantiacus*, *Sporotrichum thermophile*, *Humicola insolens*, and *Chaetomium thermophile*. They can degrade a wide range of cellulosic matters with minimal risk of contamination by pathogens. It is still difficult to effectively utilize these microorganisms and enzymes to treat cellulosic materials. Advanced biological techniques are still needed to improve their cellulolytic rates.

3.3. *Bacterial cellulosome*

Some bacterial cellulase systems differ from fungal ones by forming multienzymes (cellulosomes) from cell-associated aggregates (Demain *et al.* 2005). During hydrolysis, bacteria attach to cellulose particle surface, and the enzymes within the cellulosome synergistically degrade the cellulose into glucose and cellulodextrans, which are transported into the cells for metabolism. Many anaerobic bacteria, such as *C. thermocellum* and *Clostridium cellulovorans*, could produce cellulosome (Fig. 13.2) containing multienzyme complex for cellulose hydrolysis in an organized, concerted, and synergistic manner (Desvaux 2005). Each individual cellulase contains

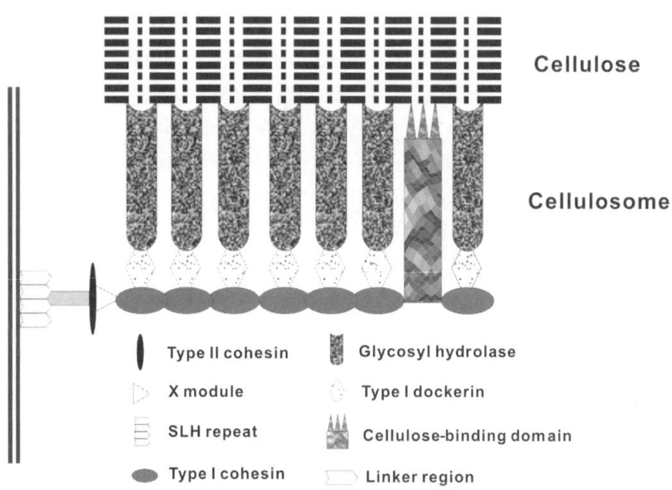

Fig. 13.2. Schematic description of a bacterial cellulosome complex.

a carbohydrate-binding module (CBM) joined by a flexible linker peptide to the catalytic domain. Some anaerobic bacteria, such as *C. thermocellum*, are known to produce cellulosomes on the surface of the cell. These studies suggest that the components of cellulosomes are strongly bound to one another via dockerin module (duplicated, non-catalytic segment of 22 amino acid residues) (Lynd *et al.* 2002). This dockerin module binds specifically to the non-catalytic scaffolding protein (CipA) consisting of nine cohesins, four X-modules, and CBM. The scaffolding can bind to the cell through type II cohesion domains. Many of the catalytic modules exhibit endoglucanases, exoglucanases, xylanases, and cellodextrinases enzyme activity. Cellulosome can efficiently hydrolyze both amorphous and crystalline cellulose, but individual peptides cannot. Recent studies on the genome sequences of aerobic *Cytophaga hutchinsonii* and anaerobic *Fibrobacter succinogenes* suggested a new mechanism, i.e., cellulose hydrolysis by these two bacteria is aided by outer membrane proteins which bind and transport individual cellulose molecules into the periplasmic space and depolymerize them by endoglucanases. This is because in *C. hutchinsonii* no genes were found to code for CBM like other aerobic cellulolytic bacteria, and in *F. succinogenes* no genes were identified to encode dockerin and scaffolding like other anaerobic cellulolytic bacteria. Also, there are no genes that encode for processive cellulases (Demain *et al.* 2005).

4. Cellulolytic Enzymes

Cellulolytic microorganism and enzymes have been well studied, and several microbial applications have been developed for textile, food, and paper-pulp processing. Table 13.2 summarizes various types of cellulolytic enzymes, their microbial sources, and industrial applications.

4.1. *Cellulose-degrading enzymes*

Bioconversion of cellulose into fermentable sugars is a biorefining process that has attracted enormous research efforts. The formation of soluble sugars from cellulose proceeds by synergistic action of at least three major types of hydrolytic enzymes and in some organisms oxidative and phosphorolytic enzymes also participate (Bhat and Bhat 1997; Lynd *et al.* 2002). The cellulose-hydrolyzing enzymes are divided into three groups: endoglucanases (endo-1,4-[3-D-glucan-4-glucanohydrolase, EC 3.2.1.4), exoglucanases (also known as cellobiohydrolases) (exo-1,4-[3-D-glucan-4-cellobiohydrolase, EC 3.2.1.91), and β-glucosidases (β-D-glucoside glucohydrolase, EC 3.2.1.21). Endoglucanases cleave intramolecular β-1,4-glycosidic linkages randomly and releases reducing sugars in the reaction mixture, while exoglucanases act on the accessible ends of cellulose molecules to liberate glucose and cellobiose. On the other hand, β-D-glucosidases hydrolyze cellooligosaccharides and soluble cellobiose and other cellodextrins to produce glucose in the aqueous phase (Zhang *et al.* 2006).

Some cellulolytic microorganisms can produce different arrays of cellulases having different modes of action and substrate specificity. Heterogeneities of the endoglucanase and cellobiohydrolase components from *Trichoderma reesei* have been studied. The fungus can produce three different types of endoglucanases (EG I, EG II, and EGIII) and two immunologically distinct cellobiohydrolases (CBH I and CBH II) extracellularly which act on the reducing and non-reducing cellulose chain ends having more applications in the textile and detergent industries (Zhang *et al.* 2006). The differences in the activities of these enzymes might be due to differences

Table 13.2. Different types of cellulolytic enzymes with their microbial sources and industrial applications.

Type of cellulolytic enzymes	Chemical name	Microbial source		Applications
		Bacteria	Fungi	
Cellulose-degrading enzymes	**Endoglucanases** (endo-l,4-[3- D-glucan-4-glucanohydrolase, EC 3.2.1.4), **Exoglucanases** (exo-l,4-[3-D-glucan-4-cellobiohydrolase, EC 3.2.1.91) **β-glucosidases** (β-D-glucoside glucohydrolase, EC 3.2.1.21) **Cellobiose phosphorylase** (EC 2.4.1.20)	*Clostridium acetobutylicum* *Clostridium cellulolyticum* *Clostridium thermocellum* *Cellulomonas fimi* *Bacillus circulans* *Cellulomonas uda* *Ruminococcus albus* SY 3 *Streptomyces* sp.		1. Hydrolysis of cellulose into fermentable sugars for the production of biofuels. 2. Useful to increase the nutritive quality of fermented foods. 3. Used as a constituent in detergent industry, also found useful for finishing cloth and other fiber. 4. Acts as an additive in the preparation of digestive enzymes. 5. Useful for the extraction of protein, starch from fruits and vegetables. 6. Useful as feed additive and silage quality improvement. 7. Useful in bioremediation such as wastewater treatment.
Hemicellulose-degrading enzymes	**Endoxylanases** (1,4-[3-D-xylan xylanohydrolase, EC 3.2.1.8) **Exoxylanase** (1,4-[3-D-xylan xylohydrolase, EC 3.2.1.37) **Xylosidase** (1,4-[β-D-xylan xylohydrolase, EC 3.2.1.37) **α-L-arabinofuranosidase** (EC 3.2.1.55)	*Bacillus pumilus* *Bacillus subtilis* *Clostridium thermocellum* *Clostridium acetobutylicum* *Streptomyces xylophagus* *Ruminococcus albus* *Cellulomonas* sp.		1. Depolymerization of hemicellulose to monomeric sugars for biofuels and other valuable chemicals production. 2. Useful in pulp industries for debarking prebleaching, and pulp fiber refining. 3. In food industries these enzymes are useful for the extraction of plant oils, starch, and coffee, useful for the production of food thickeners for bakery products, for the processing of cereal flour process for clarification of fruit juices and wines.

(*Continued*)

Table 13.2. (*Continued*)

Type of cellulolytic enzymes	Chemical name	Microbial source: Bacteria	Microbial source: Fungi	Applications
	1,4-β-D-Mannanase (1,4-β-D-mannan mannanohydrolase, EC 3.2.1.78) **1,4- β-mannosidases** (β-D-1,4-mannoside mannohydrolase, EC 3.2.1.25) α-**galactosidase** (α-galactoside galactohydrolase, EC 3.2.1.22)			4. Feed supplementation to improve nutritional quality of agricultural silagee
Lignin-degrading enzymes	**Lignin peroxidase** (ligninase, EC 1.11.1.14) **Manganese peroxidase** (EC 1.11.1.13) **Versatile peroxidases** (EC 1.11.1.16) **Laccases** (benzenediol: 02 oxidoreductase, EC 1.10.3.2) **NAD(P)H: quinine oxidoreductase** **Aryl alcohol dehydrogenase**	*Streptomyces* sp. *Cellulomonas* sp. *Cellulomonas biazotea*	*Phanerochaete chrysosporium* *Trichoderma reesei* *Pleurotus ostreatus* *Aspergillus niger* *Fomitopsis palustris*	1. Useful biological tool for the degradation of lignin, polyaromatic hydrocarbons textile dyes, and other recalcitrant organic compounds 2. For the delignification of wood and agricultural residues to increase the digestibility of ruminant feed 3. Useful for food and beverage industries like to enhance or modify the color appearance of food or beverages. Important to remove undesirable phenolics, responsible for the browning, haze formation, and turbidity development in clear fruit juice, beer, and wine

(*Continued*)

Table 13.2. (*Continued*)

Type of cellulolytic enzymes	Chemical name	Microbial source: Bacteria	Microbial source: Fungi	Applications
				4. Useful as a biosensor for clinical biochemistry and enzyme immunoassays 5. Useful in the synthesis of useful organic and polymer 6. Most importantly acts as a useful system to remove lignin from lignocellulosic biomass for easy saccharification and simultaneous biorefinery products
Pectin-degrading enzymes	**Endo polygalacturonase** (pectin depolymerase, pectinase, EC 3.2.1.15), **Exopolygalacturonase** (EC 3.2.1.67) **Exopolygalacturanosidase** (EC 3.2.1.82) α-**L-rhamnosidases** (EC 3.2.1.40) α-**L-arabinofuranosidases** (EC 3.2.1.55) **Endo-arabinase** (EC 3.2.1.99)	*Aspergillus sojae* *Aspergillus niger* *Aspergillus japonicus* *Penicillium paxilli* *Colletotrichum lindemuthianum* *Rhizoctonia solani* *Fusarium oxysporum* *Cystofilobasidium capitatum* *Penicillium canescens* *Penicillium expansum* *Penicillium italicum* *Penicillium viridicatum* *Pythium splendens* *Rizopus oryzae*	*Aspergillus niger* *Fusarium oxysporum* *Aspergillus wentii* *Trichoderma koningii* *Neurospora crassa* *Penicillium wortmanni* *Polyporus sulphureus* *Pleurotus ostreatus* *Rhizopus chinensis* *Phanerochaete chrysosporium* *Humicola insolens* *Schizophyllum commune*	(1) Useful in the conversion of pectin into fermentable sugar for the production of biogas and ethanol (2) Useful in the degradation of highly esterified pectins of fruits without producing methanol. They are useful in the food industry as a thickener, texturizer, emulsifier, stabilizer, filler in confections, dairy products, and bakery products (3) Useful for clarification of juices In the textile industries they are used for retting of plant fibers such as ramie, sunn hemp, jute, flax and hemp.

of amino acids in the active centers of these enzymes. The repeating cellobiose unit of cellulose present in two distinct stereochemical forms. It is supposed that the two stereospecific forms of endoglucanase and of exoglucanases might exist and are required to bring about optimal cellulose hydrolysis (Wood and Bhat 1988). Moreover, many wood-degrading fungi species of *Aspergillus*, *Phanerochaete*, and *Trichoderma* produce two types of oxidative enzymes, namely cellobiose dehydrogenase (CDH; EC 1.1.9.18) and cellobiose:quinone oxidoreductase (CBQ; EC 1.1.5.1), which oxidize the reducing end group in cellobiose or higher cellodextrins or even cellulose in the presence of a suitable electron acceptor (Rabinovich *et al.* 2002a). Instead of β-D-glucosidases, some aerobic and anaerobic bacteria produce cellobiose phosphorylase (EC 2.4.1.20), which hydrolyzes cellobiose but not cellotriose or higher cellodextrins. In addition, lactonase (EC 3.1.1.17) also involved synergistically with cellulases in the degradation of cellulose (Kuhad *et al.* 1997).

4.2. *Hemicellulose-degrading enzymes*

Hemicellulose is a heterogeneous polymer consisting of pentoses, hexoses, and sugar acids requires several different enzymes for effective degradation. Thus, hemicellulose also acts as an important source of fermentable sugars for biorefining applications. A variety of enzymes are responsible for the degradation of hemicellulose, such as glucuronidase, acetylesterase, xylanase, β-xylosidase, galactomannanase, and glucomannanase (Duff and Murray 1996). Microorganisms producing hemicellulolytic enzymes, including fungal species such as *Aspergillus* and *Fusarium* (Kuhad *et al.* 1997), bacteria species such as *Bacillus*, *Streptomyces*, *Cellulomonas*, and *Thermomonospora* (Zhang *et al.* 2007), as well as anaerobic bacteria such as *Clostridium*, *Thermoanaerobacter*, and *Acetivibrio* (Rabinovich *et al.* 2002a). Xylan and mannose form the backbone of the hemicellulose polymers. Complete degradation of branched xylans requires the concerted synergistic action of different enzymes. Endoxylanases (1,4-[3-D-xylan xylanohydrolase, EC 3.2.1.8) could hydrolyze the 1,4-[3-D-xylopyranosyl linkages of xylans. The existence of two types of fungal endoxylanases, such as arabinose-releasing xylanases and xylotriose-cleaving xylanases, has been reported. *Sporotrichum dimorphosporum* found to be efficient degrader of redwood arabinoglucuronoxylan, producing different glycoxylan oligosaccharides. Moreover, exoenzyme (1,4-[3-D-xylan xylohydrolase, EC 3.2.1.37) has been reported in some studies. Xylanase production was widely studied in species of *Bacillus* by using pure and agricultural raw materials (Adsul *et al.* 2007). *Cellulomonas biazotea* NCIM-2550 can produce effective cellulase and xylanase in the presence of pure cellulosic substrates (CMC and xylan) and sugar cane bagasse (Saratale *et al.* 2010). *Streptomyces* can produce endoxylanase, having higher performance where the degree of branching of arabinose is low. In addition, β-D-xylosidase (1,4-[β-D-xylan xylohydrolase, EC 3.2.1.37) has the ability to degrade xylooligosaccharides to xylose (de Lima *et al.* 2005). The xylosidase activity was observed in the fungal species of *Aspergillus*, *Trichoderma*, *Sclerotium*, *Penicillium*, and some bacterial species, such as *Bacillus*. In addition, α-L-arabinofuranosidase (EC 3.2.1.55) hydrolyzes non-reducing α-L-arabinofuranosyl groups of arabinoxylans and arabinogalactans. Some fungi, mainly *Agaricus bisporus*, *P. ostreatus*, and *T. reesei*, were found to produce α-D-glucuronidase (Kuhad *et al.* 1997) for the hydrolysis of xylo-oligomers. 1,4-β-D-Mannanase (1,4-β-D-mannan mannanohydrolase, EC 3.2.1.78) is involved in the hydrolysis of D-mannans and D-galacto-D-mannans. A variety of microorganisms including rumen bacteria and fungi have ability to produce endo and exo types of D-mannanases mainly at extracellular location; however, some bacteria, such as *Aerobacter mannanolyticus* and *Xanthomonas campastris*, produce endomannans at intracellular location (Eriksson *et al.* 1990). Some fungi such as *Polyporus sulphurius* and mushroom, *Tremella fuciformis*, can produce 1,4-β-mannosidases (β-D-1,4-mannoside mannohydrolase, EC 3.2.1.25) could hydrolyze 1,4-1inked-[β-D-mannosyl groups from the non-reducing end of mannose-containing glycopepetides. In addition, α-galactosidase (α-galactoside galactohydrolase, EC 3.2.1.22) catalyzes the hydrolysis of melibiose, methyl-, ethyl-, and phenyl derivatives of D-galactosides (Rabinovich *et al.* 2002a, b).

4.3. *Lignin-degrading enzymes*

Due to its amorphous, insoluble nature, and lack of steroregularity, lignin is resistant to hydrolytic attack. The oxidative biodegradation of lignin by white-rot fungi, mainly *Phanerochatete chrysosporium*, is widely studied. Mainly three types of phenoloxidases, such as lignin peroxidase (LiP), manganese-dependent peroxidase (MnP), and laccase, are found important in the ligninolytic enzyme system (Kumar *et al.* 2008).

Lignin peroxidase (LiP, ligninase, EC 1.11.1.14) was first discovered based on the H_2O_2-dependent C_α–C_β cleavage of lignin model compounds and subsequently shown to catalyze depolymerization of methylated lignin *in vitro* (Glenn and Gold 1983; Gold *et al.* 1984). LiP are glycoproteins with molecular weight estimated of 38–46 kDa having a heme as prosthetic group. White-rot fungus *P. chrysosporium* is an efficient producer of lignin peroxidase. Mainly lignin peroxidase (LiP) catalyzes the oxidation of nonphenolic aromatic lignin moieties and similar compounds. The enzymes have the ability to depolymerize lignin and oligomers structurally related to lignin *in vitro* (Tien and Kirk 1983) by one-electron abstraction to form L-form reactive radicals (Kersten *et al.* 1985). Also the cleavage of aromatic ring structures has been reported (Umezawa and Higuchi 1987). LiP has been used to mineralize a variety of recalcitrant aromatic compounds, such as three- and four-ring polycyclic aromatic hydrocarbons (PAHs), polychlorinated biphenyls (Gunther *et al.* 1998) and dyes (Saratale *et al.* 2009). In some white-rot fungi a natural metabolite 2-chloro-1,4-dimethoxybenzene found to be redox mediator in the LiP-catalyzed oxidations (Teunissen *et al.* 1998).

Manganese peroxidase (MnP, EC 1.11.1.13) was first discovered by Gold *et al.* (1984) in the nitrogen-depleted ligninolytic cultures of *P. chrysosporium*. The reaction mechanism of MnP is similar with LiP as it requires H_2O_2 as a cosubstrate and contains heme as a prosthetic group. This enzyme is abundantly found in wood-degrading white-rot fungi and ectomycorrhizal fungi (Arora *et al.* 2002). Organic acids, including oxalic acid, malonic acid, and pyruvic acids, are found to be inducers for the enzyme. This fungal enzyme has ability to oxidize the phenolic structure of lignin and other aromatic compound to carbon dioxide (Hofrichter 2002). Recently a third group of peroxidases, i.e., versatile peroxidases (VP; EC 1.11.1.16), has been found in species of *Pleurotus* and *Bjerkandera*; it can be considered as a hybrid between MnP and LiP, since it can oxidize not only Mn^{2+} but also phenolic and other aromatic compounds (Heinfling *et al.* 1998).

Laccases (benzenediol: oxygen oxidoreductase, EC 1.10.3.2) are copper-containing enzymes belonging to a small group known as blue oxidase enzymes. Laccase is a diphenol oxidase that catalyzes the oxidation of several aromatic and inorganic substances (particularly phenols) with the concomitant reduction of oxygen to water (Durán *et al.* 2002). Laccases have four copper atoms distributed among binding sites. They are classified into three types, i.e., Copper I, II, and III, which are differentiated by specific characteristic properties that allow them to play an important role in the catalytic mechanism of the enzyme (McGuirl and Dooley 1999). The molecular weight of laccase is in the range of 60–390 kDa (McGuirl and Dooley 1999). Wood-degrading white-rot fungi such as *Coriolus versicolor*, and *Phanerochaete chrysosporium* are found to be effective producers of laccases. Laccases have been intensively studied for their industrial applicability (Yaropolov *et al.* 1994; Bajpai 1999), molecular genetics, cloning (Cullen 1997; Collins and Dobson 1997; Hatamoto *et al.* 1999) and in the degradation of recalcitrant compounds, such as chlorophenols, PAHs (Fahr *et al.* 1999), organophosphorus compounds (Amitai *et al.* 1998) and lignin-related structures (Widsten and Kandelbauer 2008). Moreover, several processes using laccases have been developed for the treatment of phenolic effluents and PAHs. In addition, NAD(P)H:quinine oxidoreductase, aryl alcohol dehydrogenase are also involved in the degradation of lignin. *P. chrysosporium* producing at least two different intracellular NAD(P)H:quinone oxidoreductases that can reduce methoxyquinone using either NADH or NADPH as electron donors (Kuhad *et al.* 1997).

4.4. *Pectin-degrading enzymes*

The pectin is a major component of plant cell wall consisting of backbone of homo-galacturonic acid. Pectin has widespread applications in the textile and food industries and is also a good carbohydrate source for bioenergy production (Kumar *et al.* 2008). Many enzymes, including polymethylgalacturonase, (endo-) polygalacturonase (pectin depolymerase, pectinase, EC 3.2.1.15), exopolygalacturonase (EC 3.2.1.67), and exopolygalacturanosidase (EC 3.2.1.82), are known to be involved in the hydrolysis of the polygalacturonic acid chain of pectin (Bhat and Bhat 1997; Jayani *et al.* 2005). Usually α-L-rhamnosidases (EC 3.2.1.40) degrade rhamnogalacturonan in the pectic backbone, while α-L-arabinofuranosidases (EC 3.2.1.55) hydrolyze the L-arabinose side chains, and endo-arabinase (EC 3.2.1.99) acts on arabinan side chains in pectin (Takao *et al.* 2002).

5. Production and Utilization of Cellulolytic Enzymes

5.1. *Recent development in the productions of cellulolytic enzymes*

Production cost usually contributes heavily to price of the commercial cellulolytic enzymes. Therefore, it is of great importance to develop efficient and cost-effective technology for the enzyme production. The factors influencing the efficiency of cellulolytic enzyme production include the producing strains, choice of substrate, substrate concentration, and cultivation/fermentation conditions. The producers of cellulolytic enzymes include fungal species, such as *Trichoderma*, *Humicola*, *Penicillium*, and *Aspergillus*, as well as bacterial species, such as *Bacillus*, *Pseudomonas*, *Cellulomonas*, and *Actinomycetes* such as *Strepomyces*, *Actinomucor*, and *Streptomyces*. Among them, fungus *T. reesei* is the most commonly used microorganism for cellulolytic enzyme production. Moreover, genetically engineered fungal or bacterial species were often used for cellulolytic enzymes production to enhance the productivity, reduce the production cost, and simplify the downstream processing. Although the commercial-scale cellulolytic enzyme production technology has been established since 1980s, there are still some new developments in recent years on the production of cellulolytic enzymes. In particular, several novel cellulases-producing microbial strains were isolated from the environment; their cellulolytic enzymes appear to possess special features, such as high thermostability, high alkaline tolerance, and multifunction. In addition, molecular biology-based technology was applied to improve the enzyme-producing strains as well as the enzymes themselves.

A novel cellulose-degrading *Paenibacillus* sp. strain was recently isolated from poultry manure compost in Taiwan through 16S rRNA gene sequencing and phylogentic analysis (Wang *et al.* 2008). This isolated strain possesses a high molecular weight (148 kDa) cellulase, possessing both CMCase and Avicelase activities. The CMCase activity of the newly isolated cellulolytic microorganism was much higher than the activity on Avicel or filter paper and this cellulase was found to have maximum CMCase activity at 60 °C and pH 6.5. Due to the promising thermostability and slight acidic tolerance of this enzyme, it has good potential for industrial use in the hydrolysis of soluble cellulose. Ko *et al.* (2007) also isolated a novel cellulase-producing *Paenibacillus campinasensis* BL11 from black liquor of brownstock at washing stage of the Kraft pulping process. Since this black liquor is strongly alkaline, the enzymes produced could be tolerant to the alkaline pretreatment of lignocellulosic biomass. *P. campinasensis* BL11 is a thermophilic, spore-forming bacterium growing at 25–60 °C over a wide range of pH, with optimal growth at neutral pH at 55 °C. This isolate could use a variety of saccharides and polysaccharides to produce multiple extracellular saccharide-degrading enzymes (e.g., a xylanase, two cellulases, a pectinase, and a cyclodextrin glucanotransferase), which have high potential to apply in biorefining industry.

More recently, a thermostable endocellulase CelDR was found in an isolated *Bacillus subtilis* DR from a hot spring (Liang *et al.* 2009) with an optimum growth temperature of 50 °C. It retained 70% of its maximum CMCase activity at 75 °C after incubation for 30 minutes. This strain offers a potentially valuable thermostable enzyme for the biorefining industry due to extreme heat tolerance. Also recently, a novel thermophilic, cellulolytic bacterium was isolated from swine waste and identified as *Brevibacillus* sp. strain JXL (Liang *et al.* 2009), capable of hydrolyzing a broad spectrum of substrates, such as crystalline cellulose, CMC, xylan, cellobiose, glucose, and xylose. The enzymes are highly thermostable, retaining 50% of their activity after 1 h at 100 °C. Furthermore, a salt-activated endoglucanase was also discovered from alkaliphilic *Bacillus agaradhaerens* JAM-KU023. With the addition of 0.2 M NaCl, the enzymes produced from this strain showed an increased optimal thermostability from 50 °C to 60 °C and optimal pH range from 7 to 9.4 (Hirasawa *et al.* 2006). In addition, a bifunctional endoglucanase/endoxylanase was obtained from *Cellulomonas flavigena* providing potential for use in different industrial processes, such as biofuel production. This bifunctional enzyme has optimum cellulase and xylanase activity at pH 6 and 9, respectively, at the optimum temperature of 50 °C (Pèrez-Avalos *et al.* 2008). Similarly, a multifunctional enzyme was produced by *Terendinibacter turnerae* T7902, which is a bacterial symbiont isolated from the wood-boring marine bivalve *Lydrodus pedicellatus* (Ekborg *et al.* 2007). This enzyme is marginally acid tolerant at the optimum conditions of pH 6 at 42 °C and is able to reduce viscosity of CMC by 40% after 25 minutes, displaying promising characteristics for the biofuel industry. All of these recently isolated enzymes suggest a promising future for the development of effective hydrolysis systems to be used in the biorefining industry.

Despite the broad spectrum of cellulases being isolated, no single enzyme is completely suitable for the hydrolysis of cellulose in the biorefining industry. However, these enzymes offer a good starting point for the improvement of cellulases in steps toward enhancing the overall economics of biofuel production. Mahadevan *et al.* (2008) focused on the amino acids around the active site of endoglucanase Cel5A from *Thermotoga aritima*, creating the N147E mutant which displayed 10% higher activity than the wild-type Cel5A. This group also showed a correlation between binding ability and the activity of the enzyme. By binding two CBDs, one from *T. reesei* and the other from *Clostridium stercorarium*, to Cel5A this CBD-engineered Cel5A displayed 14- to18-fold higher hydrolytic activity toward Avicel (Mahadevan *et al.* 2008). Furthermore, the enzyme activity increased by 80% for a mutant Cel5Z endoglucanase of *Pectobacterium chrysanthemi* when compared to the wild-type enzyme. However, this mutant enzyme was created by the use of a nonsense mutation which removed the C-terminal region creating a truncated Cel5Z containing 280 amino acids compared to the native Cel5Z which has 426 amino acids. Without the CBD, this enzyme would not be efficient for hydrolysis of crystalline cellulose but could offer potential for solubilized cellulose (Lim *et al.* 2005). Likewise, the Cel5Z::Omega mutant of *P. chrysanthemi* hydrolyzed CMC with 1.7-fold higher activity than the intact Cel5Z cellulase (Park *et al.* 2002). Similarly, the production efficiency of a complex multifunctional enzyme Cel44C-Man26A secreted by *Paenibacillus polymyxa* GS01 was enhanced by using protein engineering approaches (Cho *et al.* 2008).

Directed evolution was also used to improve the thermal stability of *C. cellulovorans* cellulosomal endoglucanase (EngB) *in vitro* by DNA recombination with non-cellulosomal endoglucanase EngD. The screening was done using CMC agar and staining with Congo red (Murashima *et al.* 2002). Further, DNA shuffling was used to create a library of mutated endoglucanases from *B. subtilis*. Interestingly, a bacterial surface display method was used to selectively screen for variants with improved activity on CMC agar with Congo Red staining. By fusing the genes with the ice nucleation protein (Inp) the resulting fusion proteins would be displayed on the bacterial cell surface for easy screening (Kim *et al.* 2000). The catalytic activity of 1,4-β-D-glucan glucohydrolase A from *Thermotoga neapolitana* was improved using error-prone polymenase chain reaction (PCR) to generate the gene variant library (McCarthy *et al.* 2004). While catalytic activity of a hyperthermostable β-glucosidase CelB from *Pyrococcus furiosus* was improved, by family shuffling, by 3- to 5-fold compared to the wild-type (Kaper *et al.* 2000).

Although many cellulose-degrading microorganisms grow rapidly, only a few produce extracellular cellulases capable of converting crystalline cellulose into glucose *in vitro*. Of these, culture filtrates from *Trichoderma viride* are the most active. The cellulase activity in *Trichoderma filtrates* is stable and may be stored for months under refrigeration, or as lyophilized, or acetone extract powders, without any significant loss of activity. Cellulase enzyme of *Trichoderma* is produced when cellulose is present in the medium, as the cellulase is an inducible enzyme. The nature of inducer, is not certain. Cellulases are repressible, which means that if the rate of carbohydrate catabolism exceeds that required for cellular biosynthesis, cellulase synthesis decreases or stops. By comparing the kinetics of cellulase repression by glucose with inhibition of RNA and protein synthesis (by actinomycin D and puromycin, respectively), Nisizawa *et al.* (1972) inferred that cellulase synthesis is repressed at the level of protein synthesis. Because the cellulases are repressible, the addition of glucose, glycerol, or any other rapidly metabolized carbon source to a cellulose culture will interfere with cellulase synthesis. The major advance in improvement of cellulase yield was the generation and isolation of hyper-producing mutants (Mandels *et al.* 1971). By irradiating spores with a linear accelerator and screening survivors for enhanced cellulase production, a strain (QM 9123) having 2-fold increase in enzyme production was found. By taking this strain and subjecting it to the same irradiation and screening procedure, strain QM9414 was isolated and produces four times the amount of cellulase as the parent.

5.2. *Immobilization of cellulolytic enzymes for repeated uses*

The cost of cellulase production represents as high as 60% of the total cost of enzymatic hydrolysis of cellulose. Therefore, every effort to reuse the cellulase should be justified. Membrane filtration and enzyme immobilization are two of most often applied techniques for the reuse of enzymes in a biochemical process. For the former, the enzymes are kept in the bioreactor by using batch or continuous ultrafiltration systems so that they can be completely recycled until their activity is declined after long-term operations. In the latter, enzymes are either entrapped inside porous matrixes or attached to the surface of a high-surface-area support via covalent bonding or adsorption so that the enzymes can be easily retained inside the bioreactor for repeated uses. Membrane filtration could be effective, but often suffers fouling problems during continuous operations. Also, the cost of ultrafiltration membrane is quite high when the process is scaled up. In contract, the enzyme immobilization method is cheaper and easier to operate, thereby being considered a better approach in general. Although immobilized enzyme system has been successfully used in a variety of biochemical processes and is shown to be technically feasible and commercially viable for the reuse of enzymes. However, since the substrates of cellulolytic enzymes are often insoluble or in solid form, the poor contact and mass transfer restriction between immobilized cellulolytic enzymes and their substrates could be a major drawback of using an immobilized enzyme system. Nevertheless, as early as 1997, immobilized cellulase was already applied to hydrolyze cellulosic materials (Karube *et al.* 1977). The cellulases were immobilized inside collagen beads and saccharification by employing a fluidized-bed reactor. Some researchers reported the adsorption of the enzyme on charcoal and on alumina, and demonstrated that these immobilized enzymes could retain their activity for a long time. Hanaee *et al.* (1997) also reported immobilization of cellulolytic enzymes by covalent attachment to organic copolymers.

A number of methods have been recently developed to immobilize enzymes. These include: (1) covalent bonding to a solid phase, (2) covalent bonding to soluble polymers, (3) physical adsorption to a solid phase, (4) crosslinking at solid surfaces, (5) crosslinking with difunctional reagents, (6) inclusion in a gel phase, and (7) encapsulation. The largest amount of work by far has been done on the covalent attachment to solid phases. The covalent attachment procedure involves the formation of an activated carrier, followed by the reaction, of the activated carrier with an enzyme to form a composite. This may take a single reaction or several steps of reaction. The most widely used technique for placing reactive organic groups on inorganic surfaces is by using silane coupling agents (Ko *et al.* 2005). Silane coupling agents have inorganic functionality at one end and organic functionality at another. The inorganic functional groups, which can be esters, halides, or silanols, condense with hydroxyl groups on inorganic surfaces. A reactive aldehyde intermediate is readily prepared by reacting glutaraldehyde with an amine group on surface of the carrier (Busto *et al.* 1997). Aldehyde groups on an inorganic carrier react with primary amines to form an imine coupling. The amine groups on the carrier and the carboxyl groups on enzyme react directly to produce amide linkages with help of carbodiimide. A substituted urea is produced from each amide linkage as by-products.

In the last few years, attempts have been made to immobilize a sequence of enzymes in close proximity to each other on the same carrier particle. The carriers can be made of organic or inorganic compounds. The morphology and size of the carrier are extremely important with respect to surface area and pore parameters, both of which in turn will affect the loading of enzyme. A sample of the diversity and variety of carriers currently available for immobilization includes glass particles (Weibel *et al.* 2004), controlled-pore glass, aluminum oxide (Heilmann *et al.* 2003), nickel oxide (Salimi *et al.* 2007), hydroxyapatite (Salman *et al.* 2008), iron oxide (Konwarh *et al.* 2009), cellulose, agarose (Patoomporm *et al.* 1986), polyarcrylamide, nylon, and collagen. Albert *et al.* (1989) commented on the potential use of magnetic materials as carriers. A wide range of magnetizable materials as carriers, including iron, iron oxide, steels, and ferrites, are available with properties which can be used to tailor materials for different reactor applications. Kondo and Fukuda (1997) reported the preparation of ferromagnetic catalyst carriers and magnetic filtration of small particles of ferro-, ferri-, and paramagnetic catalyst carriers.

Immobilized enzyme reactors fall into several general categories, including batch reactors, continuous stirred tank reactors, fixed-bed reactors, and fluidized-bed reactors. Other variations include a stirred tank reactor in which the immobilized enzyme is enclosed in mesh containers attached to a stirrer to give adequate agitation with minimal attrition of immobilized enzymes. Closset *et al.* (1974) studied a tubular membrane reactor for the hydrolysis of starch by α-amylase. Enzyme and starch were contained in the membrane, which was permeable to the produced maltose, but not to the starch and enzyme. Olanoff and Venkatasubramanian (2004) used a device consisting of alternate collagen-enzyme membrane and backing layers wound around a feed distributor. A new approach to immobilize enzymes on magnetizable particles has been carried out a step further (Zhang *et al.* 2008) by using a fluidized-bed reactor containing papain bound to magnetic carriers. The particulate-immobilized enzymes were retained in the column by a circular magnet encircling the upper part of the column. The aforementioned immobilization techniques and processes could be applied to enzymatic hydrolysis of cellulosic substrates. However, since cellulosic substrates are usually insoluble, the mass transfer between the immobilized cellulase and substrates is of particular concern. To enhance the efficiency of mass transfer, the immobilized cellulase should be of small particle size (e.g., using magnetite nanoparticles as carrier) and of high accessibility (i.e., low mass transfer hindrance).

6. Enzymatic Pretreatment of Cellulosic Materials

As pretreatment of lignocellulose waste is the rate-limiting step and has a major influence on the costs of both prior operation (e.g., lignocellulose particle size reduction) and subsequent operations (e.g., enzymatic hydrolysis and fermentation) (National Research Council 1999). Conventional physicochemical methods for cellulose hydrolysis not only require large inputs of energy but also make secondary pollution. Biorefineries, on the other hand, utilize the activities of whole microbial cells and their enzymes to convert cellulose into fermentable sugars. Biorefinery is the overwhelming choice for lignocellulosic waste treatment due to their ease of use, high efficiency and low energy requirement, as compared to conventional physicochemical methods. Biological pretreatment of cellulosic materials could be achieved by using cellulolytic enzyme directly or using cellulolytic microorganisms for a combined microbial and enzymatic hydrolysis.

6.1. *Direct application of cellulolytic enzymes*

Enzymatic hydrolysis of cellulose is carried out by the cellulose-hydrolyzing enzyme cellulases, a mixture of several enzymes that hydrolyze crystalline/amophorous cellulose to fermentable sugars (Duff and Murray 1996). The interaction between hydrolytic enzymes and cellulosic substrates is complex, in part due to the significant number of possible interactions in the system involving a multienzyme complex that adheres to a multicomponent insoluble biomass substrate and acts catalytically upon it (Rabinovich *et al.* 2002a; Zhang *et al.* 2006). There are several advantages of enzymatic hydrolysis, including little energy requirement and mild reaction conditions, high substrate specificity, high yield of sugars, and high hydrolysis efficiency. However, compared to chemical processes, enzymatic hydrolysis has certain disadvantages, including low hydrolysis rate and high cost. Enzymatic hydrolysis of cellulose consists of three steps: adsorption of cellulase enzymes onto the surface of the cellulose, biodegradation of cellulose to fermentable sugars, and desorption of cellulase. Retardation of cellulase activity during hydrolysis may be because of the irreversible adsorption of cellulase on cellulose (Converse *et al.* 1988; Zhang *et al.* 2006).

Conventional chemical pretreatment using acid or alkali could disrupt the crystalline structure of lignocelluloses and make cellulose more accessible to the enzymes for the conversion of the polysaccharides into fermentable sugars (Mosier *et al.* 2005). Conducting such treatment at high temperature with thermostable enzymes has the advantages of enhancing enzyme penetration and disorganization of lignocellulosic cell walls (Turner *et al.* 2007). Supplementation of surfactants (e.g., Tween 20 and Tween 80) during hydrolysis is capable of modifying the cellulose surface property and minimizing the irreversible binding of cellulase on cellulose (Kaar and Holtzapple 1998; Eriksson *et al.* 2002). The addition of polymers such as polyethylene glycol (PEG) can also effectively increase enzymatic hydrolysis of lignocelluloses due to a higher availability of enzymes for cellulose degradation (Borjesson *et al.* 2007). In lignin-containing substrates, addition of bovine serum albumin (BSA) reduced adsorption of cellulase on lignin resulting in an increase in the activity (Yang and Wyman 2006; Ferreira *et al.* 2009). Recently, Saratale *et al.* (2010) reported that the addition of certain metal additives, such as Mn^{2+}, could effectively enhance the multicomponent cellulase enzyme system of *C. biazotea* NCIM-2550. Additional research efforts have been taken to improve the cellulase enzyme system by studying the cellulase structure and mechanism of action, the reconstitution of cellulase mixtures (cocktails), enzyme immobilization, random mutagenesis as well as genetic engineering approaches for cost-effective cellulase enzyme production (Cherry and Fidantsef 2003; Himmel *et al.* 1996; Tao and Cornish 2002; Zhang *et al.* 2006).

6.2. *Microbial hydrolysis of cellulosic materials*

Effective degradation of lignocellulosic biomass could be achieved by the use of efficient cellulolytic microorganisms which can produce effective cellulolytic enzymes during hydrolysis. Filamentous fungi are found ubiquitous in the environment, inhabiting ecological niches such as soil, living plants, and lignocellulosic waste material. The fungi rapidly adapt their metabolism to varying carbon and nitrogen sources which is achieved through the production of a large set of intra- and extra-cellular enzymes able to degrade various complex organic pollutants. Several studies have shown that white-rot fungi are the most effective microorganisms for the pretreatment of lignocelluloses, such as wood chips, wheat straw, bermuda grass, and softwood (Hatakka 1983; Akin *et al.* 1995). Some reports suggested that application of mixed bacterial culture is also useful for efficient hydrolysis of lignocellulosic waste (Datar *et al.* 2007). Processes of using fungal and bacterial systems for cellulose hydrolysis are inexpensive and easy to operate. However, they suffer the drawback of the consumption of the hydrolyzed products (such as reducing sugars) due to the cell growth.

To overcome these problems, some studies suggest a two-stage process using mixed or pure microbial culture for hydrolysis and the subsequent fermentative bioenergy production (Lo *et al.* 2008, 2009; Saratale *et al.* 2010). Recently some investigators observed that using temperature shift strategy the efficiency of two-stage process was enhanced. In this process, the bacterium was first cultivated at suitable temperature for cell growth and cellulases production. After certain bacterial growth, the temperature was increased to increased to a higher level which is favorable for enzyme activity of cellulase system but inhibits the bacterial growth. In this way, more reducing sugar was produced, thereby achieving higher biohydrogen production. In addition, the selected microbial communities used to harvest energy from lignocellulosic waste must be resilient to fluctuations in environmental conditions, variations in nutrient and energy inputs, and intrusion by microbial invaders that might consume the desired energy product. It was observed that different microorganisms can grow on lignocellulosic biomass under anaerobic condition and convert this abundant organic matter into useful forms of energy such as methane, hydrogen, or even electricity. Fermentation is the essential step in any process to recover energy from lignocellulosic biomass. The great advantage of fermentation is fast degradation of solids and other complex organics found in the wastes and agricultural products into mixture of simpler molecules, such as organic acids, alcohols, and hydrogen. Thus, desirable bioenergy production could be achieved by developing a novel and effective bioreactor design and by optimization and improvement of various physicochemical and operation conditions for pretreatment.

7. Anaerobic Treatment and Bioenergy Production from Lignocellulosic Wastes

Today, there is a rekindled worldwide interest in the development of new and cost-efficient processes for converting plant-derived biomass to bioenergy (Kuhad *et al.* 1997; Gong *et al.* 1999). However, the task of hydrolyzing lignocelluloses to fermentable monosugars is still technically problematic because of their low porosity, crystallinity of cellulose fiber, and the lignin and hemicellulose content which prevents cellulase from accessing the substrate (Pan *et al.* 2005; Saratale *et al.* 2008). Hence, it would be helpful to apply physicochemical or biological pretreatment to increase the biodegradability of lignocellulosic materials and enzymatic hydrolysis to convert them into sugars. These techniques, however, usually involve complicated procedures and require large amount of energy and chemicals (Lo *et al.* 2009; Dale 1999). Thus, there is an urgent need to develop more effective and environmentally friendly treatment methods for the hydrolysis and utilization of lignocellulosic biomass. Anaerobic fermentation is an effective process to recover bioenergy from municipal, agricultural, food-processing and livestock wastes and wastewaters. The process is a proven, simple, low energy requirement, and economically feasible. It has the ability to generate biofuels and some valuable chemical products from lignocellulosic waste (Wang and Wan 2009; Hallenbeck and Benemann 2002).

7.1. *Cellulosic hydrogen production*

Hydrogen is a proposed "fuel of the future." The reason for this hope is due to the development of fuel cell technology which may generate electricity from hydrogen with high efficiency and zero emissions of air pollutants. At present, almost all of the hydrogen produced comes from non-renewable fossil sources, such as natural gas and coal. Hence, clean and sustainable production of hydrogen is of great demand. Fermentative production of hydrogen from renewable resources such as lignocellulosic wastes would become the major and attractive source of hydrogen for energy production in the future (Kapdan and Kargi 2006). Usually biohydrogen can be produced by using photo fermentation (Chen *et al.* 2008) and dark fermentation (Lo *et al.* 2009; Wu *et al.* 2008). Dark fermentation for hydrogen production using renewable biomass has been demonstrated using various organic constituents in wastewater (Lin and Chang 1999; Ren *et al.* 2006, Wu *et al.* 2006). However, only limited data on hydrogen yield from lignocellulosic waste or wastewater sludge have been reported (Saratale *et al.* 2008; Kapdan and Kargi 2006).

Cellulosic biohydrogen production could be achieved via two approaches. One is to use a direct process in which cellulose hydrolysis and hydrogen production are taken in one reactor, and the other is to use a two-stage process, in which cellulose is hydrolyzed in the first stage followed by dark hydrogen fermentation for hydrogen production in the second stage (Saratale *et al.* 2008). The direct process could be cheaper and more commercially feasible, but it may be less efficient since the preferred conditions for cellulose degradation and dark H_2 fermentation could be significantly different and the cellulose hydrolysis step could become the rate-limiting step in the system. In contrast, the two-stage process could be more efficient because the cellulose hydrolysis and dark fermentation could be optimized individually. However, the operation cost might be increased when using the two-stage process. Moreover, to enhance the biodegradation efficiency of cellulosic wastes, a thermophilic process may be preferred. In fact, some thermophilic H_2-producing bacteria, such as *C. thermocellum*, could produce cellulolytic enzymes for effective cellulose hydrolysis.

Significant amounts of hydrogen were produced from cellulosic feedstocks (straw, wood chips, grass residue, paper waste, saw dust, etc.) using conventional anaerobic dark fermentation technology and natural mixed microflora under conditions that favor for hydrogen-producing acetogenic bacteria and inhibit methane-producing bacteria (Sparling *et al.* 1997; Valdez-Vazquez *et al.* 2005). Some studies reported that the application of thermophilic microorganisms, such as *Thermococcus kodakaraensis* KOD1, *Clostridium thermolacticum*, and *C. thermocellum* JN4, could be useful for the cellulosic biohydrogen production (Kanai *et al.* 2005; Liu *et al.* 2008). Recently, hydrogen production in batch culture under anaerobic dark fermentation utilizing cellulose, shredded filter paper, and delignified wood fibers (DLWs) by *C. thermocellum* 27405 has been reported (Levin *et al.* 2006). However, depending on the metabolic shift used by the organisms within the consortium, hydrogen yields varied (Hallenbeck and Benemann 2002). In this process the observation suggested that during as well as after hydrolysis some non-hydrolytic microorganisms consumed the produced reducing sugars for their growth, reducing the hydrogen production yield significantly. Also, partial hydrolysis observed makes the process less efficient (Saratale *et al.* 2008).

Some investigators demonstrated that hydrogen can be effectively produced from cellulose in a two-stage process, in which cellulose hydrolysis is carried out in the first stage, and the sugar-rich hydrolysate is effectively converted into hydrogen in the second stage (Mosier *et al.*

2005). The studies indicated that this process requires mild conditions relative to most chemical hydrolysis with increased hydrogen yield, thus the process becoming more advantageous in practical applications due to being more economically feasible and less energy intensive (Lo *et al.* 2008; Taguchi *et al.* 1996; Lo *et al.* 2009; Saratale *et al.* 2010). However, the critical challenges of the hydrogen fermentation with these two approaches were low hydrogen conversion efficiency, time consuming, unstable hydrogen production, and significant production cost. Most cellulosic biohydrogen-producing processes reported in the literature were operated on batch or semi-batch mode. This is partly due to the nature of cellulosic substrate, which is usually in solid form. However, in two-stage process, when cellulosic materials were solubilized by hydrolysis, it would be easier to operate on a continuous mode. Although little is known from literature regarding bioreactor design for cellulosic biohydrogen production, the suitable bioreactor design and optimization of the operating parameters, such as pH, hydraulic retention time (HRT), and temperature, could enhance the hydrogen conversion efficiency. In addition, enhancement in hydrogen production could be achieved by controlling the microbial ecology in the fermentation process so that electrons and energy flow to hydrogen, instead of being diverted to other end products. For that purpose, it is essential to understand the microbial ecology and to control the favorable ecology by using appropriate engineering strategies.

7.2. *Cellulosic ethanol production*

Bioethanol can be produced by using feedstocks containing sucrose (e.g., sugar cane, sugar beet, sweet sorghum, and fruits), starch (e.g., corn, milo, wheat, rice, potatoes, cassava, sweet potatoes, and barley), and lignocellulose (e.g., wood, straw, and grasses) (Goh *et al.* 2010). Most industrial ethanol production uses sugar cane molasses or enzymatically hydrolyzed starch (from corn or other grains), and yeast *Saccharomyces cerevisiae* (Balat and Balat 2009). By-products of this process are carbon dioxide, low amounts of methanol, glycerol, etc. Yeast fermentation of glucose syrups to ethanol has been well progressed in recent years but found economically infeasible. Thus, abundant and renewable lignocellulosic biomass feedstock has been considered as the low-cost feedstock for bioethanol production (Gray *et al.* 2006). However, efficient production of ethanol from lignocellulosic biomass needs pretreatment because of its recalcitrant compact structure. Some studies reported that the alcohol fermentation using lignocellulosic hydrolysates has technological problems, one of which is that the enzymatic hydrolysis reaction of cellulose is about two orders of magnitude slower than the average ethanol fermentation rate with yeast (Antoni *et al.* 2007). In addition, there is a theoretical gap in simultaneous saccharification of cellulosic biomass and ethanol fermentation as well as proportion of pentose and hexose sugars concentration (Hahn-Hägerdahl *et al.* 2007; Torney *et al.* 2007). Bacterial ethanol fermentation can use all sugars derived from cellulosic biomass; however, it suffers with catabolite repression. The widely studied *Zymomonas mobilis* is considered the work horse of bacterial ethanol fermentation (Alterthum and Ingram 1989). Recently a new recombinant *Escherichia coli* B strain LY165 developed for bioethanol plant located in the Bay of Osaka, Japan (Ohta *et al.* 1991). Cheese whey has also been used as a substrate for ethanol fermentation (Siso 1996). *Streptococcus fragilis* and *Kluyveromyces fragilis* are used widely for commercial ethanol production (Pesta *et al.* 2006). The thermophilic bacterium *C. thermocellum* could readily hydrolyze cellulosic biomass, degrading hemicellulose, and cellulose for ethanol production (Lynd *et al.* 2002). Cellulolytic *Clostridia* gives significant cellulose hydrolysis but after hydrolysis diversion toward different metabolic shifts, gives mixed gaseous, acidogenic fermentation products (H_2 and CO_2; acetate, lactate, butyrate, and ethanol) (Lynd *et al.* 2002; Demain *et al.* 2005; Wu *et al.* 2008). *C. thermocellum* is found sensitive to higher ethanol concentrations and produces a range of less favorable by-products (Tailliez *et al.* 1989). However, the metabolic engineering study of *C. thermocellum* suggests its applicability in cellulosic ethanol production (Antoni *et al.* 2007).

Some studies reported that after hydrolysis of lignocellulosic biomass, the produced pentose sugars (mainly D-xylose and L-arabinose) create problem in yeast alcohol fermentation, because yeast strains lack the xylose utilization enzymes (mainly xylose reductase and xylitol dehydrogenase) (Hahn-Hägerdahl *et al.* 2007). In agricultural raw material and hardwoods, pentose sugars are present in larger proportion, which cannot be neglected if we want to increase the yield of ethanol and complete substrate utilization. Recently, Hahn-Hägerdahl *et al.* (2007) developed pentose-fermenting yeast strains and applied for pilot projects. Bioethanol has a high octane number (108), broad flammability limits, high flame speeds, and high heats of vaporization, thereby increasing the theoretical efficiency and thus being advantageous over gasoline in an IC engine (Balat 2007). However, the production of ethanol is costly, and the bioethanol has only 66% of the energy content in the gasoline. The ethanol gasoline also has the drawback of high corrosivity, low flame luminosity, low vapor pressure (making cold starts difficult), miscibility with water, toxicity to ecosystems (MacLean and Lave 2003), and increasing exhaust emissions of acetaldehyde. Thus, many problems need to overcome when we consider using bioethanol as a future energy resource for transportation. Nevertheless, utilization of lignocellulosic wastes as the feedstock for bioethanol production is a technically feasible and commercially viable process, which is expected to contribute heavily to next generation of biofuels production.

8. Conclusion

Lignocellulosic waste could be a severe environmental pollutant, but it is the most promising feedstock or substrate for producing biofuels or other value-added chemicals via microbial fermentation processes. Therefore, to increase the economical benefits of waste treatment of lignocellulosic materials, using anaerobic fermentation to convert lignocellulosic materials into energy-rich products, such as ethanol and hydrogen, is an ideal and reasonable approach that could achieve waste minimization and energy production simultaneously. The major factor affecting the efficiency of the conversion of lignocellulosic materials into energy products is the hydrolysis/saccharification of lignocellulose. The key to a successful cellulosic bioenergy production is to develop effective technology leading to rapid and high yield hydrolysis of lignocellulose converting it into fermentable sugars for subsequent fermentative production of biofuels. A variety of cellulolytic microorganisms have been isolated from the environment and various types of cellulolytic enzymes have been developed and produced for the use in the saccharification step. Physicochemical pretreatment could be useful in enhancing the efficiency of the enzymatic hydrolysis of lignocellulose. When the lignocellulosic substrate is pretreated and hydrolyzed into fermentable sugars, the fermentation technology utilizing these sugars for the production of bioenergy is relatively well-developed. However, for the aspect of process design, how to combine or integrate the hydrolysis and fermentation processes in an efficient and economical way could also be a critical challenge. In addition, how to adapt the energy-producing process with the existing waste or wastewater treatment facility is also of great importance. Therefore, development of a successful cellulosic bioenergy production process requires collaboration by the experts from different disciplines, including biotechnologists, microbiologists, chemical engineers, environmental engineers, etc. Production of clean energy from renewable resources with an environmental-compatible way is requisite for sustainable development of human beings. Cellulosic bioenergy is a good candidate to achieve this goal but still needs more efforts to advance the technology toward commercialization.

References

Adsul, M. G., Bastawde, K. B., Varma, A. J., and Gokhale, D. V. (2007) Strain improvement of *Penicillium janthinellum* NCIM-1171 for increased cellulase production. *Bioresour. Technol.* **98**, 1467–1473.

Akin, D. E., Rigsby, L. L., Sethuraman, A., Morrison, W. H-III, Gamble, G. R., and Eriksson, K. E. L. (1995) Alterations in structure, chemistry, and biodegradability of grass lignocellulose treated with the white rot fungi *Ceriporiopsis subvermispora* and *Cyathus stercoreus*. *Appl. Environ. Microbiol.* **61**, 1591–1598.

Albert, G., Sangha, O., and Cady, R. E. (1989) Cellulase immobilization on Fe_3O_4 and characterization. *Biotechnol. Bioeng.* **33**, 321–326.

Alterthum, F. and Ingram, L. O. (1989) Efficient ethanol production from glucose, lactose, and xylose by recombinant *Escherichia coli*. *Appl. Environ. Microbiol.* **55**, 1943–1948.

Amitai, G., Adani, R., Sod-Moriah, G., Rabinovitz, I., Vincze, A., and Leader, H. (1998) Oxidative biodegradation of phosphorothiolates by fungal laccase. *FEBS Lett.* **438**, 195–200.

Antoni, D., Zverlov, V. V., and Schwarz, W.H. (2007) Biofuels from microbes. *Appl. Microbiol. Biotechnol.* **77**, 23–35.

Arora, D. S., Chander, M., and Gill, P. K. (2002) Involvement of lignin peroxidase, manganese peroxidase and laccase in degradation and selective ligninolysis of wheat straw. *Int. Biodeter. Biodegrad.* **50**, 115–120.

Bajpai, P. (1999) Application of enzymes in the pulp and paper industry. *Biotechnol. Prog.* **15**, 147–157.

Balat, M. (2007) Global bio-fuel processing and production trends. *Energy Explor. Exploit.* **25**, 95–218.

Balat, M. and Balat, H. (2009) Recent trends in global production and utilization of bioethanol fuel. *Appl. Energy* **86**, 2273–2282.

Bennett, J. W., Wunch, K. G., and Faison, B. D. (2002) Use of fungi in biodegradation. In: Hurst, C. J. (ed.), *Manual of Environmental Microbiology*, Washington DC: AMS Press, pp. 960–971.

Bhat, M. K. (2000) Cellulases and related enzymes in biotechnology. *Biotechnol. Adv.* **18**, 355–83.

Bhat, M. K. and Bhat, S. (1997) Cellulose degrading enzymes and their potential industrial applications. *Biotechnol. Adv.* **15**, 583–620.

Bisaria, V. S. (1991) Bioprocessing of Agro-residues to glucose and chemicals. In: Martin, A. M. (ed.), *Bioconversion of Waste Materials to Industrial Products.* London: Elsevier, pp. 210–213.

Borjesson, J., Peterson, R., and Tjerneld, F. (2007) Enhanced enzymatic conversion of softwood lignocellulose by poly (ethylene glycol) addition. *Enzyme Microb. Technol.* **40**, 754–762.

Brown, R. M. and Saxena, I. M. (2000) Cellulose biosynthesis: A model for understanding the assembly of biopolymers. *Plant. Physiol. Biochem.* **38**, 57–67.

Brummell, D. A. (2006) Cell wall disassembly in ripening fruit. *Funct. Plant Biol.* **33**, 103–119.

Busto, M. D., Ortega, N., and Perez-Mateos, M. (1997) Stabilisation of cellulose by cross-linking with glutaraldehyde and soil humates. *Bioresour. Technol.* **60**, 27–33.

Chen C. Y., Saratale G. D., Lee C. M., Chen P. C., and Chang J. S. (2008) Developing a solar-energy-excited optical fiber photobioreactor for phototrophic H_2 production. *Int. J. Hydrogen Energy* **33**, 6886–6895.

Cherry, J. R. and Fidantsef, A. L. (2003) Directed evolution of industrial enzymes: An update. *Curr. Opin. Biotechnol.* **14**, 438–443.

Cho, K. M., Math, R. K., Hong, S. Y., Asraful Islam, S. M., Kim, J. O., Hong, S.J., Kim, H., and Yun, H. D. (2008) Changes in the activity of the multifunctional beta-glycosyl hydrolase (Cel44C-Man26A) from *Paenibacillus polymyxa* by removal of the C-terminal region to minimum size. *Biotechnol. Lett.* **30**, 1061–1068.

Closset, G. P., Cobb, T. J., and Shah, Y. T. (1974) Study of performance of a tubular membrane reactor for an enzyme catalyzed reaction. *Biotechnol. Bioeng.* **16**, 345–360.

Collins, P. J. and Dobson, A. D. W. (1997) Regulation of laccase gene transcription in *Trametes versicolor. Appl. Environ. Microbiol.* **63**, 3444–3450.

Converse, A. O., Matsuno, R., Tanaka, M., and Taniguchi, M. (1988) A model for enzyme adsorption and hydrolysis of microcrystalline cellulose with slow deactivation of the adsorbed enzyme. *Biotechnol. Bioeng.* **32**, 38–45.

Crawford, R. L. (1981). *Lignin Biodegradation and Transformation*. New York: John Wiley and Sons. ISBN 0-471-05743-6.

Cullen, D. (1997) Recent advances on the molecular genetics of ligninolytic fungi. *J. Biotechnol.* **53**, 273–289.

Dale, B. E. (1999) Biobased industrial products: Bioprocess engineering when cost really counts, *Biotechnol. Prog.* **15**, 775–776.

Das, H. and Singh, S. K. (2004) Useful by-products from cellulosic wastes of agriculture and food industry — A critical appraisal. *Crit. Rev. Food. Sci. Nutr.* **44**, 77–89.

Datar, R., Huang, J., Maness, P. C., Mohagheghi, A., Czernik, S., and Chornet, E. (2007) Hydrogen production from the fermentation of corn stover biomass pretreated with a steam-explosion process. *Int. J. Hydrogen. Energy* **32**, 932–939.

de Lima, A. L. G., do Nascimento, R. P., Bon, E. P. D., and Coelho, R. R. R. (2005) *Streptomyces drozdowiczii* cellulase production using agro-industrial by-products and its potential use in the detergent and textile industries. *Enzyme Microb. Technol.* **37**, 272–277.

Demain, A. L., Newcomb, M., and David Wu, J. H. (2005) Cellulase, clostridia, and ethanol microbiol. *Mol. Biol. Rev.* **69**, 124–154.

Desvaux, M. (2005) The cellulosome of *Clostridium cellulolyticum. Enzyme Microb. Technol.* **37**, 373–385.

Duff, S. J. B. and Murray, W. D. (1996) Bioconversion of forest products industry waste cellulosics to fuel ethanol — A review. *Bioresour. Technol.* **55**, 1–33.

Durán, N., Rosa, M. A., Dannibale, A., and Gianfreda, L. (2002) Applications of laccases and tyrosinases (phenoloxidases) immobilized on different supports: A review. *Enzyme Microb. Technol.* **31**, 907–931.

Ekborg, N. A., Morrill, W., Burgoyne, A. M., Li, L., and Distel, D. L. (2007) CelAB, a multifunctional cellulase encoded by *Teredinibacter turnerae* T7902T, a culturable symbiont isolated from the wood-boring marine bivalve *Lyrodus pedicellatus. Appl. Environ. Microbiol.* **73**, 7785–7788.

Eriksson, K. E. L., Blanchette, R. A., and Ander, P. (1990) *Microbial and Enzymatic Degradation of Wood and Wood Components*, Berlin Heidelberg, New York: Springer.

Eriksson, T., Borjesson J., and Tjerneld, F. (2002) Mechanism of surfactant effect in enzymatic hydrolysis of lignocellulose. *Enzyme Microb. Technol.* **31**, 353–364.

Fahr, K., Wetzstein, H. G., Grey, R., and Schlosser, D. (1999) Degradation of **2**, 4-dichlorophenol and pentachlorophenol by two brown rot fungi. *FEMS Microbiol. Lett.* **175**, 127–132.

Ferreira, S., Duarte, A. P., Ribeiro, M. H. L., Queiroz, J. A., and Domingues, F. C. (2009) Response surface optimization of enzymatic hydrolysis of *Cistus ladanifer* and *Cytisus striatus* for bioethanol production. *Biochem. Eng. J.* **45**, 192–200.

Fromm, J., Rockel, B., Lautner, S., Windeisen, E., and Wanner, G. (2003) Lignin distribution in wood cell walls determined by TEM and backscattered SEM techniques. *J. Structural Biol.* **143**, 77–84.

Glenn, J. K., Gold, M. H., (1983) Decolorization of several polymeric dyes by the lignin-degrading basidiomycete *Phanerochaete chrysosporium. Appl. Environ. Microbiol.* **45**, 1741–1747.

Goh, C. S., Tan, K. T., Lee, K. T., and Bhatia, S. (2010) Bioethanol from lignocellulose: Status, perspectives and challenges in Malaysia. *Bioresour. Technol.* **101**, 4834–4841.

Gold, M. H., Kuwahara, M., Chiu, A. A., and Glenn, J. K. (1984) Purification and characterization of an extracellular hydrogen peroxide requiring diarylpropane oxygenase from the white rot basidiomycete. *Phanerochaete chrysosporium. Arch. Biochem. Biophys.* **234**, 353–362.

Goldemberg, J. (2007) Ethanol for a sustainable energy future. *Science* **315**, 808–810.

Goldemberg, J. and Johansson, T. B. (2004) *World Energy Assessment Overview: Update*. New York, USA: United Nations Development Programme.

Gong, C. S., Cao, N. U., Du, J., and Tsao, G. T. (1999) Ethanol production by renewable sources *Adv.* Biochem. *Eng. Biotechnol.* **65**, 207–241.

Gray, K. A., Zhao, L., and Emptage, M. (2006) Bioethanol. *Curr. Opin. Chem. Biol.* **10**, 141–146.

Gunther, T., Sack, U., Hofrichter, M., and Latz, M. (1998) Oxidation of PAH and PAH-derivatives by fungal and plant oxidoreductases. *J. Basic Microbiol.* **38**, 113–122.

Hahn-Hägerdahl, B., Karhumaa, K., Fonseca, C., Spencer-Martins, I., and Gorwa-Grauslund, M. F. (2007) Toward industrial pentose-fermenting yeast strains. *Appl. Microbiol. Biotechnol.* **74**, 937–953.

Hallenbeck, P. C. and Benemann, J. R. (2002) Biological hydrogen production: Fundamentals and limiting processes. *Int. J. Hydrogen Energy* **27**, 1185–1193.

Hanaee, J., Schilling, F. R., Partzsch, G. M., Brasse, H., Schwarz, G., Macias, F. A., Molonillo, J. M. G., Torres, A., Varela, R. M., Castellano, D., and Kumakura, M. (1997) Preparation of immobilized cellulase beads and their application to hydrolysis of cellulosic materials. *Process Biochem.* **32**, 555–559.

Hatakka, A. I., (1983) Pretreatment of wheat straw by white-rot fungi for enzymatic saccharification of cellulose. *Appl. Microbiol. Biotechnol.* **18**, 350–357.

Hatamoto, O., Sekine, H., Nakano, E., and Abe, K. (1999) Cloning and expression of a cDNA encoding the laccase from *Schizophyllum commune. Biosci. Biotechnol. Biochem.* **63**, 58–64.

Heilmann, A., Teuscher, N., Kiesow, A., Janasek, D., and Spohn, U. (2003) Nanoporous aluminum oxide as a novel support material for enzyme biosensor. *J. Nanosci. Nanotechnol.* **3**, 275–279.

Heinfling, A., Ruiz-Duenas, F. J., Martınez, M. J., Bergbauer, H., Szewzyk, U., and Martınez, A. T. (1998) A study on reducing substrates of manganese-oxidizing peroxidases from *Pleurotus eryngii* and *Bjerkandera adusta. FEBS Lett.* **428**, 141–146.

Himmel, M. E., Adney, W. S., Baker, J. O., Nieves, R. A., and Thomas, S. R. (1996) Cellulases: Structure, function and applications. In: Wyman C. E. (ed.), *Handbook on Bioethanol.* Bristol, PA: Taylor & Francis, pp. 144–161.

Hirasawa, K., Uchimura, K., Kashiwa, M., Grant, W. D., Ito, S., Kobayashi, T., and Horikoshi, K. (2006) Salt-activated endoglucanase of a strain of alkaliphilic *Bacillus agaradhaerens. Antonie Leeuwenhoek* **89**, 211–219.

Hofrichter, M. (2002) Review: Lignin conversion by manganese peroxidase (MnP). *Enzyme Microb. Technol.* **30**, 454–466.

Howard, R. L., Abotsi, E., Rensburg, J. E. L., and Howard, S. (2003) Lignocellulose biotechnology: Issues of bioconversion and enzyme production. *Afr. J. Biotechnol.* **2**, 602–619.

James, W. E., Jha, S., Sumulong, L., Son, H. H., Hasan, R., Khan, M. E., Sugiyarto, G., and Zhai, F. (2008) *Food Prices and Inflation in Developing Asia: Is Poverty Reduction Coming to an End?* Manila, Philippines: Asian Development Bank.

Jayani, R. S., Saxena, S., and Gupta, R. (2005) Microbial pectinolytic enzymes: A review. *Process Biochem.* **40**, 2931–2944.

John, F., Monsalve, G., Medina, P. I. V., and Ruiz, C. A. A. (2006) Ethanol production of banana shell and cassava starch. Dyna Universidad Nacional de Colombia. *Medellin*, **73**, 21–27.

Kaar, W. E. and Holtzapple, M. T. (1998) Benefits from Tween during enzymic hydrolysis of corn stover. *Biotechnol. Bioeng.* **59**, 419–427.

Kanai, T., Imanaka, H., Nakajima, A., Uwamori, K., Omori, Y., and Fukui, T. (2005) Continuous hydrogen production by the hyperthermophilic archaeon, *Thermococcus kodakaraensis* KOD1. *J. Biotechnol.* **116**, 271–282.

Kapdan, L. K. and Kargi, F. (2006) Biohydrogen production from waste materials. *Enzyme Microb. Technol.* **38**, 569–582.

Kaper T., Lebbink J. H., Pouwels J., Kopp J., Schulz, G. E., van der Oost J., and de Vos, W. M. (2000) Comparative structural analysis and substrate specificity engineering of the hyperthermostable beta-glucosidase CelB from *Pyrococcus furiosus. Biochemistry* **39**, 4963–4970.

Karube, I., Tanaka, S., Shirai, T., and Suzuki, S. (1977) Hydrolysis of cellulose in a cellulose-bead fluidized bed reactor. *Biotechnol. Bioeng.* **19**, 1183–1191.

Keeney, R. and Hertel, T. W. (2008) *The Indirect Land Use Impacts of US Biofuel Policies: The Importance of Acreage, Yield, and Bilateral Trade Responses.* West Lafayette, Indiana, USA: Center for Global Trade Analysis, Purdue University.

Kersten, P. J., Tien, M., Kalyanaraman, B., and Kirk, T. K. (1985) The ligninase from *Phanerochaete Chrysosporium* generates cation radicals from methoxybenzenes. *J. Biol. Chem.* **260**, 2609–2612.

Kim, Y. S., Jung, H. C., and Pan, J. G. (2000) Bacterial cell surface display of an enzyme library for selective screening of improved cellulose variants. *Appl Environ. Microbiol.* **66**, 788–793.

Kim, J. and Yun, S. (2006) Discovery of cellulose as a smart material. *Macromolecules* **39**, 4202–4206.

Kimura, Y., Asada, Y., Oka, T., and Kuwahara, M. (1991) Molecular analysis of a *Bjerkandera adusta* lignin peroxidase gene. *Appl. Microbiol. Biotechnol.* **35**, 510–514.

Ko, C. H., Chen, W. L., Tsai, C. H., Jane, W. N., Liu, C. C., and Tu, J. (2007) *Paenibacillus campinasensis* BL11: A wood material-utilizing bacterial strain isolated from black liquor. *Bioresour. Technol.* **98**, 2727–2733.

Ko, I. K., Kato, K., and Iwara, H. (2005) A thin carboxymethyl cellulose substrate for the cellulose-induced harvesting of an endothelial cell sheet. *J. Biomater. Sci. Polym. Ed.* **16**, 1277–1291.

Koh, L. P. and Ghazoul, J. (2008) Biofuels, biodiversity, and people: Understanding the conflicts and finding opportunities. *Biol. Conserv.* **141**, 2450–2460.

Kondo, A. and Fukuda, H. (1997) Preparation of thermo-sensitive magnetic hydrogel microspheres and application to enzyme immobilization. *J. Ferment. Bioeng.* **84**, 337–341.

Konwarh, R., Karak, N., Rai, S. K., and Mukherjee, A. K. (2009) Polymer-assisted iron oxide magnetic nanoparticle immobilized keratinase. *Nanotechnology* **20**, 225107–225117.

Kotchoni, O. S., Shonukan, O. O., and Gachomo, W. E. (2003) *Bacillus pumilus* BpCRI **6**, a promising candidate for cellulase production under conditions of catabolite repression. *Afri. J. Biotechnol.* **2**, 143–157.

Kuhad, R. C., Singh, A., and Eriksson, K. E. (1997) Microorganism's enzymes involved in the degradation of plant fiber cell walls. *Adv. Biochem. Eng. Biotechnol.* **57**, 45–125.

Kumar, R., Singh, S., and Singh, O. V. (2008) Bioconversion of lignocellulosic biomass: Biochemical and molecular perspectives *J. Ind. Microbio. Biotechnol.* **35**, 377–391.

Lagaert, S., Beliën, T., and Volckaert, G. (2009) Plant cell walls: Protecting the barrier from degradation by microbial enzymes, *Seminars in Cell & Developmental Biology* **20**, 1064–1073.

Leonowicsz, A., Matuszewska, A., Luterek, J., Ziegenhagen, D., Wojtas-Wasilewska, M., and Cho, N. S. (1999) Biodegradation of lignin by white rot fungi. *Fungal. Genet. Biol.* **27**, 175–185.

Levin, D. B., Islam, R., Cicek, N., and Sparling, R. (2006) Hydrogen production by *Clostridium thermocellum* 27405 from cellulosic biomass substrates. *Int. J. Hydrogen Energy* **31**, 1496–1503.

Liang, Y., Yesuf, J., Schmitt, S., Bender, K., and Bozzola, J. (2009) Study of cellulases from a newly isolated thermophilic and cellulolytic *Brevibacillus* sp. strain JXL. *J. Ind. Microbiol. Biotechnol.* **36**, 961–970.

Lim, W. J., Hong, S. Y., An, C. L., Cho, K. M., Choi, B. R., Kim, Y. K., An, J. M., Kang, J. M., Lee, S. M., Cho, S. J., Kim, H., and Yun, H. D. (2005) Construction of minimum size cellulase (Cel5Z) from *Pectobacterium chrysanthemi* PY35 by removal of the C-terminal region. *Appl. Microbiol. Biotechnol.* **68**, 46–52.

Lin, C. Y. and Chang, R. C. (1999) Hydrogen production during the anaerobic acidogenic conversion of glucose. *J. Chem. Technol. Biotechnol.* **74**, 498–500.

Liu, Y., Yu, P., Song, X., and Qu, Y. (2008) Hydrogen production from cellulose by co-culture of *Clostridium thermocellum* JN4 and *Thermoanaerobacterium thermosaccharolyticum* GD17. *Int. J. Hydrogen Energy* **33**, 2927–2933.

Lo, Y. C., Bai, M. D., Chen, W. M., and Chang, J. S. (2008) Cellulosic hydrogen production with a sequencing bacterial hydrolysis and dark fermentation strategy. *Bioresour. Technol.* **99**, 8299–8303.

Lo, Y. C., Saratale, G. D., Chen, W. M., Bai, M. D., and Chang, J. S. (2009) Isolation of cellulose-utilizing bacteria for cellulosic biohydrogen production. *Enzyme Microb. Technol.* **44**, 417–425.

Lo, Y. C., Su, Y. C., Chen, C. Y., Chen, W. M., Lee, K. S., and Chang, J. S. (2009) Biohydrogen production from cellulosic hydrolysate produced via temperature-shift-enhanced bacterial cellulose hydrolysis. *Bioresour. Technol.* **100**, 5802–5807.

Lynd, L. R., Weimer, P. J., Vanzyl, W. H., and Pretorius, I. S. (2002) Microbial cellulose utilization: Fundamentals and biotechnology. *Microbiol. Mol. Biol. Rev.* **66**, 506–577.

MacLean, H. L. and Lave, L. B. (2003) Evaluating automobile fuel/propulsion system technologies. *Prog. Energy Combus. Sci.* **29**, 1–69.

Mahadevan, S. A., Wi, S. G., Lee, D. S., and Bae, H. J. (2008) Site-directed mutagenesis and CBM engineering of Cel5A (*Thermotoga maritima*). *FEMS Microbiol. Lett.* **287**, 205–211.
Mandels, M., Weber, J., and Parizek, R. (1971) Enhanced cellulase production by a mutant of *Trichoderma viride*. *Appl. Microbiol.* **21**, 152–154.
Martinez, G., Larrondo, N., Putman, N., Gelpke, M. D. S., Huang, K., and Chapman, J. (2004) Genome sequence of the lignocellulose degrading fungus *Phanerochaete chrysosporium* strain RP78. *Nat. Biotechnol.* **22**, 1–6.
McCarthy, J. K., Uzelac, A., Davis, D. F., and Eveleigh, D. E. (2004) Improved catalytic efficiency and active site modification of 1,4-beta-D-glucan glucohydrolase A from *Thermotoga neapolitana* by directed evolution. *J. Biol. Chem.* **279**, 11495–11502.
McGuirl, M. A. and Dooley, D. M. (1999) Copper-containing oxidases. *Curr. Opin. Chem. Biol.* **3**, 138–144.
McKendry, P. (2002) Energy production from biomass: Overview of biomass. *Bioresour. Technol.* **83**, 37–43.
Mosier, N., Wyman, C., Dale, B., Elander, R., Lee, Y. Y., Holtzapple, M., and Ladisch, M. (2005) Features of promising technologies for pretreatment of lignocellulosic biomass. *Bioresour. Technol.* **96**, 673–686.
Murashima, K., Kosugi, A., and Doi, R. H. (2002) Thermostabilization of cellulosomal endoglucanase EngB from *Clostridium cellulovorans* by in vitro DNA recombination with non-cellulosomal endoglucanase EngD. *Mol. Microbiol.* **45**, 617–626.
National Research Council, Committee on Biobased Industrial Products (1999) *Biobased Industrial Products: Priorities for Research and Commercialization*, National Academy Press.
Niehaus, F., Bertoldo, C., Kahler, M., and Antranikian, G. (1999) Extremophiles as a source of novel enzymes for industrial application. *Appl. Microbiol. Biotechnol.* **51**, 711–729.
Nisizawa, T., Suzuki, H., and Nisizawa, K. (1972) Catabolite Pepression of cellulase formation in *Trichoderma viride*. *J. Biochem.* **71**, 999–1007.
Ohmiya, K., Maeda, K., and Shimizu, S. (1987) Purification and properties of endo-β-1,4-glucanase from *Ruminococcus albus*. *Carbohydr. Res.* **166**, 145–155.
Ohta, K., Beall, D. S., Mija, J. P., Shanmugam, K. T., and Ingram, L. O. (1991) Genetic improvement of *Escherichia coli* for ethanol production: Chromosomal integration of *Zymomonas mobilis* genes encoding pyruvate decarboxylase and alcohol dehydrogenase II. *Appl. Environ. Microbiol.* **57**, 893–900.
Olanoff, L. S. and Venkatasubramanian, K. (2004) Perfusion trails with a collagen-immobilized enzyme in an extracorporal reactor: Activity, stability, and biocompatibility. *J. Biomed. Mat. Res.* **11**, 125–136.
Pan, X., Xie, D., Gilkes, N., Gregg, D. J., and Saddler, J. N. (2005) Strategies to enhance the enzymatic hydrolysis of pretreated softwood with high residual lignin content. *Appl. Biochem. Biotechnol.* **124**, 1069–1080.
Park, S. R., Cho, S. J., Kim, M. K., Ryu, S. K., Lim, W. J., An, C. L., Hong, S. Y., Kim, J. H., Kim, H., and Yun, H. D. (2002) Activity enhancement of Cel5Z from *Pectobacterium chrysanthemi* PY35 by removing C-terminal region. *Biochem. Biophys. Res. Commun.* **291**, 425–430.
Patoomporm, C.-A., Yutaka, K., Yumi, M., Teruo, O., and Takashi, S. (1986) Properties of cellulose immobilized on agarose gel with spacer. *Biotechnol. Bioeng.* **28**, 1876–1878.
Pérez-Avalos, O., Sánchez-Herrera, L. M., Salgado, L. M., and Ponce-Noyola, T. (2008) A bifunctional endoglucanase/endoxylanase from *Cellulomonas flavigena* with potential use in industrial processes at different pH. *Curr. Microbiol.* **57**, 39–44.
Pesta, G., Meyer-Pittroff, R., and Russ, W. (2006) Utilization of whey. In: Oreopoulou, V. and Russ, W. (eds.), *Utilization of By-products and Treatment of Waste in the Food Industry*. New York: Springer. Vol. 1, pp. 1–11.
Prasad, S., Singh, A., and Joshi, H. C. (2007) Ethanol as an alternative fuel from agricultural, industrial and urban residues. *Resour. Conserv. Recycl.* **50**, 1–39.
Puppan, D. (2002) Environmental evaluation of biofuels. *Period Polytech. Ser. Soc. Man. Sci.* **10**, 95–116.
Qu, Y., Zhu, M., Liu, K., Bao, X., and Lin, J. (2005) Studies on cellulosic ethanol production for sustainable supply of liquid fuel in China. *Biotechnol. J.* **1**, 1235–1240.
Rabinovich, M. L., Melnik, M. S., and Bolobova, A. V. (2002a) Microbial cellulases: A review. *Appl. Biochem. Microbiol.* **38**, 305–321.
Rabinovich, M. L., Melnik, M. S., and Bolobova, A. V. (2002b) The structure and mechanism of action of cellulolytic enzymes. *Biochemistry* **68**, 850–871.

Ragauskas, A. J., Williams, C. K., Davison, B. H., Britovsek, G., Cairney, J., Eckert, C. A., Frederick, W. J. J.R., Hallett, J. P., Leak, D. J., and Liotta, C. L. (2006) The path forward for biofuels and biomaterials. *Science* **311**, 484–489.

Rajaram, S. and Verma, A. (1990) Production and characterization of xylanase from *Bacillus thermoalkalophilus* growth on agricultural wastes. *Appl. Microbiol. Biotechnol.* **34**, 141–144.

Ren, N. Q., Li, J. Z., Li, B. K., Wang, Y., and Liu, S. R. (2006) Biohydrogen production from molasses by anaerobic fermentation with a pilot-scale bioreactor system. *Int. J. Hydrogen Energy* **31**, 2147–2157.

Rifiin, J. (2002) *The Hydrogen Economy*. New York: Tarcher Putnam.

Rowell, M. R. (1992) Opportunities for lignocellulosic materials and composites. In: *Emerging Technologies for Material and Chemicals from Biomass: Proceedings of Symposium.* Washington, DC: American Chemical Society, pp. 26–31.

Saha, B. C. (2000) Alpha-L-arabinofuranosidases — biochemistry, molecular biology and application in biotechnology. *Biotechnol. Adv.* **18**, 403–423.

Saka, S. and Goring, D. A. I. (1985) In: Higuchi, T. (ed.), *Biosynthesis and Biodegradation of Wood Components.* Orlando, FL: Academic, p. 51.

Salimi, A., Sharifi, E., and Noorbakhsh, A. (2007) Immobilization of glucose oxidase on electrodeposited nickel oxide nanoparticles: Direct electron transfer and electrocatalytic activity. *Biosensors Bioelectronics* **22**, 3146–3153.

Salman, S., Soundararajan, S., Safina, G., Satoh, I., and Danielsson, B. (2008) Hydroxyapatite as a novel reversible in situ adsorption matrix for enzyme thermistor-based FIA, *Talanta* **77**, 490–493.

Sánchez, C. (2009) Lignocellulosic residues: Biodegradation and bioconversion by fungi. *Biotechnol. Adv.* **27**, 185–194.

Saratale, G. D., Chen, S. D., Lo, Y. C., Saratale, R. G., and Chang, J. S. (2008) Outlook of biohydrogen production from lignocellulosic feedstock using dark fermentation — A review. *J. Sci. Ind. Res.* **67**, 962–979.

Saratale, G. D., Humnabadkar, R. P., and Govindwar, S. P. (2007) Presence of mixed function oxidase system in *Aspergillus ochraceus* (NCIM-1146). *Indian J. Microbiol.* **47**, 304–309.

Saratale, G. D., Saratale, R. G., Lo, Y. C., and Chang, J. S. (2010) Multicomponent cellulase production by *Cellulomonas biazotea* NCIM-2550 and their applications for cellulosic biohydrogen production. *Biotechnol. Prog.* **26**, 406–416.

Saratale, R. G., Saratale, G. D., Chang, J. S., and Govindwar, S. P. (2009) Ecofriendly decolorization and degradation of Reactive Green 19 using *Micrococcus glutamicus* NCIM-2168. *Bioresour. Technol.* **110**, 3897–3905.

Schubert, C. (2006) Can biofuels finally take center stage? *Nat. Biotechnol.* **24**, 777–784.

Shary, S., Ralph, S. A., and Hammel, K. E. (2007) New insights into the ligninolytic capability of a wood decay *Ascomycete. Appl. Environ. Microbiol.* **73**, 6691–6694.

Siso, M. I. G. (1996) The biotechnological utilization of cheese whey: A review. *Bioresour. Technol.* **57**, 1–11.

Slade, R., Bauen, A., and Shah, N. (2009) The commercial performance of cellulosic ethanol supply-chains in Europe. *Biotechnol Biofuels*. **2–3**, 1–20.

Sparling, R., Risbey, D., and Poggi-Varaldo, H. M. (1997) Hydrogen production from inhibited anaerobic composters. *Int. J. Hydrogen Energy* **22**, 563–566.

Taguchi, F., Yamada, K., Hasegawa, K., Takisaito, T., and Hara, K. (1996) Continuous hydrogen production by *Clostridium* sp. strain No. 2 from cellulose hydrolysate in aqueous two phase system. *J. Ferment. Bioeng.* **82**, 80–83.

Tailliez, P., Girard, H., Millet, J., and Beguin, P. (1989) Enhanced cellulose fermentation by an asporogenous and ethanol-tolerant mutant of *Clostridium thermocellum*. *Appl. Environ. Microbiol.* **55**, 207–211.

Takao, M., Akiyama, K., and Sakai, T. (2002) Purification and characterization of thermostable endo-1,5-α-L-arabinase from a strain of *Bacillus thermodenitrificans*. *Appl. Environ. Microbiol.* **68**, 1639–1646.

Tao, H. and Cornish, V. W. (2002) Milestones in directed enzyme evolution. *Curr. Opin. Chem. Biol.* **6**, 858–864.

Terashima, N. and Fukushima, K. (1989) In: Lewis, N. G., and Paice, M.G. (eds.), *Plant Cell Wall Polymers: Biogenesis and Biodegradation.* Washington, DC: American Chemical Society, p. 160.

Teunissen, P. J., Sheng, D., Reddy, G. V., Moenne-Loccoz, P., Field, J. A., and Gold, M. H., (1998) 2-Chloro-1,4-dimethoxybenzene cation radical: Formation and role in the lignin peroxidase oxidation of anisyl alcohol. *Arch. Biochem. Biophys.* **360**, 233–238.

Tien, M. and Kirk, T. K., (1983) Lignin-degrading enzyme from the hymenomycete *Phanerochaete chrysosporium* burds. *Science* **221**, 661–663.

Torney, F., Noeller, L., Scarpa, A., and Wang, K. (2007) Genetic engineering approaches to improve bioethanol production from maize. *Curr. Opin. Biotechnol.* **18**, 1–7.

Turner, P., Gashaw, M., and Karlsson, E. N. (2007) Potential and utilization of thermophiles and thermostable enzymes in biorefining. *Microb. Cell Fact.* **6**(9) (doi:10.1186/1475-2859-6-9).

Umezawa, T. and Higuchi, T. (1987) Mechanism of aromatic ring cleavage of β-O-4 lignin substructure models by lignin peroxidase. *FEBS Lett.* **218**, 255–260.

Valdez-Vazquez, I., Sparling, R., Risbey, D., Rinderknecht-Seijas, N., and Poggi-Varaldo, H. M. (2005) Hydrogen generation via anaerobic fermentation of paper mill wastes. *Bioresour. Technol.* **96**, 1907–1913.

Wang, C. M., Shyu, C. L., Ho, S. P., and Chiou, S. H. (2008) Characterization of a novel thermophilic, cellulose-degrading bacterium *Paenibacillus* sp. strain B39. *Lett. Appl. Microbiol.* **47**, 46–53.

Wang, J. L. and Wan, W. (2009) Factors influencing fermentative hydrogen production: A review. *Int. J. Hydrogen Energy* **34**, 799–811.

Weibel, M. K., Roberto, B., Delotto, R., and Humphrey, A. E. (2004) Immobilized enzymes: Pectin esterase covalently coupled to dorous glass particles. *Biotechnol. Bioeng.* **17**, 85–98.

Widsten, P. and Kandelbauer, A. (2008) Laccase applications in the forest products industry: A review *Enzyme Microb. Technol.* **42**, 293–307.

Wood, T. M. and Bhat, K. M. (1988) Methods for measuring cellulase activities. *Methods Enzymol.* **160**, 87–112.

Wu, K. J., Saratale, G. D., Lo, Y. C., Chen, S. D., Chen, W. M., Tseng, Z. J., and Chang, J. S. (2008) Fermentative production of 2, 3 butanediol, ethanol and hydrogen with *Klebsiella* sp. isolated from sewage sludge. *Bioresour. Technol.* **99**, 7966–7970.

Wu, S. Y., Hung, C. H., Lin, C. N., Chen, H. W., Lee, A. S., and Chang, J. S. (2006) Fermentative hydrogen production and bacterial community structure in high-rate anaerobic bioreactors containing silicone-immobilized and self-flocculated sludge. *Biotechnol. Bioeng.* **93**, 934–946.

Yang, B. and Wyman C. E. (2006) BSA treatment to enhance enzymatic hydrolysis of cellulose in lignin containing substrates. *Biotechnol. Bioeng.* **94**, 611–617.

Yaropolov, A. I., Skorobogatko, O. V., Vartanov, S. S., and Varfolomeyev, S. D. (1994) Laccase: Properties, catalytic mechanism, applicability. *Appl. Biochem. Biotechnol.* **49**, 257–279.

Yoon, J. J., Cha, C. J., Kim, Y. S., Son, D. W., and Kim, Y. K. (2007) The brown-rot basidiomycete *Fomitopsis palustris* has the endo-glucanases capable of degrading microcrystalline cellulose. *J. Microbiol. Biotechnol.* **5**, 800–805.

Zhang, P., Berson, Y. H., Sarkanen, E., and Dale, S. (2009) Pretreatment and biomass recalcitrance: Fundamentals and progress. *Appl. Biochem. Biotechnol.* **153**, 80–83.

Zhang, R., Vigneswaran, S., Ngo, N., and Nguyen, H. (2008) Fluidized bed magnetic ion exchange as pretreatment process for a submerged membrane reactor in wastewater treatment and reuse, *Desalination* **227**, 85–93.

Zhang, Y. H. P., Ding, S. Y., Mielenz, J. R., Elander, R., Laser, M., Himmel, M., McMillan, J. D., and Lynd, L. R. (2007) Fractionating recalcitrant lignocellulose at modest reaction conditions. *Biotechnol. Bioeng.* **97**, 214–223.

Zhang, Y. H. P., Himmel, M. E., and Mielenz, J. R. (2006) Outlook for cellulase improvement: Screening and selection strategies. *Biotechnol. Adv.* **24**, 452–481.

Zhang, Y. H. P. and Lynd, L. R. (2003) Quantification of cell and cellulose mass concentrations during anaerobic cellulose fermentation: Development of an ELISA-based method with application to *Clostridium thermocellum* batch cultures. *Anal. Chem.* **75**, 219–227.

Chapter 14

Biohydrogen Production by Fermentation and Microbial Electrolysis Cells

N. Q. Ren*, A. J. Wang, K. Zhang†, B. F. Liu and D. Sun

State Key Laboratory of Urban Water Resource and Environment, Harbin Institute of Technology, Harbin 150090, P.R. China

**rnq@hit.edu.cn*

†waj0578@hit.edu.cn

This chapter is to review recent research progress on the three technologies for biohydrogen production from wastes and wastewater, i.e., dark fermentation, photo-fermentation, and microbial electrolysis cells (MEC). Among the three technologies, the development of dark fermentation is the most advanced, as evidenced by the installation of a full-scale demonstration plant, whereas the development of the other two still remains at infancy stage. For dark- and photo-fermentation, discussions are mainly focused on the development in China. Discussion of dark fermentation includes reactor types, startup strategies, and factors affecting the hydrogen production process, such as pH, partial pressure of hydrogen, hydraulic retention time, organic loading rates, sludge pretreatment, and metal toxicity, as well as pilot and full-scale production experiences. Discussion of photo-fermentation includes inoculums, light intensity, carbon and nitrogen sources, pH, metal ions, and the process of coupling with dark fermentation. Discussion of MEC includes its architecture, startup, membrane and electrode materials, anode potential, methane production, etc. Examples of producing hydrogen from MEC using fermentation effluent of molasses, lignocellulose, and cellobiose are also reviewed. Lastly, a conceptual flow diagram for a proposed biohydrogen production system treating wastes and wastewater is discussed.

1. Introduction

The excessive use of fossil fuels has caused serious environmental problems, making it critical to develop low-cost, clean, and renewable energy resources to meet growing environmental concerns and energy demands. Among the potential energy alternatives being considered, hydrogen appears to be one of the most promising energy carriers for the following reasons: it is a clean fuel producing only water when combusted; it has a high energy yield of $122\,kJ \cdot g^{-1}$, which is much greater than that of hydrocarbon fuels (e.g., $46\,kJ \cdot g^{-1}$ for gasoline); and it can be converted to electricity directly through fuel cells. Currently, there are four available methods for producing hydrogen, including water electrolysis, thermo-chemical process, radiolysis, and biological process. The biological process is potentially most attractive since biohydrogen production is environmentally friendly and less energy intensive from renewable sources when fermentative consortia are employed. In fermentative processes, wastes and wastewater rich in carbohydrates can be used as substrates to achieve the purpose of reducing pollution from uncontrolled waste degradation and the generation of hydrogen as a valuable fuel. However, the hydrogen yield of dark fermentation has a strict theoretical yield of $4\,mol\text{-}H_2 \cdot (mol\text{-}glucose)^{-1}$ (only 1/3 of the available electrons are converted to hydrogen), while in practice yields of only

2–3 mol-H_2·(mol-glucose)$^{-1}$ could be achieved. The remaining 2/3 of chemical energy is wasted in the form of acetate and other fermentation products (including butyrate and alcohols). Photo-fermentation and microbial electrolysis cell (MEC) are the alternative ways to convert these wastes by-products into hydrogen.

This chapter highlights the recent advances of biohydrogen production by dark and photo-fermentations, as well as the development of MEC technology. For dark and photo-fermentation, discussions are mainly focused on the development in China. The section of dark fermentation focuses the achievements of pilot and full-scale hydrogen production from wastewater. Reactor types, startup strategies, process controlling factors, use of complex organic wastes as feedstock, and hydrogen–methane co-generation are elaborated. The section of photo-fermentation focuses on the processes and corresponding controlling factors, as well as coupling dark- and photo-fermentation. The section of MEC includes the principle, architecture, and operation of MEC, as well as the coupling dark fermentation and MEC system. Finally, a schematic flow diagram is given to visualize the integrated system for cascade-producing hydrogen and to indicate its further development.

2. Dark-fermentative Hydrogen Production Process and Controlling Factors

2.1. *Dark-fermentative hydrogen production process*

A comprehensive review on dark-fermentative hydrogen-producing process was recently reviewed by Li and Fang (2007a). In China, the most commonly used anaerobic reactors for wastewater treatment are continuously stirred tank reactors (CSTRs), upflow anaerobic sludge blanket (UASB) reactors, expanded granular sludge bed (EGSB) reactors, and anaerobic baffled reactors (ABRs). Many of these reactors have been applied for biohydrogen production. Among them, CSTRs are the most commonly used. CSTR has a simple configuration and operation. With continuous stirring, hydrogen can rapidly released from the mixed liquor reducing the feedback inhibition caused by the increase of hydrogen partial pressure. Ren *et al.* patented a CSTR design in 1992 (Patent number: ZL9211474.1) for continuous biohydrogen production from wastewater. The reactor could operate at high sludge concentrations with mixed liquor volatile suspended solids (MLVSS) ranging 10–30 g·l^{-1}, and have a maximum hydrogen yield of 10.4–11.4 m^3·m^{-3}·d^{-1} and a hydrogen production rate of 36–40 ml·(g-VSS)$^{-1}$·h^{-1} (VSS: volatile suspended solids).

Compared to CSTRs, UASB reactors may retain large amounts of granular sludge with super settleability. They may be operated at shorter hydraulic retention times (HRTs) and thus higher organic load rates (OLRs). Mu and Yu (2006) found that the hydrogen-producing granular sludge in a UASB reactor had diameters of 1.0–3.5 mm, an average density of 1.04 g·ml^{-1} and a settling velocity ranging 32–75 m·h^{-1}. Using glucose as substrate, the hydrogen yield was 1.00–1.44 mol-H_2·(mol-glucose)$^{-1}$ at 18 h of HRT. The maximum hydrogen production rate was 202 ml·l^{-1}·h^{-1} when treating 10.7 g·l^{-1} of glucose.

Guo *et al.* (2008a) used an EGSB reactor for fermentative hydrogen production from molasses-containing wastewater using granular activated carbon (GAC) as carrier. At 35 °C, the reactor had a hydrogen yield of 3.47 mol-H_2·(mol-sucrose)$^{-1}$, and a specific hydrogen production rate of 3.16 mmol-H_2g-VSS^{-1}h^{-1}. The average attached biomass concentration was 17.1 gl^{-1}. Stable ethanol-type fermentation was carried out with ethanol and acetate accounted for 89.1%

of the organic residues. When treating starch-containing wastewater, Guo *et al.* (2008b) found a maximum hydrogen production rate of $1.64\,l{\cdot}l^{-1}{\cdot}d^{-1}$ at pH 4.4 under an OLR of 1.0 g-starch$\cdot l^{-1}\cdot d^{-1}$ and an HRT of 4 h.

Li *et al.* (2007) used a three-compartment ABR to produce hydrogen at 35 °C from a molasses-containing wastewater having a COD (chemical oxygen demand) of 5,000 mg $\cdot\, l^{-1}$. The ABR showed a specific hydrogen production rate of 0.13 l-$H_2 \cdot$ (g-MLVSS)$^{-1} \cdot d^{-1}$ and a substrate conversion rate of 0.13 l$\cdot$(g-COD)$^{-1}$.

2.2. *Startup and controlling factors of dark-fermentative hydrogen production*

2.2.1. *Reactor startup strategies*

During the startup period, hydrogen-producing bacteria in the seed sludge are acclimated and enriched in the reactor. Conventionally, methanogenic sludge is used as the seed sludge after a heat treatment to suppress the bioactivity of those bacteria that cannot produce hydrogen (Li and Fang 2007a). HRT and pH are the key controlling factors for the startup. Ren and Gong (2006) developed a new startup strategy using the feed-to-microorganism ratio (F/M) as the key controlling parameter to enrich the hydrogen-producing sludge in a CSTR fed with molasses. With an initial biomass concentration of 6.24 g-VSS $\cdot\, l^{-1}$, the reactor was operated at 6 h of HRT and an initial OLR of 7.0 kg-COD$\cdot m^{-3}\cdot d^{-1}$ (corresponding to an F/M ratio of 2–3 g-COD$\cdot$ (g-VSS)$^{-1}\cdot d^{-1}$), resulting in an average hydrogen yield of 2.01 mol$\cdot$(mol-hexose)$^{-1}$ and an average hydrogen production rate of 276 ml$\cdot$(g-VSS)$^{-1}\cdot d^{-1}$.

2.2.2. *Controlling operational parameters*

Of all parameters which may affect hydrogen production, the reactor pH is found to be the most crucial (Fang and Liu 2002). Some researches showed that the pH may affect the fermentation pathway of hydrogen production. Ren *et al.* (1997) were the first to observe an ethanol-type fermentation with ethanol as the main constituent of the fermentation residue in a CSTR at pH 4.5. Ren *et al.* (2007) later found that three CSTRs operated at pH 6.0–6.5, 5.5–6.0, and 4.0–4.5 had, respectively, propionate-type, ethanol-butyrate type, and ethanol-type of fermentation. Wang *et al.* (2007a) found that acetate was a dominant by-product at pH < 4.0 and butyrate-type fermentation was predominated at pH 4.5–6.0 while seeded with methanogenic sludge. For the reactor that seeded with heat-pretreated sludge, butyrate was dominant at pH > 3.6, and ethanol-type fermentation was dominant at pH < 3.6.

In UASB reactors, Mu *et al.* (2006a) found that the H_2 partial pressure (ranging 2.8–5.2×10^4 Pa), production rate (ranging 61–145 ml$\cdot l^{-1}\cdot h^{-1}$), and yield (ranging 0.68–1.61 mol$\cdot$ (mol-glucose)$^{-1}$) were all related to pH. The maximum H_2 partial pressure was found at pH 3.4, whereas the maximum H_2 production rate and yield were at pH 4.2. Zhao and Yu (2007) found in a similar experiment that the maximum H_2 partial pressure was at pH 7.5, and the maximum H_2 production rate and yield were both at pH 6.5–7.5.

HRT is the other important parameter. Chu *et al.* (2009) found that, in a CSTR operated at 30 °C and pH 5.5, the number of hydrogen-producing bacteria decreased gradually with increasing HRT, i.e., 9.2×10^8, 8.2×10^8, 2.8×10^8, and 6.2×10^7 cells$\cdot ml^{-1}$ at HRT 6, 8, 12, and 14 h, respectively. At the optimum HRT of 6–8 h, the hydrogen yield was 1.4–1.5 mol$\cdot$(mol-glucose)$^{-1}$.

Yu *et al.* (2002) studied biohydrogen production from a high-strength rice winery wastewater on the effects of HRT (2–24 h), COD concentration (14–36 g-COD·l^{-1}), pH (4.5–6.0), and temperature (20–55 °C). They found the maximum hydrogen production rate of 9.33 l·(g-VSS)$^{-1}$·d^{-1} and hydrogen yield of 1.37–2.14 mol·(mol-hexose)$^{-1}$ at HRT of 2 h, COD of 34 g·l^{-1}, pH 5.5, and 55 °C. Zhao *et al.* (2008) used the response surface methodology (RSM) with a central composite design to optimize the performance of a UASB reactor. Results showed that a maximum hydrogen yield of 1.62 mol·(mol-hexose)$^{-1}$ when treating an wastewater containing 14.5 g·l^{-1} of sucrose at an HRT of 16.4 h.

2.2.3. *Factors affecting microbial hydrogen production*

A number of parameters may affect the microbial hydrogen production, most related literature of which was reviewed by Li and Fang (2007a). Among those affecting parameters, the four most significant ones include pretreatment of seed sludge for the selection of hydrogen-producing bacteria, temperature, pH, plus trace metals and nutrients.

In most studies, hydrogen-producing bacteria are acclimated and enriched from anaerobic sludge, and some used compost and soil. Since these sources contain substantial amount of hydrogenotrophic and non-hydrogen-producing bacteria, a pretreatment process is deemed essential to inhibit their activities. The pretreatment processes include heat, acid, alkali, aeration, microwave, freeze–thaw, and chemical methanogen inhibitors, since many hydrogen-producing bacteria are spore formers and thus can survive such treatments while others cannot. Cai *et al.* (2009) evaluated the effectiveness of acid, alkali, heat, KNO_3 pretreatments, and found the heat pretreatment was most effective. Mu *et al.* (2007) compared heat, acid, and alkali pretreatment, and also found the highest H_2 yield of 2.0 mol-H_2·(mol-glucose)$^{-1}$ was obtained from the heat-treated sludge, whereas the lowest yield of 0.48 mol-H_2·(mol-glucose)$^{-1}$ from the alkali-treated sludge. Wang and Wan (2008d) compared five pretreatment methods using acid, alkali, heat, aeration, and chloroform, respectively, at 35 °C and an initial pH of 7.0. They also found that the heat treatment was most effective, resulting in a maximal hydrogen production rate of 120.4 ml-H_2·h^{-1} and a yield of 221.5 ml-H_2·g^{-1} and a substrate degradation efficiency of 97.2%.

Mu *et al.* (2006b) studied H_2 production from glucose by a mixed culture at various temperatures within the mesophilic range. Results showed that glucose degradation rate and efficiency, H_2 yield, and growth rate of H_2-producing bacteria all increased with the temperature increase from 33 to 41 °C. Similarly, Wang and Wan (2008c) also found that hydrogen production from glucose increased with temperature from 20 to 40 °C. Treating wastewater from a cattle farm, Tang *et al.* (2008) found the maximum hydrogen yield of 319 ml-H_2·g-COD^{-1} at 45 °C and pH 5.5.

Fang and Liu (2002) found that pH 5.5 was the optimal pH for biohydrogen production from glucose by a mixed culture. Zheng and Yu (2004) conducted a similar experiment but found hydrogen yield fluctuated at pH 4.0–5.0, and an increase of pH beyond this range led to a drastic decrease in hydrogen yield. Fan *et al.* (2004), on the other hand, also found the optimal pH at 5.4 ± 0.2, whereas, Li *et al.* (2008) studied the effect of pH (5.0–7.0) and glucose concentrations (5.0–20 g·l^{-1}) on biohydrogen production using a river bed sludge as seed. Results showed a maximum H_2 production rate of 0.22 mol-H_2·(mol-glucose)$^{-1}$·h^{-1} and a cumulative H_2 yield of 1.83 mol-H_2·(mol-glucose)$^{-1}$ at pH 6.0 treating 7.5 g·l^{-1} of glucose.

Biohydrogen production is also affected by trace metals and nutrients. Li and Fang (2007b) tested the toxicity of six heavy metals on fermentative hydrogen production, and found the $C_{I,50}$ values, at which the bioactivities of sludge was reduced by 50% from the controls, as follows: 30 mg·l^{-1} for Cu^{2+}; 1,600 mg·l^{-1} for Ni^{2+} and Zn^{2+}; 3,000 mg·l^{-1} for Cr^{3+}; 3,800 mg·l^{-1} for Cd^{2+}; and >5,000 mg·l^{-1} for Pb^{2+}. As compared to the other acidogenesis or methanogenesis granular sludge, hydrogen-producing sludge in general showed a higher resistance to heavy metal toxicity. Wang and Wan (2008a) investigated the effect of Fe^{2+}, which is essential to hydrogenase, with concentrations ranging from nil to 1,500 mg·l^{-1}, on the fermentative hydrogen production at 35 °C and the initial pH of 7.0. Results showed that the maximum hydrogen yield of 311.2 ml·(g-glucose)$^{-1}$ was obtained at Fe^{2+} concentration of 350 mg·l^{-1}. Zhang *et al.* (2005a) obtained a maximum hydrogen yield of 2.73 mol·(mol-sucrose)$^{-1}$ at the Fe^{2+} concentration of 1,600 mg·l^{-1}. Zhang and Shen (2006) found the optimal Fe^{2+} concentrations were 800, 200, and 25 mg·l^{-1}, respectively, at 25, 35, and 45 °C. Wang *et al.* (2007b) found the optimal Fe^{2+} and Mg^{2+} concentrations for hydrogen production by fermentative bacteria strain B49 were 80 mg·l^{-1} and 200 mg·l^{-1}, respectively.

In addition to Fe^{2+}, Ni^{2+} is another element essential to hydrogenase, and Mg^{2+} is an activator of many enzymes related to glycolysis. Wang and Wan (2008b) found the optimal Ni^{2+} concentration was 0.1 mg·l^{-1} for fermentative hydrogen production from glucose. Hao *et al.* (2006) found that the specific hydrogen production yield and rate were 28.04–28.97 ml·g^{-1} and 7.52–7.83 ml·g^{-1}·h^{-1} at the optimum Na^{+} concentration of 1,000–2,000 mg·l^{-1}. On the other hand, metals could also inhibit microbial hydrogen-producing capability. Zheng and Yu (2005) found that specific hydrogen production rate was enhanced by Cu at 50–100 mg·l^{-1} and Zn at 10–250 mg·l^{-1}, but inhibited by 200 mg·l^{-1} of Cu or 500 mg·l^{-1} of Zn.

Wang *et al.* (2009a) studied the effect of ammonia concentration (0–10 g-N·l^{-1}) on fermentative hydrogen production by a mixed culture at 35 °C and an initial pH of 7.0. They found that the maximum hydrogen production yield and rate were 298.8 ml-H_2·g^{-1} and 8.5 ml-H_2·h^{-1}, respectively, at 0.1 g-N·l^{-1}. Yuan *et al.* (2008) studied the effect of L-cysteine (0.1–1.0 mmol·l^{-1}), and found maximum hydrogen yields of 3.10 mol-H_2·(mol-sucrose)$^{-1}$ with 0.6 mmol of L-cysteine at 35 °C, and 3.28 mol·(mol-sucrose)$^{-1}$ at 0.1 mM L-cysteine at 30 °C.

Liu and Shen (2004) examined the interaction of several parameters, i.e., initial pH (ranging 4.0–9.0), Fe^{2+} (1.2–100 mg·l^{-1}), NH_4 (NH_4HCO_3, 0.56–11.28 g·l^{-1}), and starch (2–32 g·l^{-1}) on hydrogen production. They found the maximum specific hydrogen production rate of 146 ml-H_2·(g-VSS)$^{-1}$·d^{-1} and the yield of 178 ml-H_2·(g-starch)$^{-1}$ were obtained at pH 7.0–8.0, starch concentration of 15 g·l^{-1}, Fe^{2+} 10 mg·l^{-1}, and NH_4HCO_3 5.64 g·l^{-1}. Xu *et al.* (2008) found the yeast powder concentration of 0.3–0.5 g·l^{-1} and phosphate concentration of 100–150 mmol·l^{-1} were optimal for *Ethanoligenens harbinense* B49 (Lin 2002) with a maximum H_2 yield of 2.26 mol·(mol-glucose)$^{-1}$ and a rate of 27.74 mmol·(g-VSS)$^{-1}$. Guo *et al.* (2009), using the response surface method (RSM), estimated the maximum hydrogen yield as 2.21 mol·(mol-glucose)$^{-1}$ treating a solution containing 14.6 g·l^{-1} of glucose, 177 mg·l^{-1} of Fe^{2+}, and 692 mg·l^{-1} of Mg^{2+}.

2.3. *Pilot and full-scale hydrogen production from wastewater*

In 1999, Ren *et al.* conducted a pilot-scale study of continuous biohydrogen production in a 1.48 m^3 fermentor for more than 200 days. Using molasses as substrate, the system was operated at OLRs ranging 3.1–85.6 kg-COD·m^{-3}·d^{-1}. The hydrogen yield increased with OLR in the range of 3.1–68.2 kg-COD·m^{-3}·d^{-1}, and decreased at higher OLRs. The maximum hydrogen production rate was 5.57 m^3·m^{-3}·d^{-1} with a specific production rate of 0.75 m^3-H_2·(kg-MLVSS)$^{-1}$·d^{-1}. At the OLR of 35–55 kg-COD·m^{-3}·d^{-1}, the hydrogen yield was 26.1 mol·kg-COD^{-1} (Ren *et al.* 2007). This achievement was regarded as one of the Ten Top Science and Technology Breakthroughs in China in 2000.

In 2005, Ren *et al.* conducted the first full-scale demonstration project of fermentative hydrogen production for one year, costing RMB 21 million. The complete system consisted of facilities for not only fermentative hydrogen production, but also seed sludge acclimation and enrichment, hydrogen purification, fermentative methane production from residues of hydrogen production, as well as wastewater treatment. A 100 m^3 CSTR (effective volume of 64.5 m^3) was used for fermentative hydrogen production from molasses wastewater. Under the operation parameters of OLR 14–45 kg-COD·m^{-3}·d^{-1}, pH 4.4–5.0, and alkalinity more than 500 mg·l^{-1} ($CaCO_3$), the reactor performed the ethanol-type fermentation with average hydrogen and biogas yields of 285 m^3·d^{-1} and 588 m^3·d^{-1}, respectively. After bioaugmentation with more efficient hydrogen-producing strains, the average hydrogen yield further enhanced to 334 m^3·d^{-1}, corresponding to a volumetric hydrogen production rate of 5.26 m^3·m^{-3}·d^{-1} and a specific rate of 0.83 m^3-H_2·(kg-MLVSS)$^{-1}$·d^{-1}. Over 95% of the total sugar in the feedstock was degraded.

The hydrogen purification system, which consisted of a desulfurization tower and pressure swing adsorption tower for CO_2 removal, improved the hydrogen content in the biogas from 40–54% to 99.99%. The estimated cost for producing H_2 of 99.9% purity was RMB1.37 per cubic meter. The produced hydrogen was further used to generate 8.8–10.8 kW of electricity via fuel cells. The effluent of the process was further used for methane fermentation followed by a wastewater treatment process. The final effluent was able to meet the national discharge standards.

2.4. *Biohydrogen production from complex organic wastes*

Table 14.1 shows the operational conditions and performance of several studies that have been conducted on hydrogen production from complex organic wastes at 36–37 °C. The feedstock included beer lees, wheat straws, cornstalk, municipal solid wastes, and sewage sludge at concentrations ranging 5–50 g·l^{-1}; and the seed sludge included cow manure, sewage sludge and digested sludge after heat, acid, or alkali pretreatment. Optimal pH was mostly found at around pH 7, with an exception at pH 11 for a process using alkaline-treated sludge. Optimal feedstock concentrations are mostly about 20 g·l^{-1}. Results showed that cornstalk had the highest hydrogen yield 150 ml·g-VS^{-1}, nearly 10-fold higher than the 16 ml·g-VS^{-1} produced by the sewage sludge. Rice and potato wastes, both are carbohydrate rich, had a relatively high hydrogen yield of about 106–134 ml·g-VS^{-1}. Beer lees and wheat straw on the other hand produced about 68 ml·g-VS^{-1} of hydrogen. Overall, these studies demonstrated that biohydrogen could be produced by actual agricultural and municipal solid wastes, and more hydrogen was produced from the carbohydrate-rich feedstock. Further studies are warranted for broadening the application of this technology.

2.5. *Hydrogen production coupling with methane production*

Dark fermentation could theoretically convert only about one-third of the chemical energy in the feedstock into hydrogen, leaving the remaining two-third in the form of VFAs and alcohols

Table 14.1. Fermentative hydrogen production from complex substrates in batch process.

Feedstock	Inoculum	Pre-treatment	pH (optimal)	Concentration (optimal)	Maximum H_2 yield	Maximum H_2 production rate ($ml \cdot g\text{-}TVS^{-1} \cdot h^{-1}$)	Reference
Beer lees	Cow dung	HCl	4.5–9.0 (6.0–7.0)	5–50 $g \cdot l^{-1}$ (20 $g \cdot l^{-1}$)	68.6 $ml \cdot g\text{-}VS^{-1}$	NA	Fan *et al.* (2006a)
Wheat straw wastes	Cow dung	HCl	4.0–9.0 (7.0–9.0)	5–35 $g \cdot l^{-1}$ (25 $g \cdot l^{-1}$)	68.1 $ml \cdot g\text{-}VS^{-1}$	10.14	Fan *et al.* (2006b)
Cornstalk wastes	Cow dung	Acid	4.0–9.0 (7.0)	5–40 $g \cdot l^{-1}$ (15 gl^{-1})	150 $ml \cdot g\text{-}VS^{-1}$	NA	Zhang *et al.* (2007)
Organic fraction of municipal solid wastes	Anaerobic sludge	Heat	5.5	20 $g\text{-}VS \cdot l^{-1}$	134 $ml \cdot g\text{-}rice^{-1}$ 106 $ml \cdot g\text{-}potato^{-1}$ 50 $ml \cdot g\text{-}lettuce^{-1}$	NA	Li *et al.* (2009)
Sewage sludge	Sewage sludge	Alkaline	3–12.5 (11)	NA	16.6 $ml \cdot g\text{-}TS^{-1}$	NA	Cai *et al.* (2004)
Sewage sludge	Sewage sludge	Heat	6.81	NA	16.26 $ml \cdot g\text{-}VS^{-1}$	NA	Xiao and Liu (2009)

in the mixed liquor (Li and Fang 2007a). These residues are commonly further treated in the conventional anaerobic reactors to produce methane. A number of studies have been conducted to investigate the utilization of various substrates and operational parameters that may significantly affect the two-stage process to product both hydrogen and methane.

Xie *et al.* (2008a) found that treating a 1% glucose solution in a two-phase anaerobic fermentation process showed the highest specific hydrogen yield of 2.75 mol-H_2·(mol-glucose)$^{-1}$ at pH 6. The maximum specific methane yield of 2.13 mol-CH_4·(mol-glucose)$^{-1}$ was obtained when over 90% of ethanol, acetic, butyric, valeric, and caproic acids were removed from the effluent of the hydrogen-producing reaction. This two-stage process recovered 82% of chemical energy from glucose, as compared to 23% by a single-stage process to produce hydrogen alone. Xie *et al.* (2008b) further studied treating potatoes in a two-stage process using a heat-treated seed sludge. They found that the hydrogen yield was enhanced from 200 ml-H_2·(g-TVS)$^{-1}$ to 218 ml-H_2·(g-TVS)$^{-1}$ and the production rate from 703 ml-H_2·(g-VSS)$^{-1}$·d^{-1} to 801 ml·(g-VSS)$^{-1}$·d^{-1}. When the potatoes was pretreated by α-amylase and glucoamylase, the hydrogen yield was further increased to 271 ml·(g-TVS)$^{-1}$ and the rate to 945 ml·(g-VSS)$^{-1}$·d^{-1}. The maximum methane yield was 158 ml·(g-TVS)$^{-1}$ and the maximum rate 103 ml·g^{-1} (TVS). The energy efficiency was about 60%, as compared to the 20% for single-stage production of hydrogen alone.

Chu *et al.* (2008) used a two-stage process treating the organic fraction of municipal solid wastes (OFMSW) in a thermophilic reactor to produce hydrogen followed by a mesophilic reactor to produce methane. The thermophilic reactor was operated at an HRT of 1.3 d and OLR of 38.4 kg-VS·m^{-3}·d^{-1} or 64.4 kg-COD_{cr}·m^{-3}·d^{-1}, whereas the methanogenic reactor was conducted at HRT of 5.0 d and OLR of 6.6 kg-VS·m^{-3}·d^{-1} or 16.3 kg-COD_{cr}·m^{-3}·d^{-1}. The process had a hydrogen yield of 205 ml-H_2·(g-VS)$^{-1}$ and a methane yield of 464 ml-CH_4·(g-VS)$^{-1}$.

Wang and Zhao (2009) used a two-stage process to produce hydrogen and methane from a food waste. The hydrogen production relied on the indigenous microbes in the food waste without adding inoculums from other sources. The optimal OLRs were 22.65 kg-VS·m^{-3}·d^{-1} with a sludge retention time (SRT) of 160 h for hydrogen production and 4.61 kg-VS·m^{-3}·d^{-1} with 26.7 d of SRT for methane production. The maximum yields were 0.065 m^3-H_2·(kg-VS)$^{-1}$ and 0.546 m^3-CH_4·(kg-VS)$^{-1}$. About 5.8% of the influent COD was converted to hydrogen and 82.2% was converted to the methane.

Lu *et al.* (2009a) used cornstalk as feedstock to produce hydrogen and methane in a two-stage process which was enhanced by an artificial microbial community. Adding *Enterobacter aerogenes* or *Clostridium paraputrificum* to the hydrogen-producing reactor was found effective. In a three-stage process, in which an extra stage of hydrolysis is conducted in the front at 55 °C and pH 5.0, adding *Clostridium thermocellum* was able to increase the soluble sugars to 0.48 kg for each kilogram of cornstalk. Overall, the hydrolysate from 1 kg of cornstalk was able to produce 2.6 mol (64 l) of hydrogen with the augmentation of *C. paraputrificum*, plus 4.7 mol (115 l) of methane. The energy recovery reached 54%.

3. Photo-fermentative Hydrogen Production Process and Controlling Factors

Unlike dark-fermentation, photo-fermentative H_2 production is still at its infancy stage. A comprehensive review on photo-fermentative H_2-producing process was recently conducted by Li and Fang (2009). Several bottlenecks remain to be overcome for large-scale production. This section discusses the photo-fermentation process, its process control, and its coupling with the dark-fermentation process to enhance the overall H_2 production.

3.1. *Photo-fermentative hydrogen production*

At present, there are three kinds of reactor configuration for photo-fermentation: column, flat panel, and tubes. Zhou (2007) used a 31.1-l cylindrical reactor for photo-H_2 production from a pretreated swine manure using immobilized photosynthetic bacteria. The experiment lasted 69 days with 63 days of stable H_2 production. The maximum H_2 production rate reached 723 ml-$H_2 \cdot l^{-1} \cdot d^{-1}$ with an average of 633 ml-$H_2 \cdot l^{-1} \cdot d^{-1}$. The H_2 conversion from COD was 173 ml-$H_2 \cdot (g\text{-}COD)^{-1}$ with a substrate conversion efficiency of 61.7%. Shi *et al.* (2008) designed a set of partially recycled system. The experimental conditions were as follows: initial OD_{660} of 0.273, temperature of 30 ± 1 °C, light intensity of 1,200 lux, culture cycle of 38 h, and effluent OD_{660} being controlled at 0.38–0.74. Zhang (2007) studied the startup performance of photo-H_2 production in a reactor packed with biofilm-coated beads, and the effects of process parameters. He found that the optimum H_2 yield of 3.45 $mg \cdot h^{-1}$ was obtained under the following conditions: light intensity 5,000 lux, wavelength 590 nm, glucose concentration 0.12 $mol \cdot l^{-1}$, pH 7.0, and temperature 25 °C. Under similar conditions, Tian *et al.* (2008) found that photo-H_2 production from a biofilm reactor had a similar H_2 yield of 3.54 $mg \cdot h^{-1}$.

3.2. *Controlling factors of photo hydrogen production*

Controlling the photo-H_2 production process needs to optimize the key parameters affecting process so as to ensure the efficient and stable H_2 production. These parameters include inoculums, light intensity, carbon and nitrogen sources, pH, biogas composition, and metal ions.

3.2.1. *Inoculums*

Yield of H_2 in general increases with the amount of inoculum, but only to certain degrees (Sasikala *et al.* 1991). An over-abundance of inoculated biomass, however, may lead to low H_2 yields, probably due to the reduction of light penetration. Yang *et al.* (2002) found that, using *Rhodopseudomonas palustris* to produce H_2 from acetate, the H_2 yield increased significantly when bacteria were at the log–growth phase, but decreased at the lag phase. Zhu *et al.* (1997), on the other hand, found that H_2 production by different strains of *Rhodopseudomonas* sp. was higher in the stable phase than in the log–growth phase. On the other hand, Xu *et al.* (1993) and Qian *et al.* (2007) found that bacteria in different growth phases had in general similar levels of H_2 production capacity, although bacteria at the stable phase may be slightly lower.

3.2.2. *Light intensity*

The capture of light energy is essential for the formation of ATP and the proton gradient. Although sunlight is the most preferred light source, most experiments so far were conducted using artificial light sources, such as incandescent lamps, halogen lamps, optical fibers, and neon tubes. Zhang *et al.* (2005b) used *Rhodobacter sphaeroides* to produce H_2 from pig manure wastewater at various light intensities. They found optimal H_2 production when the light intensity was within 1,200–2,000 lux. Ren *et al.* (2009) found that cell growth was slow at light intensities below

1,000 lux, and the H_2 yield was highest under the light intensity of 3,000–5,000 lux. Further increase of light intensity to 9,000 lux, H_2 yield decreased with light intensity, although cells grew rapidly. Similar results were observed by Tao *et al.* (2008), who found that, at 1,500–5,000 lux, H_2 production of *R. sphaeroides* strain ZX-5 increased gradually with light intensity; but gradually decreased when the light intensity was increased above 5,000 lux.

3.2.3. *Carbon and nitrogen sources*

The hydrogen conversion efficiency of photosynthetic bacteria varies greatly among bacterial species or strains. This is because the substrate is metabolized differently by the bacteria and the substrate's ability to generate H^+ is different and thus affecting the H_2 production (Barbara *et al.* 1985). Photo-H_2 production is catalyzed by nitrogenase, the activity of which may be inhibited by NH^{4+}. Generally, glutamate and ethanolamine are used as the nitrogen source for photo-H_2 production. Tao *et al.* (2008) isolated an *R. sphaeroides* strain ZX-5 which can utilize acetate, malate, butyrate, succinate, pyruvate, and lactate as substrates for H_2 production with substrate conversion efficiencies of 60–90%. They also found that 7 $mmol \cdot l^{-1}$ of glutamates was the optimal nitrogen concentration. Ammonium sulfate, however, inhibited H_2 production significantly.

Industrial wastewater, domestic wastewater, and agricultural wastes have also been used as feedstock for photo-H_2 production. Shi and Yu (2006) used mixed volatile acids as substrate for H_2 production by *Rhodopseudomonas capsulata* in continuous reactors. When the medium contains acetate of 1.8 $g \cdot l^{-1}$, propionate of 0.2 $g \cdot l^{-1}$, and butyrate of 1 $g \cdot l^{-1}$, the maximum H_2 production rate and substrate conversion efficiency were 37.8 $ml \cdot g^{-1} \cdot h^{-1}$ and 45%, respectively. You *et al.* (2005) used pig manure wastewater as substrate for *R. sphaeroides*, and achieved 27.3 $ml \cdot l^{-1} \cdot d^{-1}$ of H_2 production rate. Tao *et al.* (2008) used *R. sphaeroides* ZX-5 for wastewater treatment, and gained an H_2 yield of 500 $ml \cdot g\text{-}COD^{-1}$ with a COD removal of 80%. Ren *et al.* (2008) reported that *Rhodopseudomonas faecalis* RLD-53 could utilize both acetate and butyrate for H_2 production. They found that the maximum H_2 production rate was 32.5 $ml \cdot l^{-1} \cdot h^{-1}$ when the acetate and butyrate concentrations were both 25 $mmol \cdot l^{-1}$.

3.2.4. *Initial pH*

The optimal pH for the growth of photo-fermentative H_2-productng bacteria is species dependent. Sun *et al.* (1995) showed that *Rhodopseudomonas* strain H produced little H_2 at the initial pH of 4–5, but its H_2 yield increased sharply at pH about 7.0. Fu *et al.*, (2004) and Ren *et al.* (2009) also showed that the highest H_2 yields of *Bacillus* sp. and *R. faecalis* RLD-53 were at pH 7.0. Wei *et al.* (2007) indicated that the optimal pH for the H_2 production of *R. sphaeroides* was pH 7–8. Under this condition, the utilization rate of acetate was 99.3%, the H_2 yield 0.91 mol-$H_2 \cdot (mol\text{-}acetate)^{-1}$, H_2 production rate 34.6 $ml \cdot (g\text{-}VSS)^{-1} \cdot d^{-1}$, and energy conversion efficiency 22.8%.

3.2.5. *Gas phase in the photo-H_2 production system*

Most previous studies of photo-fermentative hydrogen production were conducted under inert atmosphere using argon (Barbosa *et al.* 2001; Chen *et al.* 2006; Yetis *et al.* 2000). This is because the hydrogen-producing activity of nitrogenase is inhibited by the presence of N_2, O_2, and possibly CO_2. In the presence of N_2, nitrogenase uses N_2 as electron acceptor to produce NH_3 during fixation, and thus reduces the electron availability for H_2 formation. In the absence of N_2, nitrogenase can resume its ability to reduce proton into H_2 (Ludden and Roberts 1995; Sasikala *et al.* 1990). Nitrogenase and hydrogenase are also sensitive to O_2, which may thus strongly inhibit the H_2 production (Weaver *et al.* 1980). Furthermore, Liu *et al.* (2009a) found that CO_2 may also inhibit the growth of H_2-producing bacteria. They showed that removing CO_2 from the reactor increased the photo-H_2 production by 12.8% and the H_2 yield by 18.9%.

3.2.6. *Metal ions*

Nickel and iron are the key components of several photo-fermentative hydrogen-producing enzymes. The reported proper concentration ranges of Ni and Fe for photo hydrogen production varied slightly: 0–4 μmol·l^{-1} Ni^{2+} and 0–80 μmol·l^{-1} Fe^{2+} (Liu *et al.* 2009b); 0–9 μmol·l^{-1} Ni^{2+} and 0–200 μmol·l^{-1} Fe^{3+} (Yang *et al.* 2003); 0–57 μmol·l^{-1} Fe^{2+} (Zhu *et al.* 2007). Outside these ranges, Ni and Fe may inhibit H_2 production. Liu *et al.* (2009b) also showed that, although Mg^{2+} is essential to the cell growth, it did not enhance H_2 production yield (Liu *et al.* 2009b).

3.3. *Combination of dark- and photo-fermentation*

Ideally, the effluent of dark-fermentation, which is rich in VFAs and alcohols, may be further treated by photo-fermentation for the production of additional hydrogen so as to maximize the overall H_2 yield. Two schemes of such a combined process have been studied: an integrated process and a sequential process. The integrated process allows dark- and photo-fermentative bacteria mixed in one system, saving space and operational cost. Liu *et al.* (2010) studied the combined process using a mixed culture of *Clostridium butyricum*, and *R. faecalis* RLD-53 and found that the maximum H_2 yield was obtained when the population ratio of the two bacteria was at 1:600. Ding *et al.* (2009) investigated the effects of operational parameters, including substrate concentration, phosphate buffer, initial pH, the ratio of bacteria and light intensity, on a mixed culture of *C. butyricum* and immobilized *R. faecalis* RLD-53, on the production of H_2 from glucose. They found a maximum H_2 yield of 4.13 mol-H_2·(mol-glucose)$^{-1}$ and a maximum H_2 production rate of 33.9 ml-H_2·l^{-1}·h^{-1}. However, due to the great difference on growth rate and acid-resistant capacity between photo- and dark-fermentative bacteria, they found that the system was very difficult to operate.

Compared to the integrated process, the sequential process is easier to operate and control, because the two groups of bacteria are in individual reactors which can be operated independently. Tao *et al.* (2007) used a sequential process for H_2 production. They used phototrophic *R. sphaeroides* SH_2C to produce H_2 from the effluent of dark-fermentation, achieving a total H_2 yield of 6.63 mol-H_2·(mol-sucrose)$^{-1}$. The effect of initial pH and C/N ratio on the dark-fermentative H_2 production was also determined. Liu *et al.* (2009c) used an ethanol-type fermentative *Ethanoligenens harbinese* B49 and an immobilized *R. faecalis* RLD-53 to study the sequential H_2 production efficiency. Their result implied that phosphate buffer could increase the H_2 yield of strain *R. faecalis* RLD-53 to 6.32 mol-H_2·(mol-glucose)$^{-1}$. Su *et al.* (2009) applied anaerobic sludge as seed sludge for dark-fermentative H_2 production from cassava starch, followed by photo-fermentation by *R. palustris*. The cumulative H_2 production was 402 ml-H_2·(g-starch)$^{-1}$, which was substantially higher than the H_2 yield by a single dark-fermentation process.

4. Microbial Electrolysis Cells for Hydrogen Production

Dark-fermentative hydrogen production converts only one-third of chemical energy into hydrogen, leaving the remaining two-third in the form of VFAs and alcohols. To overcome this barrier, Liu *et al.* (2005) developed a bioelectrolytic method, called microbial electrolysis cell (MEC), by applying a small input of voltage. MEC is previously known as biocatalyzed electrolysis cell (BEC) or bioelectrochemically assisted microbial reactor (BEAMR) (Liu *et al.* 2005; Rozendal *et al.* 2006a; Ditzig *et al.* 2007; Logan 2008). By improving the MEC materials and architecture, Cheng and Logan (2007a) obtained a hydrogen yield of 3.95 mol-H_2·(mol-acetate)$^{-1}$, which is very close to theoretical maximum of 4 mol-H_2·(mol-acetate)$^{-1}$. In this section, the principle, design, and operation of MEC process will be discussed, along with the coupling process.

4.1. *Principle and architecture of microbial electrolysis cells*

4.1.1. *Principle of hydrogen production in MECs*

MEC is very similar to microbial fuel cell (MFC) (Liu *et al.* 2005; Rozendal *et al.* 2006a). Current generation in both processes is dependent on the presence of certain bacteria on the anode surface, called exoelectrogens (Logan 2008), which oxidize organic matters and transfer electrons to the anode. Unlike MFC which produces electricity, MEC produces hydrogen by applying a small voltage in the absence of oxygen. In an MEC, exoelectrogens transfer the electrons from the oxidation of organic matters to the anode, releasing carbon dioxide and protons. The electrons flow via a circuit to the cathode where electrons and protons react to form hydrogen. The process is called electrohydrogenesis or microbial electrolysis (Rozendal *et al.* 2008a; Cheng and Logan 2007a; Logan *et al.* 2008). Compared to fermentation, MEC is an electrically driven hydrogen-producing process that overcomes the thermodynamic barrier and may achieve the theoretical hydrogen recovery of 100%. MEC requires in practice a voltage slightly greater than 200 mV (110 mV in theory), which is significantly lower than the 1.21 V theoretically needed for water electrolysis.

4.1.2. *MEC architecture*

The first MEC system developed was a two-chamber (H-type) reactor separated by a small Nafion cation exchange membrane (CEM) which was held in a tube between the chambers (Liu *et al.* 2005). Due to the large distance between the electrodes and the small size of the CEM, the internal resistance for this reactor was high and initial performance was poor. However, increasing the anode surface area and reducing the electrode spacing did not increase the performance without increasing the size of CEM (Logan *et al.* 2008), because the membrane hinder proton diffusion creating a substantial pH gradient between two electrodes (Rozendal *et al.* 2006b; Kim *et al.* 2007). Performance was later improved by increasing the size of CEM relative to the electrode-projected surface areas (Cheng and Logan 2007a; Rozendal *et al.* 2007). The performance was further improved by using a cubic reactor (small electrode spacing), graphite granules as anode (high surface area), and a relatively large anion exchange membrane (AEM) (Cheng and Logan 2007a).

A few continuous flow architectures have been investigated (Rozendal *et al.* 2006a, 2007, 2008b, c). Rozendal *et al.* (2007) moved the cathode onto the membrane, forming a gas diffusion electrode. This architecture eliminates the liquid surrounding the cathode, although a second chamber is still needed for gas collection. In the continuous flow mode, a hydrogen production rate of $0.3\,m^3\text{-}H_2\cdot m^{-3}\cdot d^{-1}$ at an applied voltage of 1.0 V was achieved. Rozendal *et al.* (2008b) tested another continuous flow architecture, in which the electrode chambers consisted of vertically orientated channels for liquid transport and a headspace for gas collection so that hydrogen could be efficiently collected without accumulating in the channels. This architecture also prevented the increase of local pH which would deteriorate the biocathode performance.

Development has also been made to simplify the MEC architecture without using membrane in order to improve hydrogen production rate and to reduce capital cost (Call and Logan 2008; Hu *et al.* 2008). In a single-chamber membrane-free MEC, electric resistance is greatly reduced and the pH gradient between electrodes is eliminated. However, in this architecture a fraction of hydrogen is consumed by the hydrogenotrophic methanogens growing in the anode chamber, resulting in not only the loss of hydrogen but also accumulation of impurities in hydrogen. Using a membrane-free MEC with a graphite fiber brush anode (high surface area) and a close electrode spacing, a maximum hydrogen production rate of $3.12\,m^3\text{-}H_2\cdot m^{-3}\cdot d^{-1}$ ($292 \pm 1\,A\cdot m^{-3}$) was obtained at an applied voltage of 0.8 V (Call and Logan 2008). Through a methanogenesis-controlling strategy, single-chamber MEC was able to achieve high coulombic efficiencies (CE) of $92 \pm 6\%$, and hydrogen recoveries of ranging from 78 ± 1 to $96 \pm 1\%$ at applied voltage of 0.3–0.8 V.

Sun *et al.* (2008) recently developed an MEC–MFC coupled system for biohydrogen production from acetate without external power input. In such a system, hydrogen was produced in an MEC using the power input from an *in situ* MFC. However, as the MFC available today can supply only a limited voltage, the system produced hydrogen at a very slow rate of $0.015\,m^3\text{-}H_2\cdot m^{-3}\cdot d^{-1}$. Further work is needed to improve the MFC voltage to above 600 mV in order to make this process feasible.

4.2. *MEC operation*

4.2.1. *MEC startup*

There are two startup strategies for MEC: one is to form anode biofilm at the MFC mode first and then change the operation mode to MEC, and the other is to start up directly at the MEC mode. The former strategy, which is more commonly used, acclimates anodophilic microbial community at the MFC mode (Liu *et al.* 2005; Call and Logan 2008; Wang *et al.* 2009b). However, the direct MEC startup strategy also performs well. Biofilm of MEC can be developed using effluent from an MFC–MEC system containing exoelectrogenic bacteria (Cheng and Logan 2007a; Rozendal *et al.* 2007, 2008c), or using biofilm scraped from the anode, or even using the activated sludge (Wang 2008). At present, it is not clear to what extent the operation of an MEC is affected by the two startup modes.

4.2.2. *Control of methane generation in MECs*

Methane may be produced during hydrogen generation in a single-chamber MEC, in which there is no separation of gases produced by the mixed microorganisms (Call and Logan 2008; Wang *et al.* 2009b; Rozendal *et al.* 2008b). Hydrogenotrophic methanogens, which are naturally enriched in the reactor from the inoculums due to high concentrations of hydrogen, not only consume hydrogen but also contaminate the produced hydrogen with methane. In a biocathode MEC, Rozendal *et al.* (2008b) found that bicarbonate in the buffer was used as carbon source by hydrogenotrophic methanogens. By removing bicarbonate from the microbial nutrient medium, hydrogenotrophic methanogenic consumption of hydrogen was substantially reduced. In a single-chamber MEC, oxygen exposure and high applied voltage are two other strategies which may be used to control methane generation. Call and Logan (2008) showed that exposing the reactors to air in between batch cycles decreased the average methane concentration from 3.5% to 0.9% at applied voltage of 0.6 V. However, air exposure could limit the growth of strict anaerobic exoelectrogens, such as *Geobacter* spp. Wang *et al.* (2009b) showed that methane could be kept below 4% by shortening the cycle time and using higher applied voltages of $\geq$ 0.6 V. More strategies need to be further investigated to inhibit methane production; these may include lowering pH, heat treating inoculums, and using methanogen inhibitor (Hawkes *et al.* 2007; Oh *et al.* 2003; Chae *et al.* 2008).

4.2.3. *Anode potential*

In order to optimize operations of MEC for hydrogen production, microbial anode potential was analyzed at various levels of operational factors in biofilm anode system, including pH, substrate concentration, and applied voltage (Wang *et al.* 2010). Although hydrogen generation generally starts at anode potential around -250 mV to -300 mV, in this study, higher current density and higher hydrogen rate were obtained when anode potential was at -400 mV or even lower. Anode biofilm worked normally between pH 6.5 and 7.0, while the lowest anode potential appeared at pH 6.8–7.0. However, the anode potential went up to -300 mV or above when pH was lowered to 6.0 or substrate concentration was below 50 $mg \cdot l^{-1}$ in the anode chamber, leading to a sharp reduction of hydrogen generation. Applied voltage did not affect anode potential much during the process of hydrogen production.

4.2.4. *MEC materials*

Same anode materials are used in both MFC and MEC systems (Logan *et al.* 2008). They include carbon cloth (Liu *et al.* 2005), carbon paper (Ditzig *et al.* 2007), graphite felt (Rozendal *et al.* 2006a; 2007), graphite granules (Cheng and Logan 2007a; Ditzig *et al.* 2007), and graphite brushes (Call and Logan 2008; Wang *et al.* 2009b; Lu *et al.* 2009b). For carbon cloth, carbon paper, and graphite felt, the electrode is either pressed or glued onto an insulated wire. The latter two materials have higher surface areas that help to have more bacteria attached to the electrode. In the case of using graphite granules, a graphite rod is inserted into the bed of granules as a current collector (Cheng and Logan 2007a). In the case of using a graphite brush, two twisted wires of titanium or stainless steel hold the carbon fibers form the brush (Logan *et al.* 2007). Cheng and Logan (2007b) found that performances of these carbon materials can be increased by an ammonia pretreatment at high temperature in MFC.

The cathodes of most MECs use platinum-catalyzed electrodes, which are commercially available and can be easily prepared in the laboratory (Cheng *et al.* 2006a,b). The paste for the catalyst layer, which is prepared by mixing commercially available platinum (e.g., 10 wt% Pt/C, E-TEK) with a chemical binder (5% Nafion solution or 2% PTFE solution), is applied to one side of the cathode, such as carbon paper or carbon cloth. Precious metal catalysts, such as platinum, are used to reduce the cathodic over-potential of hydrogen evolution reaction by lowering the activation energy and by affecting the overall rate of reaction.

However, platinum-catalyzed electrodes are costly and thus not practical for use in full-scale MECs. One alternative is to use biocathodes. Rozendal *et al.* (2008b) obtained an active biocathode by at first developing an electrochemically active biofilm (enriching hydrogen-oxidizing bacteria) on the anode and then reversing the polarity of the electrode. But current densities using biocathode are substantially lower than those in a platinum-catalyzed electrode at a given applied voltage. Another attempt is to develop novel precious-metal-free cathode. Selembo *et al.* (2009) developed nickel oxide catalysts through cathodic electrode position of $NiSO_4$ and $(NH_4)_2SO_4$ on a sheet metal and used them in MECs. This cathode performed similarly to that of platinum sheet metal in MECs. But the long-term mechanical stability of the nickel oxide layer needs to be improved. Harnisch *et al.* (2009) synthesized tungsten carbide powder through a carburization procedure and explored its electrocatalytic behavior in MECs by pasting the powder onto a graphite disc of Nafion. Hu *et al.* (2009) developed NiMo and NiW catalysts by electrodepositing NiMo and NiW on a carbon-fiber-weaved cloth. At an applied voltage of 0.6 V, similar performances were observed in MECs for NiMo catalyst ($2.0\,m^3$-$H_2 \cdot m^{-3} \cdot d^{-1}$) and platinum catalyst ($2.3\,m^3$-$H_2 \cdot m^{-3} \cdot d^{-1}$), both performed better than NiW catalyst. A brush cathode made of stainless steel without any catalyst was also investigated in MEC systems. Call *et al.* (2009) showed that high surface area stainless steel brush cathodes produced hydrogen at rates and efficiencies similar to those achieved by platinum-catalyzed carbon cloth cathodes in single-chamber MECs. Using this stainless steel brush at an applied voltage of 0.6 V, a hydrogen production rate of $1.7 \pm 0.1\,m^3$-$H_2 \cdot m^{-3} \cdot d^{-1}$ was achieved with a current density of $188 \pm 10\,A \cdot m^{-3}$ and an overall energy efficiency of $78 \pm 5\%$.

Various membrane materials have been used in MECs (Liu *et al.* 2005; Cheng and Logan 2007a; Rozendal *et al.* 2007, 2008c) including cation exchange membranes (CEM) such as Nafion 117 (Ion Power Inc., New Castle, DE) and Fumasep FKE (FuMA-Tech GmbH, Germany), anion exchange membranes (AEMs; Fumasep FAB, FuMA-Tech GmbH, Germany; AMI, Membranes International, Glen Rock, NJ), bipolar membrane (BPM; FumaSep FBM, FuMA-Tech GmbH, Germany), and charge mosaic membrane (CMM; Dainichiseika Color & Chemicals, Co. Ltd., Japan). CEM, which was first used in an MEC system, is inefficient on proton diffusion because other cations, such as Na^+, K^+, NH_4^+, and Ca^{2+}, transport charges through the membrane (Gil *et al.* 2003; Rozendal *et al.* 2006b; Kim *et al.* 2007) due to the low proton concentration in solution. This leads to a pH increase in the cathode chamber and a pH decrease in the anode chamber, resulting in a loss of voltage. By using an AEM, proton transport was substantially enhanced as proton being carried by anions such as phosphate and bicarbonate through the membrane (Cheng and Logan 2007a).

4.3. *Hydrogen production by coupling dark-fermentation and MEC*

Hydrogen can be produced by coupling fermentation and MEC. In such a system, the effluent of dark-fermentative hydrogen production, containing mostly VFAs and some alcohols, is further treated by MEC to produce hydrogen. Most of MEC studies conducted so far used pure substrates, primarily acetate. Cheng and Logan (2007a) showed that MEC could produce hydrogen at 0.6 V with yields of 3.95 mol-$H_2 \cdot$(mol-acetate)$^{-1}$, 8.01 mol-$H_2 \cdot$(mol-butyrate)$^{-1}$, and

6.25 mol-H_2·(mol-propionate)$^{-1}$, equivalent to, respectively, 91%, 80%, and 89% of the theoretical maxima.

4.3.1. *MEC coupled with ethanol-fermentation from molasses wastewater*

Using a single-chamber MEC, Lu *et al.* (2009b) showed that additional hydrogen could be produced from the effluent of a dark-fermentation reactor producing hydrogen from a molasses wastewater. This fermentation effluent consisted primarily of ethanol, acetate, propionate, butyrate, valerate, and residual sugars. Using a buffered effluent at pH 6.7–7.0, the MEC at 0.6 V had a hydrogen production rate of 1.41 ± 0.08 m^3-H_2·m^{-3}·d^{-1} and a maximum overall hydrogen recovery of 83 ± 4% (based on both electrical energy and substrate utilization). When the MEC was coupled with a fermentation system, the overall hydrogen recovery was 96%, with a hydrogen production rate of 2.11 m^3-H_2·m^{-3}·d^{-1}. The produced energy equals to 287% of the electrical energy input. However, when using the un-buffered fermentation effluent at pH 4.5–4.6, the MEC produced little hydrogen.

4.3.2. *MEC coupled with H_2 fermentation from lignocellulose and cellobiose*

Lalaurette *et al.* (2009) coupled dark-fermentation with an MEC at 0.5 V to produce hydrogen from lignocellulose and cellobiose. The fermentation effluent consisted primarily of acetate, lactate, succinate, formate, and ethanol. The single-chamber MEC treating the lignocellulose fermentation effluent had a hydrogen yield of 750 ± 180 ml·(g-COD)$^{-1}$ and a production rate of 1.00 ± 0.19 m^3-H_2·m^{-3}·d^{-1}; the corresponding values for treating the cellobiose fermentation effluent were 900 ± 140 ml·(g-COD)$^{-1}$ and 0.96 ± 0.16 m^3-H_2·m^{-3}·d^{-1}. Energy produced from lignocellulose effluent was 230 ± 50% of the electrical energy input; and the corresponding value for cellobiose effluent was 220 ± 30%. The MEC had an overall hydrogen yield of 9.95 mol-H_2·(mol-glucose)$^{-1}$ treating the cellobiose effluent.

These results demonstrate that the process of coupling fermentation and MEC could achieve high hydrogen yields and production rates. However, the fermentation effluent, which usually has a pH of 4.5–5.0, needs to be buffered at around pH 7 by phosphate before being treated by an MEC. Since buffering the effluent using phosphate would not be practical in full-scale process, a strategy on applying an acid-tolerant exoeletrogenic biofilm on electrodes in MEC warrants a further investigation.

5. Perspective

This review has shown that anaerobic technology may provide a feasible solution for the production of hydrogen from wastes and wastewater. Figure 14.1 illustrates a schematic flow diagram of such a biohydrogen production system. At present, many concerns remain on how to establish a practical and affordable waste-to-hydrogen process. One possibility is to develop an integrated system for multi-organic wastes, which could include coupling processes for production of H_2–CH_4, H_2-electricity, biofuel, and value-added bio-products such as bio-surfactants and bio-flocculants. Through this approach, a multi-choice scheme might be performed according to the needs for the future bio-energy and bio-products market. Aside from the scientific

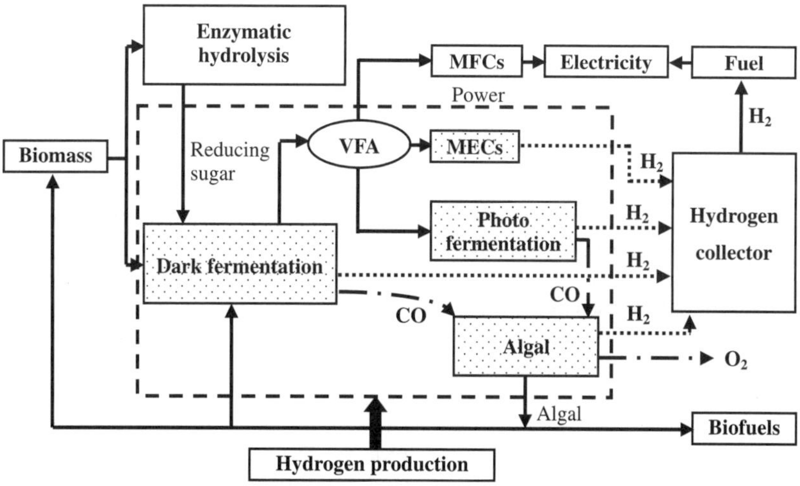

Fig. 14.1. A schematic flow diagram of biohydrogen production.

assessment, a careful and detailed economic analysis is needed in order to boost this blueprint into practice.

References

Barbara, R., Genthner, S., and Judy, D. (1985) Wall ammonium uptake in *Rhodopseudomonas capulata*. *Arch. Microbiol.* **141**, 219–224.

Barbosa, M. J., Rocha, J. M. S., Tramper, J., and Wijffels, R. H. (2001) Acetate as a carbon source for hydrogen production by photosynthetic bacteria. *J. Biotechnol.* **85**, 25–33.

Cai, M. L., Liu, J. X., and Wei, Y. S. (2004) Enhanced biohydrogen production from sewage sludge with alkaline pretreatment. *Environ. Sci. Technol.* **38**, 3195–3202.

Cai, J. L., Wang, G. C., Li, Y. C., Zhu, D. L., and Pan, G. H. (2009) Enrichment and hydrogen production by marine anaerobic hydrogen-producing microflora. *Chin. Sci. Bull.* **54**, 2656–2661.

Call, D. and Logan, B. E. (2008) Hydrogen production in a single chamber microbial electrolysis cell lacking a membrane. *Environ. Sci. Technol.* **42**, 3401–3406.

Call, D., Merrill, M., and Logan, B. E. (2009) High surface area stainless steel brushes as cathodes in microbial electrolysis cells (MECs). *Environ. Sci. Technol.* **43**, 2179–2183.

Chae, K. J., Choi, M. J., Kim, K. Y., Ajayi, F. F., Chang, I. S., and Kim, I. S. (2008) Selective inhibition of methanogens for the improvement of biohydrogen production in bioelectrochemical cells. In: *Proceedings of the 2008 Asian Bio-Hydrogen Symposium*, pp. 297–308.

Chen, C. Y., Lee, C. M., and Chang, J. S. (2006) Hydrogen production by indigenous photosynthetic bacterium *Rhodopseudomonas palustris* WP3-5 using optical fiber-illuminating photobioreactors. *Biochem. Eng. J.* **32**, 33–42.

Cheng, S., Liu, H., and Logan, B. E. (2006a) Increased performance of single-chamber microbial fuel cells using an improved cathode structure. *Electrochem. Commun.* **8**, 489–494.

Cheng, S., Liu, H., and Logan, B. E. (2006b) Power densities using different cathode catalysts (Pt and CoTMPP) and polymer binders (Nafion and PTFE) in single chamber microbial fuel cells. *Environ. Sci. Technol.* **40**, 364–369.

Cheng, S. and Logan, B. E. (2007a) Sustainable and efficient biohydrogen production via electrohydrogenesis. *Proc. Natl. Acad. Sci. U.S.A.* **104**, 18871–18873.

Cheng, S. and Logan, B. E. (2007b) Ammonia treatment of carbon cloth anodes to enhance power generation of microbial fuel cells. *Electrochem. Commun.* **9**, 492–496.

Chu, C. F., Li, Y. Y., Xu, K. Q., Ebie, Y., Yuhei, I., and Kong, H. N. (2008) A pH- and temperature-phased two-stage process for hydrogen and methane production from food waste. *Int. J. Hydrogen Energy* **33**, 4739–4746.

Chu, C. F., Ebie, Y., Inamori, Y., and Kong, H. N. (2009) Effect of hydraulic retention time on the hydrogen yield and population of *Clostridium* in hydrogen fermentation of glucose. *J. Environ. Sci.* **21**, 424–428.

Ding, J., Liu, B. F., Ren, N. Q., Xing, D. F., Guo, W. Q., Xu, J. F., and Xie, G. J. (2009) Hydrogen production from glucose by co-culture of *Clostridium butyricum* and immobilized *Rhodopseudomonas faecalis* RLD-53. *Int. J. Hydrogen Energy* **34**, 3647–3652.

Ditzig, J., Liu, H., and Logan, B. E. (2007) Production of hydrogen from domestic wastewater using a bioelectrochemically assisted microbial reactor (BEAMR). *Int. J. Hydrogen Energy* **32**, 2296–2304.

Fan, Y. T., Li, C. L., Lay, J. J., Hou, H. W., and Zhang, G. S. (2004) Optimization of initial substrate and pH levels for germination of sporing hydrogen-producing anaerobes in cow dung compost. *Bioresour Technol.* **91**, 189–193.

Fan, Y. T., Zhang, G. S., Guo, X. Y., Xing, Y., and Fan, M. H. (2006a) Biohydrogen-production from beer lees biomass by cow dung compost. *Biomass Bioenergy* **30**, 493–496.

Fan, Y. T., Zhang, Y. H., Zhang, S. F., Hou, H. W., and Ren, B. Z. (2006b) Efficient conversion of wheat straw wastes into biohydrogen gas by cow dung compost. *Bioresour. Technol.* **97**, 500–505.

Fang, H. H. P. and Liu, H. (2002) Effect of pH on hydrogen production from glucose by a mixed culture. *Bioresour Technol.* **82**(1), 87–93.

Fu, Y. J., Huang, J. L., Long, M. N., Wei, W. L., and Zhang, F. Z. (2004) Isolation and condition of a H_2-producing bacteria. *J. Nanchang Univ.* **28**, 167–169.

Gil, G. C., Chang, I. S., Kim, B. H., Kim, M., Jang, J. K., Park, H. S., and Kim, H. J. (2003) Operational parameters affecting the performance of a mediator-less microbial fuel cell. *Biosens. Bioelectron.* **18**, 327–334.

Guo, W. Q., Ren, N. Q., Wang, X. J., Xiang, W. S., Meng, Z. H., Ding, J., Qu, Y. Y., and Zhang, L. S. (2008a) Biohydrogen production from ethanol-type fermentation of molasses in an expanded granular sludge bed (EGSB) reactor. *Int. J. Hydrogen Energy* **33**, 4981–4988.

Guo, W. Q., Ren, N. Q., Chen, Z. B., Liu, B. F., Wang, X. J., Xiang, W. S., and Ding, J. (2008b) Simultaneous biohydrogen production and starch wastewater treatment in an acidogenic expanded granular sludge bed reactor by mixed culture for long-term operation. *Int. J. Hydrogen Energy* **33**, 7397–7404.

Guo, W. Q., Ren, N. Q., Wang, X. J., Xiang, W. S., Ding, J., You, Y., and Liu, B. F. (2009) Optimization of culture conditions for hydrogen production by *Ethanoligenens harbinense* B49 using response surface methodology. *Bioresour. Technol.* **100**, 1192–1196.

Hao, X. L., Zhou, M. H., Yu, H. Q., Shen, Q. Q., and Lei, L. C. (2006) Effect of sodium ion concentration on hydrogen production from sucrose by anaerobic hydrogen-producing granular sludge. *Chinese J. Chem. Eng.* **14**, 511–517.

Harnisch, F., Sievers, G., and Schroder, U. (2009) Tungsten carbide as electrocatalyst for the hydrogen evolution reaction in pH neutral electrolyte solutions. *Appl. Catal. B Environ.* **89**, 455–458.

Hawkes, F. R., Hussy, I., Kyazze, G., Dinsdale, R., and Hawkes, D. L. (2007) Continuous dark fermentative hydrogen production by mesophilic microflora: Principles and progress. *Int. J. Hydrogen Energy* **32**, 172–184.

Hu, H., Fan, Y., and Liu, H. (2008) Hydrogen production using single chamber membrane-free microbial electrolysis cells. *Wat. Res.* **42**, 4172–4178.

Hu, H., Fan, Y., and Liu, H. (2009) Hydrogen production in single-chamber tubular microbial electrolysis cells using non-precious-metal catalysts. *Int. J. Hydrogen Energy* **34**, 8535–8542.

Kim, J. R., Cheng, S., Oh, S.-E., and Logan, B. E. (2007) Power generation using different cation, anion and ultrafiltration membranes in microbial fuel cells. *Environ. Sci. Technol.* **41**, 1004–1009.

Lalaurette, E., Thammannagowda, S., Mohagheghi, A., Maness, P. C., and Logan, B. E. (2009) Hydrogen production from cellulose in a two-stage process combining fermentation and electrohydrogenesis. *Int. J. Hydrogen Energy* **34**(15), 6201–6210.

Li, C. L. and Fang, H. H. P. (2007a) Fermentative hydrogen production from wastewater and solid wastes by mixed cultures, *Crit. Rev. Environ. Sci. Technol.* **37**, 1–39.

Li, C. L. and Fang, H. H. P. (2007b) Inhibition of heavy metals on fermentative hydrogen production. *Chemosphere* **67**, 668–673.

Li, J. Z., Li, B. K., Zhu, G. F., Ren, N. Q., Bo, L. X., and He, J. G. (2007) Hydrogen production from diluted molasses by anaerobic hydrogen-producing bacteria in an anaerobic baffled reactor (ABR). *Int. J. Hydrogen Energy* **32**, 3274–3283.

Li, R. Y. and Fang, H. H. P. (2009) Heterotrophic photo-fermentative hydrogen production, *Crit. Rev. Environ. Sci. Technol.* **39**(12), 1081–1108.

Li, Z., Wang, H., Tang, Z. X., Wang, X. F., and Bai, J. B. (2008) Effects of pH value and substrate concentration on hydrogen production from the anaerobic fermentation of glucose. *Int. J. Hydrogen Energy* **33**, 7413–7418.

Li, D., Yuan, Z. H., Sun, Y. M., Kong, X. Y., and Zhang, Y. (2009) Hydrogen production characteristics of the organic fraction of municipal solid wastes by anaerobic mixed culture fermentation. *Int. J. Hydrogen Energy* **34**, 812–820.

Lin, M. (2002) *Hydrogen Production Mechanism and Ecological Law of a New Species of High Efficiency Hydrogen Production by Fermentation.* Doctoral thesis, Harbin Institute of Technology, China.

Liu, G. Z. and Shen, J. Q. (2004) Effects of culture and medium conditions on hydrogen production from starch using anaerobic bacteria. *J. Biosci. Bioeng.* **98**, 251–256.

Liu, H., Grot, S., and Logan, B. E. (2005) Electrochemically assisted microbial production of hydrogen from acetate. *Environ. Sci. Technol.* **39**, 4317–4320.

Liu, B. F., Ren, N. Q., Ding, J., Xie, G. J., and Cao, G. L. (2009a) Enhanced photo-H_2 production of *R. faecalis* RLD-53 by separation of CO_2 from reaction system. *Bioresour. Technol.* **100**, 1501–1504.

Liu, B. F., Ren, N. Q., Ding, J., Xie, G. J., and Guo, W. Q. (2009b) The effect of Ni^{2+}, Fe^{2+} and Mg^{2+} concentration on photo-hydrogen production by *Rhodopseudomonas faecalis* RLD-53. *Int. J. Hydrogen Energy* **34**, 721–726.

Liu, B. F., Ren, N. Q., Xing, D. F., Ding, J., Zheng, G. X., Guo, W. Q., and Xie, G. J. (2009c) Biohydrogen production by immobilized *R. faecalis* RLD-53 using soluble metabolites from ethanol fermentation bacteria *E. harbinense* B49. *Bioresour. Technol.* **100**, 2719–2723.

Liu, B. F., Ren, N. Q., Tang, J., Ding, J., Liu, W. Z., Xu, J. F., Cao, G. L., and Xie, G. J. (2010) Biohydrogen production by mixed culture of photo-fermentation and dark-fermentation bacteria. *Int. J. Hydrogen Energy*, doi:10.1016/j.ijhydene.2009.05.005.

Logan, B. E., Cheng, S., Watson, V., and Estadt, G. (2007) Graphite fiber brush anodes for increased power production in air-cathode microbial fuel cells. *Environ. Sci. Technol.* **41**, 3341–3346.

Logan, B. E. (2008) *Microbial Fuel Cells*, Hoboken, NJ: John Wiley & Sons, Inc.

Logan, B. E., Call, D., Cheng, S., Hamelers, H. V. M., Sleutels, T. H. J. A., Jeremiasse, A. W., and Rozendal, R. A. (2008) Microbial electrolysis cells for high yield hydrogen gas production from organic matter. *Environ. Sci. Technol.* **42**, 8630–8640.

Lu, Y., Lai, Q. H., Zhang, C., Zhao, H. X., Ma, K., Zhao, X. B., Chen, H. Z., Liu, D. H., and Xing, X. H. (2009a) Characteristics of hydrogen and methane production from cornstalks by an augmented two- or three-stage anaerobic fermentation process. *Bioresour. Technol.* **100**, 2889–2895.

Lu, L., Ren, N., Xing, D., and Logan, B. E. (2009b) Hydrogen production with effluent from an ethanol-H_2-coproducing fermentation reactor using a single-chamber microbial electrolysis cell. *Biosens. Bioelectron.* **24**, 3055–3060.

Ludden, P. W. and Roberts, G. P. (1995) The biochemistry and genetics of nitrogen fixation by photosynthetic bacteria. In: *Anoxygenic Photosynthetic Bacteria*. Netherlands: Kluwer, Academic Publishers, pp. 929–947.

Mu, Y. and Yu, H. Q. (2006) Biological hydrogen production in a UASB reactor with granules. I: Physicochemical characteristics of hydrogen-producing granules. *Biotechnol. Bioeng.* **94**, 980–987.

Mu, Y., Yu, H. Q., and Wang, Y. (2006a) The role of pH in the fermentative H_2 production from an acidogenic granule-based reactor. *Chemosphere* **64**, 350–358.

Mu, Y., Zheng, X. J., Yu, H. Q., and Zhu, R. F. (2006b) Biological hydrogen production by anaerobic sludge at various temperatures. *Int. J. Hydrogen Energy* **31**, 780–785.

Mu, Y., Yu, H. Q., and Wang, G. (2007) Evaluation of three methods for enriching H_2-producing cultures from anaerobic sludge. *Enzyme. Microb. Technol.* **40**, 947–953.

Oh, S.-E., VanGinkel, S., and Logan, B. E. (2003) The relative effectiveness of pH control and heat treatment for enhancing biohydrogen gas production. *Environ. Sci. Technol.* **37**, 5186–5190.

Qian, Y. F., Zheng, G. H., Kang, Z. H., and Wang, L. (2007) Influence of factors on hydrogen production capability of *Rhodobacter sphaeroides*. *Ind. Microbiol.* **37**, 6–12.

Ren, N. Q., Wang, B. Z., and Huang, J. C. (1997) Ethanol-type fermentation from carbohydrate in high rate acidogenic reactor. *Biotechnol. Bioeng.* **54**, 428–433.

Ren, N. Q. and Gong, M. L. (2006) Acclimation strategy of a biohydrogen producing population in a continuous-flow reactor with carbohydrate fermentation. *Eng. Life Sci.* **6**, 403–409.

Ren, N. Q., Li, J. Z., Li, B. K., Wang, Y., and Liu, S. R. (2006) Biohydrogen production from molasses by anaerobic fermentation with a pilot-scale bioreactor system. *Int. J. Hydrogen Energy* **31**, 2147–2157.

Ren, N. Q., Xing, D. F., Rittmann, B. E., Zhao, L. H., Xie, T. H., and Zhao, X. (2007) Microbial community structure of ethanol type fermentation in biohydrogen production. *Environ. Microbiol.* **9**, 1112–1125.

Ren, N. Q., Liu, B. F., Ding, J., Guo, W. Q., Cao, G. L., and Xie, G. J. (2008) The effect of butyrate concentration on photo-hydrogen production from acetate by *Rhodopseudomonas faecalis* RLD-53. *Int. J. Hydrogen Energy* **33**, 5981–5985.

Ren, N. Q., Liu, B. F., Ding, J., and Xie, G. J. (2009) Hydrogen production with *R. faecalis* RLD-53 isolated from freshwater pond sludge. *Bioresour. Technol.* **100**, 484–487.

Rozendal, R. A., Hamelers, H. V. M., Euverink, G. J. W., Metz, S. J., and Buisman, C. J. N. (2006a) Principle and perspectives of hydrogen production through biocatalyzed electrolysis. *Int. J. Hydrogen Energy* **31**, 1632–1640.

Rozendal, R. A., Hamelers, H. V. V., and Buisman, C. J. N. (2006b) Effects of membrane cation transport on pH and microbial fuel cell performance. *Environ. Sci. Technol.* **40**, 5206–5211.

Rozendal, R., Hamelers, H. V. M., Molenkamp, R. J., and Buisman, C. J. N. (2007) Performance of single chamber biocatalyzed electrolysis with different types of ion exchange membranes. *Wat. Res.* **41**, 1984–1994.

Rozendal, R. A., Hamelers, H. V. M., Rabaey, K., Keller, J., and Buisman, C. J. N. (2008a) Towards practical implementation of bioelectrochemical wastewater treatment. *Trends Biotechnol.* **26**, 450–459.

Rozendal, R. A., Jeremiasse, A. W., Hamelers, H. V. M., and Buisman, C. J. N. (2008b) Hydrogen production with a microbial biocathode. *Environ. Sci. Technol.* **42**, 629–634.

Rozendal, R. A., Sleutels, T. H. J. A., Hamelers, H. V. M., and Buisman, C. J. N. (2008c) Effect of the type of ion exchange membrane on performance, ion transport, and pH in biocatalyzed electrolysis of wastewater. *Wat. Sci. Technol.* **57**, 1757–1762.

Sasikala, K., Ramana, C. V., Rao, P. R., and Subrahmanyam, M. (1990) Effect of gas phase on the photoproduction of hydrogen and substrate conversion efficiency in the photosynthetic bacterium *Rhodobacter sphaeroides* O.U.001. *Int. J. Hydrogen Energy* **15**, 795–797.

Sasikala, K., Ramana, C. V., and Rao, P. R. (1991) Environmental regulation for optimal biomass yield and photoproduction of hydrogen by *Rhodobacter sphaeroides* O.U. 001. *Int. J. Hydrogen Energy* **16**, 597–601.

Selembo, P. A., Merrill, M. D., and Logan, B. E. (2009) The use of stainless steel and nickel alloys as low-cost cathodes in microbial electrolysis cells. *J. Power Sources* **190**, 271–278.

Shi, X. Y. and Yu, H. Q. (2006) Continuous production of hydrogen from mixed volatile fatty acids with *Rhodopseudomonas capsulate*. *Int. J. Hydrogen Energy* **31**, 1641–1647.

Shi, Y. Z., Zhang, Q. G., Wang, Y., and Jing, Y. Y. (2008) Experimental study on continuous culture of photosynthetic bacteria for hydrogen production from biomass. *Trans. Chin. Soc. Agr. Eng.* **6**, 218–221.

Su, H. B., Cheng, J., Zhou, J. H., Song, W. L., and Cen, K. F. (2009) Improving hydrogen production from cassava starch by combination of dark and photo-fermentation. *Int. J. Hydrogen Energy* **34**, 1780–1786.

Sun, Q., Xu, X. Y., and Jiao, Y. W. (1995) Studies on the conditions of hydrogen photoproduction by photosynthetic bacteria. *Acta. Microbiologica. Sinica* **35**, 65–73.

Sun, M., Sheng, G., Zhang, L., Xia, C., Mu, Z., Liu, X., Wang, H., Yu, H., Qi, R., Yu, T., and Yang, M. (2008) An MEC-MFC-Coupled system for biohydrogen production from acetate. *Environ. Sci. Technol.* **42**, 8095–8100.

Tang, G. L., Huang, J., Sun, Z. J., Tang, Q. Q., Yan, C. H., and Liu, G. Q. (2008) Biohydrogen production from cattle wastewater by enriched anaerobic mixed consortia: Influence of fermentation temperature and pH. *J. Biosci. Bioeng.* **106**, 80–87.

Tao, Y. Z., Chen, Y., Wu, Y. Q., He, Y. L., and Zhou, Z. H. (2007) High hydrogen yield from a two-step process of dark- and photo-fermentation of sucrose. *Int. J. Hydrogen Energy* **32**, 200–206.

Tao, Y. Z., He, Y. L., Wu, Y. Q., Liu, F. H., Li, X. F., Zong, W. M., and Zhou, Z. H. (2008) Characteristics of a new photosynthetic bacterial strain for hydrogen production and its application in wastewater treatment. *Int. J. Hydrogen Energy* **33**, 963–973.

Tian, X., Liao, Q., Zhang, P., Zhu, X., and Wang, Y. Z. (2008) Start-up behavior and photo-hydrogen production performance of phototrophic biofilm reactor. *J. Eng. Thermophys.* **29**, 1553–1557.

Wang, Y., Mu, Y., and Yu, H. Q. (2007a) Comparative performance of two upflow anaerobic biohydrogen-producing reactors seeded with different sludges. *Int. J. Hydrogen Energy* **32**, 1086–1094.

Wang, X. J., Ren, N. Q., Xiang, W. S., and Guo, W. Q. (2007b) Influence of gaseous end-products inhibition and nutrient limitations on the growth and hydrogen production by hydrogen-producing fermentative bacterial B49. *Int. J. Hydrogen Energy* **32**, 748–754.

Wang, A. (2008) Towards microbial electrolysis cell: Metagenomics-enabled understanding of anodophilic functional consortium. *Microbial Fuel Cell First International Symposium*, May 27–29, State College, Pennsylvania, USA.

Wang, J. L. and Wan, W. (2008a) Effect of Fe^{2+} concentration on fermentative hydrogen production by mixed cultures. *Int. J. Hydrogen Energy* **33**, 1215–1220.

Wang, J. L. and Wan, W. (2008b) Influence of Ni^{2+} concentration on biohydrogen production. *Bioresour. Technol.* **99**, 8864–8868.

Wang, J. L. and Wan, W. (2008c) Effect of temperature on fermentative hydrogen production by mixed cultures. *Int. J. Hydrogen Energy* **33**, 5392–5397.

Wang, J. L. and Wan, W. (2008d) Comparison of different pretreatment methods for enriching hydrogen-producing bacteria from digested sludge. *Int. J. Hydrogen Energy* **33**, 2934–2941.

Wang, X. and Zhao, Y. C. (2009) A bench scale study of fermentative hydrogen and methane production from food waste in integrated two-stage process. *Int. J. Hydrogen Energy* **33**, 245–254.

Wang, B., Wan, W., and Wang, J. L. (2009a) Effect of ammonia concentration on fermentative hydrogen production by mixed cultures. *Bioresour. Technol.* **100**, 1211–1213.

Wang, A., Liu, W., Cheng, S., Xing, D., Zhou, J., and Logan, B. E. (2009b) Source of methane and methods to control its formation in single chamber microbial electrolysis cells. *Int. J. Hydrogen Energy* **34**, 3653–3658.

Wang A., Liu, W., and Ren, N. (2010) Key factors affecting microbial anode potential in a microbial electrolysis cell for H_2 production. *Int. J. Hydrogen Energy*, doi: 10.1016/j.ijhydene.2009.11.125.

Weaver, P. F., Lien, S., and Senbert, M. (1980) Photobiological production of hydrogen. *Sol. Energy* **24**, 3–45.

Wei, L. F., Zheng, X. J., Zhang, Z. H., Xu, P. Y., and Wei, M. B. (2007) Influence of pH value on hydrogen production from sodium acetate by photosynthetic bacteria. *Henan Chem. Ind.* **24**, 32–34. (in Chinese).

Xiao, B. Y. and Liu, J. X. (2009) Biological hydrogen production from sterilized sewage sludge by anaerobic self-fermentation. *J. Hazard. Mater.* **168**, 163–167.

Xie, B. F., Cheng, J., Zhou, J. H., Song, W. L., and Cen, K. F. (2008a) Cogeneration of hydrogen and methane from glucose to improve energy conversion efficiency. *Int. J. Hydrogen Energy* **33**, 5006–5011.

Xie, B. F., Cheng, J., Zhou, J. H., Song, W. L., Liu, J. Z., and Cen, K. F. (2008b) Production of hydrogen and methane from potatoes by two-phase anaerobic fermentation. *Bioresour. Technol.* **99**, 5942–5946.

Xu, X. Y., Zheng, P., Yu, X. E., and Feng, X. S. (1993) Research on hydrogen production from organic waste water with immobilized cells of photosynthetic bacteria (PSB) II. The characteristics of hydrogen photo-production by *Rhodopseudomonas sp.* D. *Acta Energiae Solaris Sin.* **14**, 288–294. (in Chinese).

Xu, L. Y., Ren, N. Q., Wang, X. Z., and Jia, Y. F. (2008) Biohydrogen production by *Ethanoligenens harbinense* B49: Nutrient optimization. *Int. J. Hydrogen Energy* **33**, 6962–6967.

Yang, S. P., Zhao, C. G., Liu, R. T., Qu, Y. B., and Qian, X. M. (2002) Hydrogen photoproduction from acetate by *Rhodopseudomonas palustris*. *Chin. J. Biotechnol.* **18**, 486–491. (in Chinese).

Yang, S. P., Zhao, C. G., Qu, Y. B., and Qian, X. M. (2003) Effects of iron and nickel on hydrogen photoevolution, cell growth, and photosynthetic pigment of photosynthetic bacteria. *Acta Microbiol. Sin.* **43**, 257–263.

Yetis, M., Gündüz, U., Eroglu, I., Yücel, M., and Türker, L. (2000) Photoproduction of hydrogen from sugar refinery wastewater by *Rhodobacter sphaeroides* O.U.001. *Int. J. Hydrogen Energy* **25**, 1035–1041.

You, X. F., Zhou, J. Y., Zhang, Q. G., and Wang, Y. J. (2005) Research on hydrogen production ability of *Rhodobacter sphaeroides* with swine manure. *J. Henan Agric. Univ.* **39**, 215–217.

Yu, H. Q., Zhu, Z. H., Hu, W. R., and Zhang, H. S. (2002) Hydrogen production from rice winery wastewater in an upflow anaerobic reactor by using mixed anaerobic cultures. *Int. J. Hydrogen Energy* **27**, 1359–1365.

Yuan, Z. L., Yang, H. J., Zhi, X. H., and Shen, J. Q. (2008) Enhancement effect of L-cysteine on dark fermentative hydrogen production. *Int. J. Hydrogen Energy* **33**, 6535–6540.

Zhang, Y. F., Liu, G. Z., and Shen, J. Q. (2005a) Hydrogen production in batch culture of mixed bacteria with sucrose under different iron concentrations. *Int. J. Hydrogen Energy* **30**, 855–860.

Zhang, J. H., Zhang, Q. G., Yang, Q. F., and Wang, Y. J. (2005b) Influence of the degree of light intensity on hydrogen production of *Rhodobacter sphaeroides* from pig dejecta wastewater. *Trans. Chin. Soc. Agr. Eng.* **21**, 134–136.

Zhang, Y. F. and Shen, J. Q. (2006) Effect of temperature and iron concentration on the growth and hydrogen production of mixed bacteria. *Int. J. Hydrogen Energy* **31**, 441–446.

Zhang, P. (2007) *Study on Hydrogen Production and Biodegradation Characteristics in Biofilm Photobioreactor*. Master thesis, Chongqing University, China.

Zhang, M. L., Fan, Y. T., Xing, Y., Pan, C. M., Zhang, G. S., and Lay, J. J. (2007) Enhanced biohydrogen production from cornstalk wastes with acidification pretreatment by mixed anaerobic cultures. *Biomass. Bioenergy* **31**, 250–254.

Zhao, Q. B. and Yu, H. Q. (2007) Fermentative H_2 production in an upflow anaerobic sludge blanket reactor at various pH values. *Bioresour. Technol.* **99**, 1353–1358.

Zhao, B. H., Yue, Z. B., Zhao, Q. B., Mu, Y., Yu, H. Q., Harada, H., and Li, Y. Y. (2008) Optimization of hydrogen production in a granule-based UASB reactor. *Int. J. Hydrogen. Energy* **33**, 2454–2461.

Zheng, X. J. and Yu, H. Q. (2004) Roles of pH in biologic production of hydrogen and volatile fatty acids from glucose by enriched anaerobic cultures. *Appl. Biochem. Biotechnol.* **112**, 79–90.

Zheng, X. J. and Yu, H. Q. (2005) Biological hydrogen production by enriched anaerobic cultures in the presence of copper and zinc. *J. Environ. Sci. Health A* **39**, 89–101.

Zhou, R. Y. (2007) *Study on the Circumfluence Cylindrical Photobioreactor for Photobiological Hydrogen Production and Its Transmission of Energy*. Doctoral thesis, Henan Agricultural University, China.

Zhu, H. G., Zhao, Q. L., and Shi, J. L. (1997) Affecting factors on hydrogen production by photosynthetic bacterium *Rhodopseudomonas*. *Chin. J. Appl. Ecol.* **8**, 194–198.

Zhu, H. G., Fang, H. H. P., Zhang, T., and Beaudettec, L. A. (2007) Effect of ferrous ion on photo heterotrophic hydrogen production by *Rhodobacter sphaeroides*. *Int. J. Hydrogen. Energy* **32**, 4112–4118.

Chapter 15

Research and Development of Biohydrogen Production in Taiwan

Chiu-Yue Lin* and Chyi-How Lay

Feng Chia University, Department of Environmental Engineering and Science, 100 Wenhwa Rd., Seatwen, Taichung, Taiwan

**cylin@fcu.edu.tw*

We have studied biohydrogen production in three kinds of continuous reactors: continuous stirred anaerobic bioreactor (CSABR), carrier-induced granular sludge bed reactor (CIGSBR), and agitated granular sludge bed reactor (AGSBR). Among the three, the CSABR was the most efficient one with a hydrogen production rate (HPR) of 362 $l \cdot l^{-1} \cdot d^{-1}$ (i.e., 14.7 $mol \cdot l^{-1} \cdot d^{-1}$), using sucrose as substrate, and an optimal yield of ca. 3.5 mol-H_2·mol-sucrose^{-1} with a feeding sucrose concentration of 30–40 g-COD·l^{-1} and a short hydraulic retention time (HRT) of 0.5 h. En route toward the commercialization of this high-rate biohydrogen production process, a pilot-scale system with a 400 l fermentor was constructed and operated to identify the scale-up operating parameters. The pilot fermentor was operated at 35 °C and fed with sucrose (20 g-COD·l^{-1}). After two days of startup operating in batch mode, the fermentor was operated in continuous feeding mode with 12 h of HRT. The H_2 production rate increased from 2.5 to 15.6 $l \cdot l^{-1} \cdot d^{-1}$ with the increase of organic loading rate from 40 to 240 g-COD·$l^{-1} \cdot d^{-1}$. Results of DGGE analysis indicate that a stable hydrogen production was obtained as *Clostridium pasteurianum* was being predominant in the fermentor.

1. Introduction

Hydrogen is a promising alternative energy carrier for the future because it is clean, renewable and generates harmless by-products from combustion. Hydrogen can be produced by gasification of hydrocarbons, electrolysis of water or fermentation of organic feedstock. Wastewater containing organic matter like glucose, xylose, and sucrose, as well as starch and cellulose may be fermented to produce hydrogen.

Most hydrogen fermentation is conducted in well-mixed systems, such as the anaerobic sequencing batch reactor (ASBR), continuously stirred tank reactor (CSTR), carrier-induced granular sludge bed reactor (CIGSBR), fluidized bed reactor (FBR), membrane bioreactor (MBR), and upflow anaerobic sludge blanket (UASB) reactor (Lin *et al.* 2008c). The main operating factors affecting fermentative hydrogen production include inoculums, substrate, reactor type, nitrogen, phosphate, metals, temperature, and pH (Wang and Wan 2008). Our previous studies have identified the optimal operating factors in batch tests (Lin and Lay 2004a,b; Lin and Lay 2005; Lin *et al.* 2006a,b, 2008a). Using the optimal conditions, we operated continuously stirred tank reactors (CSTR) to enrich the hydrogen-producing sludge and to enhance hydrogen production efficiency (Lin and Chang 2004; Lin and Cheng 2006; Lin *et al.* 2008b).

However, when CSTR was operated at short hydraulic retention times (HRT) of less than 2 h, sludge washout might occur resulting in operational instability and a severe reduction of hydrogen production rate (HPR). Supported financially by Taiwan's Bureau of Energy and Feng Chia University (FCU), our research team has successfully developed a high-rate hydrogen production process using granulated sludge in several types of reactors, including CSABR, CIGSBR, and AGSBR. These reactors exhibited high HPR because they could retain a high biomass concentration in the bioreactor and allow effective contact between microorganisms and substrates (Chen *et al.* 2002; Lee *et al.* 2004; Wu *et al.* 2006).

The high-rate H_2 production processes produced a maximum HPR and hydrogen yield of 10–362 $l \cdot l^{-1} \cdot d^{-1}$ and 3.5–3.2 mol-$H_2 \cdot mol^{-1}$ sucrose, respectively. These reactors were then applied to an on-line electricity generation system via a proton-exchange membrane fuel cell (PEMFC). In this work, a "bioH_2 to electricity" demonstration system was fabricated by connecting a 3l immobilized-cell-seeded CSABR bioreactor with a PEMFC stack to generate electricity. This demonstration is expected to provide useful information for the future commercial development (Lin *et al.* 2007).

At present, the commercial viability of the AGSBR system is further assessed in a pilot-scale (400 l fermentor) operation. The pilot plant was constructed to identify the scaleup operating parameters for the practical production of biohydrogen at a commercial scale. A strategy was also developed to effectively start up this pilot system.

2. Anaerobic Hydrogen Fermentation

Fermentation is a multi-step process involving a number of serial and parallel reactions, in which H_2 is a key intermediate (Koskinen 2008). The process is carried out by microorganisms in three stages of reaction: hydrolysis, acidogenesis, and methanogenesis (Fig. 15.1). In the initial hydrolysis, complex organic matters are hydrolyzed into monomers such as glucose and amino acids by the action of extracellular enzymes produced by hydrolytic bacteria. In the stage of acidogenesis, the hydrolyzed monomers are converted into volatile fatty acids (VFAs) and alcohols. The higher VFAs, such as propionic and butyric acids, are further converted into acetic acid and H_2/CO_2 by acetogenic bacteria. Finally, acetic acid and H_2/CO_2 are converted into methane by methanogens (Brock *et al.* 1994; Malina and Pohland 1992; Schmidt and Ahring 1996).

In order to recover hydrogen, the final step of fermentation, i.e., the methanogenesis, must be blocked. The bioactivity of methanogens in the fermentor has to be suppressed. This could be achieved by a heat treatment of the seed sludge because methanogens are sensitive to heat, whereas hydrogen-producing bacteria are not (Chen *et al.* 2002; Lay *et al.* 1999; van Ginkel *et al.* 2001). Other means to limit methanogenesis in fermentor include (1) operating at short HRT to wash out the slow growing methanogens and (2) operating at an acidic environment to inhibit the pH-sensitive methanogens (Chen *et al.* 2001; Lin and Chang 1999).

3. High-Rate Biohydrogen Production Technology

High biomass concentration, up to certain degrees, would increase the treatment efficiency and the organic loading in bioreactions. Retention strategies affect both the quality and quantity of the biomass in hydrogen fermentation (Koskinen 2008). Sludge granulation is the most

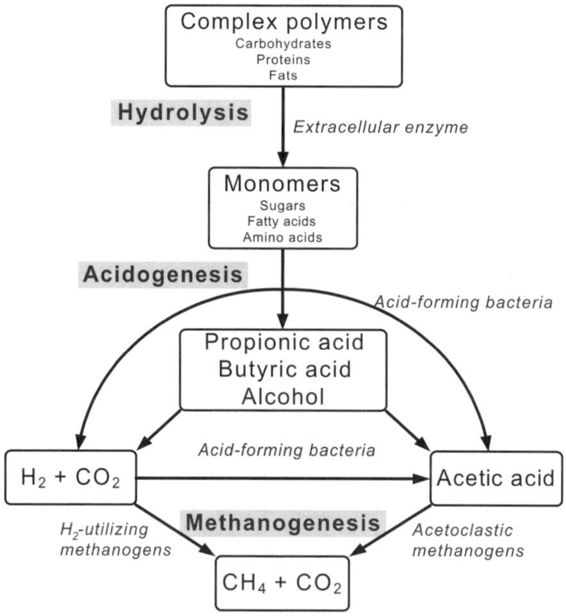

Fig. 15.1. Schematic description of anaerobic digestion process (Malina and Pohland 1992).

effective means to increase the biomass concentration and retention in bioreactor and to improve H_2 production by altering the microbial community structure (Wu *et al.* 2006, 2008a). Granulation is a complex process involving physicochemical, biological, and hydrodynamic mechanisms, in which microbial composition, extracellular polymeric substances (EPS) and hydrodynamics play important roles. Granulation may be enhanced by several means, including (1) addition of cations, e.g., Al^{3+}, Ca^{2+}, Fe^{3+}, Mg^{2+}, which may reduce the repulsive forces acting between the negatively charged bacteria cells, (2) addition of polymers which form bridges between the cells (Fig. 15.2), and (3) proper reactor design and operating strategy (Koskinen 2008).

3.1. *Culture enrichment technology*

Bacteria capable of producing hydrogen exist widely in natural environments such as soil, wastewater sludge, and compost (Wang and Wan 2008). We tested for seed sludge from those collected from sewage, paper mill wastewater, food wastewater, and beach sludge in Taiwan (Chen *et al.* 2002; Lin *et al.* 2008a; Lee *et al.* 2004). The sludge was heat treated (at 100 °C for 45 min) to inhibit methanogenic activity. Other factors affecting hydrogen-producing bacteria enrichment are substrate concentration, reactor type, nitrogen, phosphate, metal ion concentration, temperature, and pH. Batch tests were conducted to investigate the operating factor effects using various substrates such as glucose, sucrose, xylose, starch, cellulose, and actual wastewaters (Table 15.1). Optimal conditions for hydrogen production from carbohydrate-rich wastewater were identified as follows: substrate concentration 20 g-COD·l^{-1}; *C*/*N* ratio 47/1; temperature 35 °C; pH 6.0–8.0 (Lin and Chang 1999; Lin and Lay 2005; Lin and Cheng 2006; Lin *et al.* 2008b).

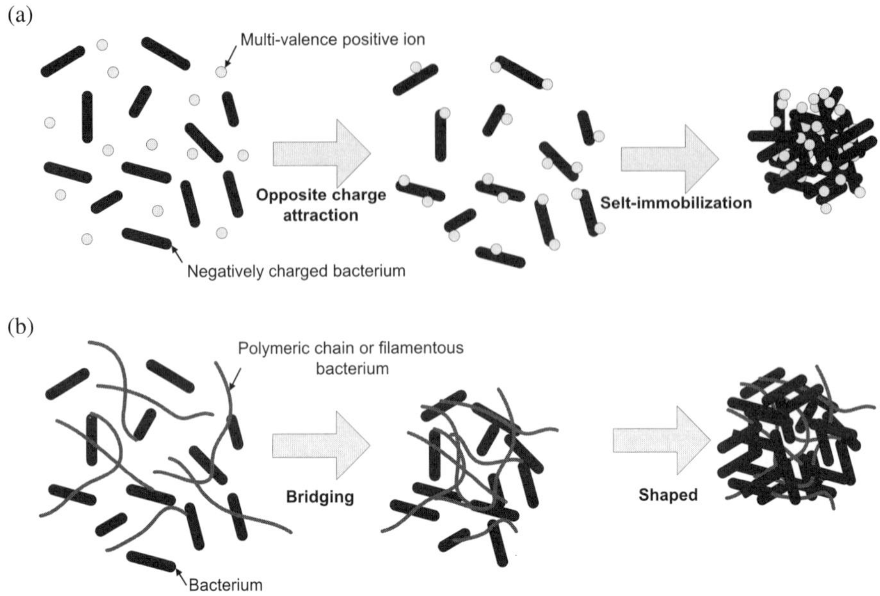

Fig. 15.2. Granulation models: (a) multi-valence positive ion bonding, and (b) polymer bonding (Liu *et al.* 2003).

Table 15.1. Hydrogen production rates and yields.

Substrate	Seed	Reactor type/ volume (l)	H_2 production Rate ($l \cdot l^{-1} \cdot d^{-1}$)	H_2 production Yield (ml-$H_2 \cdot$(g-carbohydrate)$^{-1}$)	References
Bench reactor					
Xylose	SS	CSABR/1.2	26	272	FCU (Wu *et al.* 2008b)
	SS	CSTR/4.0	2.5	95	FCU (Lin and Cheng 2006)
	C2	CSTR/NA	12	231	Japan (Taguchi *et al.* 1995)
Glucose	SS	CSABR/1.2	180	210	FCU (Shu 2005)
	SS	AGSB/0.9	191	222	FCU (Huang 2005)
	ADS	AFBR/1.0	57	158	Singapore (Zhang *et al.* 2007)
	AS	CSTR/2.0	37	272	USA (van Ginkel and Logan 2005)
Sucrose	SS	CSABR/1.2	362	216	FCU (Wu *et al.* 2006)
	SS	CSTR/4.0	27	308	FCU (Chen and Lin 2003)
	ADS	CSTR/2.3	10	651	UK (Hussy *et al.* 2005)
	SS	CSTR/3.0	13	265	Hong Kong (Fang *et al.* 2002)
Starch	BS	AGSB/1.0	48	173	FCU (Cheng *et al.* 2008)
	PS	CSTR/4.0	18	204	FCU (Lin *et al.* 2008b)
	MC	CSTR/0.26	5.6	74	Korea (Arooj *et al.* 2007)
	ADS	CSTR/2.3	3.0	255	UK (Hussy *et al.* 2003)
Pilot reactor					
CMS	SS	Batch/370	1.5	144	FCU (Chang *et al.* 2009)
Sucrose	SS	AGSB/400	15.6	286	FCU (unpublished data)

*SS, sewage sludge; C2, *Clostridium* sp. strain No. 2; ADS, anaerobic digested sludge; AS, agricultural soil; CC, cracked cereals; BS, beach sludge; and PS, paper mill sludge.

(a)

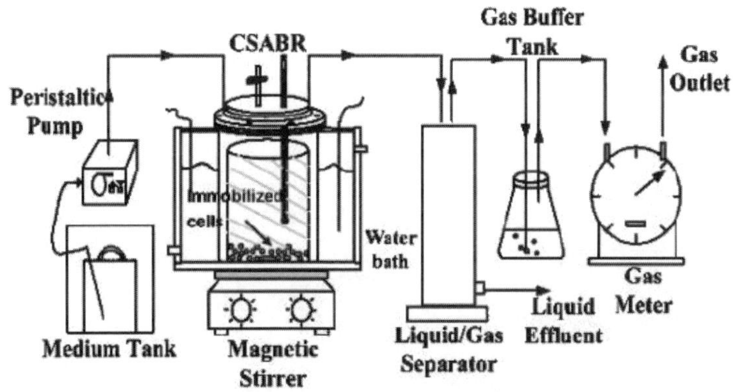

(b)

Fig. 15.3. A bench-scale hydrogen-producing CSABR: (a) schematic diagram, and (b) photo (Wu *et al.* 2006).

3.2. *High-rate hydrogen fermentation system*

3.2.1. *Continuously stirred anaerobic bioreactor (CSABR)*

Our previous study showed that CSABR had the highest hydrogen production rate (Wu *et al.* 2006) (Fig. 15.3). A CSABR seeded with silicone-immobilized sludge was tested for high-rate fermentative H_2 production at 0.5–6 h of HRT using sucrose as substrate at concentrations of 10–40 g-COD·l^{-1}. At feeding concentrations 30–40 g-COD·l^{-1} and HRT 0.5 h, this reactor had a maximum HPR of 362 l·l^{-1}·d^{-1} (14.7 mol·l^{-1}·d^{-1}) and an optimal hydrogen yield of 3.5 mol-H_2·mol-sucrose^{-1}. The formation of self-granulated sludge during short HRT operation was crucial to the high-rate H_2 production. The reactor had a high biomass concentration of 35.4 g-VSS·l^{-1} even when it was operated at an extremely low HRT of 0.5 h. Sludge granulation might also trigger a change of bacterial community resulting in a twofold increase in the specific HPR. According to denatured-gradient-gel-electrophoresis (DGGE) analysis, decreasing HRT resulted in a decrease in bacterial population diversity. The sludge which had the best PHR (when the reactor was operated at 0.5 h of HRT with sucrose concentration at 30 g-COD·l^{-1}) was dominated by *Clostridium pasteurianum*.

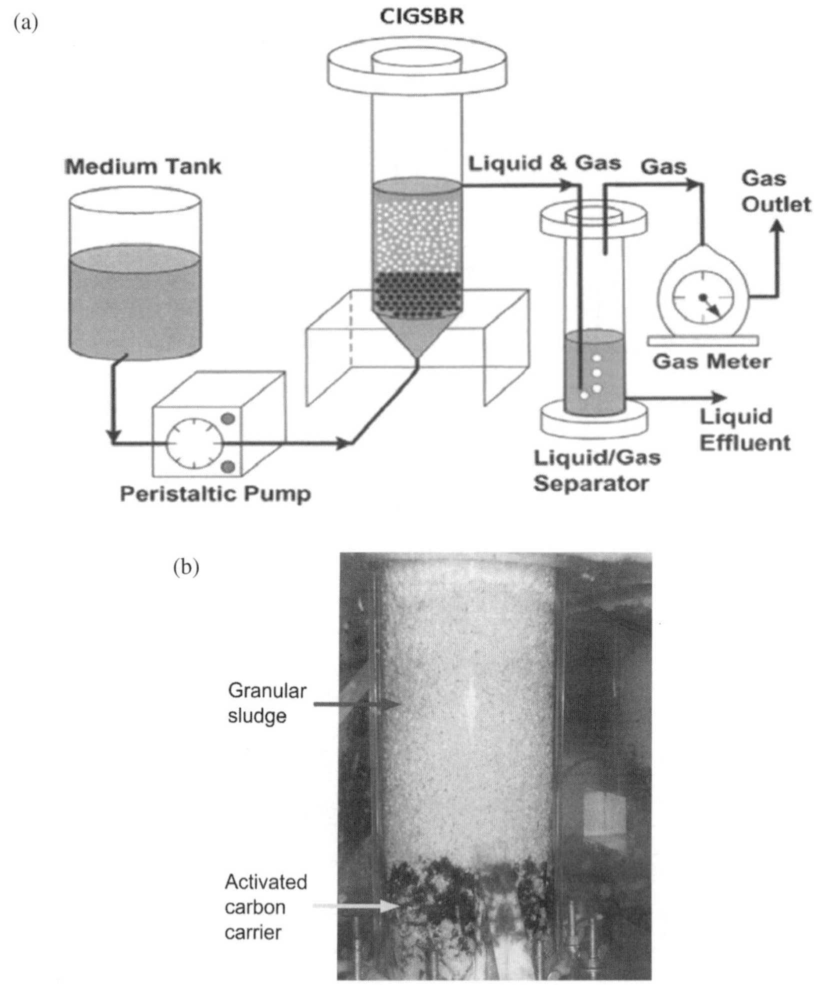

Fig. 15.4. A bench-scale hydrogen-producing CIGSBR: (a) schematic diagram, and (b) photo (Lee *et al.* 2004).

3.2.2. *Carrier-induced granular sludge bed reactor (CIGSBR)*

Long-term continuous operation of biohydrogen reactor had experienced problems with poor biomass retention at low HRTs as well as poor mass-transfer efficiency at high HRTs. A CIGSBR (Fig. 15.4) (Lee *et al.* 2004) was developed to improve biomass retention and mass-transfer efficiency. Calcium ion was supplemented to enhance the granular sludge mechanical strength. Adding 5.4–27.2 $mg{\cdot}l^{-1}$ of Ca^{2+} led to a threefold increase in biomass concentration and a fivefold increase in HPR (up to 122 $l{\cdot}l^{-1}{\cdot}d^{-1}$). Two reflux strategies were utilized to enhance the mass-transfer efficiency of the CIGSBR. Liquid reflux enhanced the HPR by 2.2-fold at an optimal liquid upflow velocity of 1.09 $m{\cdot}h^{-1}$, giving a peak biomass concentration of 22 $g\text{-}VSS{\cdot}l^{-1}$. Gas reflux at a rate of 1.0–1.49 $m{\cdot}h^{-1}$ gave similar HPR and decreased the biomass concentrations to 2–7 $g\text{-}VSS{\cdot}l^{-1}$. These strategies were effective for a stable and efficient H_2 production for 100 days (Lee *et al.* 2004).

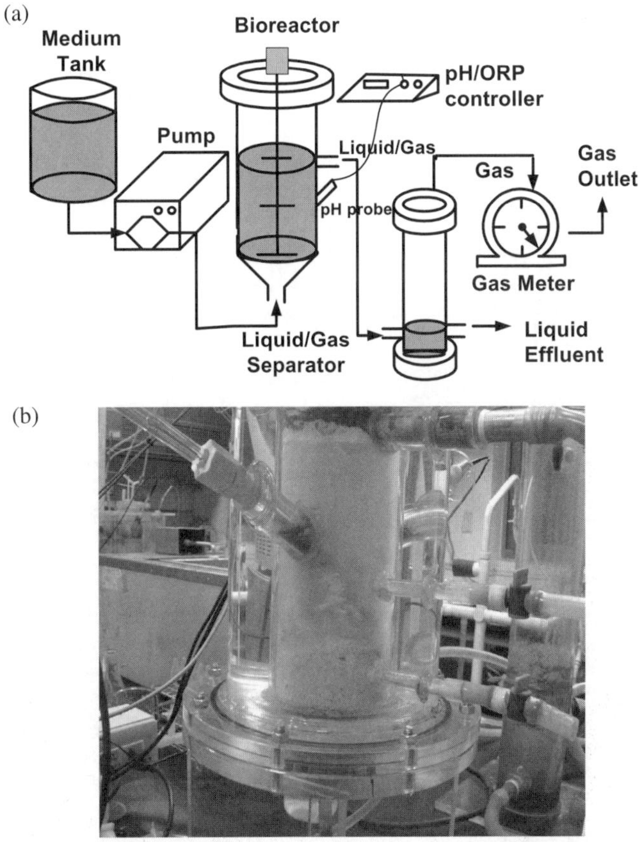

Fig. 15.5. A bench-scale hydrogen-producing AGSBR: (a) schematic diagram and (b) photo (Cheng *et al.* 2008).

3.2.3. *Agitated granular sludge bed reactor (AGSBR)*

A schematic description and photo of the AGSBR are shown in Fig. 15.5. The AGSBR had a working volume of 1 l and a paddle was arranged for agitation and carried out agitation rate of 15 rpm. The reactor was initially added with 1 $g \cdot l^{-1}$ of powdered activated carbon as carrier and inoculated with an intertidal sludge from Tai-Xi, Taiwan. The AGSBR was fed with starch wastewater, operated at pH 6.0 and 5.5 and at various HRTs. The peak HPR and hydrogen yield were achieved when the reactor was operated at 0.5 h of HRT. The HPRs were 48 $l \cdot l^{-1} \cdot d^{-1}$ at pH 6.0 and 42 $l \cdot l^{-1} \cdot d^{-1}$ at pH 5.5 (Cheng *et al.* 2008); the total sludge density was 45–48 g-VSS$\cdot l^{-1}$.

4. Integration of Hydrogen Production and Fuel Cell for on-Line Electricity Generation

The H_2 production system was also connected with a PEMFC system (Fig. 15.6) for direct electricity generation. For this purpose, a CSABR seeded with silicone-immobilized sludge was used. This reactor continuously and stably produced H_2 for over 300 days at 6 h of HRT using an influent sucrose concentration of 30 g-COD$\cdot l^{-1}$. The peak HPR and yield were 28 $l \cdot l^{-1} \cdot d^{-1}$

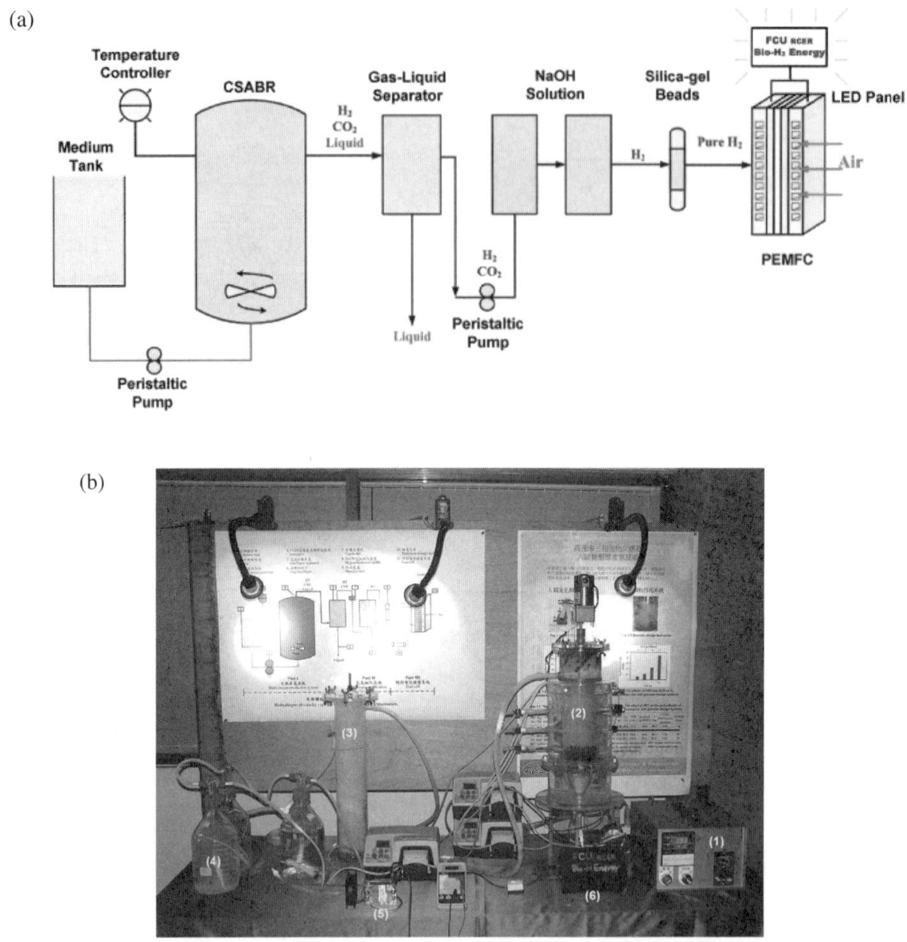

Fig. 15.6. A PEMFC electricity generation system using hydrogen produced by CSABR treating wastewater: (a) schematic diagram and (b) photograph: (1) Temperature controller, (2) CSABR, (3) gas/liquid separator, (4) CO_2 absorber, (5) PEMFC, and (6) LED panel for power demonstration (Lin *et al.* 2007).

and 3.71 mol-H_2·mol-sucrose^{-1}, respectively. The CO_2 and water vapor in the biogas were respectively removed by caustic solution in two bottles which were arranged in series (Fig. 15.4) and a column packed with silica gel. This final gas containing H_2 with >99% purity was then fed into the PEMFC system at a rate of 1.72 l·h^{-1}, generating 0.87 W of electricity at 25 °C with stable voltage of 2.28 V and a current of 0.38 A (Lin *et al.* 2007).

5. Pilot-Scale Hydrogen Fermentation System

With the financial support from Taiwan's Bureau of Energy and FCU, we built a 400 l pilot reactor for hydrogen production (Fig. 15.7), using the technology developed from our laboratory studies. The pilot system was composed of two feedstock tanks (each 0.75 m^3), a mixing tank (0.75 m^3), a fermentor (0.4 m^3 with a working volume of 0.38 m^3), a gas/liquid separator (0.3 m^3), and an automatic control system. Carbon source and nutrients were stored separately and mixed in

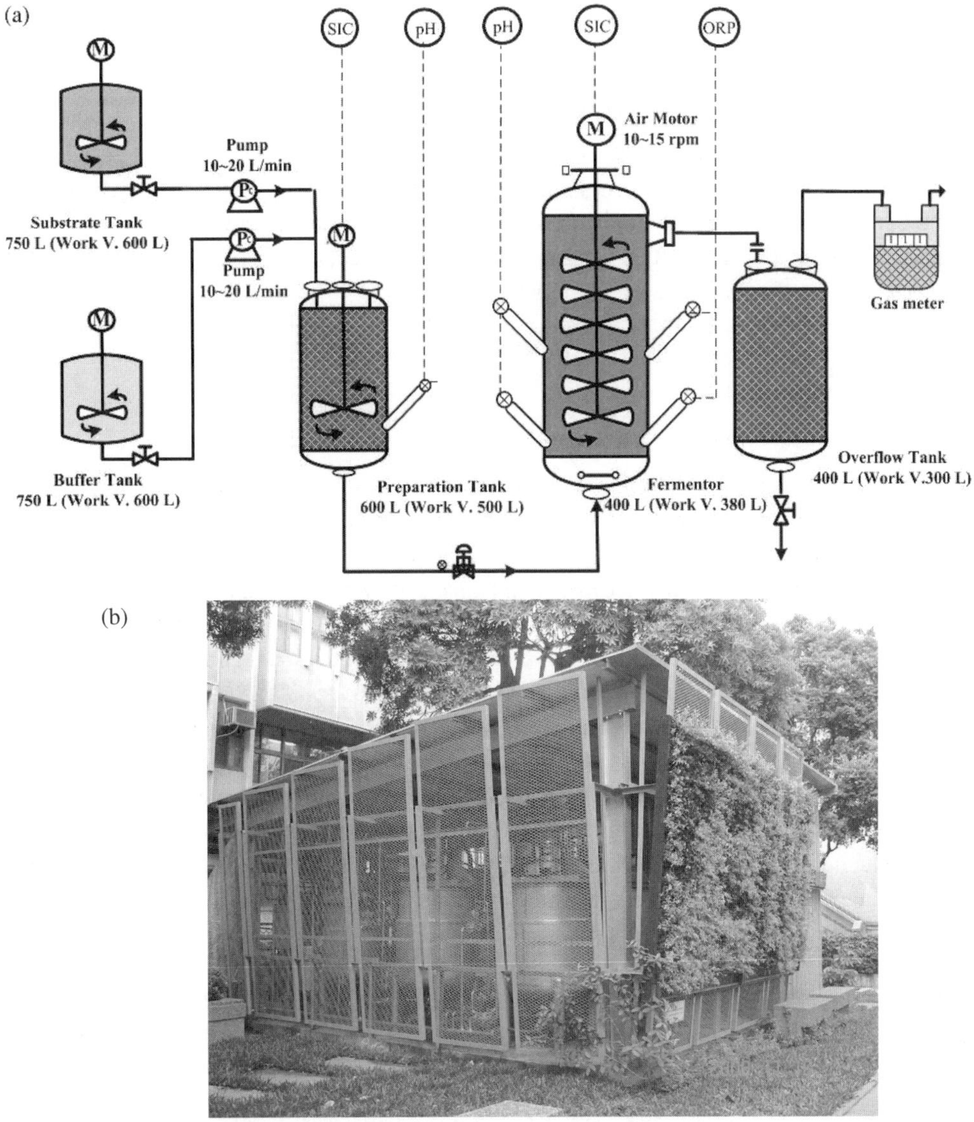

Fig. 15.7. Hydrogen production pilot plant (0.4 m^3): (a) layout (not shown), and (b) photo.

a tank before feeding into the fermentor. The fermentor was equipped with a water jacket for temperature control and a blender with a mixing rate of 10–15 rpm. The control panel is able to continuously monitor and control the mixing rate, temperature, feeding flows, pressure, and hydrogen content in biogas in the system (Lin *et al.* 2008c). The seed sludge for the pilot plant reactor operation was collected from a beach in Western Taiwan. The sludge was heat treated (at 100 °C for 45 min), and acclimated in a 1l AGSBR operated at 2 h of HRT, 37 ± 1 °C and pH 6 for two weeks using sucrose (20 g-COD·l^{-1}) as substrate. The acclimated sludge was then used to seed the pilot plant reactor which was operated at 35 °C feeding with sucrose (20 g-COD·l^{-1}). The pilot plant reactor was started up by operating in a batch mode for two days and then switched to a continuous feeding mode with 12 h of HRT. During the continuous operation, the HPR increased from 2.5 to 15.6 $l \cdot l^{-1} \cdot d^{-1}$ with the increase of organic loading rate from 40 to 240 g-COD$\cdot l^{-1} \cdot d^{-1}$ (Table 15.1).

Concentrated molasses fermentation soluble (CMS) was also tested as substrate for the pilot plant reactor. In this operation, the seed sludge was obtained from the waste activated sludge of Li-Ming Municipal Sewage Treatment Plant in Taichung, Taiwan. The screened sludge was heat treated before being inoculated into the pilot plant reactor. Batch experiments were conducted in the plant reactor at pH 6.7 and 37 °C using CMS at a COD concentration of 40 $g \cdot l^{-1}$. The HPR was 1.5 $l \cdot l^{-1} \cdot d^{-1}$ at 30 rpm of mixing rate (Table 15.1).

6. Microbial Community

6.1. *Bench-scale bioreactor with glucose, xylose, and starch as carbon source*

Table 15.2 shows the bacterial community determined via polymerase chain reaction (PCR)–DGGE for several bench-scale hydrogen-producing bioreactors feeding on various carbon sources. *Clostridium* was the dominant H_2-producing population in the bioreactors seeded with mixed sludge. *C. pasteurianum* and *Clostridium butyricum* were the main bacterial populations in reactors fed, respectively, with glucose and xylose. For the fermentor fed with starch, three *Clostridium* species, i.e., *C. pasteurianum*, *C. butyricum*, and *Clostridium tyrobutyricum* were present, suggesting that starch was hydrolyzed into sugars to support bacterial growth. Two major non-hydrogen-producing bacteria in the bioreactors were *Klebsiella* sp. and *Streptococcus* sp. The symbiosis of *Clostridium* sp. and these two bacteria in our systems was evidenced by their growth characteristics. *Klebsiella* sp. may readily consume oxygen, creating an anaerobic environment (the ORP may drop from 100 mV to -500 mV during the growth) for the growth of H_2-producing *Clostridium*. FISH images show that *Streptococcus* sp. cells were located inside the sludge granules surrounding the rod-shaped *Clostridium*-like cells. Result of Alcian Blue stain showed that most EPS that existed in the interior of the granule was produced by *Streptococcus* sp. from pure culture cultivation.

The analysis of bacterial community degrading starch for hydrogen production showed that *Bifidobacterium* sp. was abundant, as compared to those degrading glucose and xylose systems.

Table 15.2. Bacteria found in various bioreactors.

Substrate	HRT (h)	Dominant H_2-producing bacteria	Less dominant bacteria
Glucose	0.5–4	*C. pasteurianum*	*Klebsiella pneumonia* *Streptococcus sp.* *Klebsiella oxytoca*
Xylose	2–12	*C. buyricum*	*K. oxytoca* *Klebsiella pneumoniae* *Pseudomonas* sp. *Dialister* sp.
Starch	0.5–12	*C. pasteurianum* *C. butyricum* *C. tyrobutyricum*	*K. pneumoniae* *Streptococcus sp.* *Pseudomonas sp.* *Dialister sp.* *Bifidobacterium sp.*

According to the cell counts by FISHFISH, *Bifidobacterium* sp. was the only high G + C content microorganism found in the system. This suggests that non-H_2-producing microbial species such as *Bifidobacterium* sp. or other starch-hydrolytic bacteria also participated in the fermentation, probably breaking down starch into simple sugars for conversion to hydrogen by *Clostridium*.

6.2. *Pilot plant reactor*

The pH value in the pilot plant reactor was the main factor affecting the bacterial community and HPR. During days 1–13, the pH of the reactor was pH 6.0 and the bacterial composition changed without noticeable pattern, while the HPR was low and unstable (0.2–3.81l^{-1}·d^{-1}). On days 14–20, the HPR averaged 5.21l^{-1}·d^{-1} at pH 5.5–6.0 with *C. pasteurianum* being the dominant hydrogen-producing bacteria. After day 20, the HPR remained in the range of 2-41l^{-1}·d^{-1} with *C. pasteurianum* being the only dominant hydrogen-producing species. This result suggests that the pilot plant reactor should be operated at pH 5.5–6.0 for the preferred hydrogen-producing bacterial community.

6.3. *Bacterial community analysis by real-time PCR*

The bacterial community population would significantly affect the hydrogen production performance of a reactor. Five main bacterial species have been found in hydrogen-producing reactors in our studies: *Clostridium* sp., *Klebsiella* sp., *Streptococcus* sp., *Bifidobacterium* sp., and *Pseudomonas* sp. The change of bacterial community may effect the performance of the pilot plant reactor and make it unstable. However, the commonly used PCR–DGGE method takes about 12–24 h. Therefore, it is important to establish a technique that is able to analyze the variation in bacterial population within a short time. We designed the multiple primers for these five bacteria to be used for real-time PCR analysis, which could give qualitative and quantitative bacterial information in 4–6 h. Using this method, the bacterial community diversity based on PCR–DGGE and real-time PCR are shown in Table 15.3. Although the results from both methods are nearly the same, the real-time PCR, however, showed a higher sensitivity and could detect some bacterial species that could not be detected by the conventional PCR–DGGE method.

Table 15.3. Detections of bacteria by real-time PCR-based multiple primers method and by PCR–DGGE.

Samples	*C.b.*	*C.p.*	*Bifido.*	*Pseudo.*	*K.p.*	*Strepto.*	*Clost.*
1	+	-	*	+	+	+	+
2	+	-	-	+	*	*	+
3	*	+	+	-	+	+	+
4	+	-	+	+	+	-	+
5	+	*	*	-	*	*	+

Marker 1: code v.s. bacteria species showed in the table is *C.b.*, *C. butyricum*; *C.p.*, *C. pasteurianum*; Bifido., *Bifidobacterium* sp.; Pseudo., *Pseudomonas* sp.; *K.p.*, *K. pneumonia*; *Strepto.*, *Streptococcus* sp.; *Clost.*, *Clostridium* sp.

Marker 2: "+" indicates the species can be detected by the two methods; "*" detected by the multiple primer method but not by PCR–DGGE, "-" detected by neither method.

7. Conclusions

The systems developed by our team at FCU had high hydrogen production rates due to its capability of retaining high concentration of efficient H_2-producers as well as enabling effective contact between these microorganisms and their substrates. Feeding a wastewater of high sucrose concentration (40 g-COD·l^{-1}) with a short HRT (0.5 h), the CSABR produced H_2 efficiently with a peak HPR of 362 l·l^{-1}·d^{-1} and a peak yield of 3.5 mol-H_2·mol-sucrose^{-1}.

A pilot plant system with a 400 l bioreactor was constructed and operated to determine the scaleup operating parameters for possible full-scale operation. This pilot plant had a peak HPR of 15.6 l·l^{-1}·d^{-1} and a yield of 2.34 mol-H_2·mol-sucrose^{-1}. Molecular bacterial community analysis showed that *C. butyricum* and *C. pasteurianum* were the dominant species in the system. Stable HPR was obtained when the fermentor was operated at pH 5.5–6.0.

Acknowledgments

The authors would like to thank our co-workers at the biohydrogen research team from Feng Chia University (Shu-Yii Wu and Ping-Jei Lin; Drs Feng-Yuan Chang and Chen-Yeon Chu), National Cheng Kung University (Jo-Shu Chang), National Chung Hsing University (Chun-Hsiung Hung and Chin-Hung Cheng) and Central Taiwan University of Science and Technology (Kuo-Shing Lee). The authors gratefully acknowledge the financial support by the Bureau of Energy, MOEA, Taiwan (Grant No. 98-D0204-3), National Science Council of Taiwan (96-2221-E-035-033-MY3) and Feng Chia University (FCU-08G27201).

References

Arooj, M. F., Han, S. K., Kim, S. H., Kim, D. H., and Shin, H. S. (2007) Sludge characteristics in anaerobic SBR system producing hydrogen gas. *Wat. Res.* **41**, 1177–1184.

Brock, T. D., Madigan, M. T., Martinko, J. M., and Parker, J. (1994) *Biology of Microorganisms*, 7th ed. New Jersey: Prentice-Hall.

Chang, F. Y., Lay, C. H., Wu, J. H., Chen, C. H., Chang, J. S., and Lin, C. Y. (2009) Performance of lab- and pilot-scale bioreactors for fermentative hydrogen production from concentrated molasses fermentation soluble. In: *The 10th Asian Hydrogen Energy Conference 2009*, Daegu, Korea, April 8–10, p. 55.

Chen, C. C., Chen, Lin, C. Y., and Lin, M. C. (2002) Acid-base enrichment enhances anaerobic hydrogen production process. *Appl. Microbiol. Biotechnol.* **58**, 224–228.

Chen, C. C. and Lin, C. Y. (2003) Using sucrose as a substrate in an anaerobic hydrogen-producing reactor. *Adv. Environ. Res.* **7**, 695–699.

Chen, C. C., Lin, C. Y., and Chang, J. S. (2001) Kinetics of hydrogen production with continuous anaerobic cultures utilizing sucrose as the limiting substrate. *Appl. Microbiol. Biotechnol.* **57**, 56–64.

Cheng, C. H., Hung, C. H., Lee, K. S., Liau, P. Y., Yang, L. H., Lin, P. J., and Lin, C. Y. (2008) Microbial community structure of a starch-feeding fermentative hydrogen production reactor operated under different incubation conditions. *Int. J. Hydrogen Energy* **33**, 5242–5249.

Fang, H. H. P., Liu, H., and Zhang, T. (2002) Characterization of a hydrogen-producing granular sludge. *Biotechnol. Bioeng.* **78**, 44–52.

Huang, Y. S. (2005) *Anaerobic Fermentative Hydrogen Production with a Novel Agitated Granular Sludge Bed.* Master Thesis, Feng Chia University. Taiwan, pp. 47–78.

Hussy, I., Hawkes, F. R., Dinsdale, R., and Hawkes, D. L. (2003) Continuous fermentative hydrogen production from a wheat starch co-product by mixed microflora. *Biotechnol. Bioeng.* **84**, 619–626.

Hussy, I., Hawkes, F. R., Dinsdale, R., and Hawkes, D. L. (2005) Continuous fermentative hydrogen production from sucrose and sugarbeet. *Int. J. Hydrogen Energy* **30**, 471–483.

Koskinen, P. E. P. (2008). *The Development and Microbiology of Bioprocesses for the Production of Hydrogen and Ethanol by Dark Fermentation.* Doctor Dissertation, Tampere University of Technology, Finland.

Lay, J. J., Lee, Y. J., and Noike, T. (1999) Feasibility of biological hydrogen production from organic fraction of municipal solid waste. *Wat. Res.* **33**, 2579–2586.

Lee, K. S., Wu, J. F., Lo, Y. S., Lo, Y. C., Lin, P. J., and Chang, J. S. (2004) Anaerobic hydrogen production with an efficient carrier-induced granular sludge bed bioreactor. *Biotechnol. Bioeng.* **87**, 648–657.

Lin, C. N., Wu, S. Y., Lee, K. S., Lin, P. J., Lin, C. Y., and Chang, J. S. (2007) Integration of fermentative hydrogen process and fuel cell for on-line electricity generation. *Int. J. Hydrogen Energy* **32**, 802–808.

Lin, C. Y., Chang, C. C., and Hung C. H. (2008b) Fermentative hydrogen production from starch using natural mixed cultures. *Int. J. Hydrogen Energy* **33**, 2445–2453.

Lin, C. Y. and Chang, R. C. (1999) Hydrogen production during the anaerobic acidogenic conversion of glucose. *J. Chem. Technol. Biotechnol.* **74**, 498–500.

Lin, C. Y. and Chang, R. C. (2004) Fermentative hydrogen production at ambient temperatures. *Int. J. Hydrogen Energy* **29**, 715–720.

Lin, C. Y., Chen, C. H., Chung, W. T., Hung, C. H., and Cheng, L. H. (2006a) Cultivation pH effect on fermentative hydrogen production from xylose using natural mixed microflora. *Process Biochem.* **41**, 1383–1390.

Lin, C. Y. and Cheng, C. H. (2006) Fermentative hydrogen production from xylose using anaerobic mixed microflora. *Int. J. Hydrogen Energy* **31**, 832–840.

Lin, C. Y., Hung, C. H., Chen, C. H., Chung, W. T., and Chen, L. H. (2006b) Effects of initial pH on fermentative hydrogen production from xylose using natural mixed cultures. *Process Biochem.* **41**, 1383–1390.

Lin, C. Y. and Lay C. H. (2004a) Carbon/nitrogen-ratio effect on fermentative hydrogen production by mixed microflora. *Int. J. Hydrogen Energy* **29**, 41–45.

Lin, C. Y. and Lay, C. H. (2004b) Effects of carbonate and phosphate concentrations on hydrogen production using anaerobic sewage microflora. *Int. J. Hydrogen Energy* **29**, 275–281.

Lin, C. Y. and Lay, C. H. (2005) A nutrient formulation for fermentative hydrogen production using anaerobic sewage sludge microflora. *Int. J. Hydrogen Energy* **30**, 285–292.

Lin, C. Y., Wu, C. C., and Hung, C. H. (2008a) Temperature effects on fermentative hydrogen production from xylose using mixed anaerobic cultures. *Int. J. Hydrogen Energy* **33**, 43–55.

Lin, C. Y., Wu, S. Y., Lin, P. J., Chang, J. S., Hung, C. H., Lee, K. S., Chang, F. Y., Chu, C. Y., Cheng, C. H., Lay, C. H., and Chang, C. C. (2008) Start-up performance of a pilot-scale hydrogen fermentation system. *The 2009 Asian Biohydrogen Symposium*, 27–28 December, Harbin, China, pp. 1–9.

Liu, Y., Xu, H. L., Yang, S. F., and Tay, J. H. (2003) Mechanisms and models for anaerobic granulation in upflow anaerobic sludge blanket reactor. *Wat. Res.* **37**, 661–673.

Malina, J. F. and Pohland, J. F. G. (1992) *Design of Anaerobic Processes for the Treatment of Industrial and Municipal Wastes.* USA: Technomic Publishing Company.

Schmidt, J. E. and Ahring, B. K. (1996) Granular sludge formation in upflow anaerobic sludge blanket (UASB) reactors. *Biotechnol. Bioeng.* **49**, 229–246.

Shu, B. J. (2005) *Using glucose and xylose as sole substrate for fermentative hydrogen production in CSABRs.* Master Degree Thesis, Feng Chia University, Taiwan, 108–109.

Taguchi, F., Mizukami, N., Yamada, K., Hasegawa, K., and Saito-Taki, T. (1995) Direct conversion of cellulosic materials to hydrogen by *Clostridium* sp. strain no. 2. *Enzyme Microbiol. Technol.* **17**, 147–150.

van Ginkel and Logan, B. E. (2005) Increased biological hydrogen production with reduced organic loading. *Wat. Res.* **39**, 3819–3826.

van Ginkel, S., Sung, S. and Lay, J. J. (2001) Biohydrogen production as a function of pH and substrate concentration. *Environ. Sci. Technol.* **35**, 4726–4730.

Wang, J. and Wan, W. (2008) Factors influencing fermentative hydrogen production: A review. *Int. J. Hydrogen Energy* **34**, 799–811.

Wu, S. Y., Hung, C. H., Lin, C. N., Chen, H. W., Lee, A. S., and Chang, J. S. (2006) Fermentative hydrogen production and bacterial community structure in high-rate anaerobic bioreactors containing silicone-immobilized and self-flocculated sludge. *Biotechnol. Bioeng.* **93**, 934–946.

Wu, S. Y., Hung, C. H., Lin, C. Y., Lin, P. J., Lee, K. S., Lin, C. N., Chang, F. Y., and Chang, J. S. (2008a) HRT-dependent hydrogen production and bacteria community structure of mixed anaerobic microflora in suspended, granular and immobilized sludge systems using glucose as the carbon substrate. *Int. J. Hydrogen Energy* **33**, 1542–1549.

Wu, S. Y., Lin, C. Y., Lin, P. J., Hung, C. H., Chang, J. S., Lee, K. S., and Chang, F. Y. (2008b) Dark fermentative hydrogen production from xylose in different bioreactors using sewage sludge microflora. *Energy Fuels* **22**, 113–119.

Zhang, Z. P., Tay, J. H., Show, K. Y., Yan, R., Liang, D. T., Lee, D. J., and Jiang, W. J. (2007) Biohydrogen production in a granular activated carbon anaerobic fluidized bed reactor. *Int. J. Hydrogen Energy* **32**, 185–191.

Chapter 16

A Two-Stage Fermentation Process Converting Waste and Wastewater to Hydrogen and Methane

Hang-Sik Shin* and Dong-Hoon Kim

Department of Civil and Environmental Engineering,
KAIST, 373-1, Guseong-dong, Yuseong-gu,
Daejeon, Republic of Korea
**hangshin@kaist.ac.kr*

Although various means of biological H_2 production have been introduced, fermentative H_2 production has superior aspects such as fastest H_2 production rate and waste treatment. This chapter describes the principle and performance of various processes producing H_2 from wastewater, livestock wastes, and solid wastes. Of all the possible feedstock, food-processing wastes have been studied the most for H_2 production since they are rich in easily biodegradable carbohydrates. More recently, there is an increasing interest in using lignocellulosic biomass due to their abundance on earth. The existence of indigenous non H_2-producing bacteria in the waste biomass and the low carbohydrate content in the substrate are the key obstacles for enhancing H_2 yield. In order to maximize bioenergy recovery and waste reduction, a two-stage process, consisting of a fermentative H_2 production stage followed by a second fermentation stage of CH_4 production from the organic residues, was proposed. It was found that 80–90% of the energy content in the wastes could be recovered by this two-stage process with less than 10% in the first-stage. At the end of chapter, the feasibility of the two-stage process as well as further research areas needed to improve the process efficiency and applicability are discussed.

1. Introduction

1.1. *Climate change, clean energy, and biomass*

Although climate change is a natural phenomenon throughout history, the present rapid climate change is of anthropogenic nature. Since the industrial revolution, the CO_2 content in atmosphere has increased from the level of 280 ppm to 380 ppm in 2007 (IPCC 2007). This increased emission of heat-trapping greenhouse gases (GHGs) may lead to climate disasters, such as sea level rising, severe drought or precipitation, and huge storms.

In order to mitigate the effect of climate changes, the 'Kyoto Protocol' was announced in 1997. According to the Kyoto Protocol, industrialized countries should reduce their total GHGs emissions by 5.2% from the 1990 level until the end of 2012. According to the most recent 'Copenhagen Accord', the increase in global temperature should be below 2 °C to prevent dangerous anthropogenic interference with the climate system, and thereby the GHG emission reduction initiated by the 'Kyoto Protocol' was further strengthened. This target can only be met by the significant transition from fossil fuels to alternative energy sources which are cheap, renewable, and zero pollution (Saxena *et al.* 2009). Tidal, geothermal, hydroelectric, and wind power are among the proper candidates; however, they are not expected to become the dominant energy sources of the future. There are still many technological breakthroughs necessary in order to make these processes economically viable (Zidansek *et al.* 2009). Meanwhile, biomass is abundant all over the world and is a carbon-neutral resource in its life cycle. Therefore, conversion of biomass to energy has recently gained significant political and scientific momentum (Fortman *et al.* 2008).

Biomass is the third largest primary energy resource in the world after coal and crude oil. Plants are mainly composed of 40–50% cellulose by weight, followed by 20–40% of hemicelluloses (McKendry 2002). It is estimated that the energy content of plants is more than seven times the world's current energy needs. Biomass is traditionally converted to energy by direct combustion, generating dust and acid rain gases as byproducts. Now, various biochemical means have been explored to convert biomass to various forms of fuels, including H_2 (Fortman *et al.* 2008).

1.2. *Hydrogen energy*

Among various alternative energy sources, H_2 is regarded as the most promising future energy carrier since it produces only water upon combustion yielding energy of $122\,kJ \cdot g^{-1}$ that is 2.75 times higher than that of hydrocarbons. In addition, H_2 can be used in conventional internal combustion engines as well as in proton exchange membrane fuel cells to produce electricity (Momirlan and Veziroglu 2002; Neelis *et al.* 2004).

H_2 is the most common element in the universe, but not found in pure form on the earth. It was first isolated from water by the English scientist, Henry Cavendish, in 1774. The idea of using H_2 as an energy carrier first appeared in the science-fiction book 'The Mysterious Island' written by French novelist Jules Verne in 1874. One hundred years later, the International Association of Hydrogen Energy (IAHE) was established to provide a platform for the hydrogen energy community. In 2003, the International Partnership for the Hydrogen Economy (IPHE) was established not only to promote the technology, but also to implement codes and standards for H_2 infrastructure, and to arouse public safety concerns, and social and economic impacts.

Today, the terms 'Hydrogen Energy,' 'Hydrogen Economy,' and 'Hydrogen Energy System' are well accepted, and frequently appear in the media. The implementation of various 'Hydrogen Energy Systems', especially based on the renewable energy sources, such as solar, wind, and biomass (Afgan and Carvalho 2004), is an idea whose time has already arrived and it would be part of the solutions to the depletion of conventional fuels and global environmental problems. Figure 16.1 shows how introducing H_2 as energy supply may dramatically affect the CO_2 concentration in the atmosphere: the CO_2 level may reach the maximum of 520 ppm at 2050 if H_2 were introduced in 2000, 625 ppm at 2065 if the H_2 introduction is delayed until 2025, and 710 ppm at 2100 without hydrogen introduction (Momirlan and Veziroglu 2002). This suggests that an early transition to the H_2 energy system is crucial. Mason (2007) also insisted that the need for an H_2 production and distribution system is inevitable, and the sooner the process begins, the less the economic burden on future generations.

H_2 is produced by either physico-chemical or biological processes (Fig. 16.2). Over 90% of H_2 is now formed by steam reforming of hydrocarbons and coal gasification. However, these methods are shunned by the public and specialists because they use fossil fuels as raw material, emitting significant GHGs (Ewan and Allen 2005). Therefore, a number of innovative

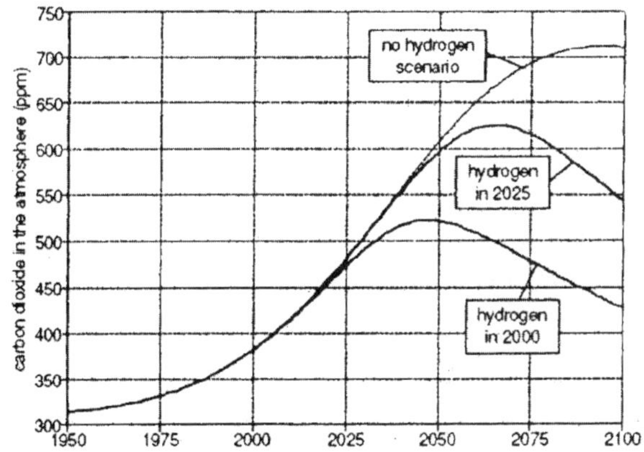

Fig. 16.1. CO_2 concentration in the atmosphere should the hydrogen energy supply be introduced in 2000 and 2025.

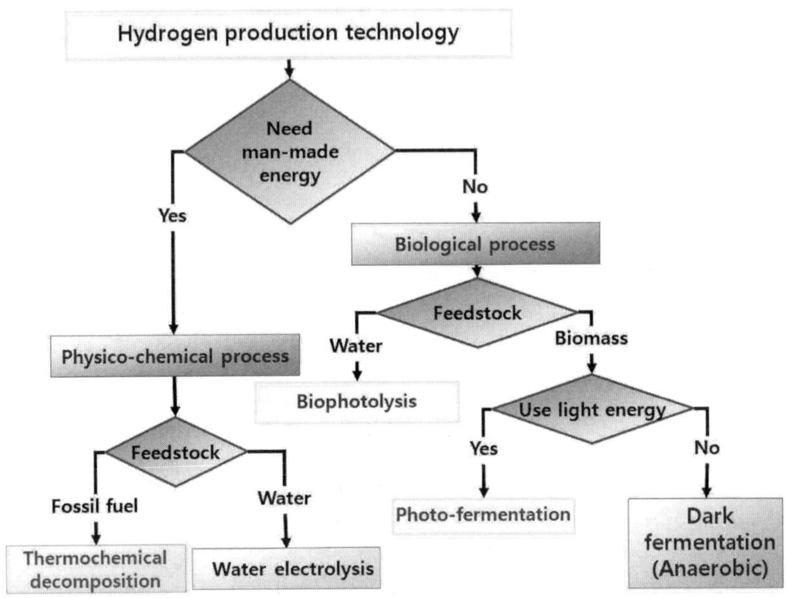

Fig. 16.2. Classification of H_2 production technology.

H_2 production means, including biological processes are being developed based upon renewable biomass (Koroneos *et al.* 2004).

2. Fermentative Hydrogen Production

As written in *The Scientist*: 'The winner in the bio-technological production of H_2 could be the founder of a "multitrillion dollar" industry,' biological means of H_2 production have been attracting much attention because of its environmentally friendly aspect (Jaffe 2003). It is a new technology that utilizes bacteria which produce H_2 during their metabolisms. There are three groups of microorganisms capable of producing H_2: algae, phototrophic bacteria, and fermentative bacteria. However, the mechanisms can be divided into two categories: photo-driven process and fermentation (Zaborsky 1998).

2.1. *Photo-driven process*

Green algae and blue green algae directly decompose water into hydrogen and oxygen using light energy as shown in Eq. (16.1). The reaction requires only water and sunlight, thus it is very attractive from the viewpoint of environmental protection.

$$2H_2O \xrightarrow{\text{Light energy}} 2H_2 + O_2 \quad \Delta G^o = 242\,\text{kJ} \tag{16.1}$$

However, those algae show rather low rates of H_2 production due to the complicated reactions involved which needed to overcome the large activation energy (+237 kJ·mol-H_2^{-1}). Another drawback is that a carrier gas is required to collect the evolved gas from the culture. In addition, separation of oxygen and hydrogen is also crucial because the hydrogen-producing enzyme, i.e., hydrogenase, is very sensitive to oxygen.

Photosynthetic bacteria such as purple non-sulfur bacteria, instead of using water as a starting compound, use simple carbohydrates and organic acids for hydrogen production, as shown below using glucose and lactate as examples in Eqs. (16.2) and (16.3).

$$\text{Glucose} + 6H_2O \rightarrow 12H_2 + 12CO_2 \quad \Delta G^o = \text{-}33.8\,\text{kJ} \tag{16.2}$$

$$\text{Lactate} + 3H_2O \rightarrow 6H_2 + 3CO_2 \quad \Delta G^o = 51.2\,\text{kJ} \tag{16.3}$$

Compared to cyanobacteria, photosynthetic bacteria require much less energy to produce H_2 from organic substances. However, activation energy must be high enough to drive nitrogenase, which is an electron transfer enzyme between ferredoxin and the hydrogen ion for the reaction shown in Eq. (16.4).

$$N_2 + 8H^+ + 8e^- + 16ATP \rightarrow 2NH_3 + H_2 + 16ADP + 16P_I \tag{16.4}$$

Nitrogenase is a cluster of iron, molybdenum, and sulfide and normally reduces N_2 to ammonia, but can also produce H_2 in the absence of N_2. Organic substances such as fatty acids, sugars, starch, and cellulose can be the starting materials for H_2 production. Photosynthetic bacteria decompose these organic substances completely, thereby they could be combined to the treatment of organic wastes (Zaborsky 1998).

2.2. *Fermentation*

H_2 can be produced by fermentative bacteria without light energy input using carbohydrate-rich substrates. The main advantage of dark fermentation is the fast reaction rate compared to the other biological processes. H_2 production through dark fermentation is related to methane fermentation which is a well-known anaerobic digestion process. During anaerobic mineralization, organic matters are converted to CH_4 and CO_2 via a series of interrelated microbial metabolic reactions, including hydrolysis, acidogenesis, acetogenesis, and methanogenesis. Fermentative bacteria hydrolyze and ferment carbohydrates, proteins, and lipids to volatile fatty acids (VFAs), which are further converted to acetate, CO_2, and H_2 by acidogenic and acetogenic bacteria. The principle of fermentative hydrogen production (FHP) is to block the methanogenesis by selective elimination of methanogens in the mixed-culture sludge and by controlling the environmental conditions such as pH and sludge retention time to avoid H_2 consumption by the methanogens (Hawkes *et al.* 2002).

Bacteria known to produce H_2 include species of *Enterobacter*, *Bacillus*, and *Clostridium*. However, most of the studies have focused on *Clostridium* sp. because of their high H_2 yield (1.61–2.36 mol-H_2·mol-glucose^{-1}) and their abundance in natural environment (Hawkes *et al.* 2002). *Clostridium* sp., obligate anaerobic bacteria, are mostly active at neutral pH, but some acid-producing species such as *Clostridium acetobutylicum*, *Clostridium beijerinckii* and *Clostridium butyricum* are active at low . Fermentation reactions can be operated at mesophilic (25–40 °C), thermophilic (40–65 °C), and extreme thermophilic (65–80 °C) conditions (Levin *et al.* 2004; Kotsopoulos *et al.* 2006). As they could form spores in harsh environments, H_2-producing inoculum was usually prepared by heat or chemical treatment of sludge from anaerobic digester, activated sludge, soil, compost, or manure (Ueno *et al.* 2001; Chen *et al.* 2002; Hawkes *et al.* 2002; Lin and Jo 2003; Morimoto *et al.* 2004; Vijayaraghavan *et al.* 2006).

Figure 16.3 depicts the metabolic pathways of fermentation products from glucose in *Clostridium* sp. Four moles of electrons obtained from glycolysis reduce two moles of NAD^+. Four moles of Fd^{2+} (ferredoxin) are reduced to Fd^+ in the next step. As a result, four moles of H_2 are generated in acetate production following reactions (2–5) and (2–6) because there is

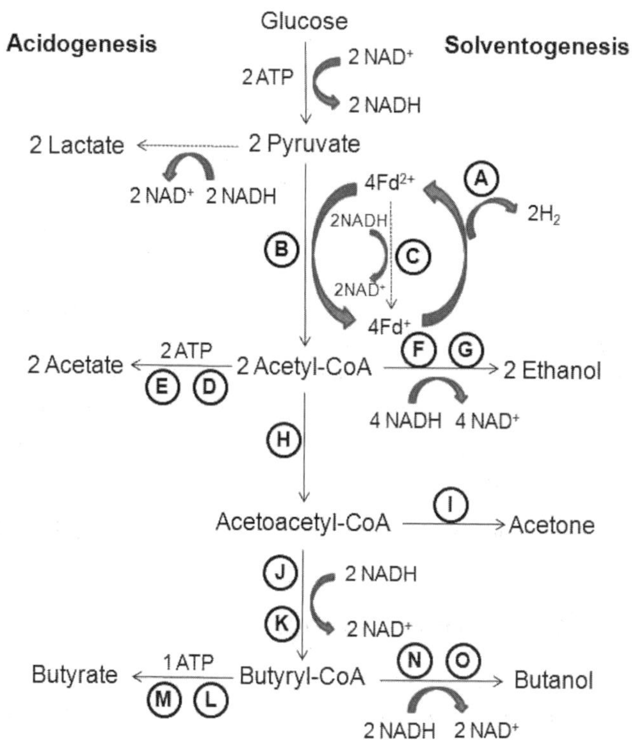

Enzymes are indicated by letters as follows: (A) hydrogenase; (B) pyruvate-ferredoxin oxidoreductase; (C) NADH-ferredoxin oxidoreductase; (D) phosphate acetyltransferase; (E) acetate kinase; (F) acetaldehyde dehydrogenase; (G) ethanol dehydrogenase; (H) thiolase; (I) acetoacetate decarboxylase; (J) 3-hydroxybutyryl-CoA dehydrogenase; (K) butyryl-CoA dehydrogenase; (L) phosphate butyryltransferase; (M) butyrate kinase; (N) butyraldehyde dehydrogenase; and (O) butanol dehydrogenase.

Fig. 16.3. Metabolic pathways of *Clostridium* sp. from glucose.

Table 16.1. Categorization of reactions involved in FHP with respect to H_2 production.

	Reactions
H_2 production	Glucose + $2H_2O \rightarrow$ 2Acetate + $2CO_2$ + $4H_2$
	Glucose $\rightarrow$ Butyrate + $2CO_2$ + $2H_2$
No relation to H_2 production	Glucose $\rightarrow$ 2Lactate
	Glucose $\rightarrow$ Butanol + $2CO_2$ + H_2O
	Glucose $\rightarrow$ 2Ethanol + $2CO_2$
H_2 consumption	Glucose + $2H_2 \rightarrow$ 2Propionate + $2H_2O$
	H_2 + $CO_2 \rightarrow$ Formate

no further consumption of NADH and Fd. Two important enzymes that participated in these reactions are NADH-Fd oxidoreductase and hydrogenase, whose activities are affected by environmental conditions such as pH, growth rate, and H_2 partial pressure. Two moles of H_2 are generated in butyrate production due to NADH reoxidization at the further step. Although more ATP could be obtained in acetate production as shown in Fig. 16.3, butyrate is the main metabolic product of *Clostridium* sp., because reaction 2–5 involves an unfavorable redox change and requires acetyl-CoA as an activator (Bahl and Dürre 2001). In addition, acetate production results in a net generation of NADH which was rapidly re-oxidized to allow glycolysis to proceed (Jones and Woods 1986). In practice, the acetate/butyrate ratio was about 0.66:1 (0.5:0.75 in molar ratio) in a normal batch fermentation (Rogers 1984). No H_2 is generated in ethanol and butanol productions.

$$NADH + H^+ + 2Fd^{2+} \xrightarrow{\text{NADHd-Fd oxidoreductase}} NAD^+ + 2Fd^+ + 2H^+ \tag{16.5}$$

$$2Fd^+ + 2H^+ \xrightarrow{\text{hydrogenase}} 2Fd^{2+} + H_2 \uparrow \tag{16.6}$$

In Table 16.1, various reactions generally found in FHP are categorized with respect to H_2 production. One mole of acetate or butyrate production is accompanied by two moles of H_2 production, while the production of propionate and formate indicates the consumption of produced H_2. Also, productions of lactate, butanol, and ethanol are unrelated to H_2 production. Thus, theoretical amount of H_2 production could be estimated based on the produced liquid-state metabolites as shown in Eq. (16.7) (Kim *et al.* 2006).

$$H_2 \text{ production} = \text{2butyrate} + \text{2acetate - propionate - formate (all in moles)} \tag{16.7}$$

2.3. *Biomass sources for FHP and their feasibility*

Carbohydrates, such as glucose, xylose, and sucrose, in wastewater are readily biodegradable for H_2 production (Li and Fang 2007). However, glycerol, the main component of a lipid, is not a suitable substrate for H_2 production (Heyndrickx *et al.* 1991). Protein is hydrolyzed to amino acids, which may be anaerobically degraded by *Clostridium* sp., producing only a small amount of H_2, or no H_2 at all if Stickland reaction is involved (Okamoto *et al.* 2000).

Most studies on FHP from wastewater use synthetic wastewater with balanced nutrients, and results may not be fully realistic. In the following sections of this chapter, only the performances of those using various actual wastes and wastewaters for FHP are discussed.

2.3.1. *Wastewater*

Most of the actual wastewater treated for FHP was from food-processing industries, such as cheese whey and palm oil mill (POME) (Table 16.2). The H_2 production from cheese whey, which is a lactose-rich wastewater from cheese manufacturing, was evaluated in batch and continuous operation (Ferchichi *et al.* 2005; Azbar *et al.* 2009). Results of these studies varied, probably due to the differences in characteristics of cheese whey, operational temperature, and process procedures.

Many studies have shown that thermophilic (50–60 °C) treatment is superior to mesophilic (30–40 °C) treatment of actual wastewaters for H_2 production (Shin *et al.* 2004; Valdez-Vazquez *et al.* 2005). This could be attributed to the simple microbial diversity under the thermophilic operation. In treating a bean curd manufacturing wastewater, H_2 production gradually decreased and totally ceased within 5 d; however, the production resumed to a normal level once the wastewater was heated to 50–90 °C (Noike *et al.* 2002). On the other hand, thermophilic treatment is costly unless waste heat is available at low or no cost.

The feasibility of POME for FHP has been studied in many Asian countries, the annual production of which is 184.6 million tons. Kelly-Yong *et al.* (2007) estimated the energy content in POME as 2.59 EJ (exajoule: 10^{18} J) per year. Studies were conducted under both mesophilic and thermophilic conditions, but the performances could not be compared due to insufficient information. Chong *et al.* (2009) investigated the effect of pH in treating POME in batch operation using pure culture. However, it is more practical to use mixed cultures from an engineering point of view, because of the ease of operation and process control plus a broad choice of feedstock (Valdez-Vazquez *et al.* 2005).

In addition to temperature, pH, hydraulic retention time (HRT), and substrate concentration are the key process parameters in continuous FHP. pH could directly affect the hydrogenase activity, metabolic pathway, and population dynamics (Dabrock *et al.* 1992; Lay 2000; Fang and Liu 2002). HRT and substrate concentration not only affect the performance but also the reactor volume required. Optimization of these operational parameters was often studied separately, but it is necessary to study the interactions of these parameters, and the optimum conditions accordingly (Fan *et al.* 2004; Wang *et al.* 2005). In a CSTR operation, the optimal HRT was generally in the range of 8–12 h and the organic loading rate (OLR) was 60 g-COD·l^{-1}·d^{-1}. However, in upflow and packed-bed reactors, the HRT could be significantly decreased down to 0.5 h (Lee *et al.* 2004). An upflow reactor treating a high-strength (34 g-COD·l^{-1}) rice winery wastewater achieved a high specific H_2 production rate of 9.33 l-H_2·g-VSS^{-1}·d^{-1} at a short HRT of 2 h, corresponding to an OLR of 408 g-COD·l^{-1}·d^{-1} (Yu *et al.* 2002).

2.3.2. *Livestock wastes*

Unlike food-processing wastewater, livestock waste is protein-rich, and thus the yield of H_2 production is limited (Table 16.3). Treating a swine waste produced a very low H_2 yield of 3 ml-H_2·g-COD^{-1} (assuming 1.2 g-COD·g-VS^{-1}), equivalent to 0.2% of the energy recovery efficiency (Kotsopoulos *et al.* 2009). Addition of glucose to the swine waste may increase H_2 production, but such a practice is impractical (Wu *et al.* 2009; Zhu *et al.* 2009). Thus, it is better to treat livestock waste at an ambient temperature for the traditional methane production, or for acid generation in a conventional two-stage process. However, the addition of livestock wastes

Table 16.2. FHP performances treating food-processing wastewaters.

Category	Substrate	Operation mode	Temperature (°C)	Main objective	Maximum H_2 yield achieved	Reference
Food-processing wastewaters	Cheese whey	Batch	35	Effect initial pH	80.14 ml-H_2·g-COD^{-1}	Ferchichi *et al.* (2005)
		Continuous	55	Effects of HRT and organic loading rate (OLR) using completely stirred tank reactor (CSTR)	492.8 ml-H_2·g-COD^{-1}	Azbar *et al.* (2009)
	Palm oil mill effluent (POME)	Continuous	60	Fed-batch H_2 production from POME	2,419 ml-H_2·l-$POME^{-1}$	Atif *et al.* (2005)
		Continuous	37	Optimizing pH, HRT, and substrate concentration	—	Vijayarahavan and Ahmad (2006)
		Batch	37	Effect of pH on H_2 production from POME using *Clostridium butyricum* EB6	3,345 ml-H_2·l-$POME^{-1}$	Chong *et al.* (2009)
	Rice winery wastewater	Continuous	55	Application of upflow reactor	2.14 mol-H_2·mol-$hexose^{-1}$	Yu *et al.* (2002)
	Citric acid wastewater	Continuous	36	Application of UASB	0.84 mol-H_2·mol-$hexose^{-1}$	Yang *et al.* (2006)
	Olive oil mill wastewater	Continuous	35	Optimizing HRT	1,651 ml-H_2·l-raw $substrate^{-1}$	Ntaikou *et al.* (2009)

Table 16.3. FHP performances treating livestock wastes.

Category	Substrate	Operation mode	Temperature (°C)	Main objective	Maximum H_2 yield achieved	Reference
Livestock wastes	Pig slurry	Continuous	70	Successful operation at hyper-thermophilic condition	3.65 ml-H_2·g-VS^{-1}	Kotsopoulos *et al.* (2009)
	Liquid swine manure + glucose	Continuous	37	Supplementation of glucose to increase H_2 production	—	Wu *et al.* (2009)
		Continuous	35	Optimizing HRT and pH	—	Zhu *et al.* (2009)

as a supplement to carbohydrate-rich wastes seems feasible as they are rich in nutrients and in buffering capacity. The pH of food-processing wastes is generally below pH 5.0, whereas the pH of livestock wastes is above pH 7.5. Therefore, the addition of livestock wastes may be beneficial to keep the operation at the optimal range of pH 6–8 for FHP (Cheng *et al.* 2002; Valdez-Vazquez *et al.* 2005).

2.3.3. *Solid wastes*

Food-processing wastes (Table 16.4) and lignocellulosic materials (Table 16.5) are the most common solid wastes treated for FHP. Results of several studies on H_2 production potential of carbohydrates, proteins, and lipids showed that carbohydrates were the most favorable substrate for FHP (Noike and Mizuno 1999; Lay *et al.* 2003; Dong *et al.* 2009). Carbohydrate-rich rice and potato produced more than 20 times more H_2 than lipid-rich fats and skins, or protein-rich eggs and lean meat (Lay *et al.* 2003). Thus, the feasibility of H_2 production from food wastes is doubtful. Two recent studies showed that the energy content in food waste converted to H_2 was only 4.43% (Kim 2008) and 1.91% (Li *et al.* 2008). The theoretical H_2 yield from hexose is 4 mol-H_2·mol-hexose^{-1} assuming all carbohydrates are converted into acetate (Table 16.1). In practice, the highest reported H_2 yield was 2.47 mol-H_2·mol-hexose^{-1} using glucose as a substrate due to the formation of other acids and alcohols (Kotsopoulos *et al.* 2006). As the energy content in one mole of glucose is equivalent to 12 moles of H_2 (Eq. (16.8)), the yield of 2.47 mol-H_2·mol-hexose^{-1} represents an energy conversion efficiency of 20.6%.

$$C_6H_{12}O_6 + 12H_2O \rightarrow 12H_2 + 6CO_2 \tag{16.8}$$

Thus, the carbohydrate content affects the feasibility of treating food wastes for FHP. On the other hand, carbohydrates could be converted to various acids during storage, resulting in a reduction of H_2 production potential (Li *et al.* 2008).

One more problem in producing H_2 from food waste is the existence of indigenous microflora. When the food waste was treated at 35 °C without any pretreatment, it naturally decomposed mainly to lactic acid (Kim *et al.* 2009). Jo *et al.* (2007) observed an unstable H_2 fermentation performance in treating food waste, and attributed it to the indigenous lactic acid bacteria (LAB)

Table 16.4. FHP performances treating solid type food-processing wastes.

Operation mode	Substrate	Temperature (°C)	Main objective	Maximum H_2 yield achieved	Reference
Batch	Bean curd manufacturing waste, rice bran, and wheat bran	35	Effects of the characteristics of substrate decomposition	61 ml-H_2·g-VS^{-1}	Noike and Mizuno (2000)
	Rice, potato, fat meat, chicken skin, egg, and lean meat	37		—	Lay *et al.* (2003)
	Rice, potato, and lettuce	37		136 ml-H_2·g-VS^{-1}	Dong *et al.* (2009)
	Food waste	35	Generation of H_2 without inoculum addition	153.5 ml-H_2·g-VS^{-1}, 2.05 mol-H_2·mol-hexose^{-1}	Kim *et al.* (2009)
	Food waste	55	Effect of pH	12.8 ml-H_2·g-COD^{-1}	Lee *et al.* (2008)
	Food waste	35/50	Find the optimum F/M ratio	57 ml-H_2·g-VS^{-1}	Pan *et al.* (2008)
Continuous	Olive pulp	55	Thermophilic H_2 production from olive pulp	35.8 ml-H_2·g-TS^{-1}	Gavala *et al.* (2006)
	Food waste	40	Effect of OLR and SRT	65 ml-H_2·g-VS^{-1}	Wang and Zhao (2009)
	Food waste	35	Test the two different sources of food waste	1.53 mol-H_2·mol-hexose^{-1}	Li *et al.* (2008)
	Food waste	35	Pretreatments of food waste for the stable H_2 fermentation	0.87 mol-H_2·mol-hexose^{-1}	Kim *et al.* (2008a)
	Food waste	35	Investigate the effect of SRT	1.12 mol-H_2·mol-hexose^{-1}	Kim *et al.* (2008b)

Table 16.5. FHP performances treating lignocellulosic wastes.

Feedstock	Pretreatment	Temperature (°C)	Maximum H_2 yield achieved	Reference
Corn straw	Stream explosion	35	63 ml-H_2·g-corn straw^{-1}	Li and Chen (2007)
Sugarcane bagasse	Acid	37	1.73 mol-H_2·mol sugar^{-1}	Pattra *et al.* (2008)
Sweet sorghum	Milling	37	3.15 mol-H_2·mol-glucose^{-1}	Ntaikou *et al.* (2008)
Cornstalk	Acid	36	143.81 ml-H_2·g-VS^{-1}	Zhang *et al.* (2007)
Grass silage	—	70	16 ml-H_2·g-VS^{-1}	Pakarinen *et al.* (2008)
Cornstalk	Acid	36	126.22 ml-H_2·g-cornstalk^{-1}	Wang *et al.* (2009)
Beer lees	Acid	36	68.6 ml-H_2·g-VS^{-1}	Fan *et al.* (2006)

originating from the food waste. Due to their unique metabolic characteristics and antibiotic function, LAB are enriched in fermentation processes of milk, meats, cereals, and vegetables (Stiles and Holzapfel 1997). Pretreatment by heat, alkali, or acid may eliminate the indigenous LAB, favoring the population of spore-forming bacteria such as *Clostridium* sp. (Kim *et al.*

2009). Kim *et al.* (2008a) also reported that alkali pretreatment of food waste at pH 12.5 for 1 day is essential for the stable H_2 production.

Lignocellulosic materials from woods, grasses, and agricultural residues are the most abundant biomass in nature. The annual yield of lignocellulosic wastes worldwide exceeds 220 billion tons worldwide, equivalent to 60–80 billion tons of crude oil (Ren *et al.* 2009). Therefore, lignocellulosic waste is an attractive, low-cost feedstock for biofuel production.

Lignocellulosic biomass is mainly composed of cellulose (32–47%), hemicelluloses (19–27%), and lignin (5–24%). Both hemicelluloses and lignin provide a protective sheath around the cellulose, and must be pretreated. The pretreatment is to remove lignin, and to partly hydrolyze hemicellulose and to alter the lignocellulosic structure for enzymes to convert the carbohydrate polymers into fermentable sugars. In order to achieve a higher H_2 yield, various pretreatment methods have been applied to enhance the H_2 yield (Fan *et al.* 2006; Zhang *et al.* 2007; Wang *et al.* 2009). The maximum yield of 149.69 ml-H_2·g-VS^{-1} was reported with a 0.2% HCl pretreatment of cornstalk; without such pretreatment, the yield was only 3.16 ml-H_2·g-VS^{-1} (Zhang *et al.* 2007).

Pretreatment may, however, produce various materials which may potentially inhibit H_2 production, including weak acids, furfural, 5-hydroxymethyl furfural (HMF), formic acid, and phenols. These byproducts may affect ATP synthesis or the intracellular redox balance of microorganisms, thereby reducing the specific growth rate. However, little is known of the actual effect of these inhibitors on H_2 production. Further studies are warranted.

Although livestock wastes and sewage sludge produced little H_2 (Wang *et al.* 2003; Yokoyama *et al.* 2007), they may be used to supplement nutrients to the carbohydrate-rich wastes (Kim *et al.* 2004) or to improve the buffering capacity (Zhu *et al.* 2008a,b). Also, the contents of Fe, P, and Ca in sewage sludge might be attributed to the enhanced H_2 production (Kim 2008).

3. Two-stage Fermentation System

Since fermentation can only convert less than 33% of the energy in substrate into H_2, further treatment should be needed to recover bioenergy and to remove the residual waste. Three possible post-treatment processes that have been studied are methane fermentation, photo-fermentation, and microbial fuel cell (MFC). The photo-fermentative post-treatment produces the maximum H_2 yield, whereas the MFC treatment produces electricity directly from the organic residues (Tao *et al.* 2007). However, these two processes are still in their infant stage, requiring more studies (Li and Fang 2007; Hawkes *et al.* 2007).

On the other hand, methanogenic treatment process has been well developed, and thus can be readily applied as the post-treatment for residual organics in the effluent of FHP. The two-stage process separates hydrogen-producing bacteria and methanogenic archaea in two reactors, each of which could be properly controlled independently for optimal operation. This may enhance process stability and overall conversion efficiencies (Ke *et al.* 2005). The organic residues from H_2 fermentation are mainly composed of acetate and butyrate, both are favorable substrates for methane conversion (Ding and Wang 2008). Han *et al.* (2005) first tested the effluent from an H_2 fermenter for a methanogenic upflow anaerobic sludge blanket (UASB) reactor, and achieved 93% of energy recovery with 5.8 h of HRT. The produced H_2 and CH_4 could be used separately, or used as a mixed fuel. Co-combustion of mixed H_2 and CH_4 could reduce nitrogen oxide emissions as compared to the combustion of CH_4 alone (Benemann *et al.* 2004).

Table 16.6. Performance of two-stage fermentation system producing H_2 and CH_4.

	Substrate	Maximum H_2 yield	Maximum CH_4 yield	Reference
Liquid-type biomass	Sucrose	1.62 mol-H_2·mol-hexose^{-1}	323 ml-CH_4·g-COD^{-1}	Kyazze *et al.* (2007)
	Sweet sorghum extract	10.4 l-H_2·kg-substrate^{-1}	29 l-CH_4·kg-substrate^{-1}	Antonopoulou *et al.* (2008)
Solid-type biomass	Food waste	310 ml-H_2·g-VS^{-1}	210 ml-CH_4·g-VS^{-1}	Han and Shin (2004)
		205 ml-H_2·g-VS^{-1}	464 ml-CH_4·g-VS^{-1}	Chu *et al.* (2008)
		65 ml-H_2·g-VS^{-1}	546 ml-CH_4·g-VS^{-1}	Wang and Zhao (2009)
	MSW[a] + Slaughterhouse waste	33.6 ml-H_2·g-VS^{-1}	Not available	Gomez *et al.* (2006)
	Olive pulp waste	35.8 ml-H_2·g-TS^{-1}	425.6 ml-CH_4·g-TS^{-1}	Gavala *et al.* (2006)
	Household solid waste	43 ml-H_2·g-VS^{-1}	500 ml-CH_4·g-VS^{-1}	Liu *et al.* (2006)
	Garbage slurry + paper	2.4 mol-H_2·mol-hexose	Not available	Ueno *et al.* (2007)
	Potato waste	30 ml-H_2·g-TS^{-1}	183 ml-CH_4·g-TS^{-1}	Zhu *et al.* (2008a,b)
	Grass silage	5.64 ml-H_2·g-VS^{-1}	467 ml-CH_4·g-VS^{-1}	Pakarinen *et al.* (2009)

[a]MSW = municipal solid waste.

Table 16.6 shows results of recent studies on the two-stage fermentation process co-producing H_2 and CH_4 from various wastewaters and solid wastes. An H_2-generating CSTR followed by an upflow anaerobic filter was operated using sucrose as a substrate at 10–30 g·l^{-1} (Kyazze *et al.* 2007). Antonopoulou *et al.* (2008) extracted liquid fraction from sweet sorghum and fed it into a two-stage CSTR fermenting system, and estimated the CH_4 potential of the solid fraction by batch tests.

In general, there was not much difference in CH_4 yield among various solid wastes. However, food wastes showed a relatively high H_2 yield probably due to their high contents of carbohydrate and other easily biodegradable constituents. Pakarinen *et al.* (2009) found that the H_2 yield of alkali-pretreated grass silage was limited. Han and Shin (2004) used a process consisting of three leaching-bed reactors for H_2 production followed by a UASB reactor for CH_4 production. Chu *et al.* (2008) operated a process consisting of a thermophilic H_2-producing reactor followed by a mesophilic CH_4-producing reactor, the effluent of which was recirculated by the H_2-producing reactor to enhance the buffer capacity, minimizing the requirement of alkaline for the H_2 fermentation. Generally, CH_4-producing reactor is operated at pH 7.5–8.5 with alkalinity concentrations of more than 5,000 mg-$CaCO_3$·l^{-1}. When methanogenic effluent was used to dilute food waste water in batch hydrogen fermentation, the alkaline requirement to keep the pH at 6.0 was reduced by 40–60%. However, special care should be taken because hydrogenotrophic methanogens in the recycled effluent could consume the produced H_2. Kraemer and Bagley (2005) reported that reducing the addition of alkaline solution by 40% resulted in a 90% drop of H_2 yield due to the H_2 consumption by methanogens. They recommended

Table 16.7. Key information needed for evaluating the efficiency of the two-stage process producing H_2 and CH_4.

Wastewater characteristics	H_2 and CH_4 fermenter
Total and soluble COD ($g \cdot l^{-1}$)	Effluent concentration (COD, VS, and carbohydrate)
TOC concentration ($g \cdot l^{-1}$)	H_2 yield (mol-H_2·mol-hexose^{-1}, ml-H_2·g-VS^{-1})
Carbohydrate content (%)	CH_4 yield (ml-CH_4·g-VS^{-1}, ml-CH_4·g-COD^{-1})
TS concentration (% or $g \cdot l^{-1}$)	HRT and SRT (d, h)
VS concentration (% or $g \cdot l^{-1}$)	Volumetric H_2 or CH_4 production rate ($l \cdot l \cdot d^{-1}$)

Table 16.8. Biogas production rate and bioenergy recovery in a two-stage fermentation system producing H_2 and CH_4.

		Biogas production rate ($l \cdot l^{-1} \cdot d^{-1}$)		Bioenergy recovery (%)			
Feedstock	HRT (d)	H_2	CH_4	H_2	CH_4	Sum	Reference
Food waste	25	0.18	0.33	8.8	66.5	75.3	Youn and Shin (2005)
	6.3	1.62	2.43	9.3	76.5	85.8	Chu *et al.* (2008)
	33.3	0.29	2.03	5.8	82.2	88.0	Wang and Zhao (2009)
Garbage and paper waste	8.0	0.81	3.90	2.0	77.3	79.3	Ueno *et al.* (2007)
Household solid wastes	17	0.19	1.81	1.0	97.8	98.8	Liu *et al.* (2006)

membrane filtration of the recycled liquid to remove hydrogenotrophs from the recycled stream.

However, the feasibility of the process depends on the energy recovery from the total energy content in the wastewater and the biogas production rate. Table 16.7 lists the key information needed to properly evaluate the efficiency of a two-stage process producing H_2 and CH_4. Most studies reported the bioenergy recovery merely in the forms of H_2 yield (as ml-H_2·g-VS^{-1} or ml-H_2·mol-hexose^{-1}) and CH_4 yield (ml-CH_4·g-VS^{-1} or ml-CH_4·g-COD^{-1}) without revealing the details of the organic matters in the influent.

From those with sufficient information, as shown in Table 16.8, 75–99% of bioenergy of food and other household wastes could be recovered; however, 67–98% of the energy was recovered as CH_4 and only 1–10% as H_2. This could be due to the low carbohydrate content in those wastes. Ueno *et al.* (2007) reported that treating solid wastes with carbohydrate content below 50% could recover no more than 10% of energy as hydrogen. In one case, the energy recovery for hydrogen was only 2% even with the high H_2 yield of 2.4 mol-H_2·mol-hexose^{-1}. Thus, based on available information so far, the process to recover hydrogen from solid wastes requires much further development effort. The initial hydrolysis is likely the bottleneck.

4. Future Research

For a two-stage process to produce H_2 and CH_4 from wastes and wastewater, the bottleneck is at the first stage. Further research is needed in order to make this process economically feasible. One is to research how to improve the hydrogen conversion efficiency of lipid and protein. H_2 yield from glycerol, the intermediate of lipid degradation was 0.41 mol-H_2·mol-glycerol^{-1} (Seifert *et al.* 2009), which is only about one-third of yield of carbohydrate. One possible approach is

Table 16.9. Nutrient composition of various marine biomass.

	Carbohydrate (%)	Protein (%)	Lipid (%)	Others (%)
Agar	61.0	16.3	3.1	19.6
Sea tangle	56.4	8.4	1.6	33.6
Gom-Pi	48.6	13.6	2.4	35.4
Red algae	40.1	15.9	2.0	42.0

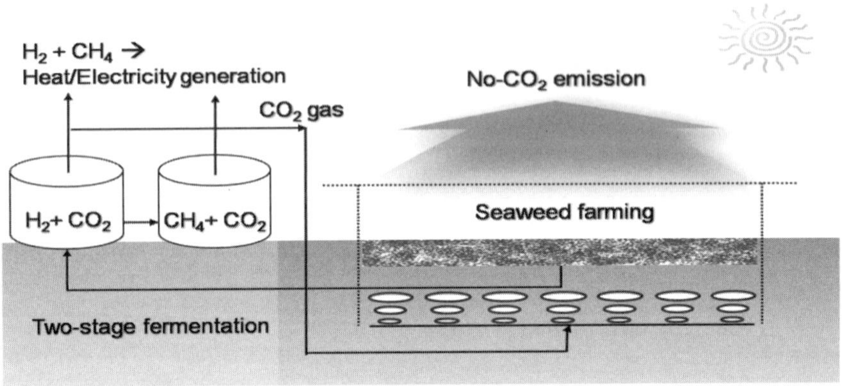

Fig. 16.4. CO_2-neutral two-stage fermentation system from aquatic biomass.

to remove bacterial genes which are responsible for degrading lipid and protein via pathways without H_2 production.

Another topic of research is using aquatic biomass to produce hydrogen. Compared to terrestrial feedstock, such as wastes from wood, corn, and sugar cane processing, using aquatic biomass as feedstock for hydrogen production has the advantages of (a) low levels of lipid, lignin, and protein (as shown in Table 16.9), (b) low cost of production, (c) high yield per unit area, and (d) high consumption of CO_2. Figure 16.4 illustrates a CO_2-neutral two-stage fermentation system, in which the by-product CO_2 is separated and fed to the aquatic culture.

At present, hydrogen-producing reactors were commonly seeded with anaerobic sludge after heat or acid/alkali pretreatment to remove the hydrogenotrophs. Cheaper and more effective alternatives should be investigated. Yokoyama *et al.* (2007) and Kim *et al.* (2009) reported that satisfactory hydrogen production was achieved by using the indigenous bacteria in food waste or cow dung without being inoculated with other anaerobic sludge.

References

Afgan, N. H. and Carvalho, M. G. (2004) Sustainability assessment of hydrogen energy systems. *Int. J. Hydrogen Energy* **29**, 1327–1342.

Antonopoulou, G., Gavala, H. L., Skiadas, I. V., Angelopoulos, K., and Lyberatos, G. (2008) Biofuels generation from sweet sorghum: Fermentation hydrogen production and anaerobic digestion of the remaining biomass. *Bioresour. Technol.* **99**(1), 110–119.

Atif, A. A. Y., Fakhru'l-Razi, A., Ngan, M. A., Morimoto, M., Iyuke, S. E., and Veziroglu, N. T. (2005) Fed batch production of hydrogen from palm oil mill effluent using anaerobic microflora. *Int. J. Hydrogen Energy* **30**(13–14), 1393–1397.

Azbar, N., Dokgoz, F. T. C., Keskin, T., Korkmaz, K. S., and Syed, H. M. (2009) Continuous fermentative hydrogen production from cheese whey wastewater under thermophilic anaerobic conditions. *Int. J. Hydrogen Energy* **34**(17), 7441–7447.

Bahl, H. and Durre, P. (2001) *Clostridia: Biotechnology and Medical Applications.* Weinheim: Wiley-VCH, Verlag, GmbH. pp. 57–63.

Benemann, J. R., Cannizzaro, C. M., and Cooney, M. (2004) Biological production of hydrogen-methane mixtures clean electricity. In: *Proceedings of the Tenth World Congress of Anaerobic Digestion.* Montreal. pp. 654–658.

Chen, C. C., Lin, C. Y., and Lin, M. C. (2002) Acid-base enrichment enhances anaerobic hydrogen production process. *Appl. Microbiol. Biotechnol.* **58**, 224–228.

Cheng, S.S., Chang, S.M., and Chen, S.T. (2002) Effects of volatile fatty acids on a thermophilic anaerobic hydrogen fermentation process degrading peptone. *Wat. Sci. Technol.* **46**(4–5), 209–214.

Chong, M. L., Rahman, N. A., Yee, P. L., Aziz, S. A., Rahim, R. A., Shirai, Y., and Hassan, M. A. (2009) Effects of pH, glucose, and iron sulfate concentration on the yield of biohydrogen by *Clostridium butyricum* EB6. *Int. J. Hydrogen Energy* **34**(21), 8859–8865.

Chu, C. F., Li, Y. Y., Xu, K. Q., Ebie, Y., Inamori, Y., and Kong, H. N. (2008) A pH- and temperature-phased two-stage process for hydrogen and methane production from food waste. *Int. J. Hydrogen Energy* **33**, 4739–4746.

Dabrock, B., Bahl, H., and Gottschalk, P. (1992) Parameters affecting solvent production by *Clostridium pasteurianum*. *Appl. Environ. Microbiol.* **58**(4), 1233–1239.

Ding, H. B. and Wang, J. Y. (2008) Response of the methanogenic reactor to different effluent fractions of fermentative hydrogen production in a phase-separated anaerobic digestion system. *Int. J. Hydrogen Energy* **33**, 6993–7005.

Dong, L., Zhenhong, Y., Yongming, S., Xiaoying, K., and Yu, Z. (2009) Hydrogen production characteristics of the organic fraction of municipal solid wastes by anaerobic mixed culture fermentation. *Int. J. Hydrogen Energy* **34**(2), 812–820.

Ewan, B. C. R. and Allen, R. W. K. (2005) A figure of merit assessment of the routes to hydrogen. *Int. J. Hydrogen Energy* **30**, 809–819.

Fan, Y., Li, C., Lay, J. J., Hou, H., and Xhang, G. (2004) Optimization of initial substrate and pH levels for germination of sporing hydrogen-producing anaerobes in cow dung compost. *Bioresour. Technol.* **91**, 189–193.

Fan, Y., Zhang, G., Guo, X., Xing, Y., and Fan, M. (2006) Biohydrogen-production from beer lees biomass by cow dung compost. *Biomass Bioenergy* **30**(5), 493–496.

Fang, H. H. P. and Liu, H. (2002) Effect of pH on hydrogen production from glucose by a mixed culture. *Bioresour. Technol.* **82**(2), 87–93.

Ferchichi, M., Crabbe, E., Gil, G. H., Hintz, W., and Almadidy, A. (2005) Influence of initial pH on hydrogen production from cheese whey. *J. Biotechnol.* **120**, 402–409.

Fortman, J. L., Chhabra, S., Mukhopadhyay, A., Chou, H., Lee, T. S., Steen, E., and Keasling, J. D. (2008) Biofuel alternatives to ethanol: Pumping the microbial wall. *Trends Biotechnol.* **26**(7), 375–381.

Gavala, H. N., Skiadas, I. V., Ahring, B. K., and Lyberatos, G. (2006) Thermophilic anaerobic fermentation of olive pulp for hydrogen and methane production: Modelling of the anaerobic digestion process. *Wat. Sci. Technol.* **53**(8), 271–279.

Gomez, X., Moran, A., Cuetos, M. J., and Sanchez, M. E. (2006) The production of hydrogen by dark fermentation of municipal solid wastes and slaughterhouse waste: A two-phase process. *J. Power Sources* **157**, 727–732.

Han, S. K. and Shin, H. S. (2004) Biohydrogen production by anaerobic fermentation of food waste. *Int. J. Hydrogen Energy* **29**(6), 569–577.

Han, S. K., Kim, S. H., Kim, H. W., and Shin, H. S. (2005) Pilot-scale two-stage process: A combination of acidogenic hydrogenesis and methanogenesis. *Wat. Sci. Technol.* **52**(1–2), 131–138.

Hawkes, F. R., Dinsdale, R., Hawkes, D. L., and Hussy, I. (2002) Sustainable fermentative hydrogen production: Challenges for process optimization. *Int. J. Hydrogen Energy* **27**, 1339–1347.

Hawkes, F. R., Hussy, I., Kyazze, G., Dinsdale, R., and Hawkes, D. L. (2007) Continuous dark fermentative hydrogen production by mesophilic microflora: Principles and progress. *Int. J. Hydrogen Energy* **32**(2), 172–184.

Heyndrickx, M., Vos, P. D., Vancanneyt, M., and Ley, J. D. (1991) The fermentation of glycerol by *Clostridium butyricum* LMG 1212t2 and 1213t1 and *C. pasteurianum* LMG 3285. *Appl. Microbiol. Biotechnol.* **32**, 637–647.

IPCC Report in the Fourth Assessment of Working Group I (2007) *Climate Change 2007 — The Physical Science Basis*. Cambridge, UK: Cambridge University Press.

Jaffe, S. (2003) Biological derived hydrogen-future fuel. *Scientist* **17**, 28–29.

Jo, J. H., Jeon, C. O., Lee, D. S., and Park, J.M. (2007) Process stability and microbial community structure in anaerobic hydrogen-producing microflora from food waste containing kimchi. *J. Biotechnol.* **131**, 300–308.

Jones, D. T. and Woods D. R (1986). Acetone-butanol fermentation revisited. *Microbiol. Rev.* **50**(4), 484–524.

Ke, S., Shi, Z., and Fang, H. H. P. (2005) Applications of two-phase anaerobic degradation in industrial wastewater treatment. *Int. J. Environ. Pollut.* **23**(1), 65–79.

Kelly-Yong, T. L., Lee, K. T., Mohamed, A. R., and Bhatia, S. (2007) Potential of hydrogen from oil palm biomass as a source of renewable energy worldwide. *Energy Policy* **35**, 5692–5701.

Kim, D. H., Han. S. K., Kim, S. H., and Shin, H. S. (2006) Effect of gas sparging on continuous fermentative hydrogen production. *Int. J. Hydrogen Energy* **31**, 2158–5169.

Kim, D. H. (2008) *Innovative Two-Stage Fermentation System Converting Organic Solid Wastes to* H_2 *and* CH_4. Ph.D thesis, Dept. of Civil and Environmental Engineering, KAIST, Republic of Korea.

Kim, D. H., Kim, S. H., and Shin, H. S. (2009) Hydrogen fermentation of food waste without inoculum addition. *Enzyme Microbiol. Technol.* **45**, 181–187.

Kim, S. H., Han, S. K., and Shin, H. S. (2004) Feasibility of biohydrogen production by anaerobic co-digestion of food waste and sewage sludge. *Int. J. Hydrogen Energy* **29**, 1607–1616.

Kim, S. H. and Shin, H. S. (2008a) Effects of base-pretreatment on continuous enriched culture for hydrogen production from food waste. *Int. J. Hydrogen Energy* **33**, 5266–5274.

Kim, S. H., Han, S. K., and Shin, H. S. (2008b) Optimization of continuous hydrogen fermentation of food waste as a function of solids retention time independent of hydraulic retention time. *Process Biochem.* **43**, 213–218.

Koroneos, C., Dompros, A., Roumbas, G., and Moussiopoulos, N. (2004) Life cycle assessment of hydrogen fuel production processes. *Int. J. Hydrogen Energy* **29**, 1443–1450.

Kotsopoulos, T. A., Fotidis, I. A., Tsolakis, N., and Martzopoulos, G. G. (2009) Biohydrogen production from pig slurry in a CSTR reactor system with mixed cultures under hyper-thermophilic temperature (70 °C). *Biomass Bioenergy* **33**(9), 1168–1174.

Kotsopoulos, T. A., Zeng, R. J., and Angelidaki, I. (2006) Biohydrogen production in granular up-flow anaerobic sludge blanket (UASB) reactors with mixed cultures under hyper-thermophilic temperature (70 °C). *Biotechnol. Bioeng.* **94**(2), 296–301.

Kraemer, J. T. and Bagley, D. M. (2005) Continuous fermentative hydrogen production using a two-phase reactor system with recycle. *Environ. Sci. Technol.* **39**, 3819–3825.

Kyazze, G., Dinsdale, R., Guwy, A. J., Hawkes, F. R., Premier, G. C., and Hawkes, D. L. (2007) Performance characteristics of a two stage dark fermentative system producing hydrogen and methane continuously. *Biotechnol. Bioeng.* **97**(4), 759–770.

Lay, J. J. (2000) Modeling and optimization of anaerobic digested sludge converting starch to hydrogen. *Biotechnol. Bioeng.* **68**(3), 269–278.

Lay, J. J., Fan, K. S., Chang, J., and Ku, C. H. (2003) Influence of chemical nature of organic wastes on their conversion to hydrogen by heat-shock digested sludge. *Int. J. Hydrogen Energy* **28**, 1361–1367.

Lee, K. S., Wu, J. F., Lo, Y. S., Lo, Y. C., Lin, J., and Chang, J. S. (2004) Anaerobic hydrogen production with an efficient carrier-induced granular sludge bed bioreactor. *Biotechnol. Bioeng.* **87**(5), 648–657.

Lee, Z. K., Li, S. L., Lin, J. S., Wang, Y. H., Kuo, P. C., and Cheng, S. S. (2008) Effect of pH in fermentation of vegetable kitchen wastes on hydrogen production under a thermophilic condition. *Int. J. Hydrogen Energy* **33**(19), 5234–5241.

Levin, D.B., Pitt, L., and Love, M. (2004) Biohydrogen production: prospects and limitations to practical application. *Int. J. Hydrogen Energy* **29**, 173–185.

Li, C. and Fang, H. H. P. (2007) Fermentative hydrogen production from wastewater and solid wastes by mixed cultures. *Crit. Rev. Environ. Sci. Technol.* **37**, 1–39.

Li, D. and Chen, H. (2007) Biological hydrogen production from steam-exploded straw by simultaneous saccharification and fermentation. *Int. J. Hydrogen Energy* **32**(12), 1742–1748.

Li, S. L., Kuo, S. C., Lin, J. S., Lee, Z. K., Wang, Y. H., and Cheng, S. S. (2008) Process performance evaluation of intermittent-continuous stirred tank reactor for anaerobic hydrogen fermentation with kitchen waste. *Int. J. Hydrogen Energy* **33**(5), 1522–1531.

Lin, C. and Jo, C. (2003) Hydrogen production from sucrose using an anaerobic sequencing batch reactor process. *J. Chem. Technol. Biotechnol.* **78**, 678–684.

Liu, D., Liu, D., Zeng, R. J., and Angelidaki, I. (2006) Hydrogen and methane production from household solid waste in the two-stage fermentation process. *Wat. Res.* **40**, 2230–2236.

Mason, J. E. (2007) World energy analysis: H_2 now or later. *Energy Policy* **35**, 1315–1329.

Mckendry, P. (2002) Energy production from biomass (part 1): Overview of biomass. *Bioresour. Technol.* **83**, 37–46.

Momirlan, M. and Veziroglu, T. N. (2002) Current status of hydrogen energy. *Renewable Sustainable Energy Rev.* **6**, 141–179.

Morimoto, M., Atsuko, M., Atif, A. A. Y., Ngan, M. A., Fakhru'l-Razi, A., Iyuke, S. E., and Bakir, A. M. (2004) Biological production of hydrogen from glucose by natural anaerobic microflora. *Int. J. Hydrogen Energy* **19**, 709–713.

Neelis, M. L., van der Kooi, H. J., and Geerlings, J. J. C. (2004) Exergetic life cycle analysis of hydrogen production and storage systems for automotive applications. *Int. J. Hydrogen Energy* **29**, 537–545.

Noike, T. and Mizuno, O. (1999) Hydrogen fermentation of organic municipal wastes. *Wat. Sci. Technol.* **42**(12), 155–162.

Noike, T., Takabatake, H., Mizuno, O., and Ohba, M. (2002) Inhibition of hydrogen of organic wastes by lactic acid bacteria. *Int. J. Hydrogen Energy* **27**, 1367–1371.

Ntaikou, I., Gavala, H. N., Kornaros, M., and Lyberatos, G. (2008) Hydrogen production from sugars and sugars sweet sorghum biomass using *Ruminococcus albus*. *Int. J. Hydrogen Energy* **33**(4), 1153–1163.

Ntaikou, I., Kourmentza, C., Koutrouli, E. C., Stamatelatou, K., Zampraka, A., Kornaros, M., and Lyberatos, G. (2009) Exploitation of olive oil mill wastewater for combined biohydrogen and biopolymers production. *Bioresour. Technol.* **100**, 3724–3730.

Okamoto, M., Miyahara, T., Mizuno, O., and Noike, T. (2000) Biological hydrogen potential of materials characteristic of the organic fraction of municipal solid wastes. *Wat. Sci. Technol.* **41**(3), 25–32.

Pakarinen, O. M., Tahti, H. P., and Rintala, J. A. (2009) One-stage H_2 and CH_4 and two-stage $H_2 + CH_4$ production from grass silage and from solid and liquid fractions of NaOH pre-treated grass silage. *Biomass Bioenergy* **33**(10), 1419–1427.

Pakarinen, O., Lehtomaki, A., and Rintala, J. (2008) Batch dark fermentative hydrogen production from grass silage: The effect of inoculum, pH, temperature and VS ratio. *Int. J. Hydrogen Energy* **33**(2), 594–601.

Pan, J., Zhang, R., El-Mashad, H., Sun, H., and Ying, Y. (2008) Effect of food to microorganism ratio on biohydrogen production from food waste via anaerobic fermentation. *Int. J. Hydrogen Energy*. **33**, 6968–6975.

Pattra, S., Sangyoka, S., Boonmee, M., and Reungsang, A. (2008) Bio-hydrogen production from the fermentation of sugarcane bagasse hydrolysate by *Clostridium butyricum*. *Int. J. Hydrogen Energy* **33**(19), 5256–5265.

Ren, N., Wang, A., Cao, G., Xu, J., and Gao, L. (2009) Bioconversion of lignocellulosic biomass to hydrogen: Potential and challenges. *Biotechnol. Adv.* **27**(6), 1051–1060.

Rogers, P. (1984) Genetics and biochemistry of *Clostridium* relevant to development of fermentation processes. *Appl. Microbiol.* **31**, 1–60.

Saxena, R. C., Adhikari, D. K., and Goyal, H. B. (2009) Biomass-based energy fuel through biochemical routes. *Renewable Sustainable Energy Rev.* **13**(1), 167–178.

Seifert, K., Waligorska, M, Wojtovski, M., and Laniecki, M. (2009) Hydrogen generation from glycerol in batch fermentation process. *Int. J. Hydrogen Energy* **34**(9), 3671–3678.

Shin, H. S., Youn, J. H., and Kim, S. H. (2004) Hydrogen production from food waste in anaerobic mesophilic and thermophilic acidogenesis. *Int. J. Hydrogen Energy* **29**, 1355–1363.

Stiles, M. E. and Holzapfel, W. H. (1997) Lactic acid bacteria of foods and their current taxonomy. *Int. J. Food Microbiol.* **36**, 1–29.

Tao, Y., Chen, Y., Wu, Y., He, Y., and Zhou, Z. (2007) High hydrogen yield from a two-step process of dark-and photo-fermentation of sucrose. *Int. J. Hydrogen Energy* **32**(2), 200–206.

Ueno, Y., Fukui, H., and Goto, M. (2007) Operation of a two-stage fermentation process producing hydrogen and methane from organic waste. *Environ. Sci. Technol.* **41**, 1413–1419.

Ueno, Y., Haruta, S., Ishii, M., and Igarashi, Y. (2001) Microbial community in anaerobic hydrogen-producing microflora enriched from sludge compost. *Appl. Microbiol. Biotechnol.* **57**, 555–562.

Valdez-Vazquez, I., Rios-Leal, E., Esparza-Garcia, F., Cecchi, F., and Poggi-Varaldo, H. M. (2005) Semi-continuous solid substrate anaerobic reactors for H_2 production from organic waste: Mesophilic versus thermophilic regime. *Int. J. Hydrogen Energy* **30**(13/14), 1383–1391.

Vijayaraghavan, K., Ahmad, D., Ibrahim, M. K. B., and Herman, H. N. B. (2006) Isolation of hydrogen generating microflora from cow dung for seeding anaerobic digester. *Int. J. Hydrogen Energy* **31**(6), 708–720.

Vijayaraghavan, K. and Ahmad, D. (2006) Biohydrogen generation from palm oil mill effluent using anaerobic contact filter. *Int. J. Hydrogen Energy* **31**(10), 1284–1291.

Wang, C. C., Chang, C. W., Chu, C. P., Lee, D. J., Chang, B. V., Liao, C. S., and Tay, J. H. (2003) Producing hydrogen from wastewater sludge by *Clostridium bifermentans*. *J. Biotechnol.* **102**, 83–92.

Wang, G., Mu, Y., and Yu, H. Q. (2005) Response surface analysis to evaluate the influence of pH, temperature and substrate concentration on the acidogenesis of sucrose-rich wastewater. *Biochem. Eng. J.* **23**, 175–184.

Wang, X. and Zhao, Y. (2009) A bench scale study of fermentative hydrogen and methane production from food waste in integrated two-stage process. *Int. J. Hydrogen Energy* **34**, 245–254.

Wang, Y., Wang, H., Feng, X., Wang, X., and Huang, J. (2009) Biohydrogen production from cornstalk wastes by anaerobic fermentation with activated sludge. *Int. J. Hydrogen Energy* **35**(7), 3092–3099.

Wu, X., Zhu, J., Dong, C., Miller, C., Li, Y., Wang, L., and Yao, W. (2009) Continuous biohydrogen production from liquid swine manure supplemented with glucose using an anaerobic sequencing batch reactor. *Int. J. Hydrogen Energy* **34**(16), 6636–6645.

Yang, H., Shao, P., Lu, T., Shen, J., Wang, D., Xu, Z., and Yuan, X. (2006) Continuous bio-hydrogen production from citric acid wastewater via facultative anaerobic bacteria. *Int. J. Hydrogen Energy* **31**(10), 1306–1313.

Yokoyama, H., Waki, M., Moriya, N., Yasuda, T., Tanaka, Y., and Haga, K. (2007) Effect of fermentation temperature on hydrogen production from cow waste slurry by using anaerobic microflora within the slurry. *Appl. Microbiol. Biotechnol.* **74**(2), 474–483.

Youn, J. H. and Shin, H. S. (2005) Comparative performance between temperature-phased and conventional mesophilic two-phased processed in terms of anaerobically produced bioenergy from food waste. *Waste Manage. Res.* **23**, 32–38.

Yu, H., Zhu, Z., Hu, W., and Zhang, H. (2002) Hydrogen production from rice winery wastewater in an upflow anaerobic reactor using mixed anaerobic cultures. *Int. J. Hydrogen Energy* **27**, 1359–1365.

Zaborsky, O. R. (1998) *Biohydrogen*. New York: Plenum Press.

Zhang, M., Fan, Y., Xing Y., Pan, C., Zhang, G., and Lay, J. (2007) Enhanced biohydrogen production from cornstalk wastes with acidification pretreatment by mixed anaerobic cultures. *Biomass Bioenergy* **31**(4), 250–254.

Zhu, H., Parker, W., Basnar, R., Proracki, A., Falletta, P., Beland, M., and Seto, P. (2008a) Biohydrogen production by anaerobic co-digestion of municipal food waste and sewage sludges. *Int. J. Hydrogen Energy* **33**, 3651–3659.

Zhu, H., Stadnk, A., Beland, M., and Seto, P. (2008b) Co-production of hydrogen and methane from potato waste using a two-stage anaerobic digestion process. *Bioresour. Technol.* **99**, 5078–5084.

Zhu, J., Li, Y., Wu, X., Miller, C., Chen, P., and Ruan, R. (2009) Swine manure fermentation for hydrogen production. *Bioresour. Technol.* **100**(22), 5472–5477.

Zidansek, A., Blinc, R., Jeglic, A., Kabashi, S., Bekteshi, S., and Slaus, I. (2009) Climate changes, biofuels and the sustainable future. *Int. J. Hydrogen Energy* **34**, 6980–6983.

Chapter 17

Bio-productions of Hydrogen and Ethanol from Sugarcane

Alissara Reungsang[1,2,*] and Pensri Plangklang[1]

[1]*Department of Biotechnology, Faculty of Technology,*

[2]*Fermentation Research Center for Value Added Agricultural Products*
Khon Kaen University, A. Muang, Khon Kaen 40002, Thailand
**alissara@kku.ac.th*

The technologies for hydrogen and ethanol productions from sugarcane juice were described based on information from laboratory-scale experiments. Bioconversions of sugarcane bagasse to hydrogen and ethanol were also discussed in terms of bagasse pretreatment and detoxification of the inhibitors in the hydrolysate. Two process configurations used in the ethanol fermentation process, i.e., separate hydrolysis/fermentation and simultaneous saccharification/fermentation, were also presented.

1. Introduction

As a sustainable energy source, hydrogen is a promising alternative to fossil fuels. It is clean and environmental friendly, producing water instead of greenhouse gases upon combustion. Hydrogen can be generated mainly from fossil fuels, biomass, and water by chemical or biological process (Han and Shin 2004; Kim *et al.* 2004). Biologically, hydrogen can be produced by either phototrophic or fermentative route. Biological processes are more environmental friendly and less energy intensive, compared to thermo-chemical and electro-chemical processes (Kapdan and Kargi 2006).

Ethanol is the most common biofuel, either used as a fuel directly or as a gasoline enhancer. It has been widely produced in many countries over the concerning of petroleum supplies shortage and global warming. Many countries have implemented programs for the blending of ethanol to gasoline. Ethanol is at present mostly produced by fermentation of sugars from sugarcane followed by distillation.

Sugarcane is one of the important industrial crops all over Thailand, except in the southern region. The total farming area is more than 960,000 ha, producing about 48 million tons of sugarcane annually (Office of the Cane and Sugar Board 2006). A national effort has been continuously made to develop new breeds of sugarcane in order to increase productivity. This may result in an excess of sugarcane product over the capacity of sugar factories. The alternative is to produce renewable biofuel, i.e., hydrogen and ethanol, using excess sugarcane juice or sugarcane bagasse as feedstock.

The use of renewable energy derived from the crop avoids many of the environmentally detrimental impact of fossil fuels such as the greenhouse effect which led to global warming.

However, producing renewable bioenergy has its own environmental costs. This includes ecological damages caused by the uses of fertilizers and pesticides, as well as soil erosion (Hill 2007). In addition, as the world population continued to increase, the conflicts of using crop for food or energy production could occur. These reasons draw the interest of using lignocellulosic materials, which are the most abundant renewable biomass in the world, as the feedstock for bioenergy productions.

Sugarcane bagasse, a waste left after sugarcane extraction process, is mostly lignocellulosic. It accounts for about 25% of sugarcane mass (Neureiter *et al.* 2002), equivalent to 12 million tons of annual production in Thailand. The most common use of sugarcane bagasse is to produce energy by combustion; only a small fraction of sugarcane bagasse is used to produce paper pulp (Carachi *et al.* 1996) and ethanol (Laser *et al.* 2002). Therefore, utilization of sugarcane bagasse as substrate to produce renewable bioenergy, such as hydrogen and ethanol, has attracted much interest. This will not only add value to sugarcane bagasse in a form of clean energy, but also provide a solution for the management of this abundant waste.

2. Composition of Sugarcane Juice

Sugarcane (*Saccharum officinarum*) is a drought-tolerant tropical and subtropical crop (Wheals *et al.* 1999). It contains 12–17% total sugars (in wet weight) and 68–72% water. The average extraction efficiency by pressing the sugarcane to obtain the sugarcane juice is 95%, and the remaining solid residue is sugarcane bagasse. The juice, primarily consisting of 90% sucrose and 10% glucose or fructose, is mainly used for producing sugar.

3. Composition of Sugarcane Bagasse

One metric ton of sugarcane generates about 280 kg of bagasse, which retains about 2.5% sugar (dry weight) (Cerqueira *et al.* 2007; Singh *et al.* 2007). Sugarcane bagasse consists of 50% cellulose, 25% hemicellulose, and 25% lignin (Pandey *et al.* 2000). Dilute acid treatment of hemicellulose fraction in sugarcane bagasse yields a solution containing mainly glucose and xylose (Aguilar *et al.* 2002) with small amount of arabinose (Gámez *et al.* 2006). Depending on the hydrolytic process, the hydrolysate may also contain the inhibitors that may be toxic to microorganisms. Such toxic effect may, however, be overcome by various treatment methods, such as base or acid treatment (Roberto *et al.* 1991; Dominguez *et al.* 1996; Alves *et al.* 1998). Table 17.1 compares the sugar compositions and inhibitor concentrations resulting from acid hydrolysis under various liquid-to-solid ratios, temperatures, and durations. Due to its high sugar content, hydrolysate of sugarcane bagasse is a very attractive raw material for bioenergy production.

4. Hydrogen Production

4.1. *Hydrogen production from sugarcane juice*

Research on biological hydrogen production has been conducted utilizing a variety of substrates, especially sugar, food waste, agro-industrial waste, and activated sludge. Bio-hydrogen production from agricultural products such as sweet sorghum (Antonopoulou *et al.* 2008; Ntaikou *et al.*

Table 17.1. Sugar compositions and inhibitor concentrations obtained from different acid hydrolysis conditions of sugarcane bagasse.

Acid	Liquid-to-solid ratio ($g \cdot g^{-1}$)	Temp (°C)	Time (min)	Concentration ($g \cdot l^{-1}$) Glucose	Xylose	Arabinose	Acetate	Furfural	Reference
HNO_3 (6%)	10	122	9.3	2.87	18.60	2.04	0.90	1.32	Rodríguez-Chong *et al.* (2004)
H_3PO_4 (4%)	8	122	300	3.00	17.60	2.60	4.00	1.20	Gámez *et al.* (2006)
HCl (2.5%)	10	140	30	5.84	21.50	2.95	1.89	5.45	Chandel *et al.* (2007)
H_2SO_4 (0.5%)	15	121	60	11.00	13.64	2.10	2.26	0.12	Pattra *et al.* (2008a)
H_2SO_4 (2%)	10	122	24	3.00	21.60	NA	3.65	0.50	Aguilar *et al.* (2002)
H_2SO_4 (7%)	5	125	120	4.50	47.2	6.2	10.7	0.83	Nigam (2000)

2008; Claassen *et al.* 2004) and sugar beet (Hussy *et al.* 2005) has also been reported. However, information on bio-hydrogen production from sugarcane juice is still limited.

The main composition of sugarcane juice is sucrose, which can be used as substrate for hydrogen production by various microorganisms such as *Clostridium pasteurianum* (Lin and Lay 2004) and *Clostridium butyricum* CGS5 (Chen *et al.* 2005) with yields of 4.80 and 2.78 mol-$H_2 \cdot$mol-sucrose^{-1}, respectively. Hydrogen production is affected by environmental factors such as pH, temperature, substrate concentration, etc. Pattra *et al.* (2007) found that suspended cells of *C. butyricum* had a maximum hydrogen production rate (141.7 mol-$H_2 \cdot l^{-1} \cdot$day^{-1}) and yield (3.05 mol-$H_2 \cdot$mol-sucrose^{-1}) at the initial pH of 6.0 and sucrose concentration of 20 $g \cdot l^{-1}$. Pattra *et al.* (2008b) also found that immobilized cell of *C. butyricum* on sugarcane bagasse gave a higher maximum hydrogen production rate (3,108 mol-$H_2 \cdot l^{-1} \cdot$day^{-1}) with a hydrogen yield of 2.66 mol-$H_2 \cdot$mol-sucrose^{-1}. In addition, the immobilized *C. butyricum* could be reused up to five batches without reduction in hydrogen yield and production rate.

When pure culture is used for hydrogen fermentation, the reactors are mostly operated under sterile condition which would be costly in industrial production. Pattra *et al.* (2009) conducted two experiments to produce hydrogen in the continuous stirred tank reactor (CSTR) under non-sterile condition: one used normal microbes in sugarcane juice and the other was augmented with *C. butyricum*. Results showed that the augmented culture had a maximum hydrogen yield (1.0 mol-$H_2 \cdot$mol-hexose^{-1}) and production rate (3.38 mmol-$H_2 \cdot l^{-1} \cdot$day^{-1}) at 4 h of hydraulic retention time (HRT). Analysis of the microbial community by DGGE indicated that the augmented microorganism, *C. butyricum*, was dominant and played the key role in hydrogen production, with support from the normal microbe, *Klebsiella pneumonia*e, under non-sterile condition. Low hydrogen content and a reduction of hydrogen production efficiency were likely resulted from the presence of *Lactobacillus harbinensis*.

Singh *et al.* (1994) found that, in addition to the fermentative bacteria, phototrophic bacteria, such as *Rhodopseudomonas,* were also able to produce hydrogen from sugarcane juice. Phototrophic bacteria have high theoretical hydrogen yields and do not produce oxygen, which may inhibit hydrogen production. In addition, they had the ability to use wide spectral light energy and convert organic substrates derived from wastes in association with wastewater treatment (Fascetti *et al.* 1998).

Feasibility of biological hydrogen production from sugarcane by fermentation was studied by Tanisho (1996). The hydrogen yield obtained was 2.4 mol-$H_2 \cdot$mol-sugar^{-1}. However, the unit cost of hydrogen energy was 30% higher than ethanol.

4.2. *Hydrogen production from sugarcane bagasse hydrolysate*

Glucose, xylose, some arabinose, and trace amounts of other sugars are products of dilute acid hydrolysis of the hemicellulose fraction of sugarcane bagasse. These sugars can be served as an alternative renewable carbon source for hydrogen production. Pattra *et al.* (2008a) found that sugarcane bagasse hydrolysate obtained from 0.5% H_2SO_4 treatment could be used for hydrogen production by *C. butyricum*. The highest hydrogen production rate (1,611 mL-$H_2 \cdot l^{-1} \cdot day^{-1}$) and yield (1.73 mol-$H_2$·mol-sugar^{-1}) were obtained at 37 °C, initial pH of 5.5, and an initial COD of 20 g·l^{-1} in sugarcane bagasse hydrolysate.

Sugarcane bagasse hydrolysate could also be produced enzymatically. Chairattanamanokorn *et al.* (2009) hydrolyzed sugarcane bagasse by using 20 U·g^{-1} of commercial cellulase derived from *Trichoderma reesei*. The hydrolysate obtained was used to produce hydrogen from thermally pretreated sludge. The optimum condition for hydrogen production was 56.5 °C and pH 5.2 with a hydrogen production yield of 1.40 mmol·g-TVS^{-1}. Interestingly, a 13-fold increase in hydrogen production yield (13.4 mmol·g-TVS^{-1}) was obtained when sugarcane bagasse was pretreated with 4 g·l^{-1} NaOH. Alkaline pretreatment was able to loose the lignin linkage with hemicellulose and cellulose within the lignocellulosic materials. Thus, the cellulase could effectively penetrate the fibers and the efficiency of enzymatic hydrolysis was distinctively increased (Chang and Holtzapple 2000; Zhao *et al.* 2008).

5. Ethanol Production

At present, fuel ethanol is mostly produced by fermentation of crops, followed by distillation. Table 17.2 lists the main feedstock used for ethanol production, depending on their availabilities, in different countries. Table 17.3 shows that annual world production of fuel ethanol reached 65,000 million liters in 2008 (Renewable Fuels Association (RFA) 2009), of which the USA was the largest producer and Brazil a close second.

Zymomonas mobilis and flocculating yeast strain, *Saccharomyces cerevisiae,* have been the focus of the research due to their great potential to be used in the large-scale ethanol production. The recombinant strains of *Z. mobilis* and *S. cerevisiae* have also been studied. Doelle and Greenfield (1985) investigated the single batch fermentation process for the commercial ethanol production by *Z. mobilis* from sucrose, sugarcane juice, and molasses at sucrose concentration of 200 g·l^{-1}. The process produced 95.5 g·l^{-1} of ethanol with an ethanol conversion efficiency of 95–98%. Thermophilic microorganisms were reported to be effective in producing ethanol as well. Thermotolerant strain, *Kluyveromyces maxianus* DMKU3-1042, was isolated from sugarcane juice. At the optimum temperature of 40 °C, this strain produced ethanol from sugarcane juice containing 22 g·l^{-1} of sugar with a maximum ethanol concentration of 6.78% (w/v) and a yield of 0.36 g-ethanol·g-sugar^{-1} (Limtong *et al.* 2007).

New strains of ethanol producers were isolated to achieve some specific requirements. The killer yeast, *Pichia* strain (CCA 510), isolated from sugarcane juice could produce ethanol and exhibit a killer activity against the fermentative yeast in the process and commercial baking yeast strains. This implies that this killer yeast could prevent the adverse effect caused by undesirable yeast presented in the system under non-sterile condition (Ceccato-Antonini *et al.* 2004).

High initial sugar concentration and the presence of ethanol in the fermentation medium would result in a long lag period for ethanol fermentation and an incomplete utilization of sugar

Table 17.2. Global ethanol blending programs (RFA 2009; Sánchez and Cardona 2008).

Country	Feed stock	Ethanol in gasoline blends (%, v/v)	Remarks
Brazil	Sugarcane	23	All gasoline must contain between 20 and 25% anhydrous ethanol.
Argentina	Sugarcane, molasses	5	Requires the use of 5% ethanol blends by 2010.
Bolivia	Sugarcane	10	Expanding ethanol blends to 25% over the next five years.
Colombia	Sugarcane	10	Requires 10% ethanol blends in cities with populations over 500,000.
Venezuela	Sugarcane	10	Phasing in 10% ethanol blending requirement.
USA	Corn	10	Oxygenation of gasoline is mandatory in dirtiest cities; tax incentives. Some states have banned MTBE and 85% blends are also available.
Canada	Corn, wheat, barley	7.5–10	By 2010, 5% of all motor vehicle fuel must be ethanol or biodiesel.
France	Sugar beet, wheat, corn	7	Set target rates for incorporation of biofuels into fossil fuels (by energy content). Calls for 5.75% in 2008 and increasing to 10% in 2010.
Germany	Sugar beet	50, 85	Mandates 8% energy content in motor fuels by 2015, 3.6% coming from ethanol.
Lithuania	Grain	7	Gasoline must contain 7–15% ETBE. The ETBE must be 47% ethanol.
Poland	Wheat	—	Mandatory "National Biofuel Goal Indicators" calling for biofuels to represent a set percentage of total transportation fuel use. 2008's standard is 3.45%, on an energy content basis.
Spain	Wheat, barley	—	Ethanol is used for ETBE production; direct gasoline blending is possible.
Sweden	Wheat	5, 85	85% blends are also available; there is no ETBE production.
Thailand	Cassava, molasses, rice	10	Gasoline in Bangkok must be blended with 10% ethanol.
India	Sugarcane	5	Requires 5% ethanol in all gasoline.
China	Corn, wheat	10	Five Chinese provinces require 10% ethanol blends — Heilongjiang, Jilin, Liaoning, Anhui, and Henan.
Philippines	Cassava, sugarcane	5	Requires 5% ethanol blends in gasoline beginning in 2008. The requirement expands to 10% in 2010.

(Tano and Buzato 2003; Jones *et al.* 1994; Sánchez and Cardona 2008). Tano and Buzato (2003) reported that the efficiency of ethanol production from sugarcane juice (152 $g{\cdot}l^{-1}$) by *Z. mobilis* ATCC31821 was decreased by 48.8% when the initial ethanol concentration in the medium was adjusted from nil to 2.5% (v/v). Cell yield and sugar consumption were also decreased by 25% and 28%, respectively. Some processes have been developed to overcome such problem, including the very high gravity technology. Jones *et al.* (1994) reported that by using the very high gravity technology, ethanol (14%, v/v) could be produced from sugarcane juice containing 224 $g{\cdot}l^{-1}$ of sugar by commercial dried yeast.

Various types of supplements such as salt, vitamin, protein and lipid were also utilized to improve yield and concentration of ethanol (Lawford and Rousseau 1997; Duarte *et al.* 1996). Saponifiable portion of soybean oil deodorizing distillate (SOD) was supplemented for

Table 17.3. Annual world production of ethanol (million liters) (RFA 2009; Sánchez and Cardona 2008).

Country	2007	2008
USA	24,600	34,069
Brazil	19,000	24,500
European Union	2,159	2,777
China	1,840	1,900
Canada	800	900
Thailand	300	340
Colombia	284	300
India	200	250
Australia	100	100
Other	312	486
Total	49,593	65,621

fermentation by *Z. mobilis* CP4 to sugarcane juice containing 152 $g \cdot l^{-1}$ of sucrose and 2.5% (v/v) of ethanol.

Addition of 5 $ml \cdot l^{-1}$ SOD produced 16 $g \cdot l^{-1}$ of ethanol with a yield of 0.35 g-ethanol·g-sucrose^{-1}, i.e., 24% higher than the yield in the batch without SOD addition (Tano *et al.* 2000). Wheat hydrolysate supplemented to the sugarcane juice could increase the ethanol concentration to 16% (v/v) (Jones *et al.* 1994). Yeast extract, micronutrients, and thiamine supplemented to the sugarcane juice for ethanol fermentation by the recombinant strains of *Escherichia coli* KO11 and *Klebsiella oxytoca* P2 produced 39.4 and 42 $g \cdot l^{-1}$ of ethanol, respectively. These concentrations were about 3.5 times higher than those without supplementation. The ingredients in LB broth supplemented to sugarcane juice could inhibit the ethanol production by KO11 but improve the ethanol production by P2 (Silva *et al.* 2005).

Different types of fermentation and reactor were employed to produce ethanol. A flask scale fed-batch fermentation of sugarcane juice was used to produce ethanol by immobilized *S. cerevisiae* AS2.1190 on sugarcane pieces. The highest ethanol concentration of 92 $g \cdot l^{-1}$ was obtained at the fifth repeated batch with sugar conversion of 99.8% while the ethanol fermentation with suspended yeast cells gave 88 g·l-ethanol^{-1} with 98.5% sugar conversion (Liang *et al.* 2008). A 10 l tower fermenter was used to continuously produce ethanol from sugarcane juice by flocculant yeast strain IR-2 with various dilution rates. The optimum dilution rate was 0.22 h^{-1}, at which the maximum ethanol concentration and yield were 72 $g \cdot l^{-1}$ and 0.47 g-ethanol·g-sugar^{-1}, respectively, with a relatively high ethanol productivity of 15.9 $g \cdot l^{-1} \cdot h^{-1}$ (Oliveira *et al.* 1999). An experiment to examine the possibility of industrial-scale ethanol production from un-supplemented non-sterile sugarcane juice was conducted. Using a 100 l fermenter *Z. mobilis* Z7 under non-sterile condition was able to yield 0.3–0.45 g-ethanol·g-sugar^{-1} at 35 °C (Lyness and Doelle 1981) over 20–29 hours of fermentation.

5.1. *Ethanol production from sugarcane bagasse*

Similar to hydrogen production from sugarcane bagasse, various pretreatment techniques using acid, alkaline, enzymes, and steam have been applied for bagasse saccharification. The resulting hydrolysate containing mixtures of pentose and hexose could be fermented to ethanol by various microorganisms. Native strains of ethanol producers, such as *S. cerevisiae* and *Z. mobilis*, cannot utilize pentose as their carbon source. *Pachysolen tannophilus*, *Candida shehatae*, and *Pichia stipitis* had been used as pentose-fermenting yeasts in the past decades (du Preez *et al.* 1986; van Zyl *et al.* 1991). However, they have a relatively low ethanol tolerance and their ethanol fermentation rates from hexose are at least five times lower than that of *S. cerevisiae*. Thus, they are not suitable for large-scale production of ethanol (Sánchez and Cardona 2008; Pasha *et al.* 2007). To overcome the problem, new technologies have been developed by using recombinant strains of ethanol producers to convert pentose and hexose in the sugarcane bagasse hydrolysate into ethanol (Prasad *et al.* 2007; Martín *et al.* 2002; Takahashi *et al.* 2000; Ho *et al.* 1998). There are three effective recombinant approaches: (1) cloning pentose-utilizing gene into *S. cerevisiae* or *Z. mobilis* (Hernández-Salas *et al.* 2009); (2) cloning genes encoding the ethanol pathway into the pentose-utilizing strains (Takahashi *et al.* 2000); and (3) cloning ethanol-resistant genes into pentose-fermenting strain (Chandel *et al.* 2007; Cheng *et al.* 2008). The use of co-cultures capable of fermenting different sugars is a simple and effective means for ethanol production has also been studied (Fu *et al.* 2009; Qian *et al.* 2006; De Bari *et al.* 2004). Some microorganism strains have been found to produce ethanol from cellulose material without saccharification, but pretreatment was required to remove lignin from sugarcane bagasse (Dawson and Boopathy 2008, 2007).

Two process configurations have been used for ethanol fermentation from bagasse. In the classical separate hydrolysis/fermentation (SHF) process, bagasse hydrolysis and hydrolysate fermentation are conducted in respective reactors, whereas in the alternative simultaneous saccharification/fermentation (SSF) process, bagasse saccharification and fermentation are conducted simultaneously in one reactor (Sánchez and Cardona 2008).

5.1.1. *SHF of sugarcane bagasse*

Physical and chemical pretreatments of bagasse would often produce hydrolate consisting of chemicals which are toxic to fermentative microorganisms. The main feature of the SHF process is that hydrolysis and fermentation are carried out separately, and thus the hydrolysate can be detoxified. Sugarcane bagasse hydrolyzed with 2.5% (v/v) HCl yielded 30.29 $g \cdot l^{-1}$ of total reducing sugars along with various fermentation inhibitors such as furans, phenols, and acetic acid. Various detoxification treatments of bagasse hydrolysate were performed before fermenting by *C. shehatae* NCIM 3501. Detoxification by ion exchange showed the best performance giving the maximum ethanol yield of 0.48 g-ethanol·g-sugar^{-1}, followed by activated charcoal, laccase, overliming, and neutralization (Chandel *et al.* 2007). Hydrochloric acid, 1.2% (v/v), pretreatment of pith and depithed bagasse yielded a high concentration of reducing sugars of 35 and 30 $g \cdot l^{-1}$, respectively. Fermentations of these hydrolysates by non-recombinant *S. cerevisiae* produced about 5 $g \cdot l^{-1}$ of ethanol with a yield of 0.15 g-ethanol·g-sugar^{-1} (Hernández-Salas *et al.* 2009).

Fermentation of an acid hydrolysate of sugarcane bagasse containing xylose, glucose, arabinose, and the inhibitor, acetic acid, by *P. stipitis* yielded 0.27 g-ethanol·g-sugar^{-1}. After removing 84% of acetic acid, fermentation of this hydrolysate resulted in a higher ethanol yield of 0.37 g-ethanol·g-sugar^{-1} (van Zyl *et al.* 1991). Cheng *et al.* (2008, 2007) investigated ethanol production by *P. tannophilus* DW06 using hydrolysate obtained from one and two hydrolysis cycles of bagasse by 1.25% (w/w) H_2SO_4 followed by an electrodialytic treatment. The fermentation of hydrolysate (containing 30 g·l-reducing-sugar^{-1}) from one cycle of acid hydrolysis produced 0.59 $g \cdot l^{-1} \cdot h^{-1}$ of ethanol with a yield of 0.35 g-ethanol·g-$sugar^{-1}$ (Qian *et al.* 2006). Two cycles of acidic hydrolysis doubled the reducing sugar concentration in the hydrolysate, producing 0.57 $g \cdot l^{-1} \cdot h^{-1}$ of ethanol with a yield of 0.34 g-ethanol·g-$sugar^{-1}$ (Cheng *et al.* 2008). *E. coli* KO11, carrying the ethanol pathway genes *pdc* (pyruvate decarboxylase) and *adh* (alcohol dehydrogenase) from *Z. mobilis* in its chromosome, has the ability to ferment pentose and hexose in hemicellulosic hydrolysates. The overlimed hydrolysate from dilute H_2SO_4 pretreatment, supplemented with tryptone and yeast extract, was readily fermented to ethanol by KO11 within 48 h. The ethanol conversion efficiency reached 91.5% of the theoretical maximum (Takahashi *et al.* 2000).

Alkaline hydrolysis, hot water, and steam have been used to pretreat sugarcane bagasse for ethanol production. However, the hydrolysates obtained from these treatments individually are unattractive due to the low concentrations of sugars and high concentrations of inhibitors (Roberto *et al.* 1991; Gong *et al.* 1993). Instead, these treatments may be combined with enzymatic treatment to produce favorable resulting hydrolysate for ethanol fermentation. Vázquez *et al.* (2007) used a response surface methodology to maximize the enzymatic conversion of cellulose in alkaline-pretreated bagasse to glucose. The hydrolysate produced under the optimum condition was then fermented by *S. cerevisiae* producing 30 $g \cdot l^{-1}$ of ethanol with a volumetric yield of 3.0 $g \cdot l^{-1} \cdot h^{-1}$. Pith and depithed bagasse were pretreated by alkaline delignification followed by enzymatic hydrolysis using 20% (w/w) each of Celluclast, Novozyme, and Viscozyme L. Cellucast is the glucanase from *Trichoderma reesei*, whereas Novozyme is the cellulase from *Aspergillus* spp., and Viscozyme L is the multi-enzyme complex of arabinase, β-glucanase, hemicellulase, cellulase, and xylanase from *Aspergillus aculeatus*. The resulting hydrolysates were used to ferment ethanol by a non-recombinant *S. cerevisiae*, producing, respectively, 12.9 and 12.5 $g \cdot l^{-1}$ of ethanol with yields of 0.26 and 0.33 g-ethanol·g-$sugar^{-1}$ (Hernández-Salas *et al.* 2009).

Bagasse was treated by dilute H_2SO_4 followed by enzymatic hydrolysis using Cellucast and Novozyme. The resulting hydrolysate containing glucose and xylose was fermented with the co-culturing of immobilized *Z. mobilis* and suspended cells of *P. stipitis* in a modified fermenter, and produced 0.49 g-ethanol·g-$sugar^{-1}$ (Fu *et al.* 2009). In another study, bagasse was pretreated by steam followed by hydrolyzed with cellulolytic enzymes. The hydrolysates were detoxified by overliming to remove fermentation inhibitors, such as phenols, acetic acid, furfural and 5-hydroxy-methyl-furfural. The hydrolysates were fermented with the recombinant *S. cerevisiae* TMB 3001 producing 0.18 g-ethanol·g-$bagasse^{-1}$ (Martín *et al.* 2002).

In a recently study, steam and wet oxidation under alkaline, neutral, or acidic condition were investigated to pretreat bagasse for SSF. Results showed that alkaline wet oxidation produced the highest ethanol conversion. Strong inhibition of ethanol fermentation was found in the steam hydrolysate, seemingly by the furan aldehydes (Martín *et al.* 2006).

Five strains of *Bacillus cereus* were identified in 5% sugarcane bagasse soaked in the substrate containing with minimum chemical requirements. The ethanol fermentation by the ethanologenic yeast strain NS6 of *B. cereus* isolate 19 and 23 expressed 12 g-ethanol·l^{-1} within 7 days of fermentation (Chaudhary and Qazi 2006).

5.1.2. *SSF of sugarcane bagasse*

In the SSF process, cellulase and/or microorganisms capable of excreting cellulose may be added to the reactor. The hydrolytic products, i.e., hexose or pentose, may be readily converted to ethanol. Compared to SHF, this process has higher ethanol yield and has less energy consumption. However, SSF process has to be operated to balance between the optimal conditions for bagasse hydrolysis and ethanol production. In addition, it requires a pretreatment step to remove lignin and higher enzyme dosages which could be costly. Among all pretreatment methods, alkaline pretreatment is most commonly applied due to its high lignin removal efficiency (Sánchez and Cardona 2008). A study was conducted to pretreat bagasse with 1% NaOH and 0.6% H_2O_2, followed by SSF using cellulase in combination with thermotolerant yeast strain, *C. krusei* ICM-Y-05. Based on the cellulose available in pretreated intermediate, the process produced up to 32 g·l^{-1} of ethanol with a 77% theoretical yield. Further increase of substrate concentration significantly reduced the ethanol yield due to poor mass transfer (Cheng and Zhang 2006).

Cheng *et al.* (2008) decreased the alkali and water consumptions by 26% and 40%, respectively, by pretreating bagasse with two recycles of NaOH/H_2O_2 treatment. Cellulase and *K. marxianus* DW08 were used in the SSF for the alkaline-pretreated bagasse producing 25 g·l^{-1} of ethanol with a yield of 0.2 g-ethanol·g-bagasse^{-1} at a rate of 0.52 g·l^{-1}·h^{-1}. Teixeira *et al.* (2000) studied the effectiveness of peracetic acid treatment of poplar wood and sugarcane bagasse, following an alkaline pretreatment. Using 6% NaOH, 15% peracetic acid, and recombinant *Z. mobilis* CP4/pZB5, the process produced 92.8% and 91.9% of theoretical ethanol yields. Acetyl compounds in the lignocelluloses were reduced by the alkaline pretreatments (Teixeira *et al.* 2000).

Hot water and steam are also commonly used to convert fractions of hemicelluloses and lignin into degradable oligomers (Sánchez and Cardona 2008). However, this may result in an inhibitory effect on ethanol fermentation SSF (Laser *et al.* 2002; van Walsum *et al.* 1996). Bagasse pretreated with hot water increased xylan recovery to 80%, and more than 90% of theoretical xylan yield could be converted to ethanol by SSF. On the other hand, the corresponding xylan recovery and yield for steam-treated bagasse were 65% and 90% (Laser *et al.* 2002). Rudolf *et al.* (2008) studied ethanol SSF treatment of steam-treated bagasse using either recombinant xylose utilizing recombinant *S. cerevisiae* TMB3400 or *P. stipitis* CBS6054 together with commercial cellulolytic enzymes. *S. cerevisiae* TMB3400 converted at pH 5.0 all glucose and nearly all xylose, whereas *P. stipitis* CBS6054 converted at pH 6.0, all glucose and over 70% xylose. Microaeration was required for *P. stipitis* CBS6054 but not for *S. cerevisiae* TMB3400.

The bagasse to enzyme ratio is one of the important parameter to a successful SSF of ethanol. Ethanol production rate and yield were limited by the slow saccharification of bagasse. Use of dilute bagasse slurry would improve the conversion efficiency but at lower rates. On the other hand, increase of bagasse concentration would require increased amount of enzyme and other combined treatments complicating the SSF process. In the study of Doran *et al.* (1994), acid-treated bagasse was converted to ethanol in an SSF process using recombinant *Kl. oxytoca* strain P2 and cellulase enzyme (Genencor Spezyme CE). At a high concentration of 160 g-bagasse·l^{-1}, a combination of 20 FPU-cellulase·g-bagasse^{-1}, pre-incubation via saccharification and reduction in particle size by grinding was required to produce 40 g-ethanol·l^{-1}. Alternatively, almost 40 g-ethanol·l^{-1} was produced with 10 FPU-cellulase·g-bagasse^{-1} by incorporating a second saccharification step followed by a second inoculation and short fermentation. In this way, over 70% of theoretical ethanol yield was achieved with the production of 20 g-ethanol·800-FPU^{-1} of commercial cellulase. Viruthagiri and Sasikumar (2008) studied the SSF process of bagasse with cellulase and thermotolerant yeast *Kluyveromyces wickerhamii* MTCC 455. Sugarcane bagasse was pretreated by 50 g·l^{-1} of NaCl and acetic acid, followed by SSF treatment at 45 °C. The process produced up to 9.5 g·l^{-1} of ethanol after 120 h of incubation.

Sasikumar and Viruthagiri (2008) optimized ethanol production from bagasse pretreated by cellulase and yeast *Kluyveromyces fragilis* using the response surface methodology. They obtained a maximum ethanol concentration of 32.6 g·l^{-1} from 180 g·l^{-1} of pretreated bagasse at 35 °C and pH 5.5 in 72 h. Kuo and Lee (2009) used the cellulose dissolving solvent, *N*-methylmorpholine-*N*-oxide monohydrate, to pretreat bagasse, followed by SSF for ethanol production by *Z. mobilis*. They achieved the yield of 0.15 g-ethanol·g-bagasse^{-1} without noticeable negative effect.

In addition to bagasse, sugarcane leaves have also been used for ethanol production through SSF process. Hari Krishna *et al.* (1998) pretreated sugarcane leaves with alkaline and oxidation, followed by SSF treatment. They were able to convert 92% of the hydrolysate containing 2.5% substrate into ethanol. Dawson and Boopathy (2007) treated sugarcane leaves with alkaline peroxide or acid to remove lignin before fermentation by *S. cerevisiae* ATCC 765. The H_2O_2 (2%) treated leaves produced 131 mg-ethanol·l^{-1}, whereas the H_2SO_4 (0.8M) treated leaves produced 336 mg-ethanol·l^{-1}.

References

Aguilar, R., Ramírez, J. A., Garrote, G., and Vázquez, M. (2002) Kinetic study of the acid hydrolysis of sugarcane bagasse. *J. Food Eng.* **55**, 309–318.

Alves, L., Felipe, M., Silva, J., Silva, S., and Prata, A. (1998) Pretreatment of sugarcane bagasse hemicellulose hydrolysate for xylitol production by *Candida guilliermondii*. *Appl. Biochem. Biotechnol.* 70–72, 89–98.

Antonopoulou, G., Gavala, H. N., Skiadas, I. V., Angelopoulos, K., and Lyberatos, G. (2008) Biofuels generation from sweet sorghum: Fermentative hydrogen production and anaerobic digestion of the remaining biomass. *Bioresour. Technol.* **99**, 110–119.

Carachi, J. C., Campana, F., and Curvelho, A. A. S. (1996) Preparation and characterization of dissolving pulps obtained from sugarcane bagasse. *Polym: Sci. Technol.* **6**, 24–29.

Ceccato-Antonini, S. R., Tosta, C. D., and Silva, A. C. D. (2004) Determination of yeast killer activity in fermenting sugarcane juice using selected ethanol-making strains. *Braz. Arch. Biol. Technol.* **47**, 13–23.

Cerqueira, D. A., Filho, G. R., and Meireles, C. D. S. (2007) Optimization of sugarcane bagasse cellulose acetylation. *Carbohydr. Polym.* **69**, 579–582.

Chairattanamanokorn, P., Penthamkeerati, P., Reungsang, A., Lo, Y. C., Lu, W. B., and Chang, J. S. (2009) Production of biohydrogen from hydrolyzed bagasse with thermally preheated sludge. *Int. J. Hydrogen Energy* **34**, 7612–7617.

Chandel, A. K., Kapoor, R. K., Singh, A., and Kuhad, R. C. (2007) Detoxification of sugarcane bagasse hydrolysate improves ethanol production by *Candida shehatae* NCIM 3501. *Bioresour. Technol.* **98**, 1947–1950.

Chang, V. and Holtzapple, M. (2000) Fundamental factors affecting biomass enzymatic reactivity. *Appl. Biochem. Biotechnol.* 84–86, 5–37.

Chaudhary, N. and Qazi, J. I. (2006) Microbiological saccharification and ethanol production from sugarcane bagasse. *Biotechnol.* **5**, 517–521.

Chen, W. M., Tseng, Z. J., Lee, K. S., and Chang, J. S. (2005) Fermentative hydrogen production with *Clostridium butyricum* CGS5 isolated from anaerobic sewage sludge. *Int. J. Hydrogen Energy* **30**, 1063–1070.

Cheng, K. and Zhang, J. A. (2006) Ethanol from sugarcane bagasse by a simultaneous saccharification and fermentation process (SSF) with *Candida krusei* Icm-Y-05. In: *Proceeding: AIChE*

Annual Meeting <http://aiche.confex.com/aiche/2006/preliminaryprogram/abstract_73730.htm> (November, 2006).

Cheng, K. K., Ge, J. P., Zhang, J. A., Ling, H. Z., Zhou, Y. J., Yang, M. D., and Xu, J. M. (2007) Fermentation of pretreated sugarcane bagasse hemicellulose hydrolysate to ethanol by *Pachysolen tannophilus*. *Biotechnol. Lett.* **29**, 1051–1055.

Cheng, K. K., Zhang, J. A., Ping, W. X., Ge, J. P., Zhou, Y. J., Ling, H. Z., and Xu, J. M. (2008) Sugarcane bagasse mild alkaline/oxidative pretreatment for ethanol production by alkaline recycle process. *Appl. Biochem. Biotechnol.* **151**, 43–50.

Claassen, P. A. M., de Vrije, T., and Budde, M. A. W. (2004) Biological hydrogen production from sweet sorghum by thermophillic bacteria. *Proceeding: 2nd World Conference on Biomass for Energy, Industry and Climate Protection* 1522–1525 (in Italy).

Dawson, L. and Boopathy, R. (2007) Use of post-harvest sugarcane residue for ethanol production, *Bioresour. Technol.* **98**, 1695–1699.

Dawson, L. and Boopathy, R. (2008) Cellulosic ethanol production from sugarcane bagasse without enzymatic saccharification. *Bioresour.* **3**, 452–460.

De Bari, I., Cuna, D., Nanna, F., and Braccio, G. (2004) Ethanol production in immobilized-cell bioreactors from mixed sugar syrups and enzymatic hydrolysates of steam-exploded biomass. *Appl. Biochem. Biotechnol.* **114**, 539–557.

Doelle, H. W. and Greenfield, P. F. (1985) The production of ethanol from sucrose using *Zymomonas mobilis*. *Appl. Biochem. Biotechnol.* **22**, 405–410.

Dominguez, J., Gong, C., and Tsao, G. (1996) Pretreatment of sugarcane bagasse hemicellulose hydrolysate for xylitol production by yeast. *Appl. Biochem. Biotechnol.* 57–58, 49–56.

Doran, J. B., Aldrich, H. C., and Ingram, L. O. (1994) Saccharification and fermentation of sugarcane bagasse by *Klebsiella oxytoca* P2 containing chromosomally integrated genes encoding the *Zymomonas mobilis* ethanol pathway. *Biotechnol. Bioeng.* **44**, 240–247.

du Preez, J. C., Bosch, M., and Prior, B. A. (1986) Xylose fermentation by *Candida shehatae* and *Pichia stipitis*: Effects of pH, temperature and substrate concentration. *Enzyme Microb. Technol.* **8**, 360–364.

Duarte, J. C., Celligoi, M. A. P. C., and Buzato, J. B. (1996) Alcoholic fermentation by *Zymomonas mobilis* CP4 or *Saccharomyces cerevisiae* in sugarcane syrup with and without lipid supplement. *Arch. Biol. Technol.* **39**, 285–293.

Fascetti, E., D'Addario, E., Todini, O., and Robertiello, A. (1998) Photosynthetic hydrogen evolution with volatile organic acids derived from the fermentation of source selected municipal solid wastes. *Int. J. Hydrogen Energy* **23**, 753–760.

Fu, N., Peiris, P., Markham, J., and Bavor, J. (2009) A novel co-culture process with *Zymomonas mobilis* and *Pichia stipitis* for efficient ethanol production on glucose/xylose mixtures. *Arch. Biol. Technol.* **45**, 210–217.

Gámez, S., González-Cabriales, J. J., Ramírez, J. A., Garrote, G., and Vázquez, M. (2006) Study of the hydrolysis of sugarcane bagasse using phosphoric acid. *J. Food Eng.* **74**, 78–88.

Gong, G. S., Chen, C. S., and Chen, L. F. (1993) Pretreatment of sugarcane bagasse hemicellulose hydrolysate for ethanol production by yeast. *Appl. Biochem. Biotechnol.* **39**, 83–88.

Han, S. K. and Shin, H. S. (2004) Biohydrogen production by anaerobic fermentation of food waste. *Int. J. Hydrogen Energy* **29**, 569–577.

Hari Krishna, S., Prasanthi, K., Chowdary, G. V., and Ayyanna, C. (1998) Simultaneous saccharification and fermentation of pretreated sugarcane leaves to ethanol. *Process Biochem.* **33**, 825–830.

Hernández-Salas, J. M., Villa-Ramírez, M. S., Veloz-Rendón, J. S., Rivera-Hernández, K. N., González-César, R. A., Plascencia-Espinosa, M. A., and Trejo-Estrada, S. R. (2009) Comparative hydrolysis and fermentation of sugarcane and agave bagasse. *Bioresour. Technol.* **100**, 1238–1245.

Hill, J. (2007) Environmental costs and benefits of transportation biofuel production from food- and lignocellulose-based energy crops. A review. *Agron. Sustain. Dev.* **27**, 1–12.

Ho, N. W., Chen, Z., and Brainard, A. P. (1998) Genetically engineered *Saccharomyces* yeast capable of effective cofermentation of glucose and xylose. *Appl. Environ. Microbiol.* **64**, 1852–1859.

Hussy, I., Hawkes, F. R., Dinsdale, R., and Hawkes, D. L. (2005) Continuous fermentative hydrogen production from sucrose and sugarbeet. *Int. J. Hydrogen Energy* **30**, 471–483.

Jones, M. A., Kolothumannil, C. T., and Ingledew, W. M. (1994) Ethanolic fermentation of blackstrap molasses and sugarcane juice using very high gravity technology. *J. Agri. Food Chem.* **42**, 1242–1246.

Kapdan, I. K. and Kargi, F. (2006) Bio-hydrogen production from waste materials. *Enzyme Microb. Technol.* **38**, 569–582.

Kim, S. H., Han, S. K., and Shin, H. S. (2004) Feasibility of biohydrogen production by anaerobic co-digestion of food waste and sewage sludge. *Int. J. Hydrogen Energy* **29**, 1607–1616.

Kuo, C. H. and Lee, C. K. (2009) Enhanced enzymatic hydrolysis of sugarcane bagasse by N-methylmorpholine-N-oxide pretreatment. *Bioresour. Technol.* **100**, 866–871.

Laser, M., Schulman, D., Allen, S. G., Lichwa, J., Antal, M. J., and Lynd, L. R. (2002) A comparison of liquid hot water and steam pretreatments of sugarcane bagasse for bioconversion to ethanol. *Bioresour. Technol.* **81**, 33–44.

Lawford, H. and Rousseau, J. (1997) Corn steep liquor as a cost-effective nutrition adjunct in high-performance *Zymomonas* ethanol fermentations. *Appl. Biochem. Biotechnol.* **63–65**, 287–304.

Liang, L., Zhang, Y. P., Zhang, L., Zhu, M. J., Liang, S. Z., and Huang, Y. N. (2008) Study of sugarcane pieces as yeast supports for ethanol production from sugarcane juice and molasses. *J. Indian Microbiol. Biotechnol.* **35**, 1605–1613.

Limtong, S., Sringiew, C., and Yongmanitchai, W. (2007) Production of fuel ethanol at high temperature from sugarcane juice by a newly isolated *Kluyveromyces marxianus*. *Bioresour. Technol.* **98**, 3367–3374.

Lin, C. Y. and Lay, C. H. (2004) Carbon/nitrogen-ratio effect on fermentative hydrogen production by mixed microflora. *Int. J. Hydrogen Energy* **29**, 41–45.

Lyness, E. and Doelle, H. W. (1981) Ethanol production from cane juice by *Zymomonas mobilis*. *Biotechnol. Lett.* **3**, 257–260.

Martín, C., Galbe, M., Wahlbom, C. F., Hahn-Hägerdal, B., and Jönsson, L. J. (2002) Ethanol production from enzymatic hydrolysates of sugarcane bagasse using recombinant xylose-utilising *Saccharomyces cerevisiae*. *Enzyme Microb. Technol.* **31**, 274–282.

Martín, C., González, Y., Fernández, T., and Thomsen, A. B. (2006) Investigation of cellulose convertibility and ethanolic fermentation of sugarcane bagasse pretreated by wet oxidation and steam explosion. *J. Chem. Technol. Biotechnol.* **81**, 1669–1677.

Neureiter, M., Danner, H., Thomasser, C., Saidi, B., and Braun, R. (2002) Dilute-acid hydrolysis of sugar-cane bagasse at varying conditions. *Appl. Biochem. Biotechnol.* **98–100**, 49–58.

Nigam, J. N. (2000) Cultivation of *Candida langeronii* in sugarcane bagasse hemicellulosic hydrolyzate for the production of single cell protein. *World J. Microbiol. Biotechnol.* **16**, 367–372.

Ntaikou, I., Gavala, H. N., Kornaros, M., and Lyberatos, G. (2008) Hydrogen production from sugars and sweet sorghum biomass using *Ruminococcus albus*. *Int. J. Hydrogen Energy* **33**, 1153–1163.

Office of the Cane and Sugar Board (2006) Summarize of the situation on sugar consumption, export and import in Thailand 2006, <http://www.ocsb.go.th/uploads/contents/14/attachfiles/ F5397.pdf> (February 2006).

Oliveira, S. C., De Castro, H. F., Visconti, A. E. S., and Giudici, R. (1999) Continuous ethanol fermentation in a tower reactor with flocculating yeast recycle: Scale-up effects on process performance, kinetic parameters and model predictions. *Bioprocess Biosyst. Eng.* **20**, 525–530.

Pandey, A., Soccol, C. R., Nigam, P., and Soccol, V. T. (2000) Biotechnological potential of agro-industrial residues. I: Sugarcane bagasse. *Bioresour. Technol.* **74**, 69–80.

Pasha, C., Kuhad, R. C., and Rao, L. V. (2007) Strain improvement of thermotolerant *Saccharomyces cerevisiae* VS_3 strain for better utilization of lignocellulosic substrates. *J. Appl. Microbiol.* **103**, 1480–1489.

Pattra, S., Sangyoka, S., Boonmee, M., and Reungsang, A. (2008a) Bio-hydrogen production from the fermentation of sugarcane bagasse hydrolysate by *Clostridium butyricum*. *Int. J. Hydrogen Energy* **33**, 5256–5265.

Pattra, S., Boonmee, M., and Reungsang, A. (2008b) Biohydrogen production from sugarcane juice by immobilized *Clostridium butyricum* on sugarcane bagasse. In: *Proceeding: The 2008 Asian Bio-hydrogen Symposium*, pp. 234–236 (in China).

Pattra, S., Lay, C. H., Lin, C. Y., O-Thong, S., and Reungsang, A. (2009) Performance and population analysis of hydrogen production from sugarcane juice by non-sterile continuous stirred tank reactor augmented with *Clostridium butyricum*. In: *Proceeding: The 2009 Asian Bio-hydrogen Symposium*, pp. 1–15 (in Thailand).

Pattra, S., Reungsang, A., and Sangyoka, S. (2007) Effects of initial pH and initial sucrose concentration on hydrogen production from sugarcane juice. In: *Proceeding: 2007 International Conference on Environmental Quality Concern, Control and Conservative*, pp. 207–215 (in Taiwan).

Prasad, S., Singh, A., and Joshi, H. C. (2007) Ethanol as an alternative fuel from agricultural, industrial and urban residues. *Resour. Conservat. Recycl.* **50**, 1–39.

Qian, M., Tian, S., Li, X., Zhang, J., Pan, Y., and Yang, X. (2006) Ethanol production from dilute-acid softwood hydrolysate by co-culture. *Appl. Biochem. Biotechnol.* **134**, 273–283.

Renewable Fuels Association (RFA) (2009) Growing Innovation 2009: America's energy future start at home. Ethanol industry outlook annual report. <http://www.ethanolrfa.org/objects/pdf/outlook/RFA_Outlook_2009.pdf> (February 2009).

Roberto, I. C., Felipe, M. G. A., Lacis, L. S., Silva, S. S., and de Mancilha, I. M. (1991) Utilization of sugarcane bagasse hemicellulosic hydrolysate by *Candida guilliermondii* for xylitol production. *Bioresour. Technol.* **36**, 271–275.

Rodríguez-Chong, A., Alberto Ramírez, J., Garrote, G., and Vázquez, M. (2004) Hydrolysis of sugarcane bagasse using nitric acid: A kinetic assessment. *J. Food Eng.* **61**, 143–152.

Rudolf, A., Baudel, H., Zacchi, G., Hahn-Hägerdal, B., and Lidén, G. (2008) Simultaneous saccharification and fermentation of steam-pretreated bagasse using *Saccharomyces cerevisiae* TMB3400 and *Pichia stipitis* CBS6054. *Biotechnol. Bioeng.* **99**, 783–790.

Sánchez, Ó.J. and Cardona, C. A. (2008) Trends in biotechnological production of fuel ethanol from different feedstocks. *Bioresour. Technol.* **99**, 5270–5295.

Sasikumar, E. and Viruthagiri, T. (2008) Optimization of process conditions using response surface methodology (RSM) for ethanol production from pretreated sugarcane bagasse: Kinetics and modeling. *Bioenergy Res.* **1**, 239–247.

Silva, G. P. D., Araújo, E. F. D., Silva, D. O., and Guimarães, W. V. (2005) Ethanolic fermentation of sucrose, sugarcane juice and molasses by *Escherichia coli* strain KO11 and *Klebsiella oxytoca* strain P2. *Braz. J. Microbiol.* **36**, 395–404.

Singh, S. P., Asthana, R. K., and Singh, A. P. (2007) Prospects of sugarcane milling waste utilization for hydrogen production. *Indian Energy Policy* **35**, 4164–4168.

Singh, S. P., Srivastava, S. C., and Pandey, K. D. (1994) Hydrogen production by *Rhodopseudomonas* at the expense of vegetable starch, sugarcane juice and whey. *Int. J. Hydrogen Energy* **19**, 437–440.

Takahashi, C., de Carvalho Lima, K., Takahashi, D., and Alterthum, F. (2000) Fermentation of sugarcane bagasse hemicellulosic hydrolysate and sugar mixtures to ethanol by recombinant *Escherichia coli* KO11. *World J. Microbiol. Biotechnol.* **16**, 829–834.

Tanisho, S. (1996) Feasibility study of biological hydrogen production from sugarcane by fermentation. *Proceeding: The 11th World Hydrogen Energy Conference*, pp. 2601–2606 (in Germany).

Tano, M. S. and Buzato, J. B. (2003) Effect of the presence of initial ethanol on ethanol production in sugarcane juice fermented by *Zymomonas mobilis*. *Braz. J. Microbiol.* **34**, 242–244.

Tano, M. S., Buzato, J. B., and Celligoi, M. A. P. C. (2000) Sugarcane juice fermentation by *Zymomonas mobilis* CP4 subjected to inhibition by ethanol and high initial concentration of substrate. *Braz. Arch. Biol. Technol.* 43. 4 (available online from <http://www.scielo.br/scielo.php?script=sciarttext&pid=S1516-89132000000400012& lng= en& nrm=iso>).

Teixeira, L., Linden, J., and Schroeder, H. (2000) Simultaneous saccharification and cofermentation of peracetic acid-pretreated biomass. *Appl. Biochem. Biotechnol.* 84–86, 111–127.

van Walsum, G., Allen, S., Spencer, M., Laser, M., Antal, M., and Lynd, L. (1996) Conversion of lignocellulosics pretreated with liquid hot water to ethanol. *Appl. Biochem. Biotechnol.* 57–58, 157–170.

van Zyl, C., Prior, B. A., and du Preez, J. C. (1991) Acetic acid inhibition of D-xylose fermentation by *Pichia stipitis*. *Enzyme Microb. Technol.* **13**, 82–86.

Vázquez, M., Oliva, M., Téllez-Luis, S. J., and Ramírez, J. A. (2007) Hydrolysis of sorghum straw using phosphoric acid: Evaluation of furfural production. *Bioresour. Technol.* **98**, 3053–3060.

Viruthagiri, T. and Sasikumar, E. (2008) Simultaneous saccharification and fermentation of pretreated sugarcane bagasse. *Bioscan* **3**, 65–69.

Wheals, A. E., Basso, L. C., Alves, D. M. G., and Amorim, H. V. (1999) Fuel ethanol after 25 years. *Tibtech* **17**, 482–487.

Zhao, Y., Wang, Y., Zhu, J. Y., Ragauskas, A., and Deng, Y. (2008) Enhanced enzymatic hydrolysis of spruce by alkaline pretreatment at low temperature. *Biotechnol. Bioeng.* **99**, 1320–1328.

Chapter 18

Synthesis Gas Fermentation

Shihwu Sung* and Po-Heng Lee

Department of Civil, Construction, and Environmental Engineering,
394 Town Engineering, Iowa State University, Ames, IA 50011-3232, USA
**sung@iastate.edu*

Using unsustainable fossil fuels for energy is a critical issue in today's society, and seeking a reliable and economical alternative energy source has raised a lot of attention recently. One of the potential solutions is to convert biomass, which is available in many forms, into fuels and commercial chemicals. The gasification of biomass followed by synthesis gas (syngas) fermentation is one of the novel methods for the potential commercialization necessary to the answer of fossil fuels replacement. An overview of fermentation from biomass-generated syngas is summarized in this chapter.

1. Introduction

Fossil fuels, which represent a significant portion of current energy supply for human activities, are of finite reserve. Moreover, burning fossil fuels result in severe environmental threats (releasing pollutants and green house gases, such as carbon dioxide and nitrogen oxides). Hence, an alternative way to produce energy and chemicals from renewable sources would reduce the stress of dependence on fossil fuels and alleviate environmental concerns (Worden *et al.* 1991). Lignocellulosic biomass like straw, wood, and corn stover are plentiful and inexpensive, and can serve as feedstock for fuels and commercial chemicals production. These biomasses of abundant supply, consisting of cellulose, hemicelluloses, and lignin, however, have to be pretreated to form simple sugars for further conversion into alternative fuels and chemicals. One pretreatment method is through acid/alkaline treatment, followed by enzymatic hydrolysis; the sugars are then fermented into fuel, such as ethanol. This process is usually extensive and leaves a large portion of un-degradable lignin.

Another pretreatment method is through gasification of lignocellulosic biomass to produce a mixture of CO, CO_2, and H_2 (known as synthesis gas, syngas or producer gas) followed by (a) a catalytic gas-conversion step or (b) anaerobic fermentation into fuels and derivative chemicals. The physicochemical limitations and disadvantages of the catalytic gas-conversion process are (Worden *et al.* 1991):

(1) Some contaminants in the syngas, primarily hydrogen sulfide and carbonyl sulfide, must be removed by an energy-intensive purification process before feeding into the catalytic reactor, making it costly.

(2) Syngas composition varies considerably from gasification processes and raw materials. However, the catalytic process requires a specific influent CO to H_2 ratio for a profitable product mix. Influent CO to H_2 ratio often needs to be adjusted by gas recompression and shift reaction mechanisms, adding significant operational costs.
(3) Catalytic processes usually operate under extreme conditions, such as high temperatures and pressures, not only creating safety concerns but also raising capital cost for reactors and other materials of construction.
(4) The catalytic processes may produce a variety of products, requiring extensive further separation.
(5) Some by-products from the catalytic process are hazardous, resulting in public health concerns.

Compared to catalytic conversion, the emerging technology of anaerobic fermentation appears to be more sustainable and attractive. It may minimize the negative impacts and produce non-toxic products. Its advantages are as follows:

(1) Microorganisms for syngas fermentation are able to tolerate hydrogen sulfide and other contaminants.
(2) Syngas fermentation is not sensitive to the syngas composition, and thus the costly gas depression procedure may be eliminated.
(3) The composition of fermentation products can be more consistent.
(4) Fermentation is operated at mild temperatures and pressures.
(5) Fermentation products are in general non-toxic.

Production of simple organic compounds from syngas is thermodynamically favorable, as shown by the negative changes of Gibbs free energy, $\Delta G°$, in Table 18.1 (Barik *et al.* 1988). Fisher and Tropsch were the first in 1932 to demonstrate that acetate could be produced from CO by an anaerobic sewage sludge (Diekert *et al.* 1982). Since then, microbial fermentation of syngas produced from cellulosic materials has been studied by many, and has even been demonstrated

Table 18.1. Theoretically feasible reactions using CO, CO_2, and H_2 (adapted from Barik *et al.* 1990).

Products	Equation	$\Delta G°$ (kcal·mol^{-1})
Formic acid	$CO + H_2O \rightarrow HCOOH$	+5.57
	$CO_2 + H_2 \rightarrow HCOOH$	+10.35
Methanol	$CO + 2H_2 \rightarrow CH_3OH$	-9.14
	$CO_2 + 3H_2 \rightarrow CH_3OH + H_2O$	-4.35
Ethanol	$2CO + 4H_2 \rightarrow C_2H_5OH + H_2O$	-34.56
	$2CO_2 + 6H_2 \rightarrow C_2H_5OH + 3H_2O$	-24.99
Butanol	$4CO + 8H_2 \rightarrow CH_3(CH_2)_2CH_2OH + 3H_2O$	-80.00
	$4CO + 12H_2 \rightarrow CH_3(CH_2)_2CH_2OH + 7H_2O$	-60.86
	$12CO + 5H_2O \rightarrow CH_3(CH_2)_2CH_2OH + 8C_2O$	-116.16
Acetic acid	$CO + 2H_2 \rightarrow CH_3COOH$	-22.73
	$2CO_2 + 4H_2 \rightarrow CH_3COOH + 2H_2O$	-13.15
Propionic acid	$3CO + 4H_2 \rightarrow CH_3CH_2COOH + H_2O$	-44.65
	$3CO_2 + 7H_2 \rightarrow CH_3CH_2COOH + 4H_2O$	-30.29
Acetone	$3CO + 5H_2 \rightarrow CH_3COCH_3 + 2H_2O$	-53.55
	$3CO_2 + 8H_2 \rightarrow CH_3COCH_3 + 5H_2O$	-39.20

in full-scale operation. Although difficulties had been encountered in early attempts, many now believe that microbial production of fuels and commercial products from syngas is commercially feasible through proper product formation, microbe selection, and process optimization (Barik *et al.* 1990).

Many microbes (some have been isolated but not yet identified) capable of producing fuels and chemicals from syngas have been reported. An overview of this cutting-edge technology is summarized in the following.

2. Microbiology of Anaerobic Microorganisms Capable of Converting Syngas to Biofuels and Chemicals

Anaerobic carboxydotrophic microorganisms known to grow on CO and H_2 as substrate are listed in Table 18.2. They are further discussed as follows.

2.1. *Mesophiles for organic acids production*

Many early researches were conducted to produce acetate and butyrate using CO as substrate. *Eubaterium limosum* isolated from various environments, e.g., human intestine, sewage, rumen, and soil, produces acetate and butyrate exclusively using CO as the sole energy source (Genthner *et al.* 1982, 1987). Other acetogens, including *Clostridium aceticum, Peptosteptococcus productus*, etc., were identified to be able to produce acetate from syngas gas. *Butyribacterium methylotrophicum* has demonstrated its ability to shift production from acetate to butyrate by inducing pH changes during the fermentation (Grethlein *et al.* 1991; Lynd *et al.* 1982; Shen *et al.* 1999). *Acetobacterium woodii* is also one of the mesophiles capable of producing acetate from CO. However, the research trend of syngas fermentation has shifted from volatile acids production to production of fuel or other high-value derivatives (Genthner *et al.* 1987).

2.2. *Mesophiles for alcohols production with the by-products of organic acids*

Microbial production of acetate, formate, butyrate, ethanol, butanol, and hydrogen has been considered to have industrial application in recent years. Among the well-studied microbes, *Clostridium ljungdahlii*, a Gram-positive, motile, rod-shaped anaerobe, was isolated from chicken yard waste. *C. ljungdahlii* grows on syngas in pH 4–7 producing ethanol and acetate (Tanner *et al.* 1993). It has been used in full-scale production of ethanol from syngas in a commercial process involving gasification, fermentation, and distillation. *Clostridium autoethanogenum* was also reported to be capable of producing ethanol and acetate from CO and CO_2. The conversion efficiency of *C. autoethanogenum* from syngas to alcohol and acetate, however, was much lower than that of *C. ljungdahlii* (Cotter *et al.* 2009). Another bacterium, *Clostridium carboxidovorans P7*, recently isolated from sediment of an agricultural settling lagoon, was reported to produce acetate, ethanol, butyrate, and butanol from CO and H_2 (Liou *et al.* 2005). Currently, the most attractive syngas fermentation process is using *C. carboxidovorans P7* to produce butanol, which is of higher energy content, lower vapor pressure, and less corrosive than ethanol.

Table 18.2. Anaerobic carboxydotrophic microorganisms (adapted from Henstra *et al.* 2007).

Species	T_{opt} (°C)	pH_{opt}	t_d (h)	Products
Mesophilic bacteria				
Clostridium autoethanogenum	37	5.8–6.0	n.a.	Acetate, ethanol
Clostridium ljungdahlii	37	6	3.8	Acetate, ethanol
Clostridium carboxidovorans P7	38	6.2	6.25	Acetate, ethanol, butyrate, butanol
Oxobacter pfennigii	36–38	7.3	13.9	Acetate, *n*-butyrate
Peptostreptococcus productus	37	7	1.5	Acetate
Acetobacterium Woodii	30	6.8	13	Acetate
Eubacterium Limosum	38–39	7.0–7.2	7	Acetate
Butyribacterium methylotrophicum	37	6	12–20	Acetate, ethanol, butyrate, butanol
Rubrivivax Gelatinosus	34	6.7–6.9	6.7	H_2
Rhodopseudomonas palustris P4	30	nr	23	H_2
Rhodospirillum rubrum	30	6.8	8.4	H_2
Citrobacter sp Y19	30–40	5.5–7.5	8.3	H_2
Mesophilic archaea				
Methanosarcina barkeri	37	7.4	65	CH_4
Methanosarcina acetivorans strain C2A	37	7	24	Acetate, formate, CH_4
Thermophilic bacteria				
Moorella thermoacetica	55	6.5–6.8	10	Acetate
Moorella thermoautotrophica	58	6.1	7	Acetate
Moorella strain AMP	60–65	6.9	n.a.	H_2
Carboxydothermus hydrogenoformans	70–72	6.8–7.0	2	H_2
Carboxydibrachium pacificus	70	6.8–7.1	7.1	H_2
Carboxydocella sporoproducens	60	6.8	1	H_2
Carboxydocella thermoautotrophica	58	7	1.1	H_2
Thermincola carboxydiphila	55	8	1.3	H_2
Thermincola ferriacetica	57–60	7.0–7.2	n.a.	—
Thermolithobacter carboxydivorans (Known as*Carboxydothermus restrictus R1*)	70	7	8.3	H_2
Thermosinus carboxydivorans	60	6.8–7.0	1.2	H_2
Desulfotomaculum Kuznetsovil	60	7	n.a.	Acetate, H_2S
Desulfotomaculum thermobenzoicum subsp. thermosyntrophicum	55	7	n.a.	Acetate, H_2S
Desulfotomaculum carboxydivorans	55	7	1.7	H_2, H_2S
Thermophilic archaea				
Methanothermobacter	65	7.4	140	CH_4
Thermococcus strain AM4	82	6.8	n.a.	H_2
Archaeoglobus fulgidus	83	6.4	n.a.	Acetate, formate, H_2S

n.a.: Not available.

2.3. *Mesophiles for hydrogen and poly-β-hydroxyalkanoate (PHA) production*

A photosynthetic purple nonsulfur bacterium, *Rhodospirillum rubrum*, can use CO as the sole carbon and energy source under anaerobic conditions, in the presence or absence of light, to produce H_2 and Poly-β-hydroxyalkanoate (PHA) (Kerby *et al.* 1995; Do *et al.* 2007). PHA is a biodegradable plastic of many potential applications, including as food additive and as the material for bone reconstruction.

2.4. *Thermophiles for hydrogen and organic acid production*

Thermophilic fermentation has the advantages of less cooling requirement of syngas, high conversion rates, and a separation benefit for alcohols; however, higher temperatures would reduce the syngas solubility (Henstra *et al.* 2007).Gram-positive thermophiles that can convert CO and H_2O into H_2 and CO_2 also have been isolated recently, including *Carboxydothermus hydrogenoformans* and *Thermolithobacter carboxydivorans*. Other thermophiles, such as *Moorella thermoautotrophica* and *Moorella thermoacetic,* have been reported to use CO producing acetate. However, no thermophiles have been reported to be able to convert CO into fuels (such as ethanol or butanol).

One of the most significant factors of syngas fermentation is the CO conversion rate and superior specific products of microorganisms which could produce valuable fuels and organic compounds. Isolation of new microbes capable of converting CO into fuels and organics and research to improve efficiency will make this technology more commercially feasible.

3. Metabolic Pathways of Syngas Fermentation

3.1. *Acetyl-CoA pathway of bacteria for the formation of organic acids and alcohols*

Products from syngas fermentation are limited to these reactions that could conserve enough energy for microbes' metabolism from syngas. The Wood–Ljungdahl pathway (Fig. 18.1 adapted

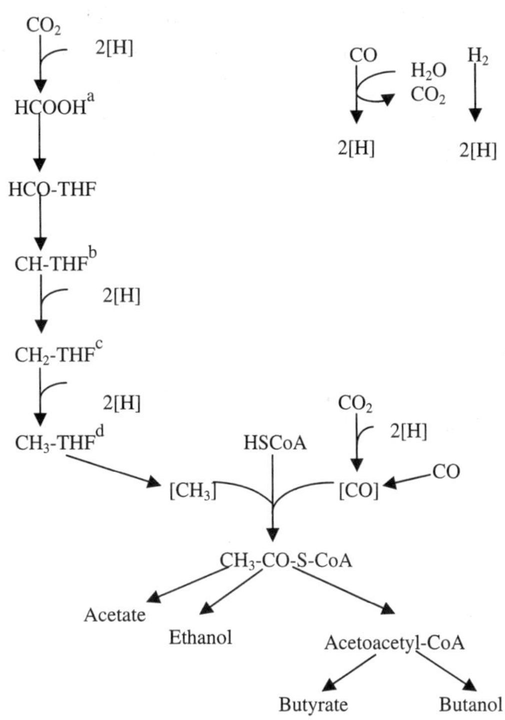

Fig. 18.1. Schematics of the reductive acetyl-CoA/Wood–Ljungdahl pathway (adapted from Henstra *et al.* 2007). [a]Formate (HCOOH), [b]methylene-tetrahydrofolate (CH-THF), [c]methenyl-tetrahydrofolate (CH_2-THF), and [d]methyl-tetrahydrofolate (CH_3-THF).

from Henstra *et al.* 2007) is the main mechanism for the production of acetate, ethanol, butyrate, and butanol from syngas fermentation by acetogens, sulfate-reducing bacteria, and archaea under strictly anaerobic condition. The carbonyl group (–CO) is converted from CO_2 by the bifunctional CO dehydrogenase complex. However, CO is abundant in syngas, so the bifunctional CO dehydrogenase complex is not manifestly required. The methyl group is reduced by CO_2 with some intermediates, i.e., formyl, methenyl, methylene, and methyl compounds bounded with a pterin cofactor. Acetyl-CoA-synthase/CO dehydrogenase complex (ACS/CODH) is responsible for forming acetyl-CoA by joining a carbonyl and a methyl group (Lindahl 2002; Ragsdale 2004; Henstra *et al.* 2007). Bacteria and Achaea have slightly different acetyl-CoA pathway. For bacteria, formate is reduced from CO_2, forming formyl compounds bounded to the pterin tetrahydrofolate with the utilization of ATP (Henstra *et al.* 2007). For Achaea, a methanofuran-bound formyl reduced from CO_2 is converted to tetrahydromethanopterin. Because the formation of acetyl-CoA from H_2 and CO_2 requires energy, acetate is transferred from acetyl-CoA to recover the lost energy from the formation of acetyl-CoA. Ethanol is produced from the further reduction of acetate. Two acetyl-CoA molecules yield an acetoacetyl-CoA which further produces butanol and buyrate (Henstra *et al.* 2007; Ragsdale and Pierce, 2008; Fischer *et al.* 2008).

3.2. *Acetyl-CoA pathway of bacteria for the formation of hydrogen*

Hydrogen production from syngas is metabolized by hydrogenogenic carboxydotrophs (Fig. 18.2 adapted from Henstra *et al.* 2007). The acetyl-CoA pathway with the energy-balanced conservation governs this bacterial metabolism (Henstra *et al.* 2007). A monofunctional CO dehydrogenase releases electrons by oxidizing CO to CO_2. Energy-converting hydrogenase (ECH) receives electrons and reduces protons to yield hydrogen (Hedderich 2004; Singer *et al.* 2006). ECH, which forms H_2 and translates protons or sodium ion, produces ATP from an ATP-synthase driven by a chemiosmotic ion gradient (Hedderich 2004; Maness *et al.* 2005).

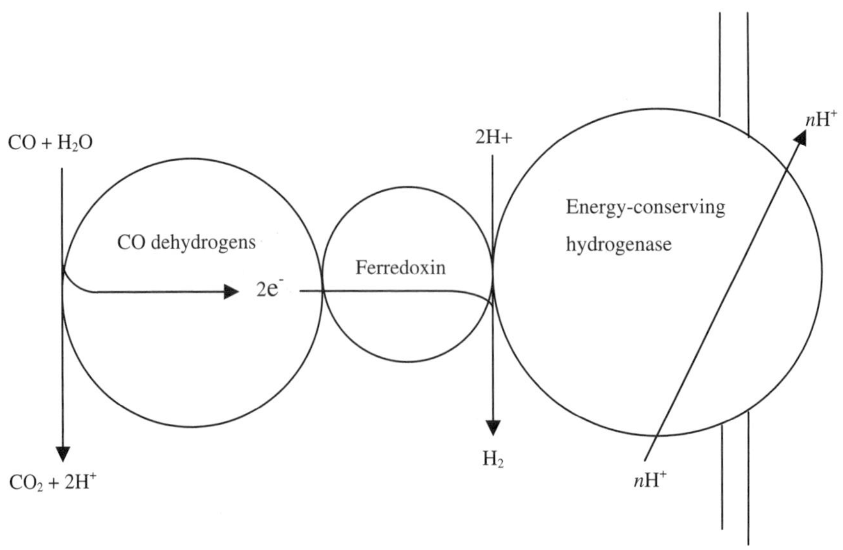

Fig. 18.2. Schematics of membrane-bound CO-oxidizing, hydrogen-evolving enzyme complex of carboxydotrophic hydrogen (adapted from Henstra *et al.* 2007).

4. Carbon Monoxide Gas–Liquid Mass Transfer

Syngas predominantly consists of CO and H_2, both of which have low water solubilities. Mass transfer rates are thus crucial to syngas fermentation. The gas–liquid mass transfer rates of CO and H_2 are only 77% and 65%, respectively, of that of oxygen, (Bredwell *et al.* 1999). The low concentrations of dissolved CO and H_2 result in insufficient supply of substrates for the microbes. This is the bottleneck for the full-scale application of syngas fermentation. This section will discuss reactor designs for improving the solubility of CO.

4.1. *Determination of CO gas–liquid mass transfer*

The mass transfer of a gas into liquid phase can be modeled using the following equations:

$$dC/dt = K_L a(C_s\text{-}C) \tag{18.1}$$

$$\ln(C_s\text{-}C)/(C_s\text{-}C_o) = (\text{-}K_L a)t \tag{18.2}$$

where $K_L a$ = CO gas–liquid mass transfer coefficient (s^{-1})
C_s = saturated dissolved CO concentration ($mg \cdot l^{-1}$)
C_o = dissolved CO concentration at the initial time 0 ($mg \cdot l^{-1}$)
t = reaction time (s)

The gas–liquid mass transfer coefficient, $K_L a$, depends not just on the gas species, but also the fermenter configuration and the mixing pattern.

4.2. *Stirred tank rector*

Stirred tank reactors (STRs) are commonly used in the full-scale fermenters, in which bacterial cells are suspended in the mixed liquor. Riggs and Heindel (2006) reported that the gas–liquid mass transfer coefficient of CO ranged from 0.003 s^{-1} to 0.043 s^{-1} (i.e., 10.8 h^{-1} to 154.8 h^{-1}) in a reactor of 7l working volume with stirrer speeds ranging 200–600 rpm, and CO flowrates ranging 1–6 $l \cdot min^{-1}$ (corresponding to a gas retention time of 1.2–7 min). A syngas consisting of 20% CO, 18% CO_2, 52% N_2, and 10% H_2 was tested to decide the CO gas–liquid mass transfer coefficient in the same reactor of 7l working volume. Results showed the CO gas–liquid mass transfer coefficient ranging from 0.02 s^{-1} to 0.08 s^{-1} (i.e., 72 h^{-1} to 288 h^{-1}) with stirrer speeds ranging 400–700 rpm and CO flowrates ranging 5–15 $l \cdot min^{-1}$ (Kapic *et al.* 2006). Six impellers (i.e., Rushton-type, Philadelphia Mixing concave 9 hollow blade turbine, Philadelphia Mixing pitched blade turbine, Lightnin A315 fluidoil, Lightnin A310, and Philadelphia Mixing LS hydrofoil) were used to improve the CO gas–liquid mass transfer in the same reactor of 7l working volume. Among the six impellers tested, the dual impeller showed the highest CO gas–liquid mass transfer with the most economical power consumption in all operational parameters (Ungerman and Heindel 2007).

4.3. *Other fermenters*

Many fermenters other than STRs have different configurations and operational modes. Some have bacteria grow on surface of carriers resulting in a higher cell density, instead of dispersed in the mixed liquor. The gas–liquid mass transfer characteristics of oxygen in attached-growth

Table 18.3. Oxygen gas–liquid mass transfer coefficients for different attached-growth reactors (Charpentier 1981; Bredwell *et al.* 1999).

Reactor type	Gas void fraction (Gas volume/liquid volume)	$K_L a$ (h^{-1})
Packed columns		
Countercurrent	0.02–0.25	0.001–0.2
Cocurrent	0.60–0.98	0.001–2.8
Plate columns		
Bubble cap	0.10–0.95	0.027–0.56
Sieve plates	0.10–0.95	0.027–1.1
Bubble columns	0.60–0.98	0.014–0.67
Packed-bubble columns	0.60–0.98	0.014–0.33
Tube reactors		
Horizontal and coiled	0.05–0.95	0.014–1.94
Vertical	0.05–0.95	0.056–2.8
Spray columns	0.20–0.95	0.002–0.042
Submerged and plunging jet	0.94–0.99	0.0008–0.017

reactors are listed in Table 18.3 (Charpentier 1981; Bredewell *et al.* 1999). Results in Table 18.3 conclude that gas–liquid transfer rates of the different attached-growth reactor designs. The CO-liquid mass transfer characteristic of the attached-growth fermenters was not highly dependent on the gas flowrates (Bredewell *et al.* 1999). Carbon monoxide was dissolved at a faster rate due to a higher CO consumption rate occurred with higher cell of density. Consequently, the attached-growth fermenter could generally achieve a higher CO transfer rate than STR for a beneficial carbon monoxide conversion at a lower gas flow (Bredewell *et al.* 1999).

4.4. *Hollow fiber membrane fermenter*

Studies have been conducted using hollow fiber membrane (HFM) to deliver oxygen and hydrogen for wastewater treatment and bio-remediation (Kim *et al.* manuscript in progress). Application of microporous HFM as a syngas diffuser has also been developed in the authors' research lab at Iowa State University (Fig. 18.3). While syngas passes through the interior of the hollow fiber, CO, and H_2 continuously permeate through the membrane into the aqueous phase of the reactor. Microbes grow on the exterior membrane surface with an enhanced biomass density for advancing the CO conversion. The polypropylene membrane bundles composed of 100, 250, and 400 fibers (with an outside diameter of 0.042 cm, a wall thickness of 50 μm, a nominal pore size of 0.2 μm and a length of 13 cm) were used to study the CO gas–liquid mass transfer rates for a fermenter with a 2l working volume. For a syngas consisting of 48.5% CO, 35.5% H_2, and 16% CO_2, the CO gas–liquid mass transfer coefficients were found ranging from 0.011 to 0.177 s^{-1} for bundles consisting of 100–400 fibers and syngas flowrates 0.025–0.140 $l \cdot min^{-1}$. Results of the study indicate that HFM fermenter is feasible for commercialization (Kim *et al.* manuscript in progress).

5. Gas–Liquid Mass Transfer Rates in Syngas Fermentation

The standard tests on gas–liquid mass transfer, however, cannot fully simulate the actual syngas fermentation condition. In syngas fermentation, the mixed liquor is of complex composition

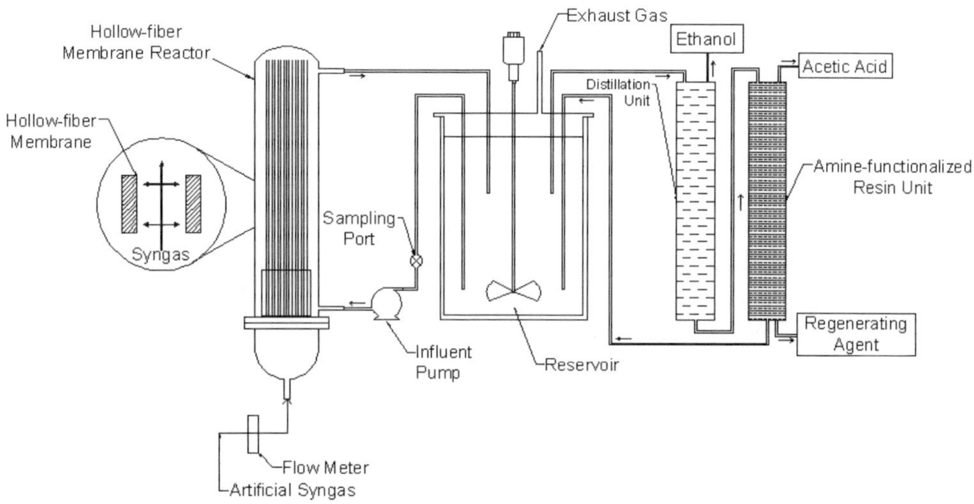

Fig. 18.3. Schematic diagram of microporous hollow fiber membrane fermenter set-up.

Table 18.4. Syngas gas–liquid mass transfer coefficients for wastewater treatment and fermentation (adapted from Bredwell *et al.* 1999).

Reactor type	Biological system	Feedstock	K_La (h^{-1})
Stirred tank –200 rpm	*B. methylotrophicum*	CO	14.2
Stirred tank –300 rpm	SRB mixed culture	Syngas mix	31 for CO 75 for H_2
Stirred tank –300 rpm	*C. ljungdahlii*	Syngas mix	35 for CO
Stirred tank –300 rpm	*R. rubrum, Methanosarcina formicum, M. barkeri*	Syngas mix	28.1 for CO
Stirred tank –450 rpm	*R. rubrum, M. formicum, M. barkeri*	Syngas mix	101 for CO
Stirred tank –200 rpm	*B. methylotrophicum*	CO	90.6 for CO
Stirred tank –200 rpm	*B. methylotrophicum*	CO	90.6 for CO
Stirred –300 rpm Microbubble sparging	SRB mixed culture	Syngas mix	104 for CO 190 for H_2
Packed-bubble column	*R. rubrum, M. formicum, M. barkeri*	Syngas mix	2.1
Trickle bed	*R. rubrum, M. formicum, M. barkeri*	Syngas mix	55.5
Trickle bed	SRB mixed culture	Syngas mix	121 for CO 335 for H_2
Trickle bed	*C. ljungdahlii*	Syngas mix	137 for CO

and the dissolved CO and H_2 are continuously consumed by the biomass. Thus, meaningful data of syngas gas–liquid mass transfer should be obtained using the specific microbes and the fermenting mixed liquor. Table 18.4 summarizes the syngas gas–liquid mass transfer rates for wastewater treatment and fermentation (Bredwell *et al.* 1999). Even though some of these studies were not used for fuels and chemicals production, the results of these studies are applicable to reactor design for syngas fermentation. CO conversion in a trickle-bed reactor by a thriculture fermentation was more than twofold of that in an STR (Klassson *et al.* 1992; Bredwell *et al.* 1999). Other studies used CO and H_2 as substrates for sulfate and nitrate reduction showing that higher gas–liquid mass transfer rates in the fermenters with bacteria in attached growth, compared to those in STR (van Houten *et al.* 1994; du Preez *et al.* 1992, 1994; Selvaraj *et al.* 1996).

6. Performances of Syngas Fermentation for Fuels and Chemicals

6.1. *Syngas fermentation for ethanol production*

Ethanol is a renewable fuel and is currently used as a supplement to gasoline (10% ethanol blend). Ethanol is primarily produced from the fermentation of readily degradable carbohydrate substrates, such as corn starch and sugarcane, causing a negative impact on global food market. Ethanol can also be produced from lignocellulosic biomass which is locally abundant. The production of fermentable sugars from lignocellulosic biomass, however, faces several challenges such as costs of pretreatment and enzymes. Another limitation is the availability of robust microbial communities capable of fermenting mixed sugars (hexose and pentose) to ethanol. Gasification of lignocellulosic biomass to produce a syngas followed by anaerobic fermentation to produce ethanol is a pioneering alternative method. Table 18.5 summarizes the performances of syngas fermentation to produce ethanol and acetate by *C. ljungdahlii* and *C. autoethanogenum*.

Enrichment of enough ethanol-producing biomass in a fermenter is the first task in the syngas fermentation process. Growth of *C. ljungdahlii* and *C. autoethanogenum* on solely syngas is extremely slow in comparison to growth on sugars with syngas supplement (Cotter *et al.* 2008; Lee *et al.* manuscript in progress). Yeast extract tended to favor the production of acetate over ethanol by *C. ljungdahlii*; removing yeast extract from a medium resulted in a threefold increase of ethanol production (Barik *et al.* 1988; Berberich *et al.* 2000). Phillips *et al.* (1994) further developed a growth media based on the elemental composition of *E. coli* for optimizing ethanol production in the *C. ljungdahlii* culture. With the designed media, *C. ljungdahlii* could produce up to 23 $g{\cdot}l^{-1}$ ethanol and 27 $g{\cdot}l^{-1}$ acetate; the corresponding values were only 1.5 $g{\cdot}l^{-1}$ ethanol and 3.5 $g{\cdot}l^{-1}$ acetate in a basal medium of ATCC 1754 PETC in a STR with cell recycle (Phillips *et al.* 1994). While *C. ljungdahlii* was able to maintain its metabolic activity with little (0.005%) to no yeast extract in the medium, the growth of *C. autoethanogenum* showed drop in media completely deficient of yeast extract (Barik *et al.* 1988; Cotter *et al.* 2009). Investigation of the effects of nitrogen limitation in the media has concluded that even though *C. ljungdahlii* could maintain its culture density in nitrogen-limited media with needed trace elements, vitamins and salts, it cannot produce ethanol and acetate under nitrogen-limited media (Cotter *et al.* 2009). Similarly, *C. autoethanogenum* could not retain its culture density if the nitrogen was not provided in the media (Cotter *et al.* 2009). A further study on high ethanol production from metabolic activity is expected for the scale-up application of syngas fermentation.

In the syngas fermentation process, pH is also one of the critical control parameters. Previous studies suggest that higher ethanol production may be achieved by operating at a low pH environment, because the microbes would produce less acid which may prevent a further pH drop and thus result in higher ethanol production (Fordyce *et al.* 1984; Datar *et al.* 2004). Another study reported that ethanol may not be related to the cell growth and suggested that high ethanol yield (over acetate) could be achieved by controlling the cells in the resting phase (Gaddy and Clausen 1992; Klasson *et al.* 1992). However, Cotter *et al.* (2008) showed the opposite result that low ethanol and acetate production was exhibited in the resting cell phase during batch fermentation.

Table 18.5. Performances of syngas fermentation for ethanol and acetate production.

Species	Reactor type	Feedstock	pH	Temp (°C)	Media	Maximum production	References
C. ljungdahlii	STR with cell recycle (13.5 l)	55% CO, 20% H_2, 10% CO_2, and 15% Ar (agitation rate: 300–500 ppm)	4.5	37	ATCC 1754 PETC medium	1.5 g·l^{-1} ethanol 3.5 g·l^{-1} acetate	Phillips *et al.* (1994)
C. ljungdahlii	CSTR with cell recycle (13.5 l)	55% CO, 20% H2, 10% CO2, and 15% Ar (flow rare: 14 ml·min^{-1}, agitation rate: 300–500 rpm)	4.5	37	Designed medium based on *E. Coli*	23 g·l^{-1} ethanol 27 g·l^{-1} acetate	Phillips *et al.* (1994)
C. ljungdahlii	CSTR (20 l)	70% CO, 15% H_2, and 15% Ar (flow rare: 14 ml·min^{-1}, agitation rate: 550 rpm)	4.5	37	ATCC 1754 PETC medium (flow rate: 0.55 ml·min^{-1})	11 g·l^{-1} ethanol and acetate	Younesi *et al.* (2006)
C. ljungdahlii	Batch (125 ml)	55% CO, 20% H_2, 10% CO_2, and 15% Ar (initial syngas pressure: 1.6 and 1.8 atm)	—	37	ATCC 1754 PETC medium	0.60 g·l^{-1} ethanol 1.11 g·l^{-1} acetate	Younesi *et al.* (2005)
C. ljungdahlii	Batch (163 ml)	55% CO, 20% H_2, 10% CO_2, and 15% Ar (initial syngas pressure: 1.8 atm)	—	37	ATCC 1754 PETC medium	0.012 g·l^{-1} ethanol 0.024 g·l^{-1} acetate	Najafpour and Younesi (2006)
C. ljungdahlii (growth phase)	Batch (250 ml)	20% CO, 10% H_2, 20% CO_2, and 50% N_2 (flow rare: 7.5 ml·min^{-1})	6.8	37	Modified reinforced Clostridial medium	0.23 g·l^{-1} ethanol 2.10 g·l^{-1} acetate	Cotter *et al.* (2008)
C. ljungdahlii (growth phase)	Batch (80 ml)	n.a.	6.8	37	Modified ATCC 1754 PETC medium	0.60 g·l^{-1} ethanol 5.34 g·l^{-1} acetate	Cotter *et al.* (2009)
C. ljungdahlii (resting phase)	Batch (80 ml)	n.a.	4.5	37	Modified nitrogen limitation media	0.02 g·l^{-1} ethanol > 0.06 g·l^{-1} acetate	Cotter *et al.* (2009)
C. autoethanogenum (growth phase)	Batch (250 ml)	20% CO, 10% H_2, 20% CO_2, and 50% N_2 (flow rare: 10 ml·min^{-1})	6.0	37	DSMZ 640	0.66 g·l^{-1} ethanol 1.40 g·l^{-1} acetate	Cotter *et al.* (2008)
C. autoethanogenum (growth phase)	Batch (80 ml)	n.a.	6.0	37	DSMZ 640 supplemented with 5 g·l^{-1} xylose	0.24 g·l^{-1} ethanol 2.40 g·l^{-1} acetate	Cotter *et al.* (2009)
C. autoethanogenum (resting phase)	Batch (80 ml)	n.a.	6.0	37	Modified DSMZ 640	0.44 g·l^{-1} ethanol 2.51 g·l^{-1} acetate	Cotter *et al.* (2009)

Microbial fermentation for alcohol production may produce volatile fatty acids as by-products. Bioactivity inhibition by these by-products may be a crucial factor in syngas fermentation. To overcome this problem, the authors' team has developed an innovative process (Fig. 18.3): syngas fermentation using hollow fiber membranes with simultaneous recovery of acetate by amine-functionalized resin or absorbent. The process utilizes the enhanced syngas–liquid mass transfer feature of the HFM, and allows the amine-functionalized resin to recover acetate as a separate saleable product and at the same time to eliminate its inhibition effect on ethanol production (Lee *et al.* manuscript in preparation).

6.2. *Syngas fermentation for butanol production*

Butanol is another biofuel which has a high potential to replace gasoline, either partially or completely. Compared to ethanol, butanol has higher energy content, but lower vapor pressure, hygroscopicity, and corrosivity. Butanol is traditionally produced from a sugar fermentation process called the ABE (acetone, butanol, and ethanol). The process is not economical due to the high cost of sugar. The alternative is to produce butanol using lignocellulosic wastes as raw material. Lignocellulosic materials may be first converted enzymatically into sugars which are then followed by the ABE process. However, the process requires extensive pretreatments for the lignocellulose, such as acid/alkaline lignin removal and enzymatic hydrolysis in order to produce fermentable sugars. On the other hand, butanol could also be produced from syngas fermentation by a new microbial species *Clostridium carboxidivorans* P7 (Liou *et al.* 2005). Most of the studies conducted on this microbe were still limited mainly on the cell growth. Ahmed *et al.* (2006) showed that tars and nitric oxide from biomass derived syngas had an adverse influence on *C. carboxidivorans* P7. A 4.51 syngas bubble column fermenter produced ethanol, butanol, and acetic acid with respective yields of 0.15, 0.075, and 0.025 mole of C for each mole of CO consumed (Rajagopalan *et al.* 2002).

References

Ahmed, A., Cateni, B. G., Huhnke, R. L., and Lewis, R. S. (2006) Effects of biomass-generated producer gas constituents, on cell growth, product distribution and hydrogenase activity of *Clostridium carboxidivorans* P7. *Biomass Bioenergy* **30**, 665–672.

Barik, S., Prieto, S., Harrison, S. B., Clausen, E. C., and Gaddy, J. L. (1988) Biological production of alcohols from coal through indirect liquefaction. *Appl. Biochem. Biotechnol.* **28**, 363–378.

Barik, S., Prieto, S., Harrison, S. B., Clausen, E. C., and Gaddy, J. L. (1990) Biological production of ethanol from coal synthesis gas. In: Wise, D. L. (ed.), *Bioprocessing and Biotreatment of Coal*, lst ed., Chapter 8 New York, Marcel Dekker Inc., pp. 1231–1154.

Berberich, J. A., Knutson, B. L., Strobel, H. J., Tarhan, S., Nokes, S. E., and Dawson, K. A. (2000) Product selectivity shifts in *Clostridium thermocellum* in the presence of compressed solvent. *Ind. Eng. Chem. Res.* **39**, 4500–4505

Bredwell, M. D., Srivastava, P., and Worden, R. M. (1999) Reactor design issues for synthesis gas fermentations. *Biotechnol. Prog.* **15**, 834–844.

Charpentier, J. C. (1981) Mass transfer in gas-liquid absorbers and reactors. *Adv. Chem. Eng.* **11**, 1–103.

Cotter, J. L., Chinn, M. S., and Grunden, A. M. (2009) Influence of process parameters on growth of *Clostridium Ljungdahlii* and *Clostridium autoetahanogenum* on synthesis gas. *Enzyme Microb. Technol.* **44**, 281–288.

Cotter, J. L., Chinn, M. S., and Grunden, A. M. (2008) Ethanol and acetate production by *Clostridium ljungdahlii* and *Clostridium autoethanogenum* using resting cells. *Bioprocess Biosyst. Eng.* **32**, 369–380.

Datar R. P., Shenkman, R. M., Cateni, B. G., Huhnke, R. L., and Lewis, R. S. (2004) *Biotechnol. Bioeng.* **86**, 587–594

Diekert, G. B. and Thayer, R. K. (1982) Carbon monoxide oxidation by *Clostridium thermoaceticum* and *Clostridium formicoaceticum*. *J. Bacteriol.* **136**, 597–606.

Do, Y. S., Smeenk, J., Broer, K. M., Kisting, C., Brown, R., Heindel, T. J., Bobik, T. A., and Dispirito, A. A. (2007) Growth of *Rhodospirillum rubrum* on synthesis gas: conversion of CO to H_2 and poly-beta-hydroxyalkanoate. *Biotechnol. Bioeng.* **97**, 279–86.

Drake, H. L. (1994) *Acetogenesis,* 1st ed. New York: Chapman and Hall.

du Preez, L. A., Opendaal, J. P., Maree, J. P., and Ponsonby, M. (1992) Biological removal of sulfate from industrial effluents using producer gas as energy source. *Environ. Technol.* **13**, 875–882.

du Preez, L. A. and Maree, J. P. (1994) Pilot-scale biological sulfate and nitrate removal utilizing producer gas and energy source. *Wat. Sci. Technol.* **30**, 275–285.

Fischer, C. R., Klein-Marcuschamer, D., and Stephanopoulos, G. (2008) Selection and optimization of microbial hosts for biofuels production. *Metabolic Eng.* **10**, 295–304.

Fordyce, A. M., Crow, V. L., and Thomas, T. D. (1984) Regulation of product formation during glucose or lactose limitation in nongrowing cells of streptococcus. *Appl. Environ. Microbiol.* **48**, 332–337.

Gaddy, J. L. and Clausen, E. C. (1992) US Patent 612:221.

Genthner, B. R. S. and Bryant, M. P. (1987) Additional characteristics of one-carbon-compound utilization by *Eubacterium limosum* and *Acetobacterium woodii*. *Appl. Environ. Microbiol.* **53**, 471–476.

Genthner, B. R. S. and Bryant, M. P. (1982) Growth of *Eubacterium limosum* with carbon monoxide as the energy source. *Appl. Environ. Microbiol.* **43**, 70–74.

Grethlein, A. J., Worden, R. M., Jain, M. K., and Datta, R. (1991) Evidence for production of n-butanol from carbon monoxide by *Butyribacterium methylotrophicum*. *J. Ferment. Bioeng.* **72**, 58–60.

Hedderich, R. (2004) Energy-converting [NiFe] hydrogenases from archaea and extremophiles: Ancestors of complex I. *J. Bioenerg. Biomembr.* **36**, 65–75.

Henstra, A. M., Sipma, J., Rinzema, A., and Stams, A. J. M. (2007) Microbiology of synthesis gas fermentation for biofuel production. *Curr. Opin. Biotechnol.* **18**, 200–206.

Kapic, A., Jones, S. T., and Heindel, T. J. (2006) Carbon monoxide mass transfer in a syngas mixture. *Ind. Eng. Chem. Res.* **45**, 9150–9155.

Kerby, R. L., Ludden, P. W., and Roberts, G. P. (1995) Carbon monoxide dependent growth of *Rhodospirillum rubrum*. *J. Bacteriol.* **177**, 2241–2244.

Kim, S. H., Lee, P. H., Chang, H. I., and Sung, S. Carbon monoxide mass transfer characteristics of hollow-fiber membranes for syngas fermentation, manuscript in progress.

Klasson, K. T., Ackerson, M. D., Clausen, E. C., and Gaddy, J. L. (1992) Bioconversion of synthesis gas into liquid or gaseous fuels. *Enzyme Microb. Technol.* **14**, 602–608.

Lee, P. H., Ni, S. Q., Choi, D. W., and Sung, S. pH effect on the growth of *Clostridium ljungdahlii* for ethanol production from syngas fermentation, manuscript in progress.

Lindahl P. A. (2002). The Ni-containing carbon monoxide dehydrogenase family: Light at the end of the tunnel? *Biochem.* **41**, 2097–2105.

Liou, J. S. C., Balkwill, D. L., Drake, G. R., and Tanner, R. S. (2005) *Clostridium carboxidivorans* sp. nov., a solvent-producing clostridium isolated from an agricultural settling lagoon, and reclassification of the acetogen *Clostridium scatologenes* strain SL1 as *Clostridium drakei* sp. nov. *Int. J. Syst. Evol. Microbiol.* **55**, 2085–2091.

Lynd, L., Kerby, R., and Zeikus, J. G. (1982) Carbon monoxide metabolism of the methylotrophic acidogen *Butyribacterium methylotrophicum*. *J. Bacteriol.* **149**, 255–263.

Maness, P. C., Huang, J., Smolinski, S., Tek, V., and Vanzin, G. (2005) Energy generation from the CO oxidation-hydrogen production pathway in *Rubrivivax gelatinosus*. *Appl. Environ. Microbiol.* **71**, 2870–2874.

Najafpour, G. and Younesi, H. (2006) Ethanol and acetate synthesis from waste gas using batch culture of *Clostridium ljungdahli*. *Enzyme Microb. Technol.* **38**, 223–228.

Phillips, J. R., Clausen, E. C., and Gaddy, J. L. (1994) Synthesis gas as substrate for biological production of fuels and chemicals. *Appl. Biochem. Biotechnol.* **45**, 145–157.

Ragsdale, S. W. (2004) Life with carbon monoxide. *Crit. Rev. Biochem. Mol. Biol.* **39**, 165–195.

Ragsdale, S. W. and Pierce, E. (2008) Acetogenesis and Wood-Ljungdahl pathway of CO_2 fixation. *Biochim. Biophys. Acta* **1784**, 1873–1898.

Riggs, S. S. and Heindel, T. J. (2006) Measuring carbon monoxide gas-liquid mass transfer in a stirred tank reactor for syngas fermentation. *Biotechnol. Prog.* **22**, 903–906.

Rajagopalan, S., Datar, R. P., and Lewis, R. S. (2002) Formation of ethanol from carbon monoxide via a new microbial catalyst. *Biomass Bioenergy* **23**, 487–493.

Selvaraj, P. T., Little, M. H., Bredwell, M. D., and Kaufman, E. N. (1996) Bioreactors for flue gas biodesulfurization using coal synthesis gas as feedstock. In: *Proceedings of the American Institute of Chemical Engineers Fall Annual Meeting*, pp. 10–15, 1996, (in Chicago, IL).

Shen, G. J., Shieh, J. S., Grethlein, A. J., Jain, M. K., and Zeikus, J. G. (1999) Biochemical basis for carbon monoxide tolerance and butanol production by *Butyribacterium methylotrophicum. Appl. Microbiol. Biotechnol.* **51**, 827–832.

Singer S. W., Hirst, M. B., and Ludden, P. W. (2006) CO-dependent H_2 evolution by *Rhodospirillum rubrum*: Role of CODH:CooF complex. *Biochim. Biophys. Acta. Bioenerg.* **1757**, 1582–1591

Tanner, R. S., Miller, L. M., and Yang, D. (1993) *Clostridium ljungdahlii* sp. nov., an acetogenic species in clostridial ribosomal-RNA homology group-I. *Int. J. Syst. Bacteriol.* **43**, 232–236.

Ungerman, A. J. and Heindel, T. J. (2007) Carbon monoxide mass transfer for syngas transfer for syngas fermentation in a stirred tank reactor with dual impeller configurations. *Biotechnol. Prog.* **23**, 613–620.

van Houten, R. T., Hulshoff Pol, L. W., and Lettinga, G. (1994) Biological sulfate reduction using gas-lift bioreactors fed with hydrogen and carbon dioxide as energy and carbon source. *Biotechnol. Bioeng.* **44**, 586–594.

Worden, R. M., Grethlein, A. J., Jain, M. K., and Datta, R. (1991) Production of butanol and ethanol from synthesis gas via fermentation. *Fuel* **70**, 615–619.

Younesi, H., Najafpour G., and Mohamed, A. R. (2005) Ethanol and acetate production from synthesis gas via fermentation processes using anaerobic bacterium. *Clostridium ljungdahlii. Biochem. Eng. J.* **27**, 110–119.

Younesi, H., Najafpour G., and Mohamed, A. R. (2006) Liquid fuel production from synthesis gas via fermentation process in a continuous tank bioreactor (CSTR) using *Clostridium ljungdahlii. Iranian J. Botechnol.* **4**, 45–53.

Index